LINDA EMORY
770-587-1418

Dragons and Tigers

A Geography of South, East and Southeast Asia

Dragons and Tigers

A Geography of South, East and Southeast Asia

Barbara A. Weightman
California State University, Fullerton

John Wiley & Sons, Inc.
New York • Chichester • Weinheim • Brisbane • Singapore • Toronto

ACQUISITIONS EDITOR *Ryan Flahive*
PROJECT MANAGER *Joan Petrokofsky*
SENIOR MARKETING MANAGER *Clay Stone*
PRODUCTION DIRECTOR *Pam Kennedy*
PRODUCTION EDITOR *Leslie Surovick*
SENIOR DESIGNER *Kevin Murphy*
COVER AND INTERIOR DESIGNER *Laura Ierardi*
ILLUSTRATION EDITOR *Anna Melhorn*
ILLUSTRATION STUDIO *Maps.com/Magellan*
COVER PHOTO CREDITS *©Tom Van Sant/Geosphere Project, Santa Monica/Science Photo Library*
INSET PHOTO *©1999 Mike Wilbur/Black Star*
BACK COVER PHOTO *Barbara A. Weightman*

This book was set in 10/12 Berkeley Book by TechBooks and printed and bound by Quebecor/Fairfield. The cover was printed by The Lehigh Press.

This book is printed on acid-free paper. ∞

The paper in this book was manufactured by a mill whose forest management programs include sustained yield harvesting of its timberlands. Sustained yield harvesting principles ensure that the numbers of trees cut each year does not exceed the amount of new growth.

Library of Congress Cataloging in Publication Data:
Weightman, Barbara A.
Dragons and tigers: a geography of South, East and Southeast Asia / Barbara Weightman.
p. cm.
Includes bibliographical references and index.
ISBN 0-471-25358-8 (cloth : alk. paper)
1. South Asia—Geography. 2. East Asia—Geography. 3. Asia, Southeastern—Geography. I. Title.

DS335 .W37 2001
915—dc21 00-043823

Printed in the United States of America

10 9 8 7 6 5 4 3 2

Preface

I have designed this book to meet the need for a comprehensive text on the Geography of Asia. Knowledge of Asian landscapes is fundamental to academic and practical learning in the twenty-first century. Virtually half of the world's people live in Asia, 40 percent in China and India alone. With some of the fastest growing economies in the world, Asian nations, notably of the Pacific Rim, are critical components of North American and other countries' economic and political strategies. Australia, Canada, and the United States are key destinations for Asian migrants whose presence intensifies networks of trans-Pacific interrelationships. Some Asian nations are global leaders in environmental damage, and several are riddled with internecine strife. I think that learning about this dynamic and multicultural region is imperative.

Dragons and Tigers explores and illustrates conditions, events, problems, and trends of both larger regions and individual nations. I use a cross-disciplinary approach to discuss evolving physical and cultural landscapes. Nature–society interrelationships provide the foundation for my emphasis on social, economic, political, and environmental problems.

I have also attempted to provide plenty of material for instructor and student use. The chapter bibliographies are multidisciplinary, referencing various approaches to the subject. I offer sufficient historical background for the grounding of current conditions and for the understanding of contemporary issues. Main point summaries are intended to sharpen the focus of the reader, and term lists highlight major geographic concepts. Furthermore, I use close-up, focus boxes to harbor interest and offer tables of demographic and quality of life data for comparative purposes. A glossary of terms is provided at the end of the text. I have tried to write a book rich and fascinating in content. My hope is that both instructors and students will want to read it.

Acknowledgments

I owe a debt of gratitude to several individuals. I especially thank my parents, Claudia Lowe, Louise Scheffer, and Eva Voigt, for their undying support over the years. I also thank the geography faculty at the University of British Columbia and the University of Washington for showing me the way to success in my academic endeavors.

I gratefully acknowledge the contributions of those who reviewed various segments of the book. Thanks to:

Patrick Buckley, *Western Washington University*
George Demko, *Dartmouth College*
Christine Drake, *Old Dominion University*
Cindy Fan, *U.C.L.A.*
Philippe Foret, *University of Oklahoma*
Melinda Meade, *University of North Carolina, Chapel Hill*
Norman Moline, *Augustana College*
David Nemeth, *University of Toledo*
Mary Lee Nolan, *Oregon State University*
Bimal K. Paul, *Kansas State University*
Yda Schreuder, *University of Delaware*
Richard Ulack, *University of Kentucky*

Any errors or deficiencies that remain in the book are solely my responsibility.

The people at John Wiley & Sons, Inc., also deserve credit. Special thanks go to Nanette Kauffman for seeing potential in my proposal and encouraging me to pursue it. Other individuals worthy of mention are:

Anne E. Smith, Publisher
Ryan Flahive, Acquisitions Editor
Karen Ayoub, Editorial Assistant
Joan Petrokofsky, Project Manager
Clay Stone, Senior Marketing Manager
Pam Kennedy, Production Director

Leslie Surovick, Production Editor
Kevin Murphy, Designer

Credits

All the photos are mine with the following exceptions: Figure 11-2 is by Risa Palm, Figure 11-4 is by John Carroll, and Figure 11-6 is by Roman Cybriwsky, courtesy of the American Geographical Society.

The following maps are reprinted with permission from Prentice Hall: Figures 3-2 and 10-18 from David L. Clawson and James S. Fisher eds., 1998, *World Regional Geography: A Development Approach,* 6th edition, Upper Saddle River, NJ: Prentice Hall, and Figure 7-3 from James S. Fisher, 1995, *Geography and Development,* 5th edition, Upper Saddle River, NJ, Prentice Hall. Figures 10-4, 10-6, 10-12, 13-1, 13-2, 13-6, 13-10, 14-12, 14-14, and 14-22 are from Harm deBlij and Peter Muller, 2000, *Geography: Realms, Regions & Concepts 2000,* New York, John Wiley & Sons, Inc. and reprinted with permission of John Wiley & Sons, Inc.

Contents

Chapter 12 3 - 24 + 25

Southeast Asia: Transition Among the Nagas 315

Chapter 13 4, 16

Mainland Southeast Asia: Turmoil and Peace 342

Chapter 1

Major Influences

"The idea of tolerance and mutual concession is based on admitting the compatibility of many different philosophical views of the world."

HAJIME NAKAMURA (1964)

Introduction

Welcome to South, East, and Southeast Asia, a region of unparalleled diversity that is home to more than half of humankind. Close to 3.5 billion people inhabit a land area of about 8 million square miles (20,761 sq km). Here arose some of the world's most admirable and powerful civilizations. Here, too, are the world's highest mountains and several of its longest rivers. Vast, densely populated plains contrast with expanses of sparsely populated mountains, deserts, and tropical forests. Hundreds of ethnic groups, religions, and languages blend in a kaleidoscopic array of lifestyles and cultural landscapes. Democracies exist alongside socialist and militarist states. The region contains some of the globe's poorest countries as well as the ascendant **dragon and tiger economies** to which we are increasingly connected.

Think of the number of Japanese, Korean, or Taiwanese cars and electronic gadgets that you have. Check your clothing labels; where were they assembled? Thailand? Indonesia? Do you eat Chinese, Indian, or Malaysian food? Is there a Buddhist or Hindu temple near your community? We are all linked in some fashion to South, East, and Southeast Asia.

Diversity makes the pursuit of multiple themes necessary for exploring and appreciating this part of the world. However, in light of contemporary geographic thought, some themes will be more apparent than others. An overarching theme will be **landscape,** a concept that goes far beyond that which we observe. It encompasses our entire sensorial experience; it is the content of our lived world. Individuals or groups do not function independently of other people or the forces of nature. Moreover, political, economic, social, and environmental systems are increasingly intertwined from local to global scales. Under this rubric, we will investigate both cultural landscapes and or physical landscapes, in context of their interrelationships.

I will stress the **social construction** of landscapes as well as **quality of life** issues. How have internal and external forces affected and effected ways of life? How well are people's needs being met? Who determines what these needs are? I will also focus on **environmental transformation.** How have local and externally imposed values, attitudes, policies, and practices altered physical, biological, and zoological realms? What are the ramifications for human beings? In addition, I will explore the meanings and processes of *development* in the context of **globalism** and **regionalism.** Is the world becoming increasingly similar? Or, is the world becoming more dissimilar? How global *is* the global economy? Is it becoming more fragmentary with the formation of competitive regional organizations?

Dragons and Tigers

Dragons and tigers represent the dynamic forces of power and change. They are symbols long associated with South, East, and Southeast Asia. In China's *I Ching,* or *Book of Changes,* the light and dark aspects of *yin* and *yang* are often expressed as dragon and tiger. In Daoism, *yang* or goodness is controlled by a green dragon; *yin* or evil is controlled by a white tiger. And, many a sage sought the dragon–tiger elixer, key to immortality. The Chinese Emperor wore *dragon robes* and sat on the *dragon throne.* The dragon is used as a metaphor for rivers and mountains—to cut a pass or install water control works is to "tame the dragon." Dragons bring luck and protection. They continue to be incorporated as architectural detail in temples, hotels, houses, and many other structures.

The tiger, aside from being a real, although increasingly rare, animal, is fabled in stories throughout these realms. In Korea, the tiger is a protective figure, watching over the land. Indonesia is rich in tiger lore with legends of tiger spirits, families, and villages. Tiger teeth and claws are said to induce prowess and potency. In South Asia, tiger images appear on seals used by early Indus Valley civilizations more than 4,000 years ago. Betrothed couples in India wear red and yellow shawls that symbolize the tiger's stripes of fertility.

Today, dragon or tiger status is ascribed to growing economies. Japan, Singapore, South Korea, and Taiwan have all been described as *dragon economies* or *economic tigers.* India is struggling for "tigerhood." Vietnam is "Asia's baby tiger."

These beasts are frequently seen in service and product names. Dragonair flies out of Hong Kong. Nine Dragon Beer is sold in China. Singapore's Tiger Beer is exported around the world. Tiger Balm ointment is a trusted remedy for headaches.

Dragons and tigers have pervaded Asian landscapes in both subtle and obvious ways. That is why they have been given a place in the title of this book.

Defining the Region

Asia's expanse and complexity require its division into more manageable and meaningful areas of study. This text will concentrate on three geographic realms of Asia: South, East, and Southeast Asia (Figure 1-1).

According to geographer Harm deBlij (1997), **geographic realms** have three sets of spatial criteria. First, physical and cultural characteristics are the largest units in which we can divide the **ecumene** (i.e., the inhabited world). Second, geographic realms are founded on similarities of **functional interactions** between people and their natural environments. These are expressed in farmscapes, transportation networks, cityscapes, environmental damage, and countless other landscape features. Third, geographic realms focus on and incorporate the world's major population clusters. South, East, and Southeast Asia encompass the two largest clusters in India and China and the fourth largest in Indonesia. You will learn about the growth and distribution of these populations in a later chapter.

South Asia comprises India, Pakistan, and Bangladesh; the Himalayan states of Nepal and Bhutan; and the island nations of Sri Lanka and the Maldives. *East Asia* includes China, Mongolia, North and South Korea, Taiwan, and Japan. *Southeast Asia* encompasses the mainland countries of Myanmar, Thailand, Lao (Laos), Cambodia, and Vietnam, in addition to the maritime states of Malaysia, Singapore, Brunei, Indonesia, and the Philippines.

"ASIA" AS A MYTH

These countries and their respective realms are parts of a larger entity called *Asia.* While we might think of Asia as a definable geographic region, actually it is

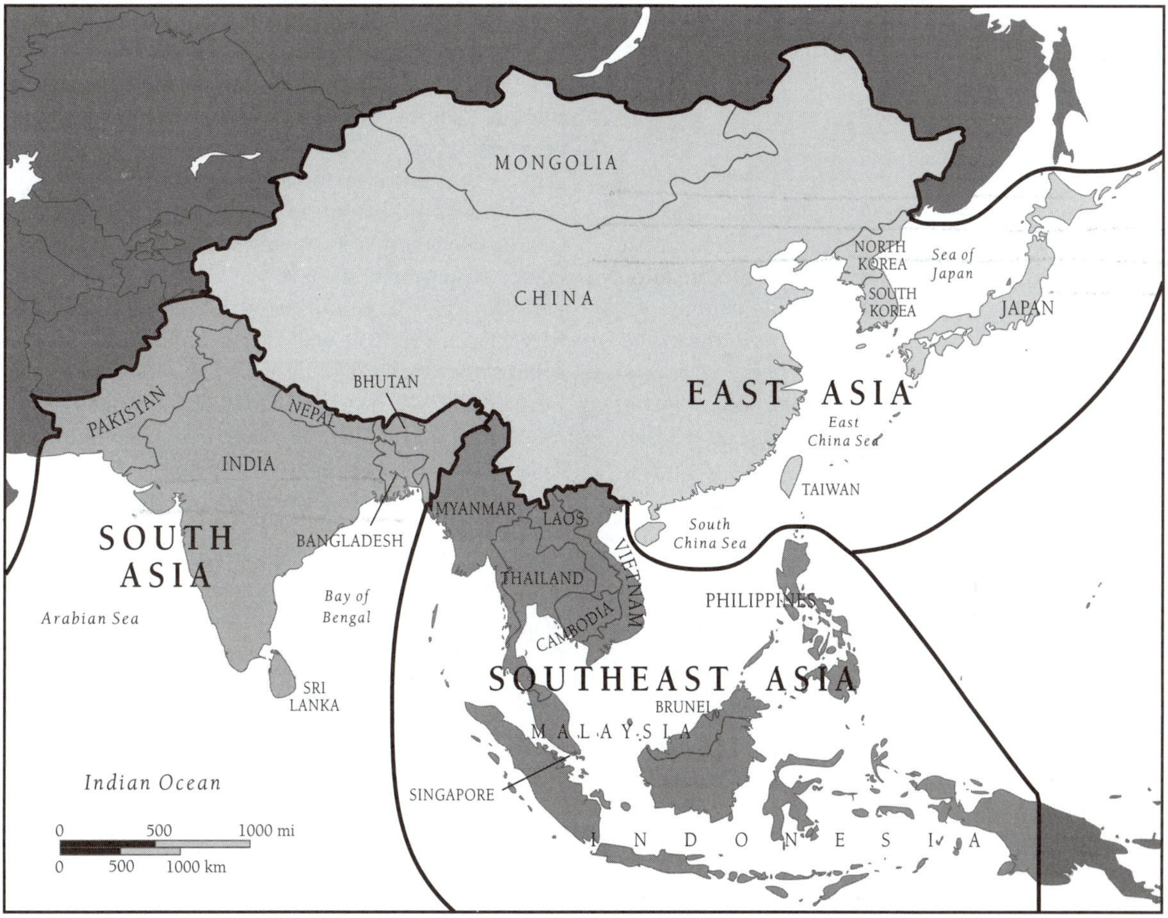

Figure 1-1

Geographic realms. These three realms are based on similarities of functional interrelationships between people and their natural environments.

more of an idea than a reality. More importantly, the evolution of the *idea of Asia* in Europe and the West shaped the nature and tenor of perceptions and interactions between Asian and non-Asian cultures throughout history. To understand this, we must explore what sometimes is called the **myth of Asia.**

There is no cultural or historical entity that equates Asia. It is more of a literary and psychological construct imposed from the outside than a geographical reality. Designations of "the Orient" and "the Far East" are also associated with the idea of Asia. Where do these terms come from, and what do they mean? What is Asia, and how did it come to be? How can it be a myth if it actually exists?

Asia, in fact, was defined by Europeans. From the time of ancient Greece, Europeans visualized the world as being divided into two or three continental parts: Asia, Europe, and sometimes Africa. Religiously inspired ***mappaemundi*** (maps of the world), portraying the earth as flat, often depicted Asia as half the known world! This *understood world* could also be divided into east and west: **Orient** meaning *east,* derived from sunrise, and **Occident,** meaning *west,* from sunset. For Europeans, *Orient* meant east and Asia. However, the initially blurry concept of Asia was filled in over time and concurrently expanded spatially.

CONCEPTUALIZING ASIA

The term *Asia* came from the Greek word for sunrise and originally referred to the Anatolian Peninsula (Turkey) in the seventh century B.C., a region later known as Asia Minor. By the fifth century B.C., Asia had acquired continental meaning. Throughout this evolutionary process, comparisons and contrasts were drawn between east and west.

Centuries later, Karl Marx (1818–1883) drew distinctions between Asia and Western Europe. He subscribed to the concept of "An Asiatic Mode of Production." In this exploitative, economic system, the urban rich and powerful sucked all profits from an undifferentiated rural peasantry laboring in virtual enslavement.

Meanwhile, Asia's colonizers had developed their own frameworks of thought. You will learn more details in later chapters. A general model is provided by geographer James L. Blaut (1994), who refers to a **colonizer's model** of the world: civilization and progress diffuse from the European center to the culturally barren periphery.

Two things are important here. First, since classical times, Western writers have commonly contrasted their own civilizations with those of Asia. Asia has been perceived as a single entity, its internal diversity homogenized in false unity, its characteristics and institutions portrayed as alien and inferior to those in Occidental realms. As the British writer Rudyard Kipling (1865–1936) said in the first line of his poem "The Ballad of the East and West": "The East is east and the West is west and never the twain shall meet."

Second, environmental determinism colored historic and geographical accounts well into the twentieth century. Ellen Churchill Semple (1863–1932), an American geographer who taught at the University of Chicago and Clark University, was complimentary to China and Japan, yet described the people of India as ignorant and superstitious. She saw tropical peoples as servile and inferior due to the high temperatures in which they lived. According to Semple, anyone who went to tropical areas would soon be reduced to the lethargic state characteristic of those residing in these debilitating regions.

Derogatory assessments not only generated **xenophobia** (fear of foreigners) but also contributed to **ethnocentrism** and the **we–they syndrome.** Ethnocentric people judge foreign cultures by their own standards and often speak in *we* (better) versus *they* (not as good) terms. Economically developed, democratic Western nations have viewed themselves as superior to those in less developed, politically alternate Eastern ones. As you will see, this notion of cultural superiority continues to pervade East–West relations. Furthermore, the history of cartography reveals a long-established practice of centering the world on Europe. This promoted a **Eurocentric** worldview, the antithesis of China's **Sinocentric** (China centered) view (see Chapter 9).

In 1998, I purchased a school notebook printed in Vietnam. On the back cover is a Eurocentered world map captioned by the words, "The World of [the] Future." What sort of maps are in your classrooms and academic books? Examine some world geography, world history, or world civilization textbooks and note how many pages are devoted to Eastern versus Western civilizations.

European notions of Asia were also colored by religious beliefs. After all, the Garden of Eden was there, the Three Wise Men came from the east, and the sun, the light of salvation, rises in the east. Prester (Presbyter) John, a wise Christian king, was believed to have ruled somewhere in Asia. In the Middle Ages, countless pilgrimages were launched to find his kingdom. In that vein of thought, the Orient was a source of light and goodness and, conversely, the Occident was dark and evil.

Beyond these scripturally inspired conceptions, Asia was portrayed as a prodigious place housing dog-headed men and other freakish beings, as well as the wicked tribes of Gog and Magog, poised to pounce upon the unwary. More significantly, the Orient was thought to possess unbounded wealth in spices, jewels, incense, and other exotica—an idea underpinning numerous land and sea explorations.

It is important to recognize that this collective, conflicting imagery of an ill-defined Asia derived from both real and imagined experiences with a region extending eastward from Egypt into Ukraine and Russian grasslands and southward into India. At the time, China was merely a fabled land on the periphery of this vast expanse called Asia. As exploration fostered increasing contacts between East and West, more places were added to the European knowledge map. Asia, as a conceived region, ultimately stretched eastward and southward into the Pacific Ocean, incorporating Japan and such far-flung islands as Indonesia and the Philippines.

By the nineteenth century, European powers were not only entrenched in many Asian countries for political and economic gain but also were struggling with defining their own boundaries. A series of wars, the rise and demise of assorted political powers, and ever-shifting national boundaries necessitated the revision of European frontiers. This was accomplished at the **Congress of Vienna** in 1815. The division between Europe and Asia was demarcated from Russia's Ural Mountains; through the Caspian and Black Seas; and along the eastern end of the Mediterranean to the African continent.

Today, Asia is recognized as part of the geographical landmass known as **Eurasia.** And so, there is a continental Asia in a Eurasian context. However, cultural, political, economic, and environmental regions, as we know them today, are no more coincident with these boundaries than with spatial constructs of earlier times. This is why it is preferable to consider smaller, more cohesive realms. *Although South, East and Southeast Asia are the regions designated for discussion, I will use the term* Asia *when referring to these collectively, or the Asian continent in general.*

Reading Place Names

There are many ways to spell **toponyms** or place names, especially in the case of Asia. This is, in part, because of varying transliterations of regional scripts into the Roman alphabet. Some of the spellings for the Chinese capital city of Beijing are: Peiping, Pei-ching, and Peking. Westerners' inability to pronounce indigenous place names led to alternate spellings. Singapore is the English version of Singha Pura that was formerly known as Temasek. Different ethnic groups may have their own versions of toponyms. For example, Urumchi in northwestern China is a Uygur name. The Chinese call it Wu-lu-mu-chi. Sometimes places disappear and reappear as something else. For instance, East Pakistan became Bangladesh in 1970. Furthermore, place names are frequently changed. Tokyo was once known as Edo. Others are newer, as with the restoration of traditional names. Bombay is now called Mumbai, and Rangoon, Burma, is currently Yangon, Myanmar. Be forewarned: If you send a letter to "Bombay," it will likely be returned to you marked "No such place."

When speaking of a place in a historical context, I will spell it as it was referred to at that time, followed by the current spelling in parentheses. For example, "In old Siam (Thailand) . . . ." In modern context, current names will stand alone. For example, "Myanmar." This might seem confusing, but, as you will find out, different maps and atlases employ different spellings. More importantly, residents of places frequently continue to use the name that has been in use the longest, or that they have a preference for. For example, in the 1970s Saigon was renamed Ho Chi Minh City, but many people continue to call it Saigon.

Religion and Belief Systems

Now let us turn to an aspect of culture of paramount significance: **religion.** Religion, or a set of sacred beliefs, pervades everyone's life, whether consciously or unconsciously. Aside from being evident in the cultural landscape in the form of sacred structures and other phenomena, religion functions in a variety of ways in society. Religions define male and female roles; affect political systems and policies; influence agricultural and dietary practices; shape environmental values; impact economic development; and, frequently, are responsible for dissension and strife.

Hinduism, Buddhism, and Islam are the major religions in Asia, but Christianity and other belief systems such as Sikhism, Confucianism, Daoism, and animism are important in specific regional settings. At this juncture, I will discuss the origin, **diffusion** or spread, and key tenets of four religions in Asia: Hinduism, Buddhism, Christianity, and Islam. Others will be discussed in their regional contexts in subsequent chapters.

All four religions have their **hearth areas,** or places of origin, in continental Asia. Each of these diffused by one or more means: conquest, word of mouth (contact diffusion), migration of believers (relocation diffusion), and the work of missionaries (proselytization). Merchants and traders played significant roles in spreading religion, especially in the case of Buddhism and Islam (see Figure 1-2).

HINDUISM

Hinduism is the oldest of the world's major religions. Its origins reach beyond the margins of documented history. Without an apparent founder, but with a substratum of belief from Indus Valley Civilization, it emerged in northern India at least 6,000 years ago. From there, it diffused southward across the Indian peninsula. Much later, it spread eastward with the diffusion of Indian influences as early as the third century A.D. and with subsequent Indian migrations through the colonial era. Hindu communities still

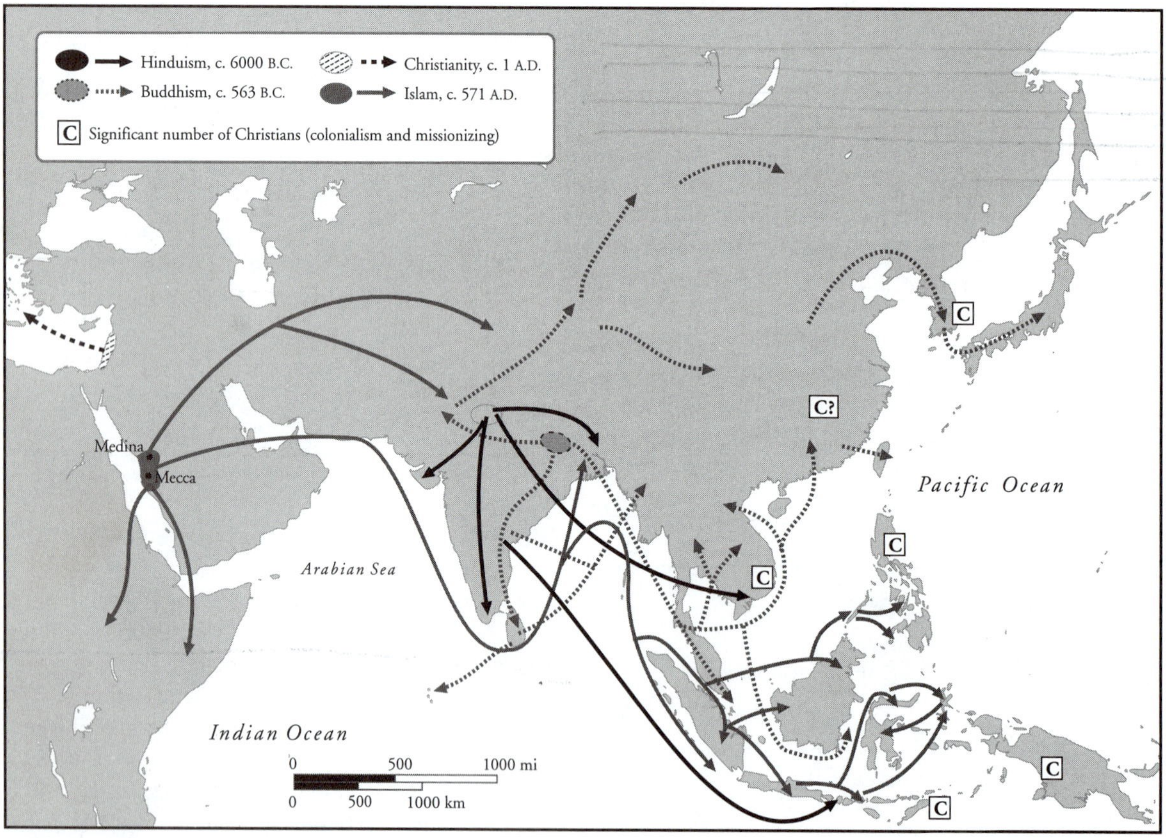

Figure 1-2

Origin and diffusion of major religions. Religions spread from hearth areas by word of mouth, migration, and proselytization.

thrive in Malaysia, Singapore, and on the island of Bali in Indonesia.

Hinduism is a **polytheistic** religion, hosting more than 333,000 (meaning countless) gods. The most important of these, in terms of daily worship and rituals, are *Shiva* and *Vishnu* (Figure 1-3). However, all deities are manifestations of a one ultimate reality known as **Brahman.** There are, in essence, as many ways to practice Hinduism as there are gods. Although there is no set creed, adherents may draw upon an enormous body of literature. Earliest are the ***Vedas,*** meaning *knowledge*. Another critical work is the **Upanishads.** This means *sitting down near* one's spiritual advisor or **guru.**

Figure 1-3

This Indian man is a saddhu—*a holy man who has divested himself of worldly possessions to seek the One Ultimate Reality. His trident indicates that he is a devotee of Shiva.* Saddhus *migrate among sacred sites and live off donations from the public. As a true* saddhu *(not a beggar), this man asked for nothing. He was merely present. Yes, I did make a donation.*

Hindus believe in the **transmigration of souls** based on accumulated **karma.** All beings possess a soul and are positioned within a hierarchy. Where one is, on what might be called a status ladder, is conditioned by the state of one's karma. Karma is the sum of an individual's deeds. Good deeds can move one up the ladder and bad deeds result in a demotion. In fact, it may take numerous lifetimes to achieve the common goal of departing the hierarchy and having one's soul unite with the One Ultimate Reality of Brahman.

Critical to the Hindu faith is the concept of **reincarnation:** the notion that a person is reborn into another cycle of existence depending on the karma of the previous life. If a woman deliberately harms or kills another person or an animal, let's say a lizard, she might be reborn as that lizard in her next life. To quote a 1960s saying: "Keep your karma clean!"

Hinduism is entrenched in all aspects of Indian culture. You will learn more about its impact on society and development in the chapters on South Asia.

BUDDHISM

Buddhism also originated in northern India, in part as a reaction to the excesses of Hinduism prevalent by the sixth century B.C. From here, it diffused with missionary monks and traders via trade routes into China, and from there, across the Korean peninsula to Japan. After becoming well established yet differentiated in Sri Lanka, it spread seaward to Southeast Asia.

Adapted to local circumstances, Buddhism diversified and evolved into many schools of thought. An extensive body of literature was created, and many temples and monasteries have rich collections of sacred books, tablets, and scrolls.

Buddhism and its various schools evolved from the teachings of Prince Siddartha Gautama, born around 563 B.C. in what today is Nepal. After years in search of an explanation of life's harsh realities, he *awakened* to a profound conclusion: The cause of suffering is desire. This realization was the key to his enlightenment as the **Buddha.**

That the world is filled with suffering and that it is caused by desire became the first two of **The Four Noble Truths.** The last two are that it is possible to overcome desire by following **The Eightfold Path.** By choosing this right path of behavior (e.g., right knowledge, aspiration, and livelihood), one can achieve **nirvana.** As inscribed on the wall of a temple in Myanmar: "*Nibbana* is not situated in any *Bhumis* [stages of being] nor is it heaven. It is a state which is dependent upon ourselves." Nirvana is an indescribable state of desirelessness and peacefulness. It is absolute bliss.

Buddha advocated withdrawal from the material world of desire into a contemplative mode, making meditation a key part of Buddhist practice. He also emphasized being generous, kind, and vowing not to harm people or animals. Consequently, vegetarianism is widely practiced in Buddhism.

As Buddhism diversified with expansion, two main schools of thought developed. **Mahayana Buddhism** is practiced in China, Korea, Japan, and, because of Chinese influence, in Vietnam. This school holds that salvation can be achieved through the intercession of divine beings. Mahayana Buddhists believe the Buddha to be their savior but may commiserate with other deities as well.

One of the most popular of these is the **Bodhisattva** Avalokitesvara (in Sanskrit, the lord who looks in every direction). A Bodhisattva is one who postpones entry into Nirvana in order to save other beings. Avalokitesvara is most popular in his female form and is enshrined in virtually all Buddhist temples. She is known as *Kuan-Yin* in China; *Kannon* in Japan; and *Kwanum* in Korea.

Theravada Buddhism developed in Sri Lanka as the monastic faith of Buddha's teachings. It places responsibility on the individual for salvation through the accumulation of merit by good behavior and religious activity. Monastic orders, or ***sangha,*** are very important, and monks and nuns are highly respected. It is Theravada Buddhism that diffused to Southeast Asia, and it is very much a part of the cultural landscapes of Cambodia, Laos, Myanmar, and Thailand (see Figure 1-4).

What happened to Buddhism in India? It rose to its zenith under Emperor Asoka in the third century B.C. But Hinduism eventually reasserted itself, and Buddhist adherents became a rarity. Aside from a few practitioners and scattered temple relics, Buddhism has faded from the Indian landscape. However, in its Himalayan refuge situation, Bhutan remains a Buddhist Kingdom. Buddhism is also important in Nepal.

Buddhist temples and other landscape elements are increasingly common in Europe, North America, and Australia. This is a result of relocation diffusion as Chinese, Laotians, Vietnamese, and others continue to migrate to other parts of the world (see Figure 1-5).

CHRISTIANITY

The spread of Christianity from its hearth region in Israel/Palestine is closely tied to European Imperialism. Carried by colonizers, administrators, soldiers, sailors, traders, and missionaries, Christian doctrine arrived in Asia. It flourished in context of colonization that began in earnest in the sixteenth century, but competition with already established religions diminished its

Figure 1-4

It was sunrise in the mountain town of Mamyo, Myanmar, where I watched Buddhist monks walk past devotees who put rice and other food into their "begging bowls." This scene exists on a daily basis throughout the world of Theravada Buddhism.

Figure 1-5

This is one of the great halls of the Hsi Lai Temple in Hacienda Heights, California. It is specifically for bodhisattvas, *persons who resist going to nirvana in order to help their fellow humans on earth. The plaza of rectangles replicates rice paddies and is called the Field of Merit. Hsi Lai Temple is similar to one in Kaohsiung, Taiwan. It appeals to multiple nationalities and accepts variation in Buddhist practices.*

Figure 1-6

I came across this Pentecostal mission in the remote Dayak community of Mencimai in Kalimantan on the island of Borneo. Most of the people of this region follow folk belief systems or are Muslim. Even Christian converts retain elements of their traditional beliefs. There are not many converts in Muslim Indonesia. The church is built on stilts because of the tropical rains and surface flooding.

impact. Although Christians can be found throughout Asia, the most significant concentrations are in Vietnam and the Philippines as a result of French and Spanish influence, respectively.

Christianity is a **monotheistic faith** postulating the idea of a single god called *God.* A Jewish man known as **Jesus Christ** is believed to be the Son of God, the Messiah that had been promised roughly 4,000 years earlier. The Messiah would restore the relationship between God and humankind. The teachings of Jesus and his disciples spread, as the Word of God, via contact conversion. Numerous versions of a book called the **Holy Bible** offer both a history of Judaism and Christianity in addition to the teachings of Jesus and his followers. Many Christians believe that the Bible *is* the Word of God. Proselytization by missionaries, using the Holy Bible, was and continues to be a powerful force in the diffusion of Christianity (Figure 1-6).

Christians believe in salvation through faith in Jesus Christ and adherence to the teachings of the Bible. Historic circumstance has seen the continual division of Christianity. Two major branches, **Roman Catholicism** and **Protestantism,** with its multiple denominations, have made the greatest headway in Asia. Christianity has grown to the point that it is self-perpetuating. Christmas is celebrated by close to 150 million Christians in Asia. People of other faiths often join in the celebrations.

ISLAM

Islam, which translates as *submission to God,* is the youngest of the foremost religions. It originated around the subsequently holy cities of Mecca (Makkah) and Medina (al Madinah) in Arabia in the sixth century A.D. Islam swept east across Asia with successive waves of conquerors, traders, and missionaries. It replaced earlier religions in many areas. For example, Buddhism and Hinduism were once important in Indonesia but were virtually eradicated under the fervor of Islam. Islam is the dominant religion in Pakistan, Bangladesh, Malaysia, Indonesia, and Brunei. It is also a potent force in several other regions such as northern India and the southern Philippines. You will learn more about Islam in its regional contexts in later chapters.

Coming from the same core area, Islam has links to both Christianity and Judaism. It, too, is a monotheistic faith. Following a series of divine revelations, **Mohammed** (571–632) began preaching in Mecca and later Medina after being forced to flee there by his enemies. His teachings about a new way of life provided welcome structure in what was then a chaotic Arab world, and converts quickly accumulated. **Muslims** (believers) formed armies to invade and conquer the lands of **infidels,** or nonbelievers. Clashes with other established faiths resulted in death to millions on all sides. Unfortunately, religious conflicts associated with Islam versus other belief systems still exist.

Muslims believe in one God, **Allah,** and recognize Mohammed (Muhammad) as His prophet. They also view Jesus Christ as a prophet. Like the Jews, they avoid pork, perceiving it as unclean. To Muslims, the will of Allah is absolute. The Word of Allah is contained in the **Koran,** or *Qur'an,* which is Arabic for "recitation." Pivotal to being a good Muslim are the **Five Pillars of Islam.** These are: acknowledgment of Allah as God and Mohammed as His Prophet; prayer five times a day; a month of daytime fasting (known as **Ramadan**); the giving of alms; and at least one pilgrimage to Mecca. Held in the last month of the lunar calendar, this yearly event is known as the ***hajj*** and involves the convergence of millions of pilgrims in the holy cities of Mecca and Medina (Figure 1-7).

Dissension over interpretation of the scriptures eventually caused division in Islam, resulting in two major branches and several other sects. The majority of Muslims are orthodox **Sunnis.** The breakaway **Shiites** account for less than 15 percent. Shiism (*Shiah*) became rooted in Persia (Iran) in the sixteenth century and subsequently spread via Iraq and Afghanistan to Pakistan. It has even influenced Sunni Islam in India. Shiites are extremely fervent in their beliefs and emotional in their practices. To them, religious leaders, or ***Imams,*** are infallible, as are their interpretations of the Koran and other religious literature.

Islam proscribes life roles, modes of dress, and social behavior. However, the degree of stricture varies greatly from place to place, especially for women. Islam is also an influential force in governments of countries like Malaysia and Indonesia, where Muslims make up the majority population.

FAITH, EXPLORATION, AND TRADE

Trade had long existed between Asia and Europe, dating back to the Roman Empire. But Arab Muslims laid the basis for a world trade system in the Mediterranean and Indian Ocean regions. As Islam is a way of life, there is no separation between religion and commerce. In fact, Islam is an ethical faith, providing even the rules of commerce. Believers who worked with each other understood rules of fair trading practices. Nonbelievers who wanted to enter Muslim trading net-

Figure 1-7

This significent mosque is at Shah Alam, the new capital of Selangor Province west of Kuala Lumpur, Malaysia. Named for the Sultan Salahuddin Abdul Aziz Masjid, it was built in 1988 and holds 16,000 people. The mosque is oriented to face the holy city of Mecca. Loudspeakers on the minarets announce the call to prayer five times a day.

works found conversion the easiest avenue. Each trading community served as a vehicle for the expansion of Islam.

In the eighth century, Arab Muslims conquered the Sind region of the Indus river delta. Until then, Buddhists had dominated trade, while Hindus focused on agriculture. Buddhist traders found it expedient to convert to Islam. Eventually, Islamic trading posts were established along India's west coast from Cambay to Calicut on the Malabar Coast (Figure 1-8). When the Muslim Mughal Empire in India expanded to the Bay of Bengal (1574–1592), access was gained to both the Coromandel coast and adjacent coasts of Southeast Asia. Muslims thereby achieved hegemony over the region's sea trade.

Prior to the appearance of Arab traders, China had been using South Asian merchants to carry their cargo. Steel had been made in China since the eighth century and was in great demand for weapons and farm implements. As a result of their overland and sea expansion, Muslims became active in the China trade and remained instrumental well into the twelfth century. Muslim trading communities existed in several Chinese cities such as Canton (Guangzhou) on the southern coast.

An international world system of commerce was now functioning via a network of trading ports and shipping lanes. By the fifteenth century, Muslim enclaves dotted the northern coasts of Java and Sumatra. Muslim-controlled Banten (Bantam) was also the focal

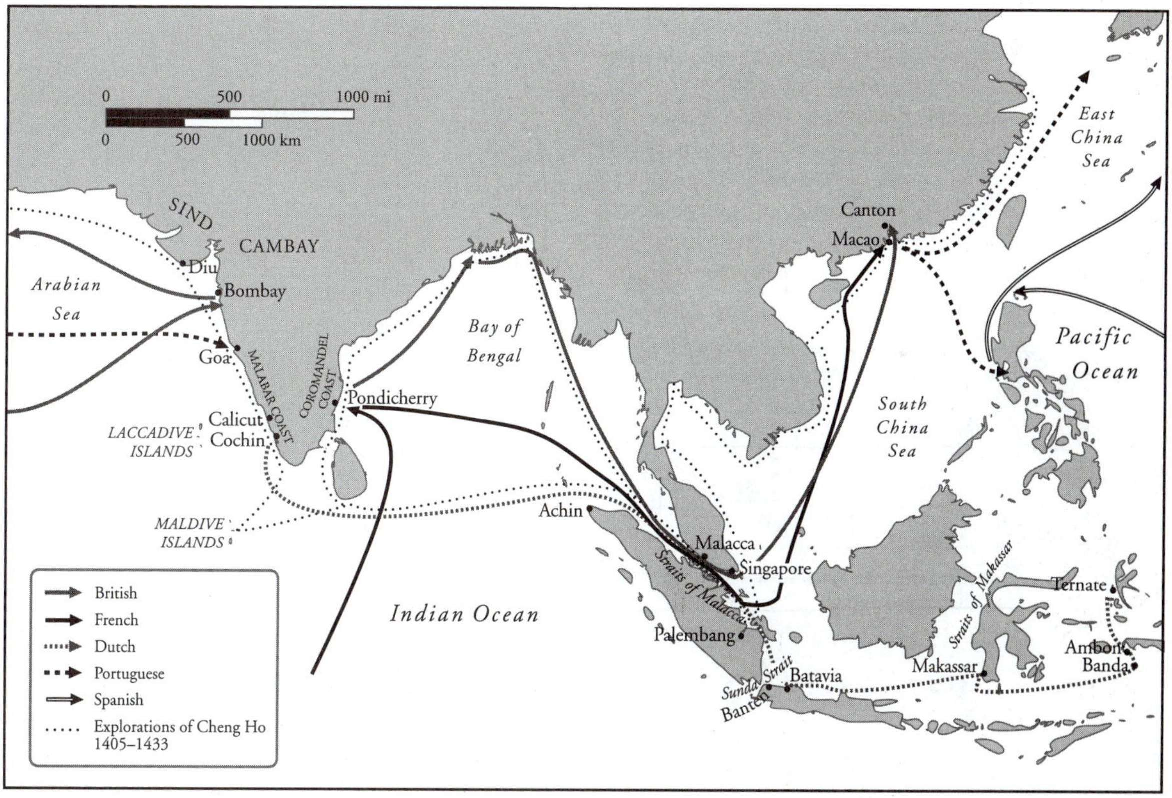

Figure 1-8

Trading centers and sea routes. Note that the Chinese explorer Cheng Ho navigated these waters long before the Europeans. The first European trading port was established by the Portuguese at Malacca (Melaka) in 1511.

point of Chinese trade in Southeast Asia. Makassar in Celebes (Sulawesi) was the key to the Spice Islands (Maluku). Achin (Aceh), Sumatra was situated at the northern entry to the Straits of Malacca. Of great importance was the founding of a transshipment port or **entrepôt** at Malacca (Melaka). Indians, Javanese, Chinese, and others migrated to these nodes of perceived opportunity where many of their descendants remain to this day.

Why this much interest? What was to be gained? Quite simply: riches. Demand was soon created for sandalwood and other aromatic woods; teak and other hardwoods; silk and porcelain; slaves and exotic animals; and countless other products. But pepper, along with cinnamon, cloves, nutmeg and a host of other aromatic and edible spices became the most revered (Figure 1-9). Arabs conjured tales of gross birds harvesting cinnamon from ferocious bat- and snake-guarded swamps to dissuade others, especially the Europeans, from discovering their spice sources.

Neither Arabs nor Europeans were first in the spice trade. Indigenous states and empires were trading such commodities long before outsiders. Roman historian Pliny told of Indonesians in outrigger canoes driven by the prevailing winds unloading cargoes of cinnamon and cassia on the east coast of Africa in the first century A.D. The great sea-borne empire of Srivijaya, centered in Palembang on the east coast of Sumatra, traversed the Malacca and Sunda Straits dominating east–west sea routes in the seventh century.

Figure 1-9

I had always wondered what mace looked like in the plant. Finally, I got a chance to see it in a market in Pinang, Malaysia. Mace is the stringy stuff covering the nut inside the outer nutmeg shell. It's amazing to think that these products changed the face of the world.

China was also involved in ocean trading. The talented Admiral Cheng Ho, in a series of voyages from 1405 to 1433, touched base at virtually every port in the region, from Southern China to the east coast of Africa (Figure 1-8). He discovered that Chinese merchants were already well established in many of these places. Europeans truly were latecomers to this dynamic, lucrative world.

Europeans: Exploration, Commerce, and Conquest

By the late 1400s and early 1500s, Europeans became familiar with what Arabs had long been aware of: the annual reversal of winds associated with Asia's monsoon climatic system (see Chapter 2). "Monsoon" comes from *mausim*, Arabic for "season." By sailing from the Mediterranean, around Africa's Cape of Good Hope and following the prevailing winds across the Indian Ocean between June and September, and doing the reverse from December to March, sailing vessels could make the round trip in less than a year.

This was the **Age of Exploration** for Europeans carrying the **White Man's Burden:** a zealous mission to convert the *heathen* to Christianity. But the real driving force was economic. And so they ventured across the globe in quest of spices, gold, and other riches. In 1488, Bartolomeu Dias of Portugal rounded the Cape of Good Hope and sailed as far as the Persian Gulf. Upon returning home he announced that the sea route to India was open! Then in 1498, Vasco da Gama sailed as far as Calicut on India's Malabar Coast and later entered Lisbon's harbor with the first cargo brought from the "Far East" to Europe by sea. Ferdinand Magellan pioneered the westward route around South America's Cape Horn in 1519. Reaching the Philippines in 1522, he initiated Spain's entry into the region. Holland's (Netherlands) 300-year involvement in the **Indies** (Indonesia) commenced with the 1595 voyage of Cornelius de Houtman, who returned to the

Netherlands with three shiploads of pepper and nutmeg.

England's Sir Francis Drake obtained a cargo of spices, plus booty plundered from a Portuguese ship, in his circumnavigation of the world (1577–1580). This accomplishment, in concert with Britain's defeat of the Spanish Armada in 1588, was a pivotal event in the sanctioning of further British expeditions. The race for territory was on, and it was inevitable that swords would cross.

Territorial gain and sea lane control became paramount concerns. The importance of reign over the Straits of Melaka cannot be underestimated. A plaque in the garden of the Sultan's former palace in Melaka reads: "Whoever is lord of Malacca has his hand on the throat of Venice."

The Portuguese established trading ports from Africa to China, and by 1520 they dominated the seas. Intolerant of Islam, and greedy for profit, they forced out the Muslims, who retreated to Aceh in Sumatra and Johor on the Malay peninsula. The Portuguese moved into Cochin, Diu, Goa, and created a new port at Bombay (Mumbai) in India. They took the critical port of Malacca in 1511, and acquired control of Macao in 1557 (Figure 1-10). The Portuguese kept Tecate in the Spice Islands until 1570, when it was wrested away by the Dutch.

Figure 1-10

This is the Porto de Santiago, main entry to the Portuguese fort of A Famosa in Melaka. The Portuguese established their fort and colony in 1511. The gate overlooks the Straits of Malacca (Melaka).

Both the Dutch and the British were anxious to oust the Portuguese from South and Southeast Asia. Backed by stockholders, with monopoly privileges and the right to raise armies and navies, their trading companies aggressively intruded into the area. **Factories,** or trading posts, were quickly set up throughout the region. The Dutch took Tenate in 1607 and founded Batavia on Java in 1619. Malacca fell into their hands in 1641, and Banten (by now British) was acquired in 1682. When the Dutch attacked the Johor port in 1784, the British responded.

Britain had initially concentrated on India but became intrigued with the economic potential of Southeast Asia, where they vied with the Dutch. They seized Melaka in 1796 and gained the island of Singapore in 1819. France entered the picture and established the first of its settlements at Pondicherry, India, in 1674. However, primarily interested in spreading Christianity, the French were not as aggressive as their contemporaries. And so the era of European colonialism was launched.

Colonialism and Imperialism

With rampant **colonization** in the fifteenth and sixteenth centuries, an extensive web of ports and shipping lanes fostered the evolution of an Atlantic economy, as well. Resources, such as silver, from the New World became integral to economic exchange between Atlantic and Pacific Realms. A world economy was now in place.

The number of imperial powers increased from two (Spain and Portugal) in the sixteenth century to five in the seventeenth century with the addition of Britain, France, and Holland. Collections of colonies worldwide were absorbed into the web of economic interchange. The only place not colonized in Southeast Asia was Siam (Thailand). It became a **buffer state** between British and French interests in Mainland Southeast Asia (Figure 1-11).

Competition was fierce during this mercantilist period during which raw materials were brought to the colonial power, where factories produced manufactured goods to be shipped back to the colonies. Under **mercantilism,** colonies were not allowed to

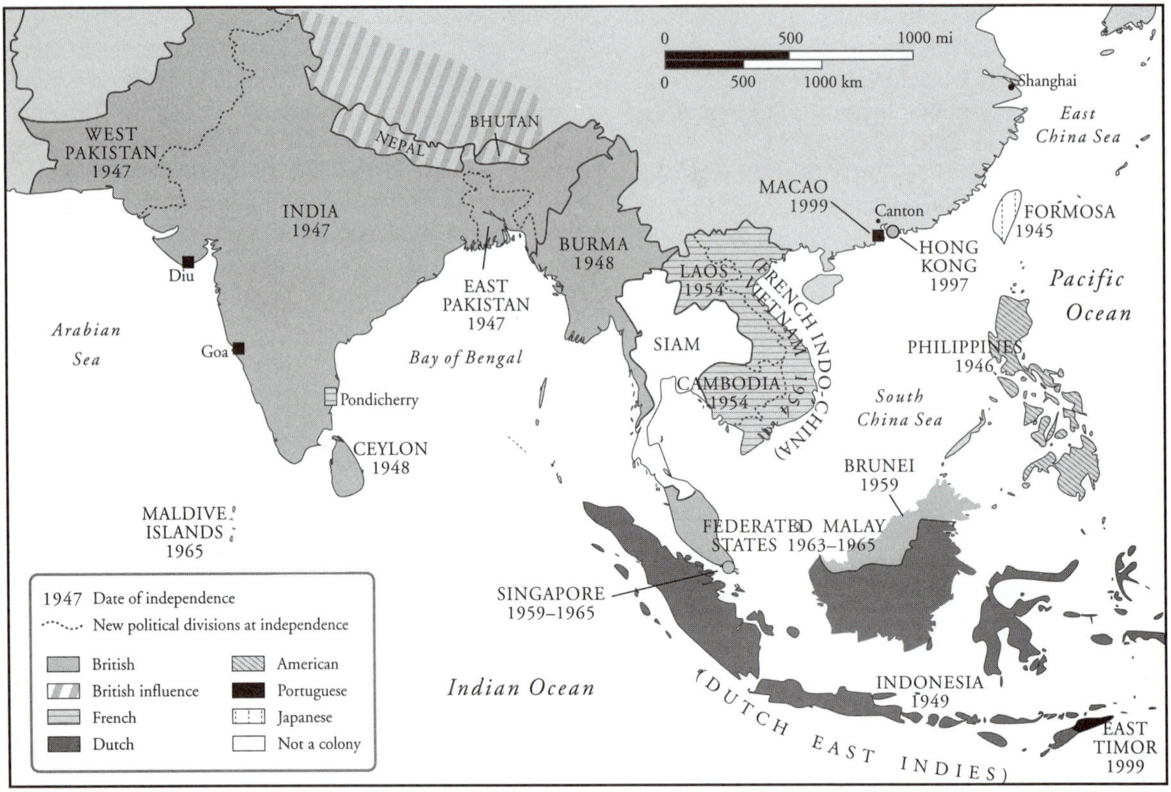

Figure 1-11

Most of Asia was either colonized or strongly influenced by foreign powers, with the exception of Thailand, which served as a buffer between English and French interests in Southeast Asia. The Philippines was a Spanish colony from 1565 to 1898 and an American one thereafter. Formosa (Taiwan) was a Japanese colony from 1895 to 1945. Although China was never a colony, many places such as Canton (Guangzhou) and Shanghai had foreign concessions.

manufacture such goods. Neither were they allowed to trade with other than their own colonial overlord. The Seven Years' War (1756–1763) between Britain and France ended French ambitions in India and North America and allowed Britain to become the world's leading power. With colonies around the globe, it was said that the sun never set on the British Empire.

Britain now expanded into India, and France reached into Indo-China. Holland tightened its grip on its Dutch East Indies (Indonesia). In order to control their monopoly on the spice trade and to jack up prices, the Dutch began destroying all the spice plants on certain islands. In fear of instant poverty, islanders resisted, with terrible consequences. For instance, almost the entire population of the Banda Islands was exterminated. By attacking and blockading Indonesian-owned ships, the Dutch managed to impoverish other islands as well.

In 1869, the British- and French-owned Suez Canal was opened, ushering in the **Age of Imperialism** (1870–1914). By reducing the sea journey to Asia by thousands of miles, interactions between Europe and Asia intensified. With increased territorial gains came friction. Rivalries, once confined to the European continent, spilled over into the colonies.

Those colonized did not accept their inequitable situation without protest. Resistance movements were countered by imperialist wars of oppression. In England these were known as *Queen Victoria's Little Wars.* In British India, French Indo-China, and Dutch Indonesia, at least 70,000 died in battle before the turn of the century. Suppression of resistance in Burma alone (1823–1826) resulted in 15,000 lives lost.

World War I (1914–1918) was a cataclysmic event in history. It spearheaded a new era for the colonial world. The Treaty of Versailles in 1919 spurred nationalistic claims on the part of colonies, and reassigned them to the victorious Western powers. But things could not be as they were; the so-called *Belle Epoch* was over.

The Colonial City

Colonial cities were intended to serve the interests of colonial powers. They were guarded outposts that functioned as military, economic, and administrative centers. Architecture was grand, meant to impress the masses with the power and glory of the empire (Figure 1-12). Symbols of colonial rule were expressed in the form of architectural styles, statues, and crests. Tudor-style apartments still exist in the former British Quarter of Shanghai, and Dutch post boxes can be found in Jakarta.

The colonial language was employed for naming places and streets and was the language of rule for administrative and educational purposes. Street signs such as Orchard Road and Rue Louis Pasteur remain in many former English and French colonial cities, and these languages continue to be spoken by educated elites.

To preserve race and class distinctions, separation was imposed on both people and landscape. Europeans were removed in their own residential quarters, segregated from the natives by physical barriers such as rail lines or waterways, or by military cantonments. Locals could only enter the foreign quarters with permission, and customarily did so in a servant capacity. Many sources refer to signs posted in China's and India's city parks stating, "No Chinese (Indians) Allowed."

Some cities were designed with a standardized plan such as Spain's 1573 *Laws of the Indies.* This called for a grid-style street layout, centered on a main plaza surrounded by government buildings and a church. Similar but smaller plazas, built as the grid expanded, were intended to ensure the dissemination of order and godliness. This influence is evident in the Philippines. Britain and France also liked the grid system because it represented order and regularity. Additionally, straight streets facilitated speedy access to trouble spots.

The colonial style city differed widely from the traditional city with its narrow, winding streets and bustling crowds. The colonial city had one focal point, but the traditional city had several, each focused on a government, religious, or market center. To the occupiers, the planned colonial city symbolized order within chaos.

As colonial power strengthened, city architecture and all that it stood for became even more imposing. Local people were relegated even further into the socioeconomic background. In Singapore and Kuala Lumpur, the British, in concert with Chinese and Indian migrants, dominated practically all activity. Indigenous Malays faded from the scene. This was also true in the Dutch East Indies. To overcome such injustices, both Malaysia and Indonesia currently have special policies favoring Malays.

Figure 1-12

This building reflects both Dutch and Malay architecture. Eventually it was used by the British as the exclusive Melaka Club. Now it is an Independence Memorial for Malaysia.

Imperialist Japan and the Rise of Nationalism

After centuries of feudalistic rule under powerful Shoguns, the Emperor Meiji was restored to the throne in 1868. Young reformers were determined to bring Japan into the Industrial Era. The Industrial Revolution had been initiated around 1750 in England and was in full swing elsewhere. Some saw potential for the island nation as the "Britain of the Pacific." Why could Japan not have a colonial empire? After all, it needed raw materials for industrialization and space for expansion. Japan flexed its muscles, attacking and defeating both China and Russia at the turn of the century. China's island of Taiwan (1895) and footholds on the mainland were the rewards of victory. Japan's empire was taking shape. Korea was annexed in 1910. Japan's territorial ambitions were fervently pursued in China, and by the mid-1930s, Japan dominated a vast empire (Figure 1-13).

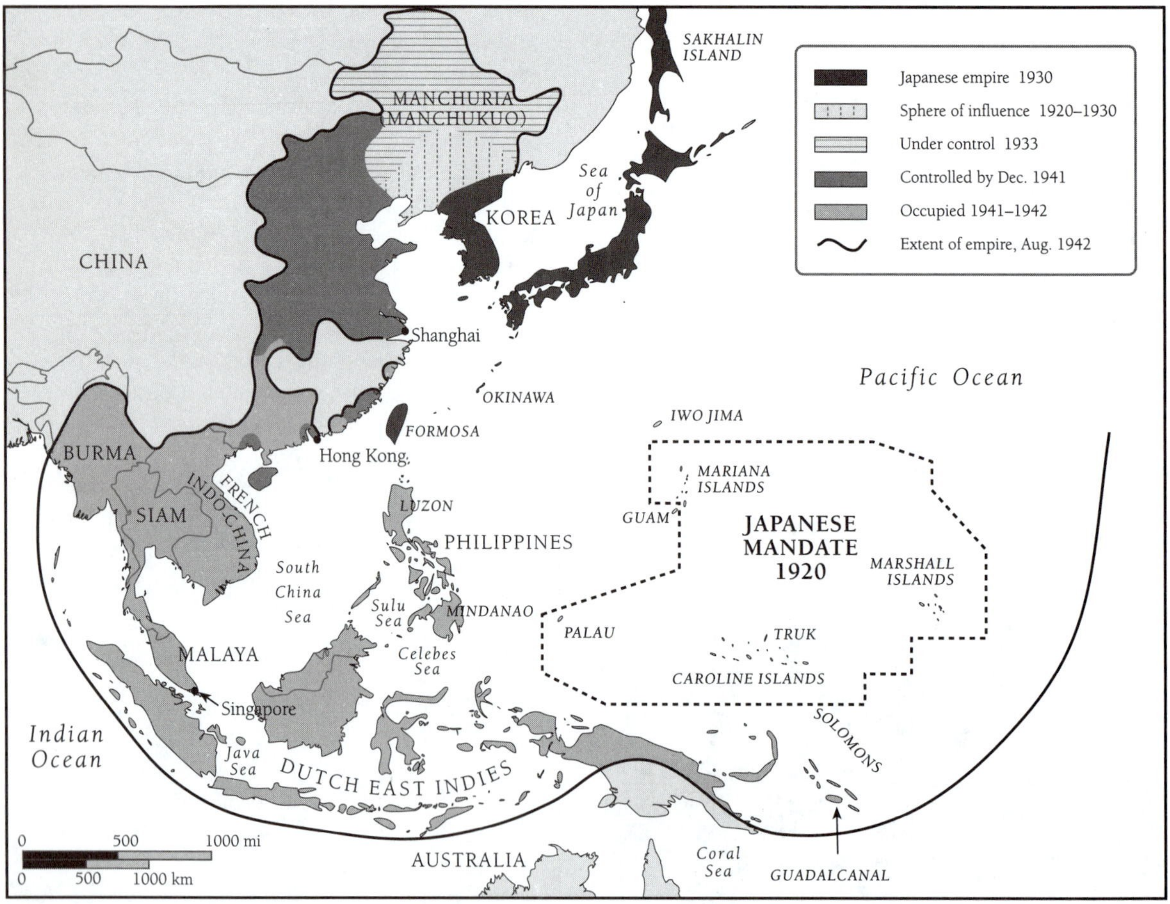

Figure 1-13

Japan's Greater East Asia Co-Prosperity Sphere. Japan won Formosa from China as a prize of war in 1985, annexed Korea in 1910, and established Manchukuo in 1933. It pressed for its goal to become "Britain of the Pacific" until its defeat by Allied Forces in 1945.

Japan's ambition was to establish a **Greater East Asia Co-Prosperity Sphere (GEACPS),** incorporating Asians minus the European imperialists. As they defeated both indigenous and colonial forces, they propagated the slogan "Asia for the Asians." Their grandiose plans were foiled when they drew the United States into the Pacific theater of war by bombing Pearl Harbor, Hawaii, on December 7, 1941. By the time of Japan's defeat in 1945, it controlled all of the northern Pacific west of the international date line. Japan then dominated 100 million people!

After 1945, Japan was relieved of all its colonies but the GEACPS had left a legacy: Japan had been the vanguard of Asian nationalism. "Asia for the Asians" became a cornerstone of independence struggles. But new socioeconomic–political theories were on the horizon to shape the entire process.

Independence and Nationalism

European imperialism began to crumble under the weight of independence movements before and after World War I: After the Second World War (1939–1945), it crashed. During the process of **decolonization,** independence was both negotiated and fought for. Countless lives were sacrificed. France, which was the most resistant to giving up its colonies between 1945 and 1954, lost 95,000 persons trying to keep its hold on Indo-China. This significant loss is overshadowed by the deaths of well over half a million Vietnamese.

Asian independence leaders, schooled in the ideals of communism or democracy, were determined to build their own independent states characterized by dignity, equality, democracy, and self-determination. China's Sun Yat-Sen (Sun Zhong-shian) (1866–1925) was regarded as the first great Asian nationalist. While China as a whole was not a colony, many parts of it were colonized, and foreign powers strongly influenced its government. The name Sun Yat-sen, along with others such as Ho Chi-Minh (1890–1969) in Vietnam and Sukarno (1901–1970) in Indonesia, soon became familiar to newsreaders and listeners. Such leaders instilled a new sense of national consciousness among their respective peoples.

Marxism and Leninism were attractive to some because they provided explanations for what was wrong with a world under imperialism and capitalism, told how to get rid of the old order, and offered a model to put in its place. Marx based his theories on the plight of an urbanizing and industrializing Europe. Lenin recognized the potential power of the rural peasantry. Asians witnessed the success of the Russian Revolution in 1917, when the rural masses overthrew an imperialistic Tsarist regime. Lenin's calls for revolution hit home with anti-imperialist, pro-nationalist groups. It is important to understand that communism was not the driving force behind independence uprisings in Asia. Nationalism and patriotism were.

Social movements of unprecedented scale and devastation brought independence, although it was achieved sporadically (refer back to Figure 1-11). Each country has its own story to be spelled out in later pages. A patchwork quilt of political systems emerged: democratic, communist, militarist. Whatever the end result, as the Chinese leader Mao Zedong (Mao Tsetung) noted in 1957, "The east wind prevails over the west wind."

Prevalence of east-wind political, economic, and social systems did not bring stability, though. Communism with its massification, lack of incentives, and mismanagement was unable to achieve its ideals. It eventually collapsed in the Soviet Union in 1991. This momentous event, coupled with new economic realities of global proportions, spearheaded profound changes in Asia and the world.

Evolving World Order

Time–space convergence is an integral part of our changing world; more space can be covered in a shorter time span. In 1850, undersea communication cables were laid around the world. In that same year, an American-built clipper ship, the *Oriental,* made a record journey from Hong Kong to London in 97 days.

In 1935, a freighter from Los Angeles, struggling in a winter storm, took 19 days to reach Tokyo. Sea-mail took weeks as well. Today, we can e-mail across the globe in mere seconds.

The number of Internet hosts has multiplied to more than 100 million. World Wide Web sites in Asia are supported and influenced by governmental or private organizations. Malaysia has the fastest-growing Internet system in Asia. Where political circumstances bar open communication (e.g., in Tibet or Myanmar), sites are sponsored by overseas communities.

Cyberspace—the sphere of information technology—is inconceivably complex and highly controversial. Geographer Paul Starrs (1998) observes that although cyberspace has been placed into reality and inured with its own personality, it has no humane substance or moral conviction. At the moment, by form, content, and accessibility, it favors the educated and more fortunate. Nevertheless, the entire nexus of human interaction is affected by its presence.

Since 1980, telecommunication traffic has averaged an increase of 20 percent a year. In fact, by 1995, the number of e-mails exceeded the number of letters sent! As of 1998, the world's most sophisticated communication infrastructure is in Singapore. The vast, intertwined, and overlapping channels of communication produced by flows of technology, hardware, and software disseminated by transnational or supranational corporations, organizations, and agencies are called **technoscapes.** Dramatic changes in technology have spurred far-reaching alterations in both national and international political, economic, and social systems. This **restructuring** is manifesting profound changes for humanity and habitat (Figure 1-14).

Figure 1-14

I was on a river trip in Borneo and was amazed at the number of satellite dishes marking river and even interior communities. This scene is from Samarinda, a port on the Mahakam River. Televisions are often shared among houses, and American programming such as MTV is available. Note the stilt houses that allow for tidal changes and rising waters in the wet monsoon.

As the world's single superpower, the United States sits in the middle of an extensive and complex, global network of flows of goods and information. The United States now exports more goods to Asia than to Europe. Money is transferred electronically among leading stock exchanges in London, New York, Hong Kong, and Tokyo. German BMWs are sold in the United States, yet they are assembled in Vietnam. Hewlett-Packard's global headquarters for microwave components and hard disk drives is in Penang, Malaysia. Designers Donna Karan and Ralph Lauren source their apparel to Taiwan, South Korea, and Singapore for manufacture. Clothing sold at The Gap, Sak's Fifth Avenue, and many other stores are produced in South Asia.

1997 was Canada's "Year of the Asia-Pacific." After the United States and Britain, the leading investors in Canada are Japan, Hong Kong, Australia, Singapore, and South Korea. Canada is not only tied economically with other Commonwealth Countries such as India and New Zealand, but also with its Pacific Rim partners. Vancouver, British Columbia, is an important port on the West Coast of North America, particularly for Canadian grain and wood products destined for Asia. It hosted the Asia Pacific Economic Cooperation Conference in 1997.

Australia is also becoming increasingly tied to this Asia-Pacific trade network. A country that shunned immigrants from Asia until the 1970s now has engagement with Asia as a major component of its foreign

Figure 1-15

Thailand's Trade Relationships Within APEC 1994. Note Thailand's connectivity to Australia, and North and South America. This trade pattern is a Pacific Rim Phenomenon. Source: Department of Foreign Affairs and Trade. 1995. The APEC Region Trade and Investment. *Canberra: Australian Government Publication Service.*

Americanization

Consider the following facts. The July 4, 1997, issue of *Asiaweek,* published in Hong Kong, sports Ronald McDonald replacing Mao Zedong's picture over the Gate of Heavenly Peace in Beijing's Forbidden City (an advertising gimmick). The Marlboro Man is riding across the back cover. There are Coca-Cola museums in China, Japan, and the Philippines, and Taiwan boasts the world's largest inflatable Coke bottle. Vietnamese versions of rap music boom in taxis and from music shops in Hanoi and Saigon. In China, Nike sells with the "Flier Qiao Dan"—Michael Jordan. Once I witnessed a group of elderly women squatting around a TV on the floor of a communal longhouse in a remote region of Indonesian Borneo. They were laughing, pointing, and shaking their heads at the apparent silliness in front of them. It was MTV.

Figure 1-16

This Kentucky Fried Chicken on the island of Java in Indonesia shows cultural integration with the inclusion of Ayam Goreng*—Fried Chicken with specific Indonesian spices. It is delicious.*

These are all examples of American **popular culture,** which is diffusing via transnationals and media into the remotest corners of Asia (Figure 1-16). What impact does this trend have on lives and landscapes?

In places such as Taiwan, Malaysia, and Thailand, some people think that American popular culture is detrimental to traditional values, especially those concerning youth behavior and familial relationships and responsibilities. In India, cooked-chicken sellers protested against Kentucky Fried Chicken for ruining their businesses. Elsewhere, small, family-run shops are being replaced by franchise chains such as 7-Eleven, and home-made ice cream parlors by such franchises as Baskin and Robbins. Avon, Amway, and other direct marketers are changing long-held relationships between customers and local merchants.

Americanization is controversial. Although it is true that such enterprises provide employment, where do most of the profits go? Televisions are becoming ubiquitous in Asia. Although television can be a way of teaching and promoting public health or family planning, it also increases expectations and promotes consumerism. This may be good for business, but is it good for society in general? How do this medium and American programming such as CNN, MTV, and various comedy, crime, and drama series influence perception and behavior? Korea and China, for example, are concerned about rising levels of obesity related to the increased availability of junk food. Malaysians and Taiwanese worry about inappropriate youth behavior.

Some Asian leaders such as Lee Kwan Yew of Singapore and Muhammed Mahathir of Malaysia, see popular culture as contributing to secularism and liberalism. Both pursue modernization and change but are convinced that national cultures should transcend such potentially harmful Western elements and "turn east" for guidance.

Main Point Summary

1. More than half of humanity inhabits South, East, and Southeast Asia. These geographic realms are part of a larger entity called Asia.
2. Asia is characterized by tremendous cultural and physical diversity. Interrelationships between people and environments generate experiential landscapes, (i.e., lived worlds or lifeworlds).
3. In the Western world, perceptions of Asia tend to be Eurocentric. Misconceptions of Asia have led to negative stereotyping, off-base broad generalizations, and a sense of superiority in attitudes and behavior.
4. Religion and belief systems pervade the lives of most people in Asia. Hinduism, Buddhism, and Islam have the most adherents, but missionary work is drawing increasing numbers to Christianity.
5. Arab Muslim traders were instrumental in charting the way for subsequent European ventures into the Indian Ocean and South China Sea. These waters were not "discovered" by Europeans, as they were already familiar to the Chinese and other sea traders. However, with God and gold in their minds, Europeans quickly gained footholds in the region.
6. Although an array of products were in demand, spices captured the European imagination in the early years of their entry into the region. As more opportunities arose for economic gain, serious competition ensued.
7. What initially were trading posts became colonial possessions, and what started out as colonies became imperial outposts of grand empires. Mercantilist policies bolstered homeland industries but depleted resources, impoverished and enslaved people, and wrought immeasurable and indelible social change on Asian cultures.
8. Two world wars and occupation by Japan altered the path of events. Indigenous, foreign-educated individuals emerged with a renewed sense of nationalism and patriotism to lead self-determination and independence movements. After significant bloodshed, the colonial powers withdrew from most of Asia.
9. In the aftermath of decolonization, a variety of political structures were established, with democracy vying with communism for dominance. The collapse of the Soviet Union in 1991 ushered in an era of significant change.
10. The new world order comprises globalization on one hand and regionalization on another. With increasing power of multinational conglomerates, individual nations band together to protect their own regional interests.
11. Americanization via business and media, while providing economic and other opportunities, is having a deleterious effect on health, behavior, and traditional cultural and social relationships.

Terms to Know

dragon and tiger economies	geographic realms	colonizer's model	religion
landscape	ecumene	xenophobia	diffusion
social construction	functional interaction	ethnocentrism	hearth area
quality of life	myth of Asia	we–they syndrome	parenthetical
environmental transformation	*mappaemundi*	Eurocentric	polytheistic
globalism	Orient	Sinocentric	Brahman
regionalism	Occident	Congress of Vienna	Vedas
	environmental determinism	toponym	Upanishads

guru
transmigration of souls
karma
reincarnation
Buddha
The Four Noble Truths
The Eightfold Path
Mahayana Buddhism
Bodhisattva
Theravada Buddhism
sangha
monotheistic
Jesus Christ
Holy Bible
Roman Catholicism
Protestantism
Mohammed
Muslim
infidel
Allah
Qur'an (Koran)
Five Pillars of Islam
Ramadan
hajj
Sunni
Shiite
Imam
entrepo[at]t
Age of Exploration
White Man's Burden
Indies
factories
colonization
buffer state
mercantilism
Age of Imperialism
Greater East Asia Co-Prosperity Sphere (GEACPS)
decolonization
timespace convergence
technoscapes
restructuring
globalization
multinational enterprise
transnational enterprise
regionalization
Asia Pacific Economic Cooperation (APEC)
Pacific Rim
Asia-Pacific
popular culture
Americanization

Bibliography

Blaut, J. 1994. *The Colonizer's Model of the World: Geographical Diffusionism and Eurocentric History*. New York: Guilford.

Borthwick, Mark. 1992. *Pacific Century: The Emergence of Modern Pacific Asia*. Boulder: Westview.

Brierley, Joanna Hall. *Spices: The Story of Indonesia's Spice Trade*. Kuala Lumpur: Oxford.

Chapman, G. P., and K. M. Baker eds. 1992. *The Changing Geography of Asia*. New York: Routledge.

Cressey, George B. 1963. *Asia's Lands and Peoples*. 3rd edition. New York: McGraw-Hill.

DeBlij, Harm J., and Peter O. Muller. 1997. *Geography: Realms Regions and Concepts*. 8th edition. New York: Wiley.

Drakakis-Smith, David. 1993. "That Was Then, This Is Now: Forty Years of Social and Economic Change in the Tropical Third World." *Singapore Journal of Tropical Geography* 14: 81–102.

Georgacas, Demetrius J. 1969. The Name Asia for the Continent; Its History and Origin. *Names* 17: 32–37.

Ginsburg, Norton. 1988. "Geography and the Pacific Century." *Asian Geographer* 7:1–11.

Hodder, R. 1992. *The West Pacific Rim: An Introduction*. London: Belhaven.

Jones, E., L. Frost, and C. White. 1993. *Coming Full Circle: An Economic History of the Pacific Rim*. Boulder: Westview.

Kitagawa, Joseph M. ed. 1989. *The Religious Traditions of Asia*. New York: Macmillan.

Knox, Paul. 1995. World Cities and the Organization of Global Space. In *Geographies of Global Change*, ed. R. J. Johnston, Peter J. Taylor, and Michael J. Watts, pp. 232–247. Oxford: Blackwell.

Lent, J. A. 1995. *Asian Popular Culture*. Boulder: Westview.

Levathes, Louise. 1994. *When China Ruled the Seas*. New York: Oxford.

Lewis, Martin W., and Karen E. Wigen. 1997. *The Myth of Continents: A Critique of Metageography*. Berkeley: University of California.

Michalak, Wieslaw, and Richard Gibb. 1997. Trading Blocs and Multilateralism in the World Economy. *Annals of the Association of American Geographers* 87: 264–279.

Nakemura Hajime. (1964) 1997. *Ways of Thinking of Eastern Peoples: India, China, Tibet, Japan*. London: Kegan Paul.

O'Meara, Molly. 1997. Riding the Dragon. *World Watch*. 10: 8–18.

Risso, Patricia. 1995. *Merchants and Faith: Muslim Commerce and Culture in the Pacific Ocean*. Boulder: Westview.

Ryan, Paul R. 1989–90. Pacific Century Dead Ahead! *Oceanus* 32: 5–12.

Spencer, Joseph E., and William L. Thomas. 1971. *Asia, East by South: A Cultural Geography*. 2nd edition. New York: Wiley.

Spencer, Joseph E. 1973. *Oriental Asia: Themes Toward a Geography*. Foundations of World Regional Geography Series. New York: Prentice-Hall.

Starrs, Paul. 1998. The Sacred, the Regional, and the Digital. *The Geographical Review* 87: 193–219.

Steadman, John M. 1969. *The Myth of Asia*. New York: Simon and Schuster.

Unwin, Tim. ed. 1996. *Atlas of World Development*. New York: Wiley.

Watanabe, Toshio. 1992. *Asia's Growth and Agony*. Honolulu: East West Center.

Chapter 2

Environments and People

"Those who would take over the earth/And shape it to their will/ Never, I notice, succeed."

TAO TE CHING (C. 3000 B.C.)

Major Topographical Features

South, East, and Southeast Asia exhibit a grand and complex display of mountains, plateaus, drainage basins, and river valleys. All of these have been modified and have influenced people's actions and interactions since the beginning of human time.

The foundation for Asia's **topography** (landforms) is an extensive zone of ancient rocks that, in early geologic time, were folded, faulted, and intruded on countless occasions. Ongoing **hydrological** (water) and **aeolian** (wind) **erosion** and **deposition** have combined to generate Asia's physical geography as we know it today (Figure 2-1). Unlike North America and Europe, Asia was not covered with massive ice sheets, with the exception of high mountain systems such as the Himalaya. In fact, **glaciation** continues to play a critical role in sculpting mountainous regions.

The most imposing topographic features are the Himalayan ranges and adjacent Tibetan Plateau. The Greater or highest Himalaya extend 1,500 miles (2415 km) from west to east with an impressive number of the world's highest peaks (check your atlas for locations). No fewer than fifteen peaks top 25,000 feet (7,000 m). Annapurna at 26,535 feet (8,090 m), Dhaulagiri at 26,689 feet (8,137m), Kanchenjunga at 28,208 feet (8,598 m), and Everest, the highest at 29,028 feet (8,848 m), are the best known mountains. Imagine a fractured rock wall around 4 miles (6 km) high and 200 miles (322 km) across. This is the majesty of the Himalaya.

Although cut by numerous passes, the Himalayan ranges have posed a formidable barrier between Indian and Chinese realms demarcating racial, linguistic, and other cultural differences. The mountains themselves form a region of **cultural convergence** where lifeways and belief systems blend in splendid landscapes (Figure 2-2).

The history of the Himalaya began 200 million years ago with the process of **continental drift** when the landmass known as Pangaea broke into pieces. At that time, Gondwanaland (Africa) and what we know now as India proceeded toward another, larger landmass called Laurasia. When the collision occurred 45 million years ago, mountains emerged in a colossal upheaval. The Tethys Sea vanished, consumed in the process. When trekking near the Tibetan border, I was

Figure 2-1

This picture of the Spiti Valley, in the Indian Himalaya, clearly illustrates the impact of water, wind, and ice in physical landscape formation. A small community with its barley fields clings to the base of the alluvial fan. The Spiti River is a tributary of the Indus River system.

Figure 2-2

There are many Tibetans living in the Himalaya of India's Himachal Pradesh. This gompa (monastery), at almost 13,000 feet (3,890 m), was built in the sixteenth century.

delighted to find two **ammonites:** flat, spiral fossil shells, evidence of life in the ancient watery world of the Tethys Sea.

The Himalaya is a product of **plate tectonics.** Plate tectonics theory postulates that the upper portion of the lithosphere comprises a patchwork of rigid plates. These are rooted in the underlying, more plastic upper layer of the earth's mantle. Movement along the fracture zones between the plates is responsible for a legion of topographic features and seismic phenomena.

Where plates diverge or move apart, molten material from the earth's core wells up into the space. Where plates converge or come together, spectacular mountains, pyroclastic events, and deep ocean trenches appear.

Did you know that Asia is considered the world's most disaster-prone region?

The Pacific Ring of Fire

Half of all major natural disasters are in the Asia-Pacific region. Because of the abundance and intensity of seismic activity, the region is often referred to as the *Pacific Ring of Fire* (Figure 2-3).

Volcanic eruptions and earthquakes have wreaked havoc on humanity for millennia. The most powerful eruption in recorded history occurred on the Indonesian island of Sumbawa in 1815 when Mount Tambora 9,348 feet (2,850 m) blew up 100 billion tons of volcanic rock and debris. The death toll was 92,000. China's Tangshan earthquake in 1976 killed more than 240,000 people, and a 1993 quake in western India's Maharashtra state took 12,000 lives. The Philippines' Mount Pinatubo eruption in 1991 and Japan's Kobe quake in 1995 are two recent catastrophes.

Tsunamis, or seismic sea waves, can be even more damaging than the seismic event. The speed of a tsunami is proportional to the depth of the water; the deeper the water, the faster the wave. As the water shallows, the fast-moving wave is slowed, causing the water to pile up on itself. Tsunamis have caused the deaths of hundreds of thousands of coastal and island people. For example, in 1976, a tsunami eliminated 8,000 people in the Philippines' Moro Bay.

The Pacific Ring of Fire also includes the world's deepest ocean trench: the Mariana Trench, which descends 36,198 feet (−11,033 m) below sea level. Another is the Indian Ocean's Java Trench at minus 23,376 feet (−7,125 m).

NORTH OF THE HIMALAYA

If you were to cross the Himalaya northward you would be standing at an elevation of approximately 15,000 feet (4,500 m) at the edge of the stupendous Tibetan Plateau. You would be gasping for air and gaping in awe at more than 400,000 square miles (1,036,001 km^2) of rugged basins and ranges. Many of the rock-strewn basins contain permanent lakes. Passes permit relatively easy access from one basin to another, but many of these are 17,000 feet (5,100 m) and higher. Here is where South Asia's four great rivers originate: the Sutlej, Indus, Ganges, and Brahmaputra (Tsang Po). The northern perimeter of the plateau is bounded by other massive ranges such as the Kunlun and Karakorum. The 28,250 feet (8,611 m) peak K2 rises in the Karakorum.

Northward again are the arid basins and mountains of northwestern China. The Tarim Basin is largest in expanse and the Turpan (Turfan) Basin lowest in elevation. The Heavenly Mountains—the Tien Shan—stand as one of this region's major divides. Although these and related basins are not important in terms of population density, they are of critical importance economically and politically.

EAST AND SOUTHEAST OF THE TIEN SHAN

Eastward is a mosaic of rugged mountains and plateaus, along with plains and more basins. The Qinling Range, Loess Plateau, North China Plain, and Sichuan Basin are only four of many significant elements of China's physical landscape. These regions are dominated by the mighty Huang He (Yellow River) and the Chiang Jiang (Central River) or Yangzi. These and other river valleys, basins, and plains make possible a livelihood for millions. Even further east are the craggy Korean peninsula and Japanese islands.

SOUTH AND SOUTHEAST OF THE HIMALAYA

South of the Himalaya, the North Indian or Indo-Gangetic Plain sweeps from Pakistan, across India, to Bangladesh. This subsiding trough of 300,000 square miles (777,001 km^2) is filled with fertile **alluvial soils**

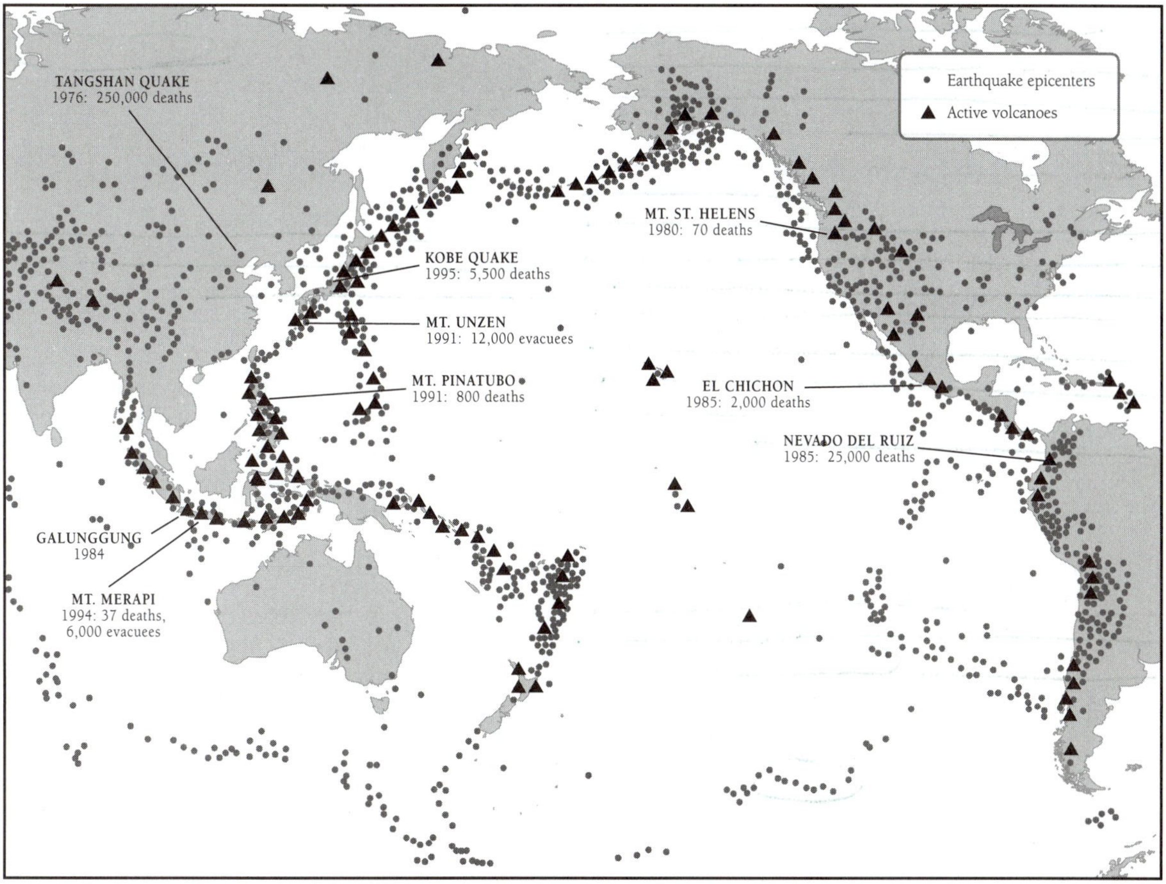

Figure 2-3

The Pacific Ring of Fire. Do you live in one of these major fault zones? How would you prepare for a seismic event?

deposited by river systems of the Indus, Ganges, and Brahmaputra. These water-deposited sediments, in concert with extensive river networks and relatively flat land, have contributed to this being one of the most densely populated regions on earth.

South of the Indo-Gangetic Plain rises India's dissected Deccan plateau. Covering most of Peninsular India, the Deccan is bordered by narrow coastal plains along the Malabar and Coromandel coasts. The tear-shaped island of Sri Lanka lies about 20 miles (32 km) to the southeast.

Mainland Southeast Asia is a washboard of mountains and **fluvial** (river-related) landforms. Mountain ranges, with elevations as high as 18,000 feet (5,400 km), are generally aligned in a north-to-south direction. One exception is the S-shaped Annamite Cordillera that follows practically the entire coast of Vietnam from southern China. Mountain chains are arranged closely in the north with virtual slots of lowland in between. Further south, the slots widen into V-shaped valleys and eventually into open plains, basins, and deltas.

Mountains have afforded cultural havens for the inhabitants of interstitial, riverine zones. They have encouraged ethno-linguistic diversity as each of hundreds of different culture groups maintained their own way of life in relative isolation from others for centuries.

Rivers are of paramount importance in mainland Southeast Asia. The Ayeyarwady (Irrawaddy), Chao Praya, Mekong, and Red River valleys have their headwaters on the Tibetan Plateau. So tortuous is the surrounding mountain terrain that the source of the Mekong was not found until 1994. River valleys not only have fostered support of millions of inhabitants but also have provided excellent sites for the growth and development of major cities, including most capitals.

Peninsular Southeast Asia extends 900 miles (1,449 km) from the Gulf of Thailand to the island of Singapore. The narrowest portion, the Isthmus of Kra, is roughly 40 miles (64 km) across. It widens to the south to form the Malay Peninsula.

ISLAND REALMS

South, East, and Southeast Asia are ringed by islands. Some, such as Sri Lanka, Singapore, and Taiwan, are politically independent states. Others are part of island chains called **archipelagoes.** Japan, the Philippines, and Indonesia are all archipelagoes. Each comprises one or more large islands and hundreds, or even thousands, of smaller ones. Japan is dominated by the large island of Honshu, but there are three other sizable and hundreds of lesser ones. The Philippines incorporates more than 7,000 islands along with the larger islands of Luzon and Mindanao. However, the biggest is not necessarily the most important. For instance, the Indonesian archipelago totals more than 13,000 islands, dominated in size by Sumatra, Sulawesi (Celebes), and Kalimantan (Borneo). But Indonesians and their political and economic centers are most heavily concentrated on the much smaller island of Java.

New Guinea, the world's second largest island (after Greenland), is politically divided between Papua New Guinea in the east and Irian Jaya in the west. Irian Jaya is occupied by and is generally considered a part of Indonesia. This island is the focus of one of Indonesia's many political and environmental controversies (Chapter 14).

There are more island groups in the Indian Ocean. Strung between Myanmar and Sumatra is a 600-mile (966 km) chain of more than 200 islands. These are known as the Andaman and Nicobar Islands. The majority are merely exposed tops of mountains protruding from the ocean floor. Southwest of India lie the Maldives, 115 square miles (298 km^2) of islands, banks, and reefs. Only about 10 percent of the roughly 2,000 of these are inhabited.

Figure 2-4

This is an example of tower karst topography in Guilin, China.

LIMESTONE AND CORAL ENVIRONMENTS

Chemically weathered limestone is the essence of **karst topography.** Fractured limestone formations, with at least 80 percent calcium carbonate, are subject to solution processes. Subsurface drainage channels are formed as water seeps through joints in the otherwise impermeable limestone. Chemical weathering takes place, resulting in pitted and lumpy surface forms and labyrinths of caves below (Figure 2-4). Karst regions in Asia are found in northwestern India, southern China, and in various parts of Southeast Asia. These formations are popular tourist attractions, drawing sightseers, photographers, rock climbers, and spelunkers from around the world.

Are you aware that the world's best dive sites are in the Asia-Pacific region?

Coral Reefs

Thriving in warm, tropical waters, coralline formations are abundant in Asia. It is estimated that more than half of the world's **coral reefs** are in the Indian Ocean, occurring along continental shelves and around islands.

Biological process is responsible for the formation of coral reefs. A coral is a small marine animal related to other marine invertebrates such as anemones and jellyfish. Corals excrete calcium carbonate from the lower half of their polyp bodies, forming a hard, calcified skeleton. Reefs are formed by colonial corals that live in a symbiotic relationship with algae. Over centuries, skeletons accumulate on raised submarine features. This biological formation makes up a coral reef.

Coral reefs protect coasts from wave erosion and contribute to the process of sandy beach formation. They provide breeding grounds and nutrients for subsistence and commercial fish, and they are habitats for other food sources like mollusks and crustaceans. Also, they are invaluable tourism assets.

Coral reefs are the marine equivalent of rainforests, the richest ecosystem on earth. Indonesia is home to some of the world's most spectacular reefs. Their biological and economic importance cannot be underestimated. Indonesian waters contain more than 5,000 species of fish, 30 species of sea snakes, and 7 varieties of turtles, among countless other creatures. However, reef populations have been significantly reduced.

Look at Sulawesi, where scientists believe that tropical marine life began. Each year, thousands of pounds of pear oysters, mother-of-pearl shells, and *trepang*—sea cucumbers—are exported. Shells such as the helmet conch and giant clams are sold as well.

Failing to recognize the intrinsic value of coral reefs, people have destroyed and endangered them in a variety of ways. Reefs have been mined for construction materials such as lime. Corals have been harvested for the aquarium trade. Dredging for sand for cement has damaged them and disrupted the marine community. Land reclamation projects have eliminated them. Reefs are ruined by heavy oil spills and the dumping of industrial waste. Speedboats, mechanical fishing devices, and fish blasting have been extremely detrimental. Fish blasters use explosives to kill large numbers of fish easily.

Since the 1970s, many nations have recognized the significance as well as the severity of loss of their coral reefs and have taken steps to protect them. Action has also been spurred by the potential for tourism dollars generated by dive sites and ancillary facilities. Nevertheless, abuses continue and now include unlimited and uncontrolled tourist activities. Unfortunately, coral reefs continue to be threatened around the world. As they are part of a larger marine ecosystem, destruction of reef formations is detrimental to many other life forms.

Physical geography can also affect political geography. We know that rivers, mountain ridges, and other physical features are used to demarcate political boundaries. Although some are disputed, most of these **natural boundaries** are stable. But bodies of water are problematic to define, presenting many opportunities for territorial disagreements.

Ocean Disputes and Conflicts

Oceans and seas are problematic because it is difficult to demarcate, in any practical sense, who owns what territory or mineral rights, or even who owns fish. In 1982, the United Nations issued The Law of the Sea Treaty, which allows each coastal state to claim a 200-

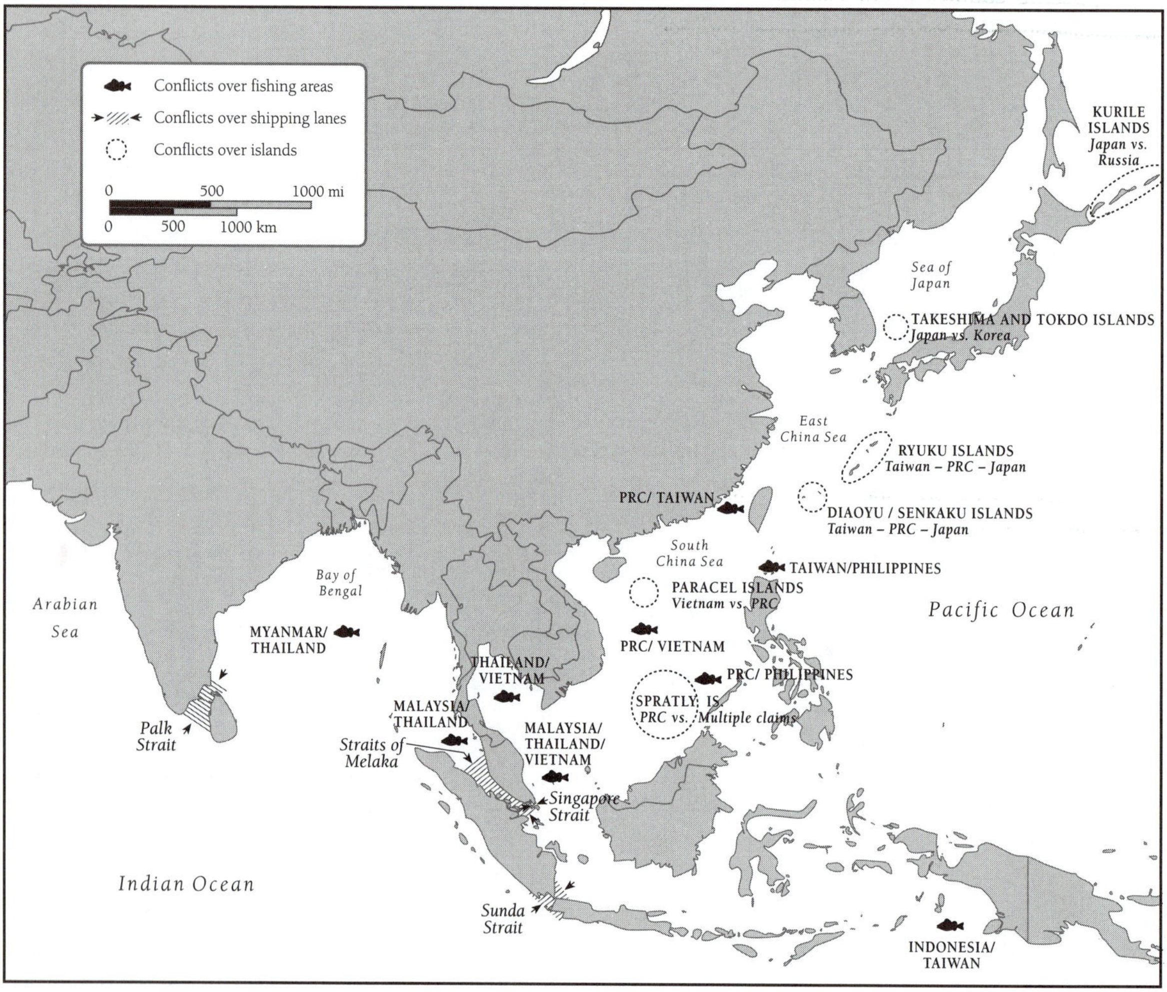

Figure 2-5

Conflicts at sea. Conflicts arise over ocean territories such as straits and islands, and over rights to resources such as fish and minerals.

nautical-mile **Exclusive Economic Zone (EEZ)** in which it controls resource rights. Countries are authorized to sell or lease rights to multinational corporations and they can control fishing and pollution. However, problems arise when fishers stray over boundaries or pollution comes from another country's EEZ.

Designation of EEZs has not solved all problems. The 200-nautical-mile EEZ does not work where distance between states is less than 400 nautical miles. To solve this dilemma, a median line is used. This line divides the waters equidistantly between the opposing countries. For example, Malaysia and Indonesia share the Straits of Melaka, the width of which ranges from 8 to 126 nautical miles. The EEZ of each country ranges from 4 to 63 nautical miles. Nevertheless, disputes are common in such narrow passages.

The case of the **Timor Gap** provides an example. In 1989, Australia and Indonesia struck a deal to explore and exploit the seabed between them. Here is thought to be one of the world's largest reserves of hydrocarbons. This perceived breakthrough followed years of controversy. In 1972, Australia and Indonesia agreed to formalize their sea boundaries along the deep Timor Trough. But East Timor was still controlled by Portugal. Both Australia and Portugal made conflicting claims to the seabed, issuing overlapping exploration licenses to oil companies. This disputed section became known as the Timor Gap.

In 1975, as East Timor was preparing for independence, Indonesia invaded the island primarily for strategic reasons. Australia, for economic reasons, had secretly approved of the invasion. Four years later, Indonesia and Australia started negotiations over the Timor Gap but did not progress until a Joint Development Zone was proposed in 1985. This divided the region into three parts, one to be jointly governed. However, Portugal argued that the treaty is invalid. It is targeting Australia at the International Court of Justice, whose jurisdiction Indonesia does not recognize. Meanwhile, several oil exploration leases have been issued.

Another point of controversy is the amount of territory that islands can claim. A small island can claim rights to fish and minerals such as manganese and oil over a disproportionate sea area. This makes possession of islands, no matter how small, critical. Jurisdiction over islands fuels tensions and conflicts in the Indian and Pacific Oceans.

The Law of the Sea Treaty also guarantees ships *innocent passage* while passing through the waters of one country to the next. Passages such as the Sunda Strait or Straits of Melaka are vital to world shipping. To close these narrow waterways would be a relatively simple task in a time of conflict.

There are those who proclaim oceans and seas to be humanity's *common heritage,* a concept introduced by the Dutch as *mare liberum* in the sixteenth century. Unfortunately, this ideal is impractical in a commercial world, and the "scramble for the seas" has intensified (Figure 2-5).

Monsoon Asia

South, East, and Southeast Asia have been termed **Monsoon Asia** by geographers and other scientists. This is because these realms are affected by intense, seasonally shifting wind systems that involve an annual cycle of wet and dry seasons. Beyond equatorial regions, summers are wet and winters are dry (Figure 2-6).

Equatorial regions experience year-round rainfall, although there are "wet" and "less wet" periods. This depends on the position of the sun, prevailing winds at the time, and topography. As wind systems shift north and south, different parts of islands receive more or less precipitation. Examine both January and July wind and precipitation patterns on a more detailed map in an Atlas. This information is critical when planning a trip. You probably do not want to arrive in Bali in the wet period.

Another feature of equatorial regions is **convectional rainfall.** The combination of solar energy and abundant moisture availability in oceans and biotic communities causes rapid evaporation. Warm, tropical air rises, cools, and condenses. The inability of cool air to hold as much moisture as warm air produces rain. This convectional rainfall occurs nearly every day

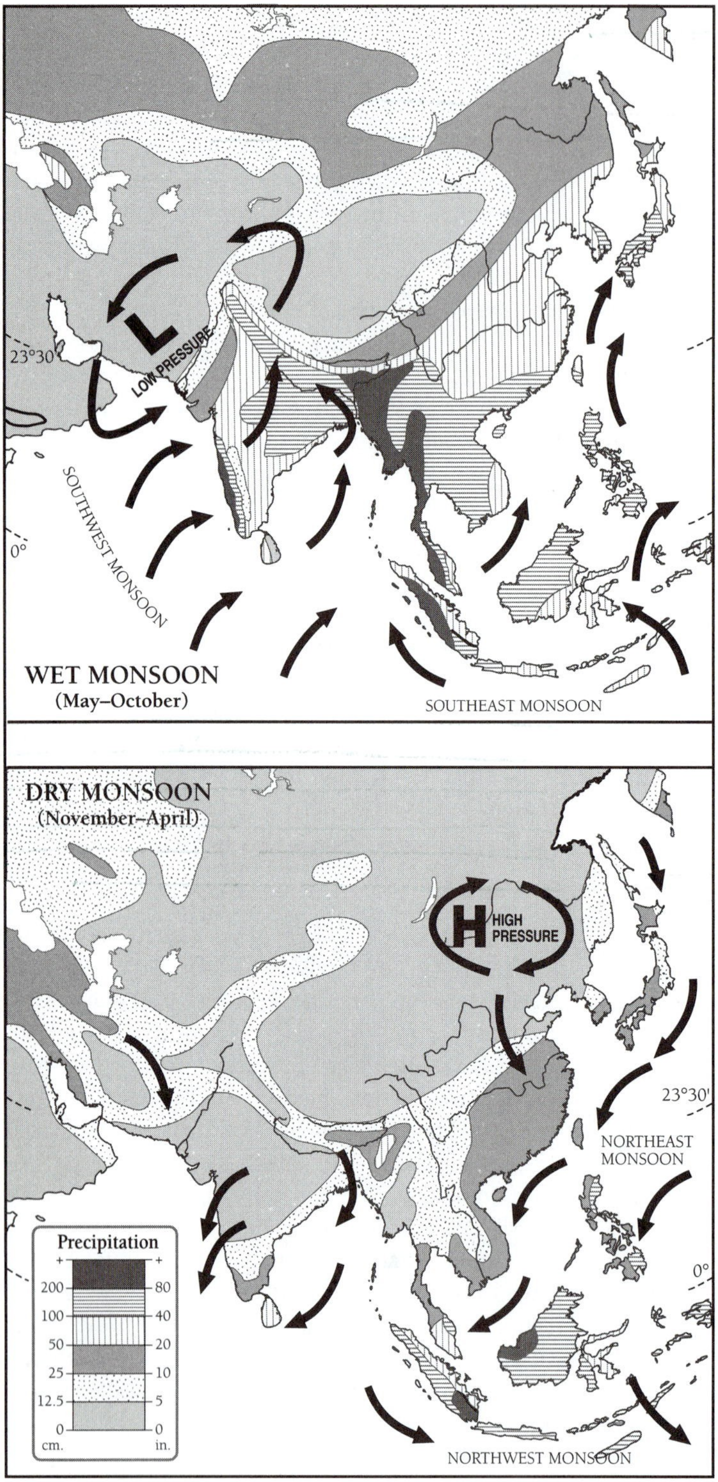

Figure 2-6

Monsoon seasons. Remember that winds blow from high to low (pressure) and are named from where they came. All of Asia is influenced by the Monsoon.

in the hotter summer months. In the less-hot months when evaporation rates are lower, several days might pass without a deluge.

Prior to rain, the humidity builds to an intolerable level. There is so much moisture in the air that 78°F (25°C) feels like 100°F (37°C). As the day wears on, storm clouds fill the sky, and in late afternoon, torrential rains wash the landscape and relieve the humidity. This is one reason why people work their farms in the early morning and late evening. Midday is a time to seek relief from the sun's intensity.

Latitudinal position is also significant, affecting both moisture availability and temperatures. In general, moisture decreases and temperature ranges increase inland and away from the equator. For example, average yearly precipitation at Guangzhou, on China's southern coast, is about 27 inches (1680 mm). At Hohhot, capital of Inner Mongolia in China's interior, it is 17 inches (426 mm). Guangzhou's yearly temperature range is 56° to 83°F (13° to 28°C). Hohhot's is 8° to 71°F (−14° to 22°C).

Both rainfall and temperatures are critical criteria for agriculture determining the need for irrigation and length of **growing season.** A growing season is the number of consecutive, frost-free days in a year.

What causes the monsoon seasons? The size of the Asian landmass or **continentality,** combined with its proximity to the Indian Ocean, are keys to explaining variations in air pressure and related upper-air circulation.

The interior of Asia is far removed from modifying maritime influences. Therefore, in winter, it is dominated by a high-pressure anticyclone while the central area of the Indian Ocean is dominated by the equatorial low-pressure trough known as the **Intertropical Convergence Zone (ITCZ).** Winds blow from zones of high-pressure to zones of relatively lower pressure. Consequently, cold, dry winds blow from the continental interior to the periphery, dessicating (drying) the landscape in the process. Temperatures average between 60° and 68°F (15° and 20°C) at lower elevations.

Cold increases with elevation. The normal lapse rate is 3.5°F/1,000 ft. (6.4°C/1,000 m). The impact of cold is worsened with the **wind-chill factor** whereby body heat is lost more rapidly with stronger winds. Further, because of diminished atmospheric density at higher altitudes, its ability to absorb and radiate heat is reduced. Insolation with its ultra-violet energy is intensified. While high elevations can produce oxygen deprivation and sunburn, exposure to extreme cold can induce *hypothermia*—frostbite with subsequent loss of body parts, or freezing to death.

Extremely cold temperatures can be experienced in lowland areas as well. In January 1983, I was in a New Delhi train station. The platforms were coated with humanity, hundreds of homeless individuals and families huddled under blankets seeking escape from the bitter cold. A newspaper reported that 17 people had frozen to death in the streets overnight.

With desiccating winds and the coming of summer, March to June can be very hot. The winds begin to shift, and people pray and prepare for the coming of the *wet monsoon.* Ideally, the wet season is between June and September, but one sure thing about the monsoon is its unpredictability. If it is early, seeds can be washed away; if it is late, crops can wither and die. This explains the many ceremonies and festivals at this time of year to propitiate various gods and spirits associated with these essential rains.

At this time, continentality effects a low-pressure zone in the interior. The ITCZ shifts northward to the Tropic of Cancer. Solar radiation heats the landmass, leaving the surrounding waters relatively cooler. A subtropical high-pressure cell emerges over the Indian Ocean. As hot, dry air sweeps outward across the ocean's surface, it quickly evaporates water. This huge, thunderous, sodden air mass moves into the lower pressure ICTZ, bringing voluminous rains to the continent. Precipitation is highest in coastal areas and on the windward slopes of mountains. As air masses move inland, available moisture diminishes and less precipitation is experienced.

Mountains and other areas of raised relief are highly influential as **topographic barriers.** The Himalaya, for example, provides a physical barrier to incoming air masses, channeling them westward across the Indo-Gangetic Plain. As moisture-loaded air meets this topographic barrier, it is forced to rise, cool, and condense, providing **orographic precipitation.** As a result, the windward slopes of the Himalaya receive abundant rainfall. The subsequently dry air masses flow over the crest and down the leeward slope, warming and drying the landscape in its path. Consequently, protected leeward slopes, or **rainshadow** areas, are comparatively dry.

Cherrapunji, India, situated at 4,309 feet (1,313 m) in the foothills of the Himalaya (just north of Bangladesh), is the record holder for most precipitation in a single year. Toronto, Canada, averages 30.1 inches (76 cm); Sydney, Australia, 44.5 inches (114 cm); Atlanta, Georgia, 50.4 inches (127 cm); and Singapore 89.3 inches (229 cm). In contrast, Cherrapunji averages 37.5 feet (1,143 cm) of precipitation a year. Such torrential rains, sweeping against the mountains from the Bay of Bengal, have caused the deaths of hundreds of thousands of people.

Another messenger of destruction is the **tropical cyclone**, termed a **typhoon** in the Pacific. Tropical cyclones typically develop between the latitudes of 8° and 15° when sea temperatures rise above 80°F (27°C). Beginning as a weak, low-pressure cell, it funnels as a deep, circular low. Moving with prevailing wind systems, it intensifies further and delivers violent storms. Depending on the location, typhoons generally strike between May and September.

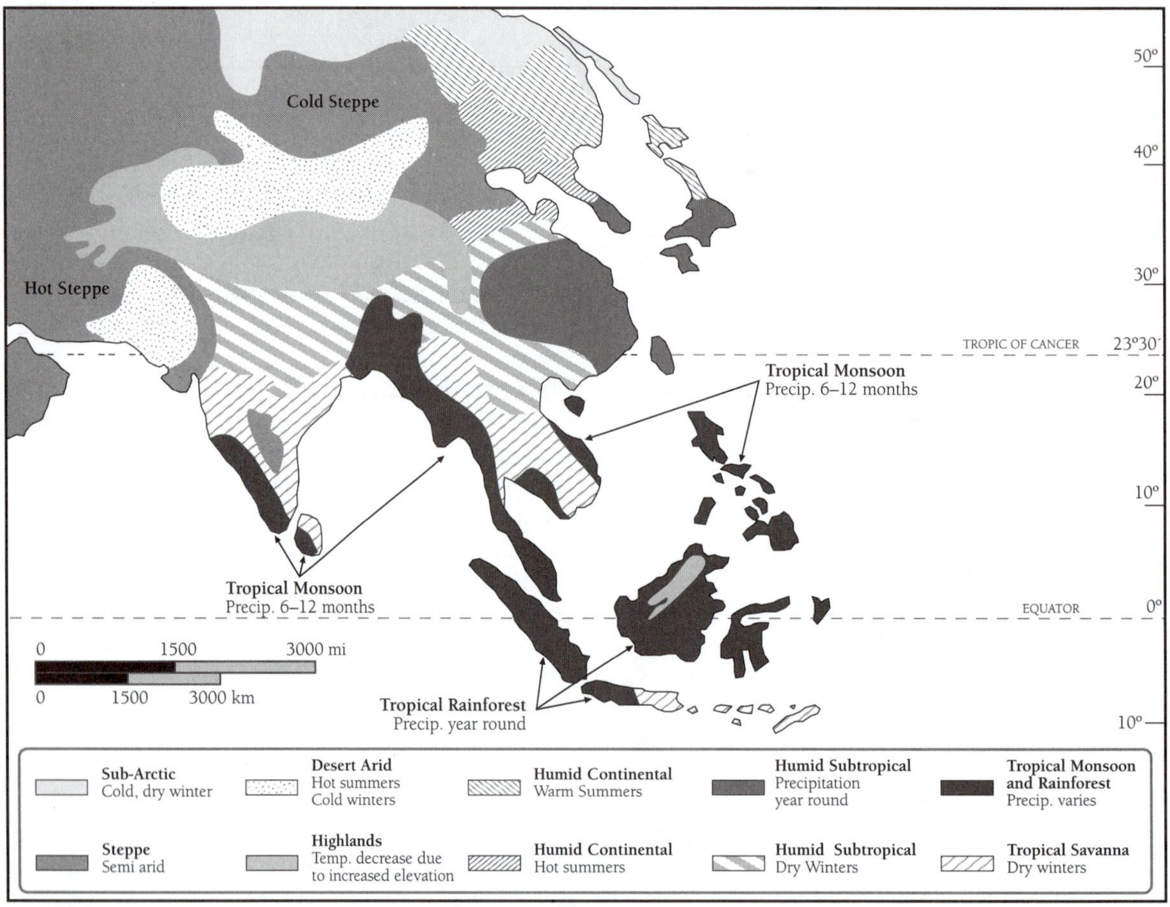

Figure 2-7

Climatic zones. Climate is a year-round phenomenon that varies with latitude, elevation, prevailing winds, and distance from moderating effects of water.

In 1997, Typhoon Tina struck the coast of South Vietnam, destroying everything in its path. Tina then moved on to wreak havoc in parts of Thailand and Bangladesh, where it degenerated into a mere tropical storm. In Vietnam, nearly 3,000 boats were sunk or swamped, and an estimated 4,000 people, mostly fishermen, perished.

REGIONAL CLIMATIC PATTERNS

Although the monsoon affects the entire region, smaller regions have their own climate characteristics. If you are interested in more detail regarding temperature ranges and precipitation amounts, be sure to consult your atlas. I will simplify here.

Climates in South, East, and Southeast Asia can be generalized into five categories: tropical; arid; humid subtropical; humid continental; and highland (Figure 2-7). To understand this pattern, you can ask the following questions. How near or far from the equator is a place? Does it experience continental extremes or moderating maritime influences? What is the elevation? Is the place influenced by the presence of a topographic barrier? Answering these questions will help you understand regional similarities and differences in climate.

CLIMATE AND PEOPLE

Quality of life in various climatic regimes depends very much on what coping mechanisms are available. Flood control works, irrigation systems, water storage facilities, healthy work animals or functional machinery, electricity, sealed (paved) roads, insurance against crop loss, and emergency food supplies are only a few.

Millions of people in Asia have minimal means to cope with disaster. An erratic monsoon, a drought, a tropical cyclone, or a lengthy cold spell can mean ruination and total despair for those living on the edge. Here is an example. In 1998, rains and pests destroyed the crops in a small area in southeastern India. Farmers unknowingly had purchased illegally diluted fertilizers and pesticides. Deeply in debt and sensing nothing but hopelessness, hundreds of farmers ingested their remaining chemicals. They believed their deaths would bring a government stipend to their wives, thereby saving their land from the moneylenders. Payment was sporadic and insufficient, and many women and their children are now landless and destitute. Those of us living in a high-tech world probably find it difficult to grasp this level of desperation. But for millions in Asia and elsewhere, desperation is a way of life.

Natural Vegetation

Natural vegetation implies undisturbed, mature plant communities. Although an undisturbed biome is a rarity, examination of natural, model biomes help to reveal the measure of human impacts. The following biomes are coincident with the climate zones of Figure 2-7.

TROPICAL RAINFORESTS

Tropical rainforests are most prominent in Southeast Asia where equatorial conditions provide plentiful year-round moisture and consistently warm temperatures. Here is the earth's most diverse biome. Of 250,000 flowering plant species in the world, 170,000 are tropical. Of these, 40,000 are in Asia, with 25,000 in Malaysia and Indonesia. Moreover, this region has tremendous **species diversity.** Western Europe, west of Russia and north of the Alps, has 50 indigenous species. Eastern North America has 171. In contrast, a single 123.5 acre research plot on the Malay peninsula exhibited 830 indigenous species.

Think of rainforests as having layers within which every plant competes for sunlight and moisture. Each layer has a different microclimate, with the upper layer having the most light, heat, and wind. The forest's biomass is in the **canopy**, above the forest floor. Leaves appear on tree trunks at approximately 15 feet (4.5 m) above ground, and this begins the canopy that reaches upward to 100 feet (30 m) or more. Poking out of the canopy are the tallest trees called **emergents.** X **Dipterocarps,** with their winged seeds, dominate the upper canopy. These are trees of grand stature, housing a wealth of flora and fauna. It is said that dipterocarps are the apex of evolution. The middle level of the canopy is densest. It has a spectacular array of mammal, bird, and other species. **Lianas** (woody climbers

X don't need to know these technical terms

and vines) loop around and between trees. These are the bane of foresters because they can so entangle a tree that even when completely cut, it remains dangling, tied to adjacent trees. Some lianas are 8 inches (20 cm) in diameter. Orchids, ferns, and other **epiphytes** dress this zone as well. Epiphytes are physically, but not nutritionally, supported by other plant life.

Beneath this exceptionally rich environment is a more humid, darker understory and the forest floor. Ferns, bamboo, shrubs, and herbs shade a relatively open floor that is strewn with decomposing plant and animal litter and fungi. This rather thin layer provides nutrients to vegetation. Most tropical soils are deceivingly infertile, leached of their nutrients by copious rainfall and high temperatures. At lower layers, aluminum and iron oxides form rock-like layers called **laterite.** Since antiquity, laterite has been cut and used as building blocks for temples and other structures. Even giant trees have shallow roots to tap the litter layer for food. Some of these have supportive bases called **buttresses** (Figure 2-8).

Where light penetrates, dense growth can be found at floor level. Dense vegetation, or **jungle,** exists along rivers, or in areas that have been cleared and abandoned. These seemingly impenetrable areas belie the more open rainforest in their vicinity.

Figure 2-8

Here is a dramatic example of a buttressed tree in the tropical rainforest. Since soil fertility is at the surface, trees develop sprawling surface rather than tap root systems. Some large species of trees develop blade-like supports like these.

TROPICAL MONSOON FORESTS

Found primarily in mainland Southeast Asia and the Indian subcontinent, the monsoon forest is a transitional biome between tropical rainforests and more arid plant realms. Unlike tropical rainforests, monsoon forests must adapt to distinct wet and dry seasons. Many trees are **semideciduous;** they lose part of their foliage in the dry months (Figure 2-9).

Tropical monsoon forests lack a continuous canopy. Their trees average only 50 feet (15 m) in height. Often cut for firewood, they make poor lumber. Teak, a highly valued hardwood for carpentry, is an exception. Trees also produce waxes, gums, resins, and lacquer. Monsoon forests possess a rich collection of wildlife.

Do your shoes have rubber soles? Do you use a palm oil–based soap? Have you used lacquer or latex paint? Have you used bamboo skewers or wooden chopsticks? Is there a coconut fiber mat at your front door? How much of the tropical forest do you use?

Riches of the Tropical Forest

Tropical forests are rife with desirable products such as fruit, tubers, plants, and other food

Figure 2-9

This monsoon forest is lining the Rapti River in southern Nepal. It is winter, dry season, and many of the trees have lost all or at least some of their leaves to compensate for the water shortage. Most of these trees are semideciduous. The Rapti is part of the Ganges River system and consequently is a sacred river.

crops; rattans and fibers; oils and resins; in addition to wood. Traditional ways of gathering forest products have little impact as compared to modern logging activity. Logging is a prime reason why more than 40 percent of Asia's tropical forests have been destroyed. Foresters seek certain types of trees to cut for local use and export. I will describe two of these here: teak (*tectona grandis*) and meranti (*shorea spp*).

Teak is regarded as a fine, medium-weight hardwood. It is noted for its strength, stability, and durability. Termite resistant, it is valued for constructing housing and furniture. It used to be in great demand for ships. Teak is native to India, Vietnam, Burma, and Thailand, but is grown commercially on Java and in Sri Lanka. Logging for export has eliminated teak forests from Thailand.

There are several species of meranti, called *seraya* in Malaysia. These are giant, buttressed trees achieving heights of over 200 feet (60 m). Meranti are among the most important export woods of Asia. Logs are sent to Australia, Europe, and Japan for use in construction and in the manufacture of plywood.

Rattans and fibers are also in demand. Used for furniture, matting, and basketry, these are collected and assembled for local use and export. Some of the commercial rattans are climbing palms that wind their way around the forest for lengths of 250 feet (75 m) and more. Rattans have been known to grow to 600 feet (180 m). The main suppliers are the Malay Peninsula, Indonesia, and China. Kapok, a waterproof fiber harvested from ripe tree pods, and coconut fiber called *coir*, serve multiple uses.

Bamboo is among the oldest materials used by humanity. Seeds and shoots are edible, as are the worms that live inside. Water pipes, containers, spears, house frames, and construction scaffolding are only some of its uses. Bamboo is a grass related to corn and wheat. A woody perennial, it can grow to 60 feet (18 m). Some fast-growing varieties can grow a foot in one day. There are at least 1,000 species of the grass, and it grows around the world in most environments. However, commercial species are tropical, and 60 percent of these are in Asia.

Tropical forests are also rich in essential oils. Vietnam, for instance, produces around 400 of these, including anise (black licorice or anisette). In India and Sri Lanka, aromatic sandalwood has a recorded history of 4,000 years and is one of the oldest known perfumery ingredients. Sandalwood oil is commercially distilled from trees at least 30 years old. Extracted from plant leaves, patchouli oil was used to scent the carpets and rugs of India. Now cultivated commercially, especially in Malaysia, it is a key ingredient in French perfumes.

Resins, known as copals and dammars, are **exudates;** they ooze out of trees as a result of injury. These can be collected in soft or hard form. Old, fossilized, hard lumps, buried beneath the ground for ages, are highly prized. Copals are used in paints and resins. Dammar is the Malay world for resin. Softer than copal and soluble, dammars are used for such things as varnishes and wallpaper finishes. Gum resins and gum turpentine are drawn from pine forests in the Himalaya foothills and elsewhere.

Forest products are primary sources of pharmaceuticals. Strychnine is but one example. Indigenous to the monsoon forests of western, coastal India, it is still widely employed there and in Southeast Asia as a stimulant, tonic, or emetic. Strychnine is also used for spear and arrow hunting and to poison vermin. You have probably read about it in mystery novels as a means of murder. Prepared from the seed of the snakewood tree, large doses of strychnine are definitely deadly.

The World of Mangroves

Mangroves are littoral plant formations of tropical and subtropical coasts where trees and shrubs have adapted to inundated and saline conditions. At least 50 million years old, they originated in Malaysia and diffused around the globe. The world's largest remaining mangrove concentration is the Sunderbans in the Bay of Bengal, home of the Bengal tiger. Mangroves comprise habitats for hundreds of living organisms from mammals, birds, crustaceans and fish, to algae, fungi, and bacteria. The entire community is called **mangal.**

Mangroves are significant for several reasons. They stabilize shorelines and decrease coastal erosion. Capable of generating over three tons of organic matter per acre per year, they build bottom sediment with their multiple roots. Animals such as water buffalo graze on them. And they act as windbreaks against coastal storms.

People find mangroves economically useful. Their wood is used in construction and provides a rich charcoal. Sugar from the nypah palm can be turned into alcohol and machinery fuel. Prawns, crabs, fish, and other delicacies are harvested for food. Unfortunately, mangroves are disappearing rapidly with population increase, urbanization, industrialization, and internationalization.

Population pressure, along with government-sponsored resettlement schemes, have instigated the clearance of vast areas of mangal. Urbanization and industrialization have seen mangroves used as waste dumps and ruined by factory effluents and oil spills. International companies cut mangroves for pulp and particleboard. They are removed to make way for plantation agriculture, fish farming, and salt evaporation vats. Grandiose damming and water diversion schemes alter the fresh–salt water balance essential to the survival of both mangrove and the creatures within.

Mangroves are capable of rapid growth. A yearly growth of 540 feet (162 m) has been recorded in Java. Palembang, Sumatra, a coastal port in the thirteenth century, is now 31 miles (50 km) inland. However, when the mangals are cleared *en masse,* they seldom regenerate. Consequently, with the twentieth century perception of mangroves as wastelands, they have been radically reduced.

TROPICAL SAVANNA

Arid parts of India, such as the eastern Deccan, are characterized by grassland with scattered trees. These tropical savannas are transitional zones between monsoon forests and even drier regimes. Vegetation is typically **xerophytic,** or drought-resistant. Spontaneous or set fires function to maintain the savanna. Early in the dry season, "cool" fires stimulate new growth. Late-occurring "hot" fires rage out of control, killing trees, seeds, and every living thing. Soils are more fertile and better drained, and savanna lands are widely used for grazing and crops such as grains and groundnuts (peanuts).

DESERT AND STEPPE ENVIRONMENTS

Xerophytic and often thorny shrubs and succulents are found in **deserts,** which are defined as regions having less than 10 inches (250 mm) of precipitation a year. Some plants bloom only sporadically, waiting perhaps years for rain. These are called **ephemerals.** Some areas of sandy or stony desert are devoid of vegetation.

Asia's deserts are primarily cold deserts. Winters can be extremely cold and summers unbearably hot. Because of sparse vegetation and openness, heat loss is rapid and nights are relatively cooler. Deserts can be made agriculturally productive with irrigation.

In the plateau regions beyond the Tibetan Plateau, such as China's Nei Mongol (Inner Mongolia), there are vast areas covered with short grasses (Figure 2-10). Referred to as **steppe,** these grasslands are the traditional home of nomads who move their herds from one pasture site to another in seasonal cycles. Steppes are China's most important grazing lands, but they are being used increasingly for cropping.

MIDLATITUDE FORESTS

Much of China, Japan, and Korea possesses broadleaf, **deciduous** forests characteristic of middle latitude locations. Deciduous trees drop their leaves in the cold season, using winter dormancy to survive until the warm days of spring. Oak, birch, and elm are three of the more familiar varieties. Needle trees such as pine, fir, and spruce exist in cooler northern regions and at higher elevations. There are nearly 2,000 woody plants in China's forests alone. These forests are important for both wood and nonwood products. Ginko, chestnuts, walnuts, tallow oil and tung oil, and birch juice are some of these. The vast majority of midlatitude forests have been eliminated or damaged for the benefit of large human populations in these parts of the world.

MOUNTAIN VEGETATION

Mountain plant communities vary vertically because of climatic change with elevation. This is known as **altitudinal zonation** of vegetation. Everything from

Figure 2-10

Here in Inner Mongolia, China, cattle graze on short grass steppe. Note the temporary holding pen in the background. After a few days, they will be moved to a fresh area.

Figure 2-11

High in the Indian Himalaya—15,000 feet (4,500 m)—a horse grazes on vegetation far above the tree line. Rich in grasses and flowers in summer, this terrain is covered with ice and snow in the winter months.

tropical, to mixed deciduous forests, to needle-leaf forests exist with increases in altitude. Forested areas give way to alpine meadows, stunted trees, and the **tree line** where neither growing season nor soil depth will support tree life. Here the sparse, stunted trees give way to the low-lying flora, mosses, and lichens of sub-arctic **tundra** (Figure 2-11). Even higher is barren rock and ice. As with other biomes, population pressure is having deleterious impacts on mountain environments, especially in the Himalaya.

Wealth of Wildlife Lost

The existence of two broad zoogeographical divisions—Gondwanan and Laurasian—can be explained by plate tectonics. About 15 million years ago, the continental plate of Antarctica, Australia, and New Guinea split apart, leaving Antarctica behind. The Australia–New Guinea section collided with southeastern Laurasia, thereby creating the Malaysian and Indonesian archipelagos. The collision took place just west of Sulawesi. This undersea fracture marks the division between Laurasian and Gondwanan faunal types. As mentioned earlier in this chapter, India, once part of Gondwanaland (Africa), crashed into Laurasia as well. In the process, India served as a raft carrying Gondwanan fauna to Laurasia.

Western and eastern **Malesia** (Malaysia and Indonesia) have strikingly different animal communities. Plant differences are not as distinct because plants more readily disperse across water. A combination of Gondwanan and Laurasian fauna inhabits the western region, while Laurasian live in the east. Here, animals such as the tree kangaroo and pygmy glider have similarities with Australian species.

The British naturalist Alfred Russel Wallace explored both South American and Southeast Asian rainforests. During his stint in Asia (1854–1862), he discovered these two distinct zoogeographic regimes. He described their fauna as having "endless eccentricities of form and extreme richness of colour." The line of demarcation between the two zones is known as the **Wallace Line.**

Thousands of wildlife species live in Asia, many not yet identified. Many are **endemic**—they occur in a given area only. Having a restricted range, these are easily wiped out. Indonesia, the Philippines, China, India, and Japan have the most endemic species of flora and fauna, and these are seriously threatened.

Since international focus is usually on tropical regions, other climes are often ignored or forgotten. Think about this: half of Japan's native mammals are in danger of being eliminated across the archipelago. Two kinds of wolf and three types of bat are now extinct. Another 22 species, including the Japanese river otter, have not been seen in the wild for a decade. More than 2,000 species of terrestrial vertebrates live in China. Many are at risk. Endemic species such as the lesser and giant panda, the golden monkey, and the Yangzi alligator are seriously endangered. Przewalski's Horse, the earth's last truly wild horse, has not been seen in the wild since 1966.

The diverse environments of South Asia provide habitats for a plethora of animals, birds, and other living things. However, Asian elephants, one-horned rhinoceros, sloth bear, and other species have sharply reduced populations (Figure 2-12). This is attributable to human pressure for land, modification or destruction of habitats, and poaching. Asian lions, once widespread and avidly sought by hunters, are now confined to India's Gir National Park in the western state of Gujarat.

Some Indian wildlife have adapted to living in human settlements. Bats, birds, mongooses, and monkeys subsist on the refuse of cities, towns, and villages. The common langur and rhesus macaque are the most prominent, especially around temples. Here they can feed off food offerings and neither Hindus nor Buddhists will harm them.

Indonesia is the most mammal rich, with close to 700 species of higher vertebrates, but it has more threatened species of mammals and birds than any other country (Figure 2-13). Nevertheless, its forests remain a spectacle of wildlife inhabiting every layer: from proboscis monkeys, gibbons, orangutans, flying foxes (bats), gliding snakes and frogs in the canopy; to tree kangaroos, and other arboreals of the understory; to clouded leopards, tigers, and other cats of the lower understory and floor; to tapirs, rhinos, and wild pigs, foraging on the floor. Birds and insects abound in countless varieties throughout.

Figure 2-12

I spotted this elephant getting a drink of water in Kerala State in southern India. The elephant had just completed a day's work hauling logs in the forests of the Western Ghats. Elephants, as work animals, are rapidly being replaced by vehicles that are responsible for much greater environmental damage.

Figure 2-14 shows the number of threatened species of mammals, birds, and reptiles in Asia. Threats come primarily from people who destroy habitats for settlement; seek economic gain from resource exploitation such as mining, forestry, and plantation agriculture; emit air and water pollution through industrialization; and hunt and trade in exotic live beasts, their skins, feathers, claws, and other body parts.

In 1989, the Philippines exported 19,000 live reptiles and close to 9,000 live primates. Indonesia exported nearly 8,000 live reptiles and more than 16,000 live primates. Countless birds are exported as well. Some are destined for medical research, some for zoos, and some for pet stores. If not properly looked after, many creatures die in transit or in captivity, or are discarded when people tire of them.

Animal trade, much of it illegal, has increased with greater demand from industrialized countries. From tropical forests around the globe, the United States purchased 266,750 live reptiles and 19,323 live

Figure 2-13

I discovered this poor creature in a small park/zoo in Banjarmassin, Kalimantan, Indonesia. It is an illegally caged proboscis monkey, known for its large nose. Given the rapid rate of deforestation in this region, the proboscis monkey is a threatened species. In another tiny cage was a baby orangutan that never stopped crying. Rescue efforts cannot keep up with the illegal poaching of wild animals.

primates in 1989. Japan and South Korea continue to be major importers. Thailand, where it is possible to purchase virtually anything, no matter how rare, is a key source region.

Animal parts also generate big profits. From 1980–1989, thousands of cat skins were exported from China, South Korea, and Mongolia. The largest buyer in Asia was Japan, with a total of more than 160,000 skins! Thousands more went to European states. Most countries now have biome protection programs, national parks, and bans or limits on wildlife trade. However, illegal trade continues, and the number of endangered species is even greater. Species destruction threatens the earth's biodiversity.

Biodiversity refers to range of variation of living organisms and is usually defined in terms of genes, species, and ecosystems. It is generally accepted that we should strive to preserve biodiversity for several reasons: present and potential uses of biological resources especially in medicine; to maintain a biosphere supportive of healthy human life; and for its own sake based on a concern for the well-being of all living things.

Recent findings reveal that tropical medicinal plants are more commonly found in "disturbed habitats"—where hunting and gathering or slash and burn agriculture is practiced, and that the virgin forest is quite devoid of them. But the *knowledge* of these plants and their uses is dying out with traditional populations.

Does this mean that we should save the people and disturbed habitats and not worry about pristine rainforest? Remember that there are unknown, potential uses for rainforest flora and that this forest, with its own complex of ecosystems, is integral to the larger biosphere in which we live.

Some environmentalists speak of **environmental criticality** when referring to environmental degradation and associated socioeconomic deterioration. Removal of forests, disturbed or not, is removal of habitat for all living creatures. Ways of life are destroyed and associated knowledge is lost. Thoughtless development reduces biodiversity of flora, fauna, and humans. Quality of life for all is surely related to the health of habitat.

Do the medicines you use derive from tropical plants or animals?

Plants, Animals, and Medicine

The United States imports more than $20 million worth of rainforest plants a year. In fact, 40 percent of medications sold in North

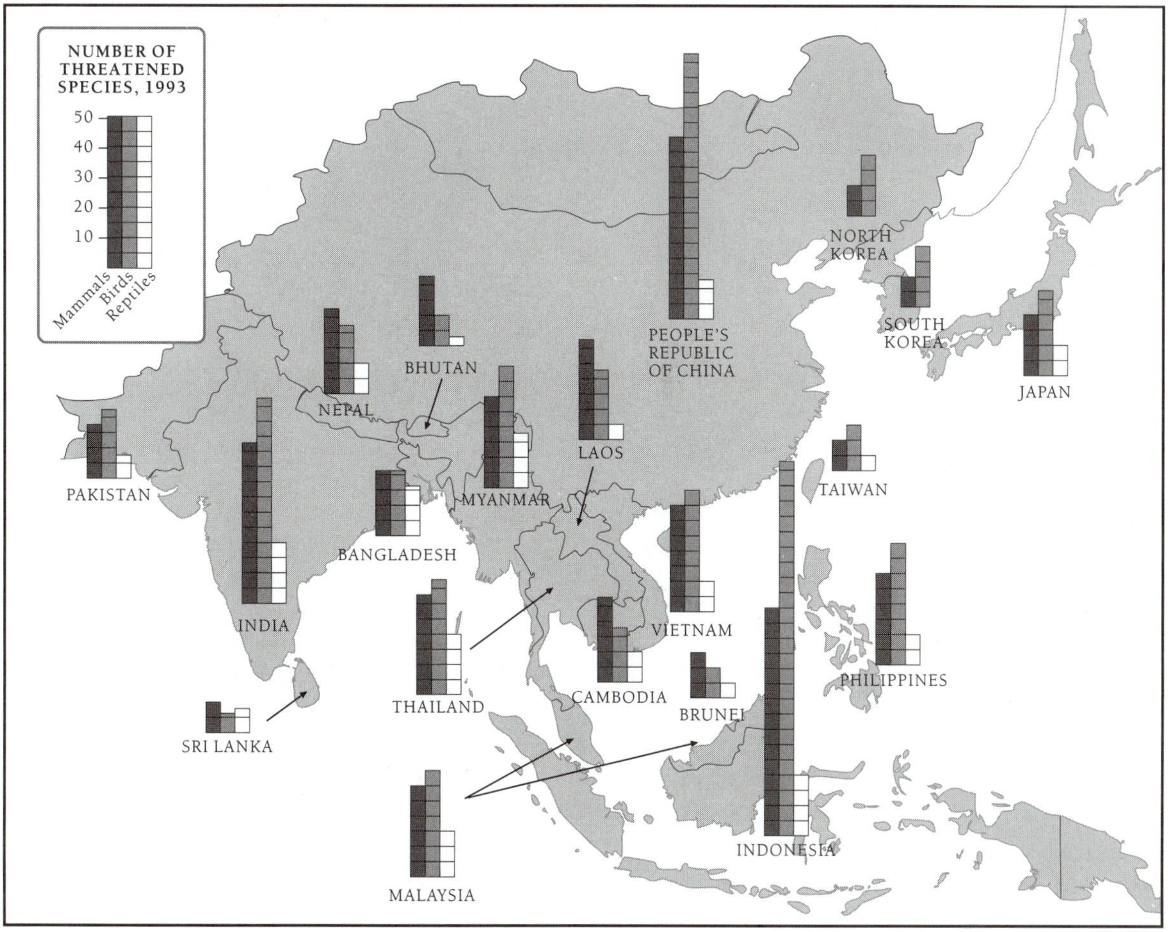

Figure 2-14

Threatened mammals, birds, and reptiles. How many species will become extinct in the twenty-first century? What are humans doing to accelerate this process? What can be done to stop or slow it down? Source: *World Resources Institute. 1996.* World Resources 1996–97. *New York: Oxford University Press.*

America derive from plants. These are either harvested from forests by local people or are commercially grown. Cinchona trees, originally from South America, are grown in India to produce the antimalarial quinine. But new, quinine-resistant strains of the disease have developed, prompting researchers to investigate antimalarial potential of new plants.

Both modern and traditional medicine impact biotic communities. In India, for example, 2,500 plants are either used by traditional healers or exported. Nepal exports 90 percent of its 700 medicinals and herbs. These countries, along with Sri Lanka, are noted for traditional Ayurvedic medical practice. Grounded in Hinduism, Ayurvedic healing relies on natural products.

Traditional medicine from India, China, and elsewhere has an enormous pharmacopia, some of which has proved invaluable to modern medical practitioners (Figure 2-15). For example, reserpine, the source of modern tranquilizers, is extracted from the serpentine root shrub. The National Cancer Institute and other organizations, intent on finding cures for cancer and HIV related diseases, are actively collecting tropical leaves, twigs, roots, and rhizomes from both forests and traditional healers. Thousands of samples have been collected yet these account for only seven to eight percent of the thousands of flowering plants in Asia. Many of these are vanishing in the wake of development.

Also useful in medicine are animal products. Snake venom derivatives can relieve phlebitis. And they can duplicate symptoms of muscular dystrophy and myasthenia gravis, enabling scientists to understand these diseases better.

Animal products are widely used in traditional practice, but Western scientists put little faith in the curative powers of most of them. A more important point is that the majority of Asian people *believe* in their efficacy. And, through experiments with placebos, scientists have shown that belief in a medicine's effectiveness can bring about actual improvement.

Faith in the mysterious workings of faunal cures drives the demand for faunal parts. In China, powdered deer antlers are employed to counter aging, and ground dried gecko is taken for respiratory problems. In Vietnam, boiled monkey bones are believed to improve circulation. Indonesia's trade in *jammu* or traditional medicinals involves millions of dollars a year. While many, better-educated urbanites are turning to Western imports, increasingly available in markets and pharmacies, rural and older populations continue to rely on long-trusted remedies.

One of the most exploited animals is the rhinoceros. Only a few hundred survive in India, Nepal, and Indonesia. A mere 70 remain on the island Java. Rhino parts are craved for their imagined powers as analgesics and aphrodisiacs. Its droppings are said to cure leech bites, its raw hide skin afflictions. Its "horn," which actually is hard, agglutinated hair, guarantees the seller instant wealth. Unfortunately, rhino horn remains in demand even with public education and legal restraints on its trade.

Figure 2-15

This is a traditional medicine pharmacy in Korea, where animal and plant products are essential. Here you can see deer antlers and packages of ginseng roots.

Thinking About the Natural Environment

Environment is a multidimensional concept, and definitions range from anthropocentric to bioethical: the environment exists to support humanity versus it has value in its own right. The former leans toward the idea that there is a sharp division between people and nature, the latter toward the notion that human and natural worlds are entwined in an all-encompassing life-world. Those who view environment in terms of its use value tend to think that problems result from mismanagement and that better management of resources will fix things. The bioethical perspective is that nonhuman phenomena have intrinsic moral significance apart from their utility. Therefore, humans have a moral responsibility to care for and protect all aspects of nature. Between these "use it" and "save it" perspectives lie numerous other interpretations.

There is a common perception that Asian culture groups, at least historically, lived closer to and had more regard for nature. Nature must be important because both East and Southeast Asians employ natural metaphors regarding social position and morality. For example, reason (*li*) is fundamental to the history of Chinese philosophy. Originally, the character for *li* meant "well distributed veins on minerals or precious stones." Later, it came to mean "principle" and then, "universal principle." Besides, nature is intrinsic to religions and belief systems. However, there is no single Asian concept of nature. Although it is true that there is less of a distinction drawn between human and natural worlds in Asia, the range of attitudes and behaviors toward it is as varied as elsewhere.

Did you know that almost-extinct tigers are descendants of the extinct sabre-tooth tigers of Siberia?

Fate of the Tiger

By the turn of the twentieth century, an estimated 100,000 tigers roamed from Siberia to Indonesia. In the dawn of the twenty-first century, there are perhaps 2,500 remaining. Big game hunting in the colonial era, modern poaching, and loss of habitat are the major causes of this catastrophe.

In India, only about half of the funds designated for "Project Tiger" ever reach it. Poaching continues because impoverished people can earn \$100 for a dead tiger. Poisoning water holes with cyanide is one method, killing tigers and all other beasts who drink. Brokers make thousands on the sale of tiger products. An Indian tiger skin can bring as much as \$15,000 in Arab nations.

Tigers are now extinct in Bali and Java, and only about 400 to 500 remain in Sumatra. Ground tiger bone can earn \$500 a gram. A 24-pound (11 kg) tiger is literally worth a fortune. From 1975 to 1992, South Korea imported nearly 4,000 kg of dried tiger bone from Sumatra—the equivalent of 338 to 620 tigers.

In southern China, an estimated 20 to 30 tigers eke out an existence in remote patches of forest. Massive forest clearing and the eradication of tigers as "pests" and "vermin" in the 1950s decimated the population. Although conservation programs now exist, these few tigers are in a precarious situation.

In mainland Southeast Asia, a few hundred tigers are rapidly meeting their demise. Aside from loss of habitat to logging concessions, plantations, and the like, tigers are starving because hunters have virtually eliminated entire game populations from range areas.

Conservation programs are largely ineffective. Lack of interest or funds to pay conservationists and game wardens, corruption, and poverty contribute to failure. Some efforts are being made to enlist the aid of local farmers in protecting "their" tigers, but the prognosis for the survival of the tiger is poor.

Although culture is frequently seen in opposition to natural forces, these remain connected in a holistic, contextual existence. While Westerners assume a direct connection between people's perception of nature and their treatment of it, this is not necessarily the case in Asia, where contextualism and pragmatism are far more important. Nature can be worshipped and harmed simultaneously; one can pay homage to the tree prior to felling it.

Certainly, Asian perceptions of nature have not stood in the way of its destruction. Yes, Western impacts, foreign companies, and foreign investment all play their roles. However, political and economic elites, the formulators of national policies, facilitate, contribute to, and benefit financially from environmentally damaging development schemes. With all the rhetoric about saving endangered beasts and habitats, the onslaught of development continues to affect the health and well-being of nature and humanity alike.

Current thinking does not dichotomize use or development from environment. Since humanity cannot survive without natural resources and these cannot exist if depleted or destroyed, development and environment are inexorably linked. **Sustainable development** is essential in light of human impacts on air, water, flora, fauna, and other natural systems. It is critical for the survival and well-being of future generations.

Conceptually, sustainable development sounds like the right thing to do. But putting these ideas into practice unearths such questions as these: What is the precise meaning of *sustainable,* and who sets the standards? Who or what should be sustained? Is a geomorphic phenomenon such as a prehistoric cave entitled to the same consideration as a tiger population? Are animals more important than fish?

In the upcoming chapters, you will read more about environmental problems and efforts to contain them. Keep in mind that because of the ephemeral characteristic of nature, local problems become regional ones, and regional problems become global ones. Keep in mind, also, that it is the poor who suffer most directly from environmental degradation. Unlike the majority of those living in the industrialized nations of the northern hemisphere, the majority of Asians live close to the land, without the mediation of air conditioners, water purifiers, and so forth. Geographer W. M. Adams (1995) says, "Sustainable development is synthetic, and constructed within the confines of current convention in northern-dominated debate about development. . . ." While debate lingers among policymakers, the full force of modern development continues to transform whatever is left of the natural world. In Chapter 3, I will discuss the dynamics of population, agriculture and food supply, and related social inequity.

Main Point Summary

1. The physiography of Asia is exceedingly complex. Ancient, massive mountain ranges and extensive, rugged plateaus offer unique environments for habitation and exploitation. The Himalaya and other ranges have channeled migration, served as barriers to cultural diffusion, and continue to hinder accessibility.
2. Fertile, alluvial lowlands such as the Indo-Gangetic and North China Plains support dense populations. Populations are generally clustered in plains, basins, and coastal lowlands.
3. The Indus, Yangzi, Mekong, and other rivers offer both food supply and power potential. Rivers are critical for settlement and as transportation arteries.
4. Oceans and seas provide problems as well as possibilities. Strategic and resource-rich areas such as the Straits of Melaka or the South China Sea are focal points of political conflict.
5. The physical and cultural diversity of fragmented island realms fosters instability. Political control is especially problematic in extensive archipelagos such as Indonesia and the Philippines.
6. South, East, and Southeast Asia are collectively known as *Monsoon Asia.* Activities are regulated by distinct wet and dry seasons, and productivity is frequently marred by climatic variance, unpredictability, and disaster.
7. Natural vegetation is coincident with climatic regimes. Biomes, increasingly destroyed or threatened by human presence, house ever-decreasing species types and numbers. Both species and genetic diversity are endangered.
8. Biomes are varied, ranging from tropical rainforests in Southeast Asia to temperate, mixed forests in East Asia; from steppe in Mongolia to deserts in Pakistan; and from mangal in Bangladesh to tundra in the high Himalaya. Both people and wildlife have wide-ranging adaptations to the environmental challenges of these different biomes.

9. Wildlife diversity is unparalleled and each biome presents an array of useful wood and resinous substances, medicinals, and food products. Population pressure, development, and illegal logging and hunting operations threaten biodiversity. These also induce environmental criticality whereby the integrity of both environmental and social life-worlds is jeopardized.

10. *Environment* is a multidimensional concept with definitions ranging from anthropocentric to bioethical. Current, idealized thinking addresses sustainable development. Unfortunately, ideas and practices are seldom convergent. Population increase, urbanization, and economic development have left few remains of the "natural world."

Terms to Know

topography
hydrological erosion
aeolian erosion
deposition
glaciation
cultural convergence
continental drift
ammonites
plate tectonics
tsunamis
alluvial soils
fluvial
archipelago
karst topography
coral reefs
natural boundary
Exclusive Economic Zone (EEZ)
Timor Gap
Monsoon Asia
convectional rainfall
growing season
continentality
Intertropical Convergence Zone (ITCZ)
wind-chill factor
topographic barrier
orographic precipitation
rainshadow
tropical cyclone
typhoon
natural vegetation
species diversity
rainforest canopy
emergents
dipterocarp
liana
epiphyte
laterite
buttress
jungle
semideciduous
exudates
mangal
xerophytic
desert
ephemerals
steppe
deciduous
altitudinal zonation
tree line
tundra
plate tectonics
Malesia
Wallace Line
endemic
biodiversity
environmental criticality
sustainable development

Bibliography

Adams, W. M. 1995. "Sustainable Development?" In *Geographies of Global Change,* ed. R. J. Johnston, Peter J. Taylor, and Michael J. Watts, pp. 354–373. Oxford: Blackwell.

Balik, Michael J., Elaine Elizabetsky, Sarah A. Laird eds. 1996. *Medicinal Resources of Tropical Forests.* New York: Columbia University.

Bayliss-Smith, Tim, and Susan Owens. 1994. "The Environmental Challenge." In *Human Geography: Society, Space, and Social Science,* ed. Derek Gregory, Ron Martin, and Graham Smith, pp. 113–145. Minneapolis: University of Minnesota.

Bruun, Ole, and Arne Kalland. 1995. *Asian Perceptions of Nature.* Richmond, Surrey, G. B.: Curzon.

Durst, Patrick B., Ward Ulrich, and M. Kashio, eds. 1993. *Non-Wood Forest Products of Asia. New Delhi:* Oxford and IBH.

Kasperson, Jeanne, Roger Kasperson, and B. L. Turner, eds. 1995. *Regions At Risk: Comparisons of Threatened Environments.* Tokyo: United Nations.

Macintosh, Donald J. 1983. Riches Lie in Tropical Swamps. *The Geographical Magazine* 55: 184–188.

Matthiessen, Peter. 1997. The Last Wild Tigers. *Audubon* 99: 54–63 & 122–125.

McNeely, Jeffrey A., and Paul S. Sochaczewski. 1988. *Soul of the Tiger.* New York: Doubleday.

Myers, Norman. 1984. *The Primary Source: Tropical Forests and Our Future.* New York: Norton.

Peissel, Michel. 1997. *The Last Barbarians: The Discovery of the Source of the Mekong in Tibet.* New York: Holt.

Saghal, Bittu. 1997. Tigerland. *New Internationalist* 288: 24–25.

Scott, Geoffrey A. 1981. Mangroves and Man in the Malay Archipelago. *Sea Frontiers* 27: 258–266.

Smith, Holly. 1997. *Adventuring in Indonesia.* San Francisco: Sierra Club.

Stone, David. 1994. *Biodiversity of Indonesia.* Singapore: Archipelago.

Tangsubkul, Phophat, and Frances Lai Fung-wai. 1983. The New Law of the Sea and Development in Southeast Asia. *Asian Survey* 23: 858–878.

United Nations Environment Programme. 1997. *Global Environmental Outlook.* New York: Oxford.

Vesilind, P. J. 1984. Monsoons. *National Geographic* 166: 712–747.

Ward, Geoffrey C. 1997. Making Room for Wild Tigers. *National Geographic* 192: 2–45.

Whitmore, T. C. 1990. *An Introduction to Tropical Rainforests.* Oxford: Clarendon.

World Conservation Monitoring Centre (WCMC). 1992. *Global Diversity: Status of the Earth's Living Resources.* London: Chapman and Hall.

WCMC. 1993. *1994 IUCN Red List of Threatened Animals.* Cambridge, U. K.: IUCN.

Chapter 3

Population, Agriculture, and Social Disparity

"The people turn in allegiance to Humanity, as surely as water flows downward or as a wild animal takes cover in the wilderness."

(MENCIUS 390–305 B.C.)

In the Beginning

It is currently believed that *Homo sapiens* (modern humans) originated in Africa and spread from there to the rest of Eurasia and ultimately the rest of the world. But new evidence from Asia and improved dating techniques have questioned this assumption.

The "Africa origin" theory holds that *Homo erectus* (predecessor of *Homo sapiens*) emerged in Africa c.1.6 million years ago and *H. sapiens* emerged about 100,000 to 200,000 years ago. No one knows precisely when *H. erectus* ceased to live and *H. sapiens* was born, but recent evidence from Indonesia indicates that both might have thrived there at the same time as late as 60,000 years ago.

Around 16,000 B.C., sea levels lowered, forming land bridges between Asia and North America, and between New Guinea, Australia, and Tasmania. The latter three formed what is referred to as Sahulland. Sundaland is the name given to the connected islands of the Philippines, Malaysia, and Indonesia (Figure 3-1). These land bridges facilitated the movement of people from Asia throughout Southeast Asia, as well as the movement of Australoids north and west. The inhabitants of New Guinea, for instance, are Australoid. The ocean gap between Bali and Lombok was an apparent barrier, but now it is clear that *H. sapiens* managed to cross the sea and traverse to Australia.

Java man was discovered in 1891 at Trinil and named *Pithecanthropus erectus,* meaning "upright ape-man." At that time, it was believed that *H. erectus* had reached Java by 800,000 B.C. But in 1994, dating techniques revealed that Java man was living on the island as early as 1.8 million years ago. This is 800,000 years before *H. erectus* was supposed to be in that region. Other recent finds confirm this time frame.

In the early 1920s, remains of what was to be called *H. erectus pekinensus* were found in a karst cave at Dragon Bone Hill in the village of Zhoukoudian near Beijing. For eons, people had collected fossils

The Wallace Line

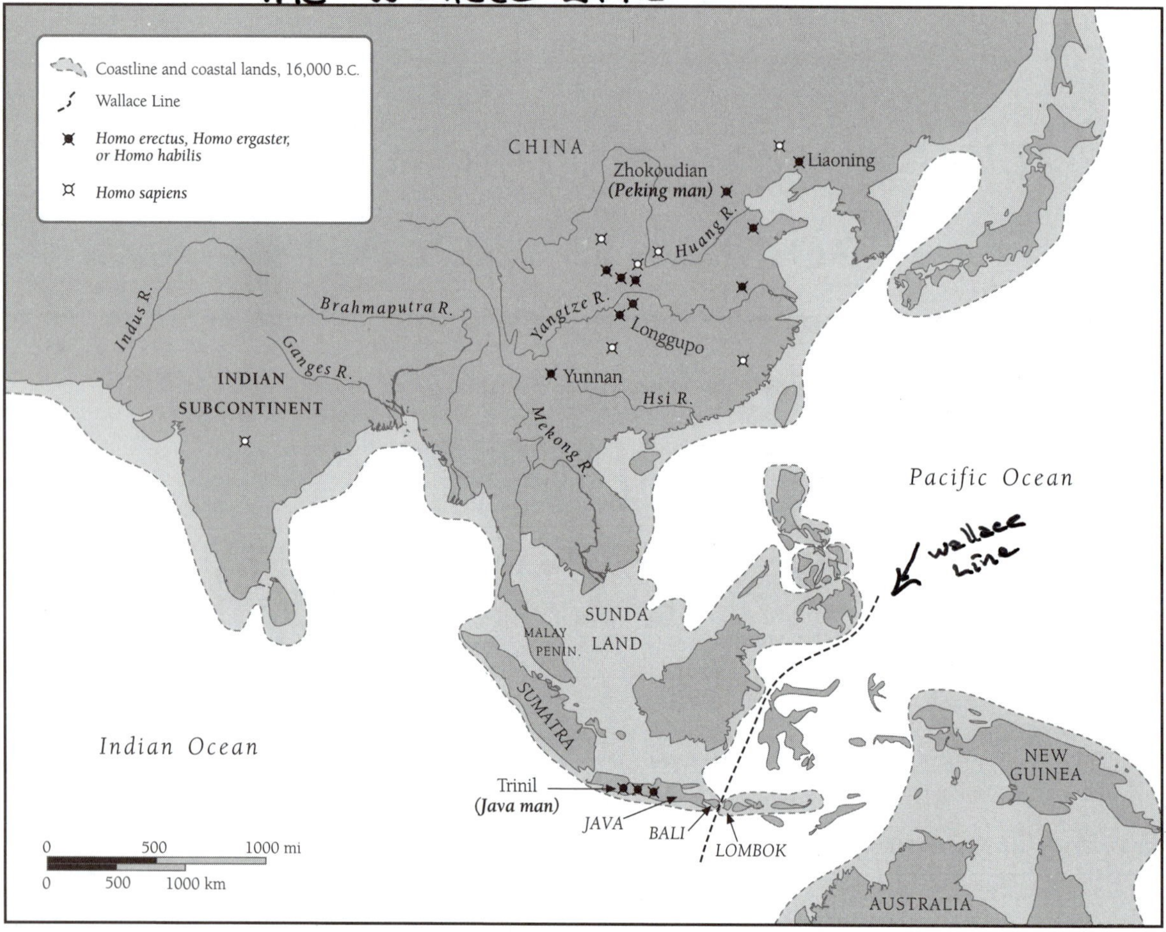

Figure 3-1

Ancient coastlines and early humans. Note how former land masses made possible the inter-realm migration of people, animals, and plants.

(dragon bones) here for use in traditional medicine. Peking man, hunter and gatherer, inhabited the cave between 460,000 and 230,000 B.C. It is now thought that the predecessor of Peking Man reached central China by 1.7 to 1.9 million years ago. Is it conceivable that humans originated in China?

When surveyed in 1984, the Longgupo (Dragon Hill) cave in Sichuan revealed a *H. erectus* fossil 1.8 to 2 million years old. Additional finds in Yunnan and elsewhere in China confirm this antiquity. Two skulls unearthed in China's Hubei province in 1989–1990 are dated at least 600,000 years old, and a 200,000 year old *H. sapiens* fossil, discovered in 1984 in a Liaoning cave, is twice as old as any other *H. sapiens* find in China. Fossil discoveries in central Java in 1996 indicate that *H. erectus* lived there 27,000 years ago, supporting the theory that *erectus* and *sapiens* coexisted. What does this mean in terms of the peopling of Asia, or even the world?

The **"out-of-Africa" theory** has been bolstered by the suggestion that *H. erectus* and *H. sapiens* were contemporaries. Discoveries in Southern Africa in 1998, as yet unanalyzed, are hypothesized to further support the African origin theory. Other scientists support a **multiregional hypothesis,** which purports that dispersed *H. erectus* populations spawned *H. sapiens* populations in several locations simultaneously. The Liaoning discovery certainly gives cause to ponder this possibility.

Growth and Dispersal

Eight thousand years ago, a mere five to ten million people inhabited the earth. Today, the earth's population is more than six billion. Most of this growth has taken place since 1800, when the world population first reached one billion. In 1997, South, East, and Southeast Asians accounted for more than four billion, or two-thirds of the global population. How and why did this happen? What are the ramifications?

Estimates of early Asian populations indicate that they were relatively small, with sporadic periods of increase and decrease. By the mid-nineteenth century, population numbers had jumped significantly. However, growth was not an even progression. Before the advent of modern medicine, population numbers fluctuated across the globe with environmental disasters, war, famine, and disease. For example, in 1279, the Mongol conquerors of China exterminated some 35 million peasants. A pandemic of plague from 1347–1351 killed 75 million of the then 300 million Eurasians.

Although worldwide high-mortality disasters have been largely quelled, regional catastrophes continue to make dents in population totals. For instance, 17,000 died in Bombay from plague in the late nineteenth century. A famine killed at least 30 million Chinese from 1959 to 1961. A million Bangladeshis perished in a hurricane in 1970. And, in one of the worst disasters in human history, at least two million Cambodians were exterminated by government policy from 1975 to 1978! Nevertheless, overall numbers remain sizable and provide large **base populations** to generate notable increments of increase.

Figure 3-2 shows the distribution of Asian populations. What accounts for this pattern? History provides the answer. Asian populations originated in and expanded from a series of core areas over a period of more than 4,000 years (Figure 3-3).

CORE AREAS OF GROWTH AND DIFFUSION

As the Neolithic revolution in primary production drew to a close, there were two major clusters of population in East and South Asia: the North China Plain and the Indus Valley of northwestern India (now Pakistan). Around 3000 B.C., these regions possessed agricultural systems productive enough to permit growth and stimulate expansion. A good measure of political control, social organization, and diversification of labor in rich environments allowed the development and maintenance of irrigation projects, as well as effective storage and distribution of agricultural surpluses. Population developments at this time derived from agrarian expansion.

Other smaller clusters existed around Japan's Inland Sea, in southern Korea, the Red River basin of northern Vietnam, northern Java, the upper Ganges valley, and elsewhere in South Asia. By the year 1 A.D. regional populations had accumulated in these places.

By the fifth century, secondary population concentrations emerged in association with the upper Mekong delta, Vietnam's central coast, central Myanmar, China's Szechuan basin, and southern India.

From the eleventh to the fourteenth century, periods of effective political control, accompanied by improvements in agricultural methods, fostered expansion into China's central Yangzi region, Cambodia's Tonle Sap basin, the wet zone of Sri Lanka, and the lower Ganges valley.

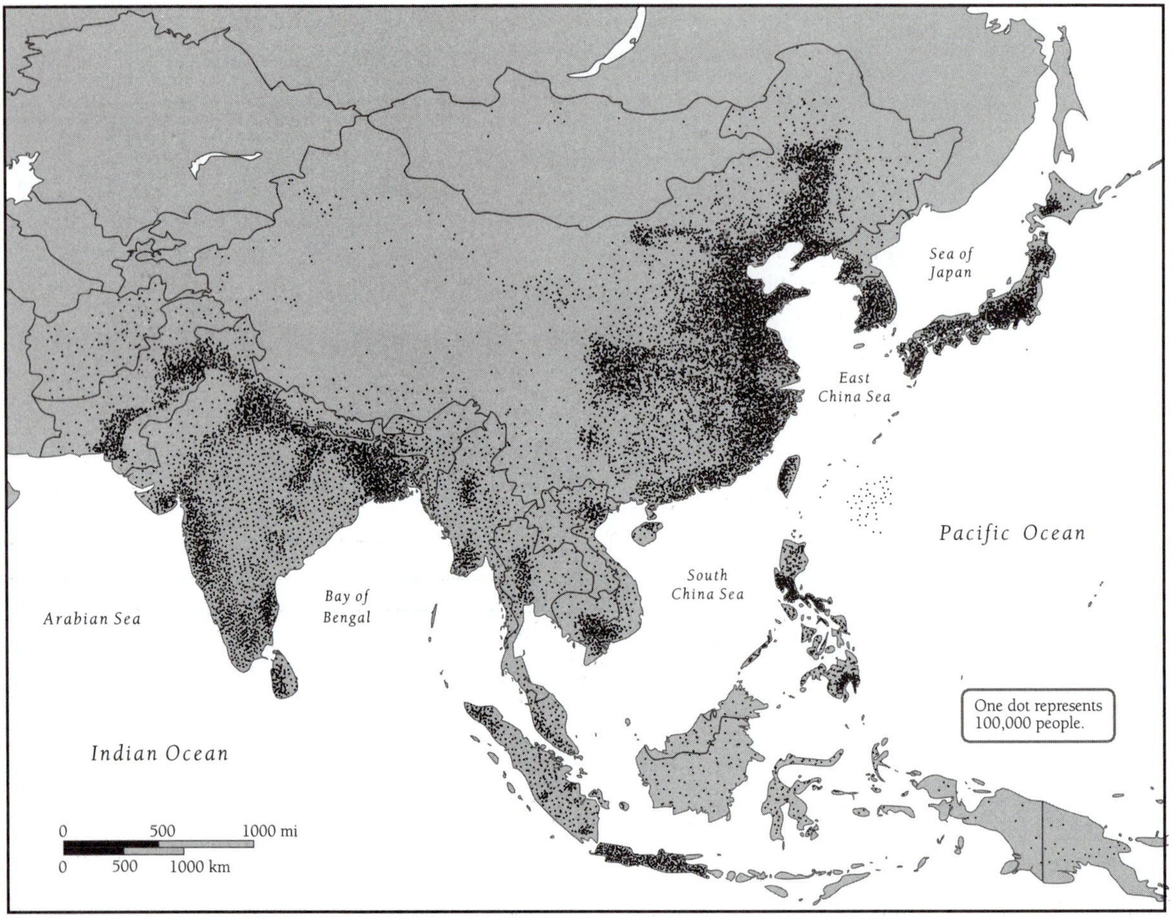

Figure 3-2

Population distribution. Compare this map with one of physical features. Note the relationship between population, plains and basins, and the significance of rivers.

Growth burgeoned in the nineteenth century under the impact of European influences. European technology, coastally oriented trade, and imposed political structures combined to effect both population increases and change in distributions. New flood control technology with large-scale drainage canals facilitated the development of river deltas. Ports and their hinterlands became focal points of development and important migration destinations. Most of Asia's dominant cities grew dramatically during this period. Tokyo, Hong Kong, Shanghai, Calcutta, Bombay (Mumbai), Bangkok, Singapore, and Jakarta are some of these.

Core areas were critical to population growth and distribution in Asia. In fact, every country except for Laos is based on one or more core areas. However, there is more to the story. The landscape of population, like any other landscape, is socially constructed; landscapes are made by people in their particular life-world context. Social organization is especially significant in the development of Asian landscapes.

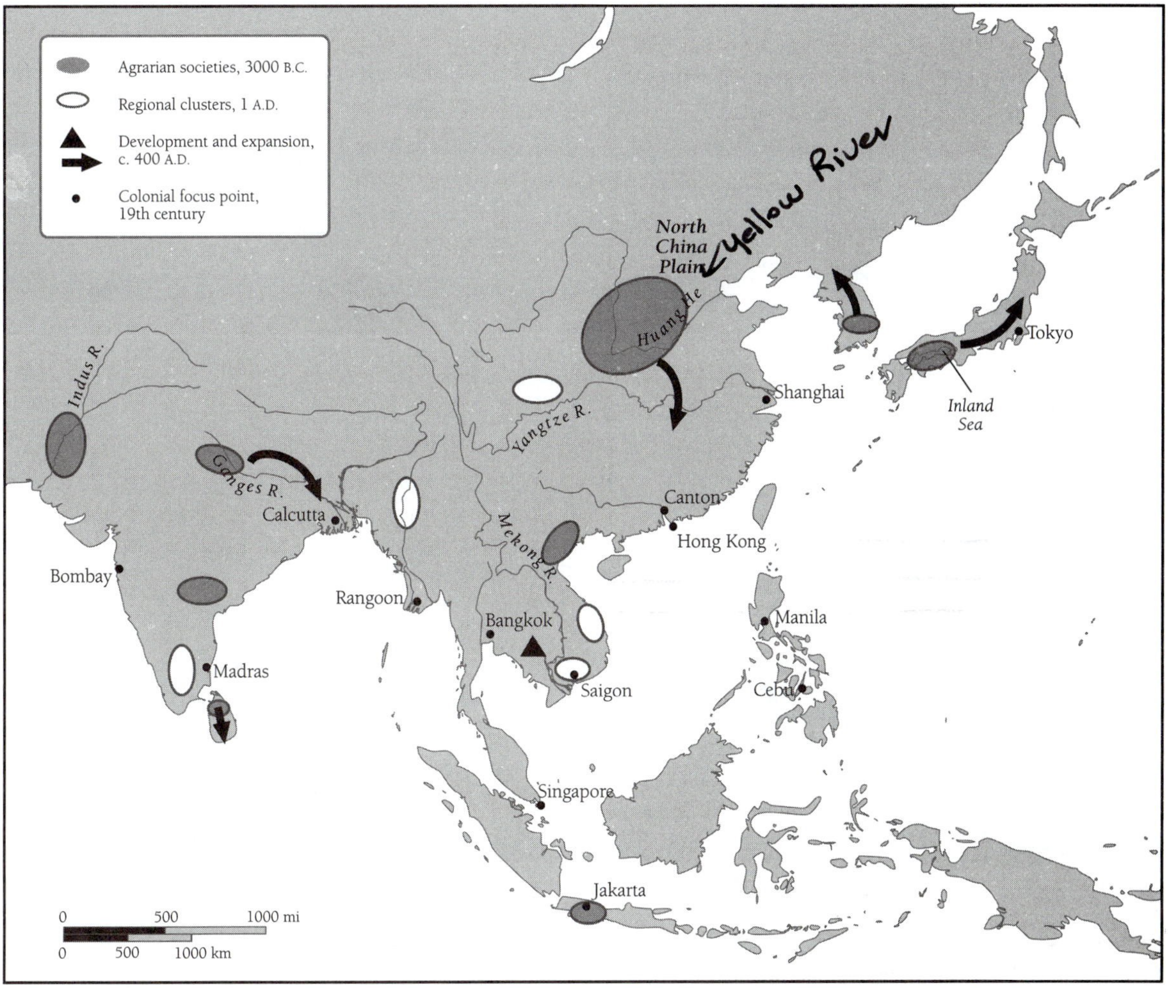

Figure 3-3

Core areas and diffusion of development. Note how alluvial lowlands fostered settlement as well as the importance of rivers and coasts in colonial settlement.

SIGNIFICANCE OF THE GROUP

Population growth and expansion were founded on intensive agriculture based on human labor. This is an undertaking requiring cooperative land management and production systems. Cooperation is encouraged because in Asia, the family is central and the group is more important than the individual. Established settlements were (and continue to be) **nucleated;** dwellings were clustered in villages (Figure 3-4). People and animals function in daily and seasonal routines of movement to and from their village and surrounding farmland. Cooperation is particularly important at planting and harvest times and in the maintenance of irrigation and flood-control systems. But this can only work in a situation of social and political stability. Consequently, long periods of stability coincided with population growth. Long-term chaos brought about population decline.

Figure 3-4

This aerial view over Pakistan shows two nucleated villages. People work the fields surrounding the villages. Much of this land is fragmented, whereby farmers do not own contiguous fields. Many farmers are tenant farmers working for landlords. Linear settlement patterns can be seen along roads and irrigation channels.

Where are you on the "I–we" spectrum of thought and action?

Thinking and Acting Collectively

In contrast to many societies where the individual is the locus of responsibility and action, Asian societies are characterized by collectivism. From birth, Asians are imbued with group consciousness extending far beyond familial and national levels to otherworldly existences. Identity derives from sets of expectations and responsibilities covering all aspects of life from emotional to economic. Long-established relationships among individuals, families, clans, work groups and so forth are regarded as both mutually beneficial and necessary to nourish and perpetuate family or organizational lineage. Thought and action revolve around "we" and collective agreement supersedes individual will.

The existence of different standards for different people is generally understood and accepted. The fact that royalty or elite groups exhibit abundant wealth, conspicuous consumption, or questionable behavior is tolerated as long as duties are performed correctly and moral responsibilities are met in ordering and providing for society at large. Authoritarian structure and order are considered prudent.

Religious beliefs frequently support a stratified status quo. Hierarchical concepts of order in the sacred realm bolster hierarchical relationships and consequential order in the secular world. *Carpe diem* and "you only live once," make little sense in context of profoundly enduring and seemingly endless continuities of collective existence.

It is true that Western influences—philosophies espousing individual rights, freedom, and social equity—are poking holes in the fabric of group identity in Asia. Change is most evident in places where values of individual achievement and entrepreneurial success are becoming prominent. Nevertheless, change does not mean erasure of tradition and Asians remain grounded, to varying degrees, in traditional collectivist consciousness.

Agricultural Foundations

Agriculture is understood as the cultivation of crops and rearing of livestock. Early humans were hunters and gatherers but at some point learned to grow their own food crops and raise their own meat supply. Plant and animal **domestication** most likely had several hearth areas. Cultural geographers Joseph Spencer and William Thomas (1976) postulate that regional crop assemblages, such as taro, yams, and bananas in Southeast Asia, encouraged agricultural developments (Figure 3-5).

Vegeculture most likely preceded seed cultivation; cultivating tuber and tree crops such as taro and bananas occurred prior to the domestication of rice and other grains. Land does not require clearing for root and tree crops, and a variety can be planted. These can be harvested at will; therefore, there is little need for storage facilities. Seed cultivation requires land clearance. Furthermore, grains have a specific growing season and must be harvested all at once, making storage necessary. Inadequate facilities invite losses to mold, insects, and rodents.

Primitive vegeculture developed in tropical regions around the world. Southern China, eastern India, and mainland and insular (island) Southeast Asia were such places. Vegeculture was practiced as early as

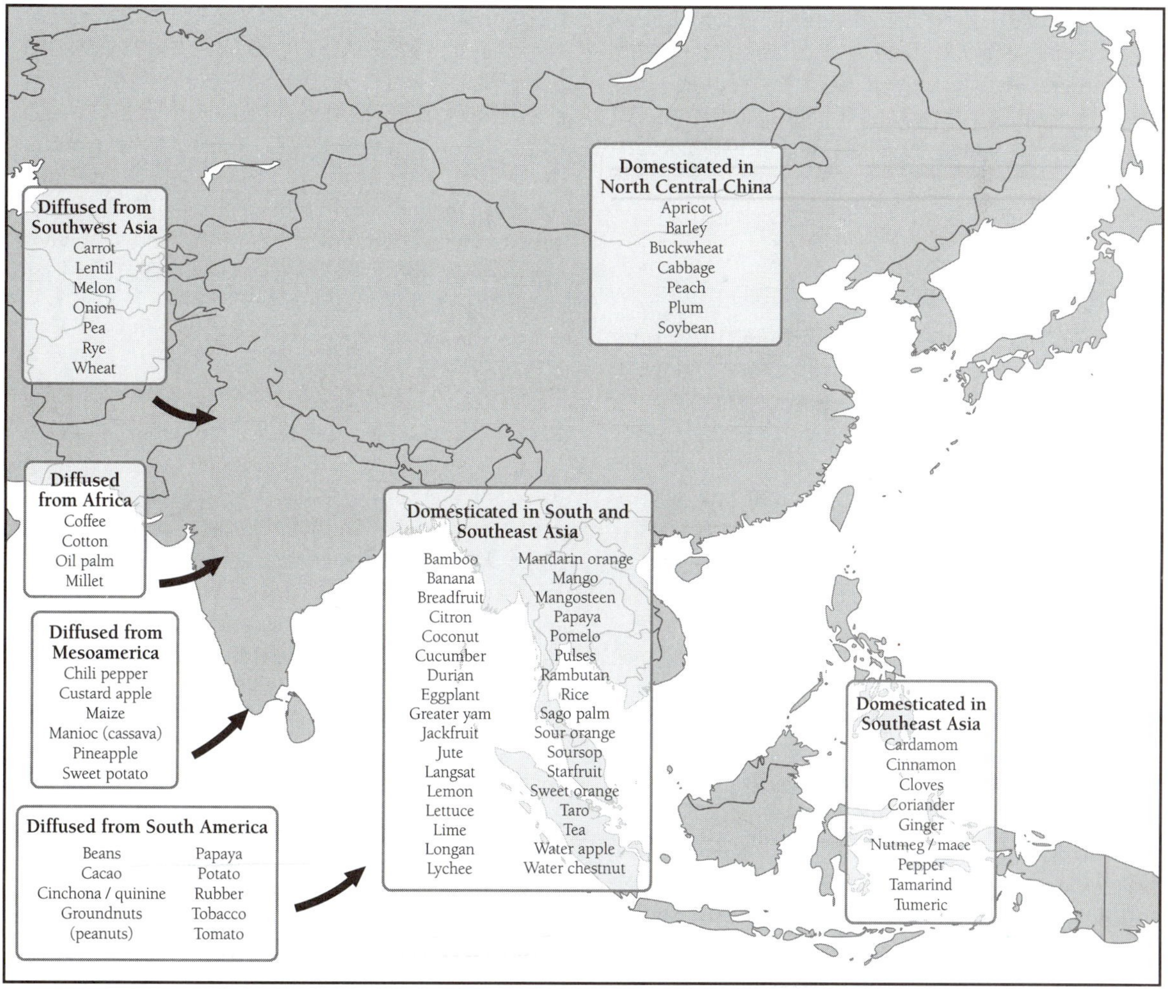

Figure 3-5

Crop assemblages, domestication, and introduction. Which of these grains, fruits, herbs, and spices are you familiar with?

9000 B.C. in Thailand and was definitely a multiregional phenomenon by 7000 B.C. Evidence of vegeculture has turned up in many areas, including Taiwan and Japan.

In previous pages I noted that Asia possesses a wealth of indigenous plant species. In fact, many "familiar" foods originated there. However, crop assemblages were altered with increased interregional trade by land and sea. Plant diffusion was slow until the fifteenth century, but by the seventeenth century, botanical gardens had been established in Europe to foster plant exchanges around the world. By the nineteenth century, the industrial world demanded nonfood, **industrial crops** such as cotton and rubber. Radical change derived from the introduction of plantation crops such as rubber and palm oil, especially in Southeast Asia. Commercial-variety rubber originated in South America and oil palms came from Africa. Today, plantation agriculture is transnational agribusiness, dependent upon the machinations of the global economy.

SHIFTING CULTIVATION

Early farming methods included **shifting cultivation,** a migratory, field rotation system found mainly in forested regions with relatively sparse populations. An area is cleared by a **slash and burn** method—cutting down all but the largest trees and the food trees and setting fire to the resultant debris. Burning brings nitrogen to the soil, which is then planted with the aid of a digging stick. Cropping is typically layered—below ground tubers, surface plants, and above surface plants and tree products. A field is used from one to three years depending on soil fertility. Then, it is abandoned and allowed to fallow naturally for ten to twenty years or more, during which it regenerates its natural flora. People move to a new area and begin again.

Shifting cultivation is an ecologically sound practice for maintaining soil fertility and encouraging forest rejuvenation. People participate in the ecosystem rather than superimpose upon it. However, this practice has drawn severe criticism as being destructive of forests, as well as being responsible for fires and ruination of soil fertility.

Problems lie not so much with the method as with the impacts of population and deforestation. Dramatic population increases, combined with the removal of forests for plantations and other uses, mean that there is insufficient land available for shifting cultivators. For example, every year in Southeast Asia, anywhere from 98,800 to 1,976,000 acres (c. 400,000 to 800,000 ha) of land used for shifting agriculture is transformed to permanent agriculture. Consequently, shifting cultivators are unable to fallow their plots long enough. Replanting too soon produces a downslide from diminished yields to total abandonment.

Shifting cultivation is still found in Southeast Asia, and in some areas of South Asia and southern China (Figure 3-6). It is known as **swidden** in Southeast Asia, but there are many local terms as well. In Indonesia, for instance, it is called *ladang*. While swidden plots are being clear-cut and permanently planted, the issue of soil fertility remains. Sustained crop yields in most tropical soils require the application of large quantities of fertilizer, essentially ensuring reliance on manufactured sources. Poorer farmers cannot afford chemical supplements, and many are forced to find subsistence elsewhere. Shifting cultivation is best described as a residual agricultural practice in Asia.

RICE CULTIVATION

Recent archaeological evidence from Zhejiang province in southern China indicates that rice was domesticated there around 5,000 B.C. Scientists think that its hearth area was in Southeast Asia, possibly Thailand, some 9,000 years ago. Gujarat in western India was another site of domestication. Swidden plots housed initial rice cultivation that was not irrigated. Wet-rice production did not become well established in China until the fourteenth century and in India even later. Rice cultivation diffused to Japan either via Korea or across the sea around 300 to 400 B.C. It was introduced into Java by sea from India, but as elsewhere, irrigated cultivation was not well organized until much later. Eventually rice replaced millets as the main food grain in the warmer parts of Asia.

Today rice is the staple food of more than half the world's population, and 90 percent of it is grown in Monsoon Asia. The inclusion of Asian colonies into the international economy stimulated population movements and the establishment of commercial, wet-

Figure 3-6

In this swidden plot in Kalimantan, Indonesia, the major crop is corn. A papaya tree stands in the foreground and a banana plant in the background. Eventually this plot will be abandoned for another.

rice systems in major river deltas. The Irrawaddy (Ayerawady), Menam (Chao Praya), and Mekong deltas were occupied between 1850 and 1930 and exported rice to Asia and the world.

Rice bowls were founded on environmental circumstances and human ingenuity regarding methodology and technology. Paddy preparation, water distribution systems, seedbed cultivation, sowing and/or transplanting techniques, and terracing and irrigating slopes are only some of the requirements (Figure 3-7). Farms are generally small, around 2 to 4 acres (1–2 ha). Plots are small as well, disallowing the use of large machines. Although the use of chemicals—fertilizers, pesticides, insecticides, and fungicides—are widespread, mechanization is not, except for some parts of Japan, South Korea, and Taiwan.

In Asia, three types of rice are sown or planted: *indica, japonica,* and *javanica. Indica* varieties have long grains that remain separated when boiled. *Japonicas* have shorter, stickier grains. *Javanicas* are somewhere in between. *Indicas* and *javanicas* dominate in tropical and subtropical areas while *japonicas* excel in cooler latitudes and at higher elevations. There are glutinous and nonglutinous varieties in all categories.

Farmers typically grow several varieties, depending on their own food preferences, market demand, and ceremonial requirements. They are also conscious of variation in maturation periods that can take from 90 to 260 days. Where climate permits, faster ripening varieties, like *indicas,* allow double or even triple cropping. Quick-ripening rice is critical where water supply is unstable.

Wet-rice cultivation relies on either rainfall or irrigation. In deeply flooded areas like delta regions, floating rice is grown. It makes up a quarter of Bangladesh's rice output, for example. Elongated stems grow in concert with rising water in the wet season, keeping the grains above the water that may be several feet (meters) high. Floating rice is tended by wading or from boats.

Since the 1960s, farmers have been planting *high-yielding varieties* (HYVs) of rice, wheat, and other crops. These were developed at research institutions around the world in an effort to sustain burgeoning populations. These stronger, higher-yielding, pest-resistant seeds, along with appropriate management techniques, generated the **Green Revolution.** Tougher seeds enabled the use of hitherto vacant land. From the mid-1960s to the mid-1980s, land under rice production increased by 15 percent in Asia. But crop yields jumped by 74 percent. Scientists are searching for a "super-rice" that will have even greater yields and be genetically protected against even more diseases and pests.

Innovations are generally adopted by the most innovative but not necessarily the biggest landholder.

Figure 3-7

Here in southern China a woman is bundling rice seedlings from the seed bed to the left. The seedlings will eventually be transplanted into a rice paddy.

However, larger operators are likely to have more connections and, therefore, easier access to loans, farming cooperatives, irrigation systems, and so forth. Female-headed households and social minorities such as low class/caste or tribal groups seldom have equal access. Other impediments relate to environmental conditions or loss of cropland to urbanization. Nevertheless, the Green Revolution has generated quantum leaps in crop yields, lowered consumer prices, and made jobs available to landless unemployed. However, in some areas labor costs have increased to the point that farmers are forced to reduce labor inputs. For instance, in Thailand, farmers are turning to broadcast seeding to counter the costs of hand-planting seedlings.

The *law of diminishing returns* dictates that increments of increase in crop yields will shrink relative to increases in inputs over time. Green Revolution agriculture calls for more knowledge in such areas as integrated pest management. And it demands more inputs, especially labor in the case of rice. With rising costs and reduction in cropland, food supply may not keep up with population growth. Geographer Jonathan Rigg (1994) notes that "the revolution is on-going," that social and other restraints shift and even disappear over time, and, "that one person's tradition-bound peasant is another's poor and exploited farmer."

"Without rice there is nothing doing."

Rice and Society

The Malay proverb above signifies the importance of rice in much of Asian society. In Chinese and Southeast Asian languages, "to eat" is the same as "to eat rice." Japan became known as *Mizumono Kuni*—the "Land of Luxurious Rice Crops." Rice is the foundation of culture. It is requisite for survival and integral to daily, weekly, monthly, and yearly routines and permeates both secular and sacred worlds.

Rice is highly digestible and nutritious when eaten with its bran intact and/or with vegetable and fish supplements. Claims of dietary deficiencies due to rice are simplistic. It is *polished* rice that has reduced protein. As part of modernization, rice mills are increasingly common, and rice is polished to a glaze. Thoroughly devoid of bran, the nutritional value of the grains is reduced.

Rice wine, beer, and liquor are widely consumed. Japanese saki is distilled from rice wine.

In some places, a common container emphasizes the role of rice as a social locus. On several occasions, I have been invited to share the rice wine pot. Sipping out of one of several reed straws protruding from large clay containers, I experienced a sense of unity with my Lao and White Tai (in Vietnam) hosts.

Rice husks are fed to animals. In the Mekong Delta, rice bran is fed to fish on fish farms. Rice bran oil is a basic ingredient of soap and many insecticides. Noodles and cosmetics are made from rice flour. The straw provides bedding for livestock and can serve as a low-grade feed. It also is the medium for cultivation of straw mushrooms.

Aside from its practical uses, rice permeates ceremonial and ritual activities. For instance, there is an annual event in Bangkok, attended by the Thai king, called *The Royal Ploughing Ceremony,* which calls for bounteous rice harvests. When Japanese Emperor Akihito was married in 1990, he carried a rice scoop to symbolize fertility. Folk dances still replicate stages in rice production, and rice motifs are frequent in weaving and basketry. Temples, goddesses, and gods are devoted to rice. It is integral to marriage, birth, and groundbreaking ceremonies. The importance of rice in Asian society can be summed up by a Balinese religious belief: rice is the "soul of man."

Population Increase and Expansion

Population numbers and distribution were tied to agricultural productivity. In stable and prosperous times, villages expanded and even coalesced. New farmlands were opened up and people moved into them. Where equal or multiple share inheritance was practiced, lands continued to be subdivided into ever-smaller plots. As populations grew and good lands filled, marginal lands became sought after. In-migration of one population forced the out-migration of another. Hunters and gathers and swidden populations, many of whom were minority peoples, were seriously affected. Southeast Asia received countless non-Chinese ethnic groups as well as Chinese in context of China's southward expansion. Repeated invasions of Aryans and later Mongols into northwestern India drove extant populations eastward and southward there as well.

Population migrations did not end in ancient history; they are ongoing. As you will learn later, Chinese movement into Manchuria is a fairly recent phenomenon and the Chinese continue to migrate into their northern, western, and southwestern frontier regions. Vietnamese migration down the Annamite coast continues to press southward into the lower Mekong delta. Population processes and patterns are constantly changing in light of social and environmental circumstances.

SIGNIFICANCE OF CHILDREN

The key dynamic of population growth is the fact that people have children. This reality draws much comment and criticism, especially when the childbearers appear not to be able to support their offspring. The necessity for having children does not rest solely on the concept of societal reproduction. Children are the core of Asian family-group culture. Children are essential for the respect and continuity of family lineage. With strong bonds between people and their lands, children (usually sons) receive and are entrusted with guardianship of family property through generations. Children sustain the group, becoming a source of pride. Men and women are truly worthy when they produce children, more so with sons. Not having children is inconceivable to the majority of Asians.

There are also practical reasons for having children. In traditional societies, children contribute their services and labor very early. Four-year-olds look after two-year-olds. Seven-year-olds work at farm tasks or help in the family business. Children make important contributions toward their own and their families' survival. They are also expected to take care of their parents. Both boys and girls expect to look after their aged parents or parents-in-law. Where there is no old-age insurance or care facilities (in most places), this cultural practice is critical.

But why do people have *so many* children? History demonstrates that life is uncertain. Natural disasters can strike at any time. Famine can quickly decimate a population. North Korea suffered a severe famine in 1998. The death toll has not been publicized. War has been all too common in the twentieth century, and millions continue to experience fighting as a way of life. Health care is available only to some and limited for all. In other words, if a family has only one or two children, the chances are not good that any will survive to carry out family responsibilities. In the War Museum of Hanoi, there is a picture of a woman who lost nine sons in wars fought in Vietnam from the 1950s to the 1970s. Who will care for her in her old age? People cannot be expected to have fewer children if many are needed to ensure an ongoing supply for labor, income, military duties, parental care, and the like. However, quality of life improvements can induce reduction in family size.

REGIONAL DYNAMICS

Table 3-1 lists statistical data used in traditional geodemographic analysis. Traditional assessment of population trends is based on information about growth and change; density and distribution; and population composition. This information is gathered by various organizations from official sources based on censuses in individual countries.

Data are not necessarily reliable. Counts are not accurate because not all people are easily accessed. Moreover, people may lie about the number of children they have or number of individuals in their household to avoid taxes, prohibitions on family size, or residency rules. Governments might skew figures for economic and political purposes. Whatever the case, look at these data not as absolute facts but rather as *indicators* of conditions and trends.

Growth and Change

Variation is a key word for understanding population dynamics in Asia. Table 3-1 demonstrates that some countries have relatively high birth rates (BRs). Bhutan, Maldives, Nepal, Pakistan, Cambodia, and Lao are among these. Nepal, Pakistan, Cambodia, Lao, and Myanmar have strikingly higher death rates (DRs) than other countries. The difference between these two measures is the natural increase rate (NIR). While Bhutan has a high BR (4%), it has a relatively low DR (.9%). This results in a very high NIR (3.1%). Maldives has a similar situation: BR 4.3% − DR .7% = NIR 3.6%. NIRs above replacement rate result in population increase. Too-low NIRs mean population decrease.

Look at the doubling time (DT) column. This indicates the number of years it will take for a population to double in size given the current NIR. The next column shows the expected population in the year 2025, given the same NIR. Notice that Japan is predicted to have fewer people in 2025 than it has now. Having both a low BR and DR, Japan's NIR is insufficient to replace its population. Does this mean that the Japanese will ultimately disappear? Read more on this topic in Chapter 13.

Although India and China do not stand out in terms of NIRs, they are notable in terms of their population totals. Because of the size of these base populations, even a small percentage increase adds many people to the total. So, while India's NI is only 1.9 percent, it will add almost 415 million people by 2025. China, with its low NIR of 1.0 percent, nevertheless will grow nearly 333 million by 2025. I will discuss India's and China's population issues separately in the regional chapters.

Density and Distribution

Density and distribution data measure spatial characteristics of populations. Refer to Figure 3-2 and note that some areas are densely populated while others are sparsely populated. Environmental and historic circumstances contribute to these patterns. People tend to avoid extreme environments such as deserts and rugged, high-elevation terrain. Access to resources is crucial to survival. For example, arable land is found on plains and along waterways, and this is where most populations reside. Past circumstances have led to a phenomenon known as **historic inertia**—people migrate to and continue to live in places where other people are already established and opportunities are perceived to exist. Consequently, densely populated

Table 3-1 Selected Population Statistics

1998	Total Pop. Mill.	Birth Rate %	Death Rate %	Nat. Inc. %	Doub. Time Years	2025 Proj. Pop.	% Urban	% −15/65+ Yrs. Old
World	**5926**	**2.3**	**0.9**	**1.4**	**49**	**8082**	**44**	**32/7**
Australia	18.7	1.4	0.7	0.7	101	23.5	85	21/12
Canada	30.6	1.2	0.7	0.5	136	40.3	77	20/12
U.K.	59.1	1.3	1.1	0.2	433	62.6	90	19/16
U.S.A.	270.2	1.5	0.9	0.6	116	335.1	75	22/13
South Asia								
Bangladesh	123.4	2.7	0.8	1.8	38	165.6	16	43/3
Bhutan	0.8	4.0	0.9	3.1	22	1.5	15	43/2
India	988.7	2.7	0.9	1.9	37	1441.2	26	36/5
Maldives	0.3	4.2	0.9	3.3	21	0.6	25	46/3
Nepal	23.7	3.3	1.1	2.2	32	39.5	10	43/4
Pakistan	141.9	3.9	1.1	2.8	25	258.1	28	41/4
Sri Lanka	18.9	1.9	0.6	1.3	53	24.1	22	35/4
East Asia	**1469**	**1.6**	**0.7**	**0.9**	**74**	**1798**	**37**	**25/7**
China	1242.5	1.7	0.7	1.0	69	1561.4	30	26/6
H. K. SAR*	6.7	0.9	0.5	0.4	161	7.8	—	18/10
Macao	0.5	1.3	0.3	1.0	71	0.6	97	25/7
Japan	126.4	1.0	0.7	0.2	330	120.9	78	15/16
N. Korea	22.2	1.8	0.9	0.9	75	26.1	59	28/6
S. Korea	46.4	1.6	0.6	1.0	68	52.7	79	22/6
Mongolia	2.4	2.4	0.7	1.6	42	3.3	57	36/4
Taiwan	21.5	1.5	0.6	1.0	73	25.4	75	23/8
S.E. Asia	**512**	**2.4**	**0.8**	**1.6**	**42**	**709**	**35**	**35/4**
Brunei	0.3	2.5	0.3	2.2	32	0.5	67	34/3
Cambodia	10.8	3.8	1.4	2.4	29	17.0	14	44/4
Indonesia	207.4	2.4	0.8	1.5	45	275.2	37	44/4
Laos	5.3	4.2	1.4	2.8	25	9.8	19	45/3
Malaysia	22.2	2.6	0.5	2.1	33	37.0	57	35/4
Myanmar	47.1	3.0	1.0	2.0	35	67.8	25	36/4
Philippines	75.3	3.0	0.7	2.3	30	116.8	47	38/4
Singapore	3.9	1.5	0.5	1.1	65	4.2	100	23/7
Thailand	61.1	1.7	0.7	1.1	64	71.6	31	27/5
Vietnam	78.5	1.9	0.7	1.2	57	109.5	20	40/5
1998	Total Pop. Mill.	Birth Rate %	Death Rate %	Nat. Inc. %	Doub. Time Years	2025 Pop. Proj.	% Urban	% −15/65+ Yrs. Old

* Hong Kong Special Autonomous Region.

† These data indicate numerical trends. They are poor indicators of quality of life.

Source: 1998 Population Data Sheet. Population Reference Bureau.

areas most likely will continue to be densely populated areas. As you will read in later chapters, some governments, like those of Indonesia and Malaysia, are trying to change their population distributions through programs encouraging migration from densely populated regions into sparsely populated regions.

Density figures are misleading. **Arithmetic density** indicates how many people exist per square mile (km) of land in a particular country. The Philippines, for instance, has a density of 638 persons per square mile. However, people are not evenly distributed throughout the islands. The northern island of Luzon is home to millions more than the southern island of Mindanao, and numerous islands are devoid of people. Arithmetic densities reveal nothing about actual spatial distribution of people.

An alternative to arithmetic density is **nutritional density** that relates numbers of people to amount of cultivated land. Nutritional density assumes a direct relationship between the number of people and land productivity potential. This information is interesting but still misleading.

Density data provide only crude measures. Laos is a relatively sparsely populated country yet it is regarded as one of the poorest in Asia. Taiwan, one of the most densely populated, is seen as one of the richest. Densely populated Singapore, also one of the richest, has virtually no arable land. Density does not account for variation in human or land productivity, differences in dietary requirements, or quality of available diet, food imports, or unequal access to food supplies.

Another indicator of distribution is **percentage urbanized.** What proportion of a country's population lives in cities? Note that definitions of *city* and *urban* vary widely. Table 3-1 shows that most of Asia is not highly urbanized. China has at least 20 cities of more than one million people, but China is only 29 percent urban; most Chinese are rural dwellers. Thailand is often perceived as "Bangkok," but Thailand is only 19 percent urban. In fact, 15 of Asia's 23 countries are less than 50 percent urbanized. This emphasizes the fact that most people in Asia are farmers. Furthermore, urbanization does not necessarily equate better living standards. For example, North Korea is 61 percent urban but has very low living standards. Some of the worst poverty is found in cities where modern skyscrapers belie other realities of **squatter settlements** and homelessness (Figure 3-8). Look for more information on urbanization in the next chapter.

Population Composition

Population composition data reveal the sex and age structure of a particular population. **Population pyramids** are visual representations of relationships be-

Figure 3-8

This is Kali Besar, one of Jakarta's several squatter settlements. Situated on a river near the sea, it is built on stilts to accommodate the tides.

tween males and females, and among different age cohorts (Figure 3-9). Most populations for Asian countries have wide bases, indicating high NIRs and youthful populations. Considering that it is female cohorts ages 15–45 that are potential child-bearers, this age group will increase dramatically as younger cohorts age. For example, a five-year-old girl in 2000 will be a potential mother in 2010 (or even earlier).

Table 3-1 lists percentages of populations under 15 and over 65 years old. These age cohorts, under traditional analysis, are perceived as being "dependent" on other age cohorts for survival. But in fact, significant numbers of these young and old people are contributors to their own survival, as well as that of other age groups.

Another measure of composition is **sex ratio:** the numerical ratio between men and women. Normally, populations have a balanced sex ratio. Although women outlive men, sex ratios at birth are biologically stable (assuming no intervention). The norm is 106 to 107 male births to every 100 females. But sex ratios in countries with long life expectancies reflect the fact that women live longer than men. In 1995, there were 105 women for every 100 men in the United States, for example.

In 1995, India and Bangladesh had sex ratios of 106, Pakistan 108, and China 105 males per 100 females. In China, among children age two or younger, there were 121 males for every 100 females. Demographers have discovered that millions of girls in these countries are missing. In light of the importance of males in the extended and generational family, there is a strong preference for boys in Asian cultures. In China, for instance, a boy is called a "big happiness" while a girl is a "little happiness." Government-sponsored and personal efforts to limit family size

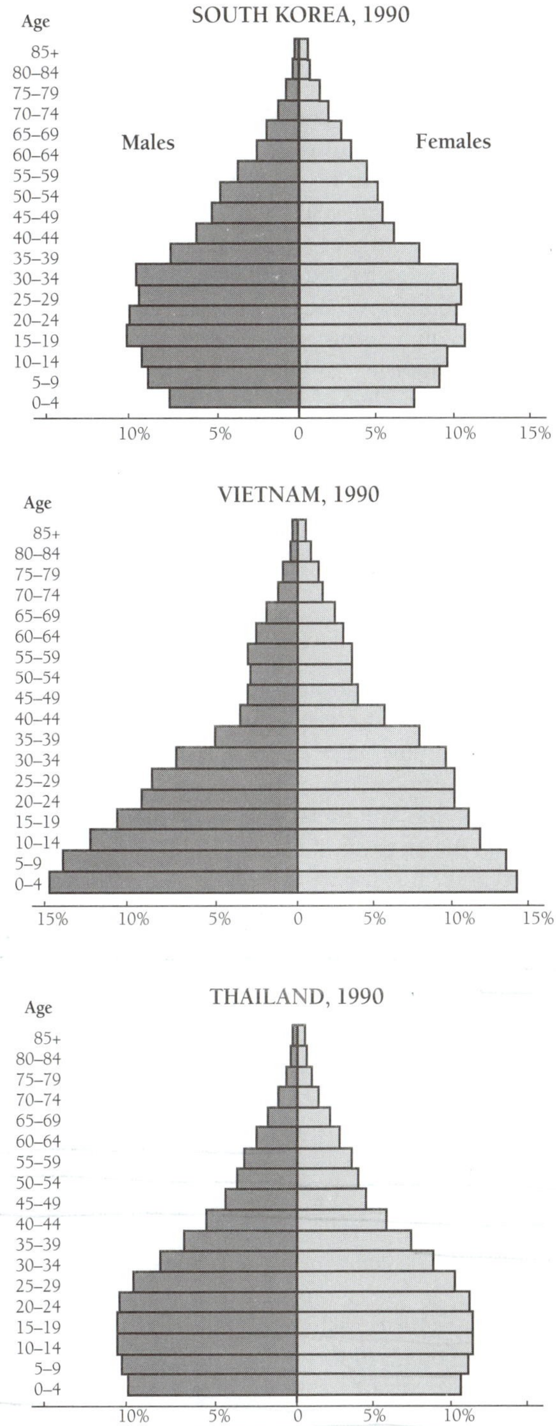

Figure 3-9

Compare these population pyramids. South Korea's narrowing pyramid base reflects its falling NI. Vietnam's broad-based pyramid indicates a relatively high NI. The impact of war from the 1950s through the 1970s is evident with the shortage of males in the 35–55 age cohorts. The narrowing base of Thailand's pyramid reflects the success of its family planning programs.

can lead to **sex-selective abortion.** Modernization has brought sex-determination devices to hospitals and clinics, enabling women to "select" the sex of their offspring. In 1996, India outlawed these practices but they persist, nevertheless.

Female infanticide, the deliberate killing of girls, has a long history in Asia (and elsewhere, including Europe), particularly in times of strife. In such times, a girl might be perceived as a liability: she will leave the family and move into her husband's household, and often require a substantial dowry payment as well. A boy will remain, support, and care for his own multigenerational household. In tough times, men and boys eat first; women and girls often do without. Such cultural practices are not simple cruelty. They are strategies for survival in stressful life-world contexts. Improvements in quality of life witness reductions in female infanticide and infant mortality in general.

POPULATION AND QUALITY OF LIFE

Selected quality of life statistics are shown in Table 3-2. While life expectancies at birth (LE) have improved over time, only Japanese, Singaporeans, and Taiwanese can expect to live as long as people in Canada or the United States. Infant mortality rates (IMR) in the first year of life and child mortality rates (CMR) in the first five years of life are excellent indicators of living conditions. Strife-torn Cambodia, Laos, and Myanmar, among the poorest countries, exhibit very high rates. Remote Bhutan also suffers very high infant and child mortality rates.

The percentage of children under five years of age suffering from malnutrition, as represented by being underweight, wasted, or stunted in growth, is also high for many countries. In seven Asian countries, 40 percent or more children are underweight. Bangladesh (67%) and India (53%) have the highest rates of these conditions.

Additional quality of life indicators are provided in Table 3-3. Access to potable (safe to drink) water is a key measure, considering its role in the occurrence and spread of infectious and parasitic diseases such as cholera and bacterial and amoebic dysentery. Note that there are sharp discrepancies between urban and rural areas. In China, 97 percent of city dwellers have safe drinking water, but only 56 percent of rural people do. Japan, South Korea, and Singapore have the highest rates of access to decent water, while Cambodia, Laos, and Vietnam have the lowest.

Access to health services and number of persons per physician figures suggest survival potential with childbirth, disease, injury, or medical complications. Again there are dramatic differences between urban and rural settings. Since the majority of people in most countries are rural, health care delivery is inadequate. In Australia, there are approximately 400 persons per doctor. Contrast this with more than 4,000 per doctor in Thailand or 7,000 per doctor in Indonesia. Both China and Mongolia have relatively low ratios of persons per doctor. This calls the definition of *doctor* into question.

In China and elsewhere, "doctors" are not necessarily fully licensed physicians. They may be little more than medical aids with a few weeks training. I met a young "doctor" in Xian, China, who told me he was a cardiologist. When I inquired about his training, he said that he had gone to "cardiology college" for four years after finishing high school. I have visited many clinics and hospitals in various parts of Asia and can assure you that conditions in most are grim. Several countries are extremely short of even minimally trained personnel, and sanitation conditions are poor. Moreover, both doctors and hospitals tend to locate in cities.

Table 3-3 also lists literacy rates and percentage of primary school pupils reaching grade five. This information is meaningful in context of population planning programs. South Asia, with the exception of Sri Lanka, has a poor record of children staying in school. In many places, girls never attend school and are the most likely to leave. This explains the male/female differential in literacy rates (Figure 3-10 p. 67).

Girls make up close to half the children attending school in China, but statistics do not show that they make up three-quarters of those enrolled but never attending. Some believe that education is wasted on girls, who will soon marry into another family's household. It is more practical that she help with chores at home as long as she lives there. It is well understood that, in more affluent circles, increasing educational achievement by women coincides with reduction in family size. An educated woman is more likely to

Table 3-2 Quality of Life Measures for Infants and Children

	Life Expect. at Birth (1998) M	Life Expect. at Birth (1998) F	% Infant Mort. Rate 0–1 (1998)	% Child Mort. Rate 0–5 (1997)	% Low Birth Weight (1994)	% of Children Under 5: Under Weight (1990–1996)	% of Children Under 5: Wasted (1990–1996)	% of Children Under 5: Stunted (1990–1996)
Australia	**75**	**81**	**0.6**	**0.8**	**6**	—	—	—
Canada	**75**	**81**	**0.6**	**0.8**	**6**	—	—	—
U.K.	74	80	0.7	1.0	7	—	—	—
U.S.A.	73	79	0.7	1.0	7	—	—	—
South Asia								
Bangladesh	58	58	8.2	11.5	50	67	17	63
Bhutan	—	—	7.1	18.9	—	38	4	56
India	59	59	7.2	11.5	33	53	18	52
Maldives	63	61	3.0	—	—	—	—	—
Nepal	55	54	7.9	11.4	—	49	6	63
Pakistan	58	59	9.1	13.7	25	38	9	50
Sri Lanka	70	74	1.65	1.9	25	38	16	24
East Asia								
China	69	73	3.1	4.7	9	16	4	32
H.K. SAR	76	82	0.4	—	—	—	—	—
Japan	77	84	0.38	0.6	7	—	—	—
N. Korea	63	69	3.9	3.0	—	—	—	—
S. Korea	70	77	1.1	0.9	9	—	—	—
Mongolia	—	—	4.9	7.4	6	12	2	26
Taiwan	72	78	0.67	—	—	—	—	—
Southeast Asia								
Brunei	70	73	0.84	1.0	—	—	—	—
Cambodia	50	53	11.6	17.4	—	40	8	38
Indonesia	60	64	6.6	7.5	14	35	—	—
Laos	52	55	9.7	13.4	18	44	10	48
Malaysia	70	75	1.0	1.3	18	44	10	44
Myanmar	60	62	8.3	15.0	16	43	8	45
Philippines	63	69	3.4	5.3	15	30	8	33
Singapore	74	80	0.38	0.6	7	—	—	—
Thailand	67	72	2.5	3.2	13	26	6	22
Vietnam	65	69	3.8	4.5	17	17	12	47

	Life Expect. at Birth (1998) M	Life Expect. at Birth (1998) F	% Infant Mort. Rate 0–1 (1998)	% Child Mort. Rate 0–5 (1997)	% Low Birth Weight (1994)	% of Children Under 5: Under Weight (1990–1996)	% of Children Under 5: Wasted (1990–1996)	% of Children Under 5: Stunted (1990–1996)

Sources: Population Reference Bureau and UNICEF.

Table 3-3 Selected Quality of Life Indicators

	% Access to Safe Water Urban (90–96)	% Access to Safe Water Rural (90–95)	% Access to Health Services Urban (90–95)	% Access to Health Services Rural (90–95)	# of Pers. Per Doctor (95)	% Literate M/F (95)	% Primary School Students Reaching Grade 5 (90–95)
South Asia							
Bangladesh	99	96	45 total		5143	49/26	47
Bhutan	75	54	65 total		10,900	56/28	82
India	85	79	100	80	2106	66/38	62
Maldives	—	—	—	—	2493	—/—	—
Nepal	88	60	—	—	12,500	41/14	52
Pakistan	82	69	99	35	1915	50/24	48
Sri Lanka	88	52	—	—	4745	93/87	92
East Asia							
China	97	56	100	83	633	90/73	88
Japan	100	85	—	—	265	100/99	100
N. Korea	—	—	—	—	—	—/—	—
S. Korea	100	76	100	100	784	99/97	100
Mongolia	100	58	—	—	401	99/97	—
Taiwan							
Southeast Asia							
Brunei	—	—	—	—	1133	—/—	—
Cambodia	65	33	80	50	8120	48/22	50
Indonesia	79	54	99	91	7236	90/78	92
Laos	60	51	67 total		—	69/44	53
Malaysia	96	66	—	—	2226	89/78	53
Myanmar	78	50	100	47	3455	89/78	—
Philippines	92	80	71 total		9528	95/94	67
Singapore	100	—	—	—	664	96/86	100
Thailand	94	88	90	90	4260	96/92	88
Vietnam	47	42	100	80	2444	97/91	—
	% Access to Safe Water Urban (90–96)	% Access to Safe Water Rural (90–95)	% Access to Health Services Urban (90–95)	Rural	# of Pers. Per Doctor (95)	% Literate M/F 95)	% Primary School Students Reaching Grade 5 (90–95)

Sources: United Nations. 1997. Statistical Yearbook for Asia and the Pacific 1996; UNICEF.

Figure 3-10

This is a boys' religious school in Pakistan's Northwest Frontier. Here the boys spend a few hours each day learning to read the Koran along with some other basic education. Note the visiting girls standing to the right. What are they doing? How will the lives of these boys and girls differ?

understand the methods and reasons for family planning, and she is in a stronger position to make choices concerning her own life.

FAMILY PLANNING PROGRAMS

Controlling family size is a high priority for most nations of South, East, and Southeast Asia. Table 3-4 charts the progress of family planning in various countries. One measure of success is the **total fertility rate (TFR)**, which refers to the number of children a woman would have if age-specific birth rates remained constant throughout her childbearing years—age 15 to 45 (some sources use 15 to 49). Populations are growing despite the fact that TFRs have fallen.

Family planning programs were established in twelve Asian countries as early as 1970. These countries in particular have seen dramatic reductions in their TFRs. Malaysia, Thailand, and Indonesia are known for their success in reducing TFRs. Bangladesh has managed to reduce its TFR from an average of 7.0 in the early 1970s to 3.6 in the late 1990s.

Other countries such as Pakistan have not taken as well to family planning. Although programs have long been in place, the secondary position of women, with a mere 24 percent literacy rate, and rural poverty in general are not conducive to female decision making regarding the number of children they bear. Only 12 percent of Pakistani women are using modern contraception methods, the lowest number in Asia.

In light of political, social, economic, and cultural variation among and within nations, success in population reduction is not easily achieved, nor is the process the same for all. The following examples will illustrate.

Myanmar

The government of low-density Myanmar ignored population growth until it began to reap the consequences of rapid population growth. Programs stress the wider spacing of births. However, abortion is common as well. The quality of medical care is indicated by the fact that 40 percent of maternal deaths result from botched abortions. Myanmar is a poor country with very limited infrastructure. Ethnic persecution and human rights violations are widespread. Contraception devices are difficult to get, especially in remote areas. Most are smuggled in from India and China.

Vietnam

Family planning has been promoted in Vietnam since the 1960s. Abortion is legal and free, making it the primary form of birth control. The government gives cash payments for sterilization or the use of IUDs but not for the use of birth control pills. In fact, only one

Table 3-4 Family Planning and the Status of Women

	Fam. Plan. Prog. Estab. Early 1970s	Total Fert. Rate (70–75)	Total Fert. Rate (98)	% Fem. Using Contraception any/mod (98)	# Fem. per 100 Males Secon. Edu. (96)	Mater. Mort. per 100,000 Live Br. (90)	% of Female Labor Force in		
							Agr. (95)	Ind. (95)	Serv. (95)
South Asia									
Bangladesh	X	7.0	3.3	49/42	50	850	65	8	27
Bhutan		5.9	5.6	8/—	40	1600	95	1	4
India	X	5.4	3.4	41/36	52	570	78	10	12
Maldives		7.0	6.4	—/23	95	548	25	63	12
Nepal		6.5	4.6	29/26	37	1500	97	1	2
Pakistan	X	7.0	5.6	18/13	41	340	15	23	62
Sri Lanka	X	4.0	2.2	66/44	106	130	50	14	36
East Asia									
China	X	4.8	1.8	83/81	73	95	74	17	10
H.K. SAR	X	—	1.1	82/75	—	—	—	—	—
Japan		2.1	1.4	64/57	—	9	7	29	65
N. Korea		5.7	1.9	—/—	94	130	41	23	35
S. Korea	X	4.1	1.7	77/66	92	70	31	28	41
Mongolia		5.8	3.1	57/41	107	65	28	21	51
Taiwan		—	1.7	—/—					
Southeast Asia									
Brunei		5.4	3.4	—/—	101	—	1	9	90
Cambodia		5.5	5.2	—/—	—	900	75	11	14
Indonesia	X	5.1	2.7	55/52	81	650	44	16	40
Laos		6.2	5.9	19/—	66	650	76	8	16
Malaysia	X	5.2	3.2	48/31	102	80	31	26	43
Myanmar		5.8	3.8	17/14	90	580	35	25	39
Philippines	X	5.5	3.7	48/30	99	280	34	13	53
Singapore	X	2.6	1.7	65/—	100	4	1	42	57
Thailand	X	5.0	2.0	72/70	93	200	64	11	25
Vietnam		5.8	2.3	75/56	99	160	57	15	28
	Fam. Plan. Prog. Estab. Early 1970s	Total Fert. Rate (70–75)	Total Fert. Rate (98)	% Fem. Using Contraception any/mod (98)	# Fem. per 100 Males Secon. Edu. (96)	Mater. Mort. per 100,000 Live Br. (90)	Agr. (95)	Ind. (95)	Serv. (95)

Sources: Population Reference Bureau Data Sheet 1998 United Nations. 1995. The World's Women 1995. UNICEF.

percent of Vietnamese women use the pill, in contrast to 30 percent of Indonesian and Thai women. Misinformation about the efficacy and side effects of the pill has discouraged its use. Moreover, 60 percent of birth control providers believe that rural women would not remember to take it daily as prescribed.

Although family planning is widespread, it is not comprehensive. Men continue to reject male sterilization, and condoms are not readily available. In 1998, the government announced the establishment of a condom factory. However, infrastructural problems continue to make both supply and delivery unreliable.

Malaysia

Malaysia has considered reducing family size only since 1984 because it has been mainly interested in creating a large domestic market for domestic goods. While programs appear successful in that TFRs have fallen from 5.2 in the 1970s to 3.3 in the 1990s, data mask ethnic group differences. For example, women of Chinese ancestry, who comprise 38 percent of the population, have a TFR of 2.0. Malay and other indigenous groups have a TFR of 4.0. Furthermore, tax breaks are provided for large families.

Bangladesh

Slowing population growth has been a priority in Bangladesh since the 1970s, when only 8 percent of women used contraceptives. A comprehensive program was developed that included training of field counselors. Regional clinics were set up throughout the nation, and 23,000 high-school-educated, female counselors, who themselves used birth control, addressed family planning issues with village women. The importance of population control was stressed via radio, television, posters, and billboards. It costs the government U.S.$62 to prevent one birth, but it saves U.S.$615 on social services.

Today, around 50 percent of Bangladeshi women use some form of birth control. The mean age at marriage has risen, the age at which women have their first child has increased, and more women stop having children earlier. Still, a large base population, in addition to a desired family size of three to four children, means that population growth remains high.

The case of Bangladesh is unique in Asia for two reasons. First, it shows that a high female literacy rate is not essential for reduced TFRs. Second, as geographer Bimal Kanti Paul (1997) points out, "This South Asian country disproves beliefs that birthrates drop only after economic conditions improve. It also refutes the contention that 'development is the best contraceptive'."

Basic population statistics should be interpreted as estimates of numerical change. They reveal nothing tangible about the well-being of people. On the other hand, projections and related measures are used to illustrate perceived population problems such as "overpopulation." The question of "too many people on earth" is a controversial one.

Are There Too Many People on Earth?

Fertility levels are down, but momentum remains for future growth as large, young cohorts move into childbearing range. Absolute numbers will continue to rise as long as the number of births per woman exceeds the replacement rate of 2.06. While fertility rates decline, the number of potential mothers increases. But, as Table 3-4 suggests, there are different rates of fertility decline. Higher fertility is typically associated with poorer people.

Scientists are reconsidering the Malthusian specter of famine in a world unable to sustain ever-increasing numbers of people. Yearly reports from the World Watch Institute (WWI) point to an array of population growth impacts on sustainability of people and environments. One indicator for this dynamic is grain production.

World grain harvests declined in the 1990s, 15 percent lower in 1994 than in 1984. Green Revolution improvements no longer bring about dramatic increments of increase in yields. Even though crop yields continue to increase, according to WWI's Lester Brown (1998), "There will eventually come a point in each country, with each grain, when farmers will not be able to sustain the rise in yields."

Soil erosion, scarcity of irrigation water, and land conversion are part of the picture. Loss of cropland to nonfarm uses has been rising in urbanizing–industrializing regions (Figure 3-11). Factories, housing, shopping malls, and even golf courses gobble up cropland. An estimated 40 courses now occupy former rice paddies in the Xi delta region in southern China. Indonesia loses enough cropland each year on Java alone to feed nearly 400,000 individuals.

Such developments defile or reduce groundwater available for irrigation. According to the WWI, irrigation water in parts of India, Indonesia, and Malaysia will fall 15 to 30 percent short if urban needs are met. In coastal areas, saltwater is drawn into depleted aquifers, rendering land saline and unusable.

Carryover grain stocks are also low, drawn down by unmet production needs. As supplies decline, prices rise, thereby reducing consumption by the poor. Concurrent rising affluence for others, in the wake of modernization, increases demand, further reducing supply. From 1991 to 1995, incomes for China rose by half. In 1994, China had a net grain export of nine million tons. In 1995, rising prices led to the import of 16 million tons. In that same year, Vietnam curtailed rice exports to control domestic prices.

Given projected population increases in Asia and worldwide, demand for grain is expected to exceed supply. Asia's grain imports are skyrocketing. By the year 2030, it is projected that China will need to import an amount of grain equivalent to the entirety of current global production!

Increased demand for grain is not simply to put food directly into people's mouths. Rising affluence stimulates greater demand for meat, milk, and eggs, which requires more grain for livestock feeding. Asia's billions are moving up the food chain, consuming more grain-intensive products. Moreover, depletion of oceanic fish supplies is requiring additional grain to feed fish raised on land-based fish farms.

Asian countries spearhead demand for nongrain products. Yet, grain provides at least twice the amount of energy when consumed directly, as opposed to being eaten in nongrain form. In other words, grain-fed meat, fish, and poultry are not as sustenance-efficient as basic cereal.

China's demand for livestock products accounted for about two-thirds of world demand for grain in the 1990s. Lester Brown (1998) believes that China exemplifies why the **demographic transition model** (four sequential stages: high birthrate (BR) & death

Figure 3-11

I took this picture from an expanse of rice paddies on the outskirts of Kyongju, South Korea. I was amazed at the urban encroachment on agricultural lands. This scene is ubiquitous in rapidly urbanizing South Korea.

rate (DR); high BR & lowering DR; lowering BR & low DR; low BR & DR), based on centuries-long growth patterns in Europe, is not viable elsewhere because there are simply insufficient resources to sustain forecast numbers.

Skeptics of Malthusian predictions of famine, disease, and pestilence trumpet technology as solving all. They argue that population pressure can lead to agricultural innovation and better productivity, as has happened in some places. Evidence also supports the fact that resource deprivation can engender self-imposed fertility control. Although this may be the case on a local scale, at the global scale, sustainability of both populations and resources remains problematic.

Does this mean that there is indeed a population crisis whereby masses of humanity face hunger, disease, and lives of misery? Geographer Alan Findlay (1995) argues that global economic structure, inequitable control over resources, and imbalanced population distributions are responsible for malnourishment and other social and environmental ills. In short, high-tech, rich nations use up far more than their fair share of resources.

According to Findlay, it is "structural forces which are producing uneven life opportunities for the world's population." The amount of food is not the problem—rather, it is access to it. Poverty and malnutrition have their roots in powerlessness. Political, social, and economic spaces vary greatly in their degree of power to exploit and control essential resources. Lack of power may derive from monetary, ethnic group, class, or gender inequities. The rich, the politically important, the ethnic dominants, the upper classes, and males in patriarchal societies benefit over the rest, especially in stressful times.

Ironically, technology is largely responsible for both the cause and cure of population crises. Technology contributed to the dramatic lowering of death rates that spurred the population explosion in the first place. Today it enables industrial and urban land encroachment and is responsible for CO_2 emissions and chemical air pollution, water pollution, and other environmental hazards. Global networks of marketing and advertising by multinational corporations promote unwise application of new technologies, in addition to increasing expectations and encouraging mass consumption. Calories wasted, lost, or spoiled increase with affluence. Calorie loss, less than 10 percent in poor nations, is 20 to 40 percent in rich nations. There is increasing divergence between rich and poor, both among and within countries. Local misery arises, in good part, from global mismanagement.

Did you know that technology can be responsible for disease?

Technology and Disease

Close to 30 new or previously unrecognized diseases have been identified since 1973. In 1995, a new strain of cholera appeared in India. Spreading with international travel and trade, it is now endemic in Asia, associated with poor sanitation. In China, Hantavirus, a new disease associated with rats and mice (which flourish in rice paddies), has affected thousands. Japanese encephalitis kills 7,000 a year in Asia. The 1990s have witnessed outbreaks of new strains of diphtheria in Laos and dengue and hemorrhagic fever in India and Malaysia.

In the 1950s, it was thought that mosquito-borne malaria would be eliminated by the use of pesticides and quinine treatment. Today, malaria is the most prominent disease in Asia and medicine cannot keep up with newly developing, virulent strains. Case levels in Sri Lanka have returned to 1950s levels. In China and India, malaria cases rose from 2 million in 1961 to 15 million in 1995. Cambodia and the Thailand–Myanmar border region have the world's highest concentration of drug-resistant malaria. This particular strain can only be treated with *qinghaosu* (arteminisin), an old Chinese remedy from the wormwood plant.

Infectious and parasitic diseases such as tuberculosis, meningitis, and diarrheal ailments are responsible for about 40 percent of deaths in poor nations (one percent in rich nations). More than 17 million people died in 1995 from such diseases that are related to poverty, malnutrition, and lack of sanitation. Medical personnel are battling new strains of these as well.

What role does technology play in this devastating situation? One of the most important causes of the evolution of new varieties of disease is environmental change. Global climatic change linked to the release of toxic emissions is one factor. Alteration of habitat forcing migration of vectors (carriers such as mosquitoes) in addition to microbial adaptation and change is another. Diseases are diffused with increased international travel and trade, as well as with improvements in transportation infrastructure. Population pressure and subsequent movement into remote areas result in exposure to diseases formerly carried only by animal populations. Unsanitary and crowded living conditions associated with intensified urbanization increase opportunities for infection, especially with TB and cholera. Overuse of pesticides and antibiotics have produced resistant strains of disease, rendering such palliatives ineffective. Increased sexual freedom and drug use, influenced by modernization, have witnessed a rise in related diseases such as AIDS.

Again, while technology can offer cures for illnesses, it can be a major cause of them. Limited, understaffed, and poorly funded medical facilities are generally unable to cope with increasing demand for services.

Gender Inequality

Gender is a culturally constructed social institution. Gender roles are expected and assumed in cultural contexts, and society is organized around gender. Institutional behaviors and relationships are structured according to understood relative positioning and roles of men and women. In most cultures, gender inequity is the norm.

To understand gender inequity is to appreciate **power relationships.** Men and women are accorded unequal roles in virtually every society. This means that women do not have equal power to access education, earn comparable wages, make economic or political decisions, or even control their own destiny regarding inheritance, marriage, child-bearing, or career. Women do not have power over land or resources, or equal if any participation in public decision making.

Violence against women is a major risk factor for women in Asia. Gender-based violence—including selective abortion, infanticide, neglect, abuse, and domestic violence—affect millions of women. In India, at least 300,000 more girls than boys die each year, and one in four deaths of women between the ages of 16 and 24 is due to "accidental burns." These and other "accidents" result primarily from the failure of a bride's family to pay an original or newly demanded dowry, or from her "inability" to produce sons (males genetically determine the sex of a child). "Dowry deaths" are estimated between 5,000 and 12,000 a year in India.

Table 3-4 shows alarmingly high rates of maternal mortality. In part, these reflect access to health care. But they also reflect gender-based ill treatment. Particularly in Muslim cultures, women often are not allowed to be seen by a male physician. Too-early marriage and pregnancy, dietary deficiency, overwork, too closely spaced births, and late-in-life pregnancies take an enormous toll. Births are frequently unassisted or aided by a midwife. Complications typically mean death, and pre- and postnatal care is scant.

Given their sheltered and uneducated circumstances, many women in patriarchal societies may be unaware of their inferior status. They accept 12 or more hours of hard labor in and out of the home, eating last and least, constant pregnancy, forced sex, beatings, and even death as their lot in life. These women do not know that they have even the potential to make a decision or control anything. Even if they are aware of their lesser position, most often there is little they can do about it.

Are you aware that millions of women and children are sex slaves in Asia?

Women and Children Enslaved: Prostitution and AIDS

Millions of women and children under 18 work as prostitutes in Asia. Roughly 60,000 girls offer sex services in the Philippines. Thailand has as many as 800,000 children working in the "sex industry," and it is estimated that 10,000 girls are bought from Myanmar each year to replace those overworked, diseased, or dead. In Phnom Penh, Cambodia, a six-year-old can be acquired for as little as three dollars. As many as 50,000 Nepalese girls work as prostitutes in India. Their average age in 1994 was 10 to 14.

Commercial sex workers are in high demand throughout Asia. Places such as Bangkok, Manila, Mumbai, Ho Chi Minh City (Saigon), and Taipei are infamous **sex cities.** Sex tourism is partly responsible. Most "sex tourists" are from Japan, North America, Europe, and the Middle East. Most are heterosexual. However, prostitutes are also deemed essential for indigenous males. Mores in several countries restrict sexual behavior for females but not for males. Millions of men work away from their home for weeks or months at a time, and prostitutes are seen as contributing to their well-being. Since local women are not readily available, imports are required.

Where do these girls come from? In Thailand, "labor agents" recruit girls from such places as Myanmar and Cambodia. Parents are paid a lump sum (this becomes a debt) and are assured that their daughter will be taken to work for a rich family, send money home, and attend school. In the face of abject poverty, sending a daughter off elsewhere might be seen as a blessing. The reality is that she will be detained in debt bondage—forced to work in a brothel until she has paid her "debt," plus the cost of food and clothing. Thailand has legal safeguards against such exploitation, but the law is rarely enforced by officials and police, who are deeply entrenched in this money-making enterprise.

In India, labor agents scour the region for children of the impoverished. Nepal is a virtual gold mine in this regard. There is also a tradition of declaring one's daughter a *devadasi,* or Hindu temple servant. Dedicated to a particular goddess, a *devadasi* participated in ritual singing, dancing, and other religious practices. However, tradition has been corrupted, and today many of these girls are forced into prostitution by the unscrupulous. The practice has been outlawed, yet it still accounts for at least 20 percent of prostitution in India.

Boys are not immune to the sex industry. Boys often see prostitution as a way out of poverty, and many pursue the urban sex trade. Both boys and girls participate in the exceedingly lucrative pornographic movie and video business. The sex industry is closely tied to the drug trade, and both are associated with the spread of sexually transmitted diseases (STDs) (Figure 3-12).

According to the World Health Organization (WHO), there are at least 18 million people infected with HIV. In 1997, more than 6,000 women a day were infected around the world. In South and Southeast Asia, 90 percent of AIDS cases are in India, Myanmar, Cambodia, and Thailand.

In Thailand, HIV infection was virtually unknown prior to 1987. Now Thailand has five times the number of cases as India, which ranks second in the world for AIDS victims. It is estimated that the year 2000 will witness 350,000 HIV-infected babies born. There will be half a million AIDS cases, more female than male, primarily in the 15–34 age cohorts. Thailand's life expectancy is predicted to fall as a result of AIDS-related deaths.

Prostitution and intravenous drug use are the main causes of HIV infection in Asia, and

Figure 3-12

No translation is needed for this Vietnamese sign. AIDS comes from drug use. Like Vietnam, many Asian countries avoid the issue of sexual transmission in their anti-AIDS programs.

both are on the increase. High-risk groups are commercial sex workers and heterosexual men. International travel and tourism as well as the trucking industry are major contributors in the diffusion of STDs. Patronizing brothels on their journeys, they bring HIV home to their unsuspecting wives and lovers.

The increase in HIV infection and AIDS cases has upped the demand for even younger sex workers. Men pay premium prices for virgins, believed to be disease-free. Once prostitutes are "too sick to work," they are left to the streets or "repatriated" to their home country to die. Some girls do manage to survive and even make money. Many of these return home as heroic "dutiful daughters" who have sacrificed themselves for their family. Their behavior is a model for the recruitment of younger sisters. However, the majority end up dead at a young age. Combating child prostitution is a major goal of women's movements in Asia.

Globalization is a contemporary and growing trend in the sex industry that has long been sustained by Europeans, North Americans, Australians, Saudi Arabians, Japanese, and other groups. This flourishing enterprise has spawned a counterpart: "sex worker/entertainer" migration streams. The Philippines, Thailand, China, and other Asian sources provide thousands of sex workers to such key destinations as Japan, Taiwan, and Saudi Arabia. Some 35,000 women leave the Philippines yearly to work as "female entertainers" in Japan. Terms such as *entertainer* or *artist* mask the reality of "body as commodity." Although some nations such as Thailand and the Philippines have enacted restrictions and regulations for employment abroad, they are largely ignored.

Prostitution is not a woman's problem. Geographer Joni Seager (1997) states: "Prostitution is not a women's institution—it is controlled by men and sustained by violence." While local-level prostitutes might have a measure of control over their own destiny, "the global sex trade is almost entirely coercive, sustained by high levels of violence and predicated on the thorough subjugation of women."

WOMEN'S AND MEN'S FARMING

It is an accepted fact that women do most of the world's agricultural work. The prevalence of women in Asian agriculture is indicated in Table 3-4. However, male-reported statistics have long shorted and devalued female inputs. For example, women's contributions at planting and harvest times may be classified as "help" and not count as "work." Historically, in the realm of labor statistics, women have been rendered practically invisible.

Just as there are prescribed life-roles for males and females, so there are gender-appropriate animals and crops. Women's animals are small, of low value, and live within range of the house where they can be fed scraps. Chickens, goats, and pigs are considered women's animals. In fact, most poultry in Asia is raised by women. Men's animals are large, of high value, beasts of burden, and require a spatially wide range for pasturage. Horses, cattle, and water buffalo are considered men's animals.

Crops are classified according to who controls their management and disposition, not who works them. Women's crops are usually vegetable and root crops suitable for subsistence or local consumption (Figure 3-13). These can be raised in spatially confined household gardens. Men's crops are typically grains and tree products designated for marketing and export, and cultivated in spatially scattered fields. These crop associations do not prevent both men and women from working in each other's domains however.

Figure 3-13

A woman sells fruit under a temple bell in Kathmandu, Nepal. The bell is rung by devotees to summon the deities. The woman is searching for lice in her daughter's hair. Mutual grooming by relatives and friends is an important mechanism of social bonding.

THE FEMINIZATION OF POVERTY

Gender inequity has drawn world attention in the latter twentieth century. The United Nations and other organizations have begun to address this issue. Nevertheless, modern developments, contributing to the impoverishment of millions, continue unabated. The **global feminization of poverty** persists with 70 percent of the world's poor being women.

The transition from subsistence farming to cash crop farming tends to favor men, as does increased commercialization. The significant role of women's gardens is downplayed or ignored, and these frequently disappear with commercialization. Studies in Sri Lanka have shown that this process has resulted in nutritional shortages. Loss of cropland to modern development forces women out of farming. In 1994, both Indonesia and Vietnam lost 20,000 hectares. From 1986 to 1992, three percent of China's cropland gave way to housing and industrial developments. In Thailand, golf courses have replaced an estimated 34,000 hectares. Remaining prime land is generally assigned to cash crops—men's crops. Moreover, men have greater access to capital, machines, water, and fertilizer, essential inputs that are controlled by men. Females are often employed as wage laborers on agricultural estates and in factories but do not receive equal remuneration. Men also tend not to share earnings with their entire family.

Mechanization can also be detrimental to women. In rice production for instance, tasks are often differentiated by gender. The introduction of rice planting machines in Taiwan took the business of transplanting seedlings out of the hands of women. Rice planting is now a man's job.

In Vietnam, men and women cooperate in rice production but women still do most of the work. Men prepare the paddies, women plant and weed. Both maintain irrigation and both do harvesting. Women hand-pound the grain before winnowing and drying it (Figure 3-14). Mechanized rice mills have recently displaced millions of women (and men) across Asia, hitting particularly hard in Indonesia, Bangladesh, and Vietnam. As rice mills eliminate perceived drudgery, they simultaneously eliminate what may be the only opportunity for women to earn income.

Figure 3-14

This Vietnamese woman is winnowing rice. Tossing it into the air, the chaff will blow away and the heavier rice will fall into the basket below. Winnowing is regarded as women's work.

The increasingly intensive use of agricultural chemicals also takes its toll. Planting and weeding call for women to spend long hours in chemically polluted water, which causes a rise in related ailments. However, use of herbicides to kill weeds eliminates the job altogether.

Because women's labor contributions are not officially apparent, women are marginalized and bypassed by development projects. Only a fraction of World Bank and other internationally funded projects target women directly. When funds are available, they are channeled to what are considered "women's activities." Yet it is known that women are more efficient money-handlers and more prudent in repaying loans. Furthermore, women, on average, are better at managing assets, generating additional income, and applying monetary increases to family well-being. Research in Malaysia and South Asia reveals that most programs favor the grooming of women as "housewives" in a nuclear family based on a perceived "American" model. These schemes emphasize female domesticity, feminine crafts, mothering and child-oriented activities, and their roles as "farmers' wives."

Gender discrimination also thrives in industrialized nations where segregation and subordination are intrinsic to patriarchal principles. In Korea, women are clustered in certain occupations such as hotel workers and are paid less than men for the same position. Women are believed to have certain talents and weaknesses associated with their femininity: nimble-fingered and obedient, they lack skills required for managerial positions. Men do not like to work with or for women. In 1993, only 4.2 percent of South Korean managers were female. Since every effort must be made to protect a woman's femininity, intense and complex business machinations and possible foreign travel in a "heartless marketplace" are viewed as detrimental to her welfare. Educational structure, content, and direction continue to foster these attitudes.

It might be assumed that increased education would broaden opportunity and boost income for women. But, this is not always the case. In fact, as women gain more education, male–female differentials may increase with persistent and pervasive discrimination and underutilization of women's talents. Moreover, educational systems continue to channel males into scientific and technological fields and women into the humanities. Textbooks portray perceived gender-appropriate roles.

In the wake of industrialization, a partnership between educational and industrial concerns has emerged. State-supported, cooperative programs have schools prepare students for employment as needed by the industrial sector. Factories recruit teenagers to work and even guarantee them a high school diploma. This is quite appealing to females whose parents are reluctant to support their schooling, especially beyond middle and high school levels.

With a deepening entrenchment of capitalism in Asian societies, inequalities' causes and solutions will be tied increasingly to foreign design and global enterprise. I am reminded of my recent introduction to the poverty-stricken landscape of Cambodia. Along the road from the airport to Phnom Penh, I was confronted with three large billboards: Coca Cola, the Marlboro Man, and Godzilla. What does this say about foreign influence and modern development? The next chapter will look more closely at the idea and practice of development in Asia.

Main Point Summary

1. Findings in China and Indonesia have questioned the "out-of-Africa" theory and support a *multiregional hypothesis* of human origins. But 1998 discoveries in Southern Africa are expected to bolster the Africa theory. Around 16,000 B.C., population migration was enabled by lowered sea levels, creating landmasses known as Sahulland and Sundaland.
2. South, East, and Southeast Asia account for two-thirds of the world's people. A sporadic growth pattern was followed by rapid increase in the nineteenth century. Core areas of population emerged in valleys, plains, and coastal areas. Developments in water control and agriculture, dependent on cooperative human efforts, sustained and nourished growth and expansion.
3. Europeans, intent on harvesting and transporting Asian resources, focused their efforts in coastal areas. Ports became both focal points of economic activity and migration destinations. Many of Asia's largest cities owe their unprecedented growth to colonial influences.
4. Vegeculture was the precursor of agriculture in Asia, and numerous plants and animals were domesticated there.
5. Shifting cultivation, originally an ecologically sound practice requiring lengthy fallow periods, is declining in importance with increasing population pressure and forest removal.
6. Rice is perceived as integral to society in most of Asia. The foundation of culture, rice was domesticated in South and Southeast Asia and diffused throughout the region. Population increase on limited arable land has rendered farm plots very small in most places. Human and material inputs are intensive, and there is little room for machinery. Rice varieties are selected in relation to maturation period, personal preference, ceremonial needs, and market demand.
7. The Green Revolution fostered the use of high-yielding plant varieties. Yet, increasing demands and the law of diminishing returns means that food supply may not meet population needs in the twentyfirst century.
8. Children are at the core of family-group culture; there is a general preference for males. People continue to have several children for both personal and practical reasons.
9. Population growth and distribution are variable, with some countries having high NI rates and others low NI rates. Countries with large base populations, such as China and India, experience sizable increases even with lowered NIRs. Family planning programs also vary in practice and efficacy.

10. Asia has several world cities such as Tokyo, Seoul, and Shanghai. It also has many cities of over a million inhabitants, with the majority of these in India and China.
11. Sex ratios reveal that India and China ascribe to practices such as sex-selective abortion and female infanticide because of strong preferences for male offspring.
12. Statistics indicate that quality of life varies both within and between countries. Malnutrition and disease continue to plague countries such as Nepal, Laos, and Cambodia, and regions of China and India. Safe water, appropriate medical care, and basic education tend to be inadequate in all but the most developed countries such as Singapore and Japan and in more developed regions around capital cities.
13. While NIRs are falling, large, young cohorts entering child-bearing years are expected to put pressure on global food supply, especially grain stocks. Even with the Green Revolution, increments of crop yields will diminish and agricultural land will continue to shrink with urbanization and development.
14. Problems are essentially structural, centering on the control of production and distribution of resources. Power is inequitable and vested in international institutions and indigenous elites. Technological solutions often invite new environmental and social problems.
15. The culturally structured institution of gender prescribes roles, behaviors, and expectations that generally ensure the subservient position of women. Women are often segregated spatially, socially, and economically, and inequitable division of labor persists from home to work place. Development schemes may further disempower women and ensure their descent deeper into poverty. Young women are often sold into the sex industry. Some are desperate enough to voluntarily sell their services. In either case, they frequently succumb to violence and disease.

Terms to Know

"out-of-Africa" theory
multiregional hypothesis
base population
nucleated villages
domestication
vegeculture
industrial crop
shifting cultivation
slash and burn agriculture
swidden
Green Revolution
historic inertia
arithmetic density
nutritional density
percentage urbanized
squatter settlements
population composition
population pyramids
sex ratio
sex-selective abortion
female infanticide
total fertility rate (TFR)
Demographic Transition Model
gender
power relationships
sex cities
global feminization of poverty

Bibliography

Asia Watch Women's Rights Project. 1993. *A Modern Form of Slavery: Trafficking of Burmese Women and Girls into Brothels in Thailand.* New York: Human Rights Watch.

Bender, William, and Margaret Smith. 1997. *Population, Food, and Nutrition.* Population Bulletin 51/4. Washington, D.C.: Population Reference Bureau (PRB).

Bray, Francesca. 1986. The Rice Economies: *Technology and Development in Asian Societies.* Berkeley: University of California.

Brown, Lester. 1998. The Future of Growth. In *State of the World 1998,* ed. L. R. Brown et al., pp. 3–20. World Watch Institute. New York: Norton.

———. 1998. Struggling to Raise Cropland Productivity. In *State of the World 1998,* ed. L. Brown et al., pp. 79–95. World Watch Institute. New York: Norton.

Brown, Lester, Cristopher Flavin, and Hal Kane. 1996. *Vital Signs 1996.* World Watch Institute. New York: Norton.

Cheng, Lucie, and Ping-chun Hsiung. 1998. Engendering the 'Economic Miracle': the Labour Market in the Asia-Pacific. In *Economic Dynamism in the Asia-Pacific,* ed. Grahame Thompson, pp. 112–36. New York: Routledge.

Cho, Hyoung, and Pil-wha Chang, eds. 1994. *Gender Division of Labor in Korea.* Seoul: Ewha Woman's University.

Findlay, Alan. 1995. Population Crises: Malthusian Specter? In *Geographies of Global Change,* ed. R. J. Johnson, P. J. Taylor, and M. Watts, pp. 152–74. Oxford: Blackwell.

Gardner, Gary. 1996. Asia is Losing Ground. *World Watch* 9: 19–27.

Gore, Rick. 1997. The Dawn of Humans. *National Geographic* 191: 84–109.

Grigg, D. B. 1974. *The Agricultural Systems of the World: An Evolutionary Approach.* New York: Cambridge.

Huke, Robert E. 1985. The Green Revolution. *Journal of Geography.* 84: 248–54.

Momsen, Janet H., and Vivian Kinnaird, eds. 1993. *Different Places, Different Voices: Gender and Development in Africa, Asia, and Latin America.* New York: Routledge.

Olshansky, S. Jay. 1997. *Infectious Diseases—New and Ancient Threats to World Health.* Population Bulletin 52/2. Washington, DC: PRB.

Paul, Bimal Kanti. 1994. Aids in Asia. *The Geographical Review* 84: 367–79.

Paul, Bimal Kanti. 1997. Changes in Reproductive Behavior in Bangladesh. *The Geographical Review* 87: 100–104.

Piper, Jacqueline. 1993. *Rice in South-East Asia: Cultures and Landscapes.* New York: Oxford.

Population Reference Bureau. 1997. *World Population Data Sheet.* Washington, D.C.: PRB.

Rigg, Jonathan. 1989. The Green Revolution and Equity: Who Adopts the New Rice Varieties and Why? *Geography* 74: 144–49.

Rigg, Jonathan. 1994. Social Influences on the Adoption of New Rice Technology in Asia: Searching for Meaning. In *Atlas of World Development,* ed. Tim Unwin, pp. 142–43. New York: Wiley.

Riley, Nancy E. 1997. *Gender, Power, and Population Change.* Population Bulletin 52/1. Washington, D.C.: PRB.

Sachs, Carolyn. 1996. *Gendered Fields: Rural Women, Agriculture, and Environment.* Boulder: Westview.

Seager, Joni. 1997. *The State of Women in the World Atlas* (2nd edition). New York: Penguin.

Spencer, Joseph E., and William L. Thomas. 1971. *Asia, East by South: A Cultural Geography.* (2nd edition) New York: Wiley.

UNDP. 1990. *Human Development Report.*

UNICEF. 1997. *The State of the World's Children.* New York: Oxford.

Chapter 4

Development, Urbanization, and Quality of Life

"If there were no contradictions and no struggle, there would be no world, no process, no life, and there would be nothing at all."

(MAO ZEDONG)

There are many ways of defining and assessing development. No two countries are exactly alike in measurable aspects of development such as per person income, calorie intake, or number of doctors. More importantly, different countries have different ideas about what development is. For example, the strength of family ties may be more important than the availability of a medical doctor. Access to sufficient food may be more important than monetary intake. Whatever the case, *development* is a concept widely used by governments and aid organizations in determining needs, investments, and project applications. In this chapter you will learn about various perceptions of development.

The Changing Meaning of Development

All nations have been assigned descriptors based on criteria such as per capita income (PCI) based on gross national or domestic product (GNP or GDP). PCI is quite misleading, in part because it does not provide information on the buying power of a currency. A more contemporary measure of income is PCI based on purchasing power parity (PPP). PPP is based on GDP, accounts for price differentials among countries, and translates local currencies into international dollars. However, PCI measures do not recognize non-monetary transactions, such as goods for services, that are very important in many Asian societies. Also, they assume an unrealistic equal (per person) distribution of money.

There is, in fact, a **rich–poor gap** not only between parts of the world, but also between countries and within individual nations (Figure 4-1). In 1992, the poorest 50 percent of the world's population had less than 15 percent of global GDP. In contrast, the richest 15 percent had 50 percent of global GDP.

Assessment of a combination of development factors positions a country on a bipolar spectrum of socio-economic development. This development spectrum

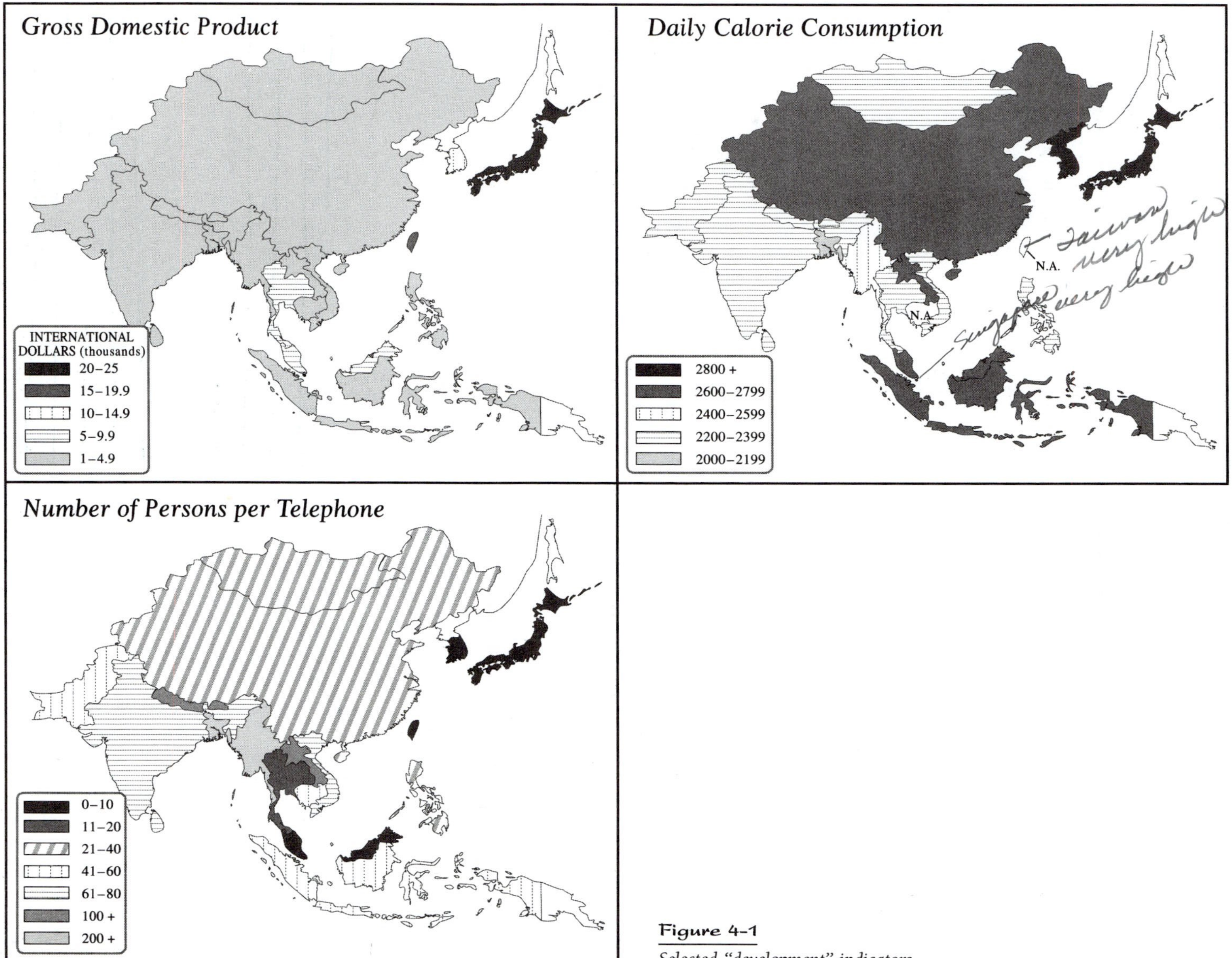

Figure 4-1
Selected "development" indicators.

is based on misleading pairs of opposites, such as less-developed or more developed; developed or underdeveloped; and rich or poor. A term that is applied to several East Asian economies, including South Korea, Taiwan, Hong Kong and Singapore, is **newly industrializing countries (NICs).** Others, such as India and Indonesia, are called **developing countries (DCs).**

Through the latter part of the twentieth century, several schools of thought prevailed. For example, development trajectory proponents purported that all nations would proceed along an evolutionary path of economic advancement according to progressive stages already experienced in the West. Marxists stressed the ills of capitalist and class exploitation and the need to alter unequal power relationships within and between nations.

Another school of thought provided countries with ideologically based designations. Technologically developed, democratic nations formed the **First World,** the communist-bloc nations were called the **Second World,** and the remainder became the **Third World.** When the Soviet Union disbanded and China began adopting vestiges of capitalism, and poorer countries such as Taiwan and Malaysia emerged as Economic Tigers, these assignations became obsolete.

A more recent theory argues that in an interdependent world, it is essential that balance be achieved to ensure mutual benefits through the control of trade between the rich, industrialized countries of the North and the poorer, Third World countries of the South. This **North–South dichotomy** partitions the world into essentially superior and inferior realms and serves the notion of socioeconomic divergence between the haves and have nots. It also connotes the notion of underdevelopment whereby the South's lack of development is at least partly a function of the North's development.

Regional inequities have always existed in Asia; they most certainly did not originate with Western impacts. But, ongoing disparities have been compounded by outside influences, especially in this century of global economic and political machinations.

Critics believe that most categorizations are pejorative, associated with paternalism and racism. Founded in "we–they" thinking, ideas of evolutionary progression along a path toward technological achievement and democratic institutions became moral justification for the application of Western social and economic principles to international relationships.

These principles have one thing in common: they begin with the notion of deprivation (judged by Western standards) and culminate with the concept of development incorporating measures of resource exploitation, urbanization, industrialization, and material consumption. Fraught with conflict and contradiction, such applications have long been contested by "recipient" nations.

Definitions and categorizations based on what people do *not* have and prescriptions as to how they *should* live, based on Western standards, have shaped perceptions and altered landscapes around the world. Moreover, students from NICs who study abroad learn nonindigenous concepts of planning and development and apply these within their homelands.

Crowded, winding "unplanned" streets, which were once functioning, multifaceted communities, have been seen (through newly trained eyes) as inappropriate spatial geometries that must be rearranged for greater efficiency and visual appeal. Squatter settlements, home to millions, perceived as unhealthy warrens of poverty and criminality, have been razed. These are rapidly rebuilt; most governments now realize that it is more expedient to implement self-help programs within existing squatter settlements. The social costs of such upheaval cannot be measured.

The vision of nations speeding along a trajectory of technological advancement and material consumption has been shattered. Things that were supposed to happen under this rubric did not. China, for example, has initiated broad economic reform and has dramatically reduced its population growth, thereby ending the specter of hordes of Communist Chinese overrunning the planet. Singapore and Thailand have become regional economic powerhouses. Social and economic protest movements have been on the rise. Common issues such as global resource depletion and environmental crises are drawing international attention. Asian leaders speak of preserving Asian values and charting their own paths. As you will learn in the following chapters, Asian development models are rapidly and dramatically reshaping landscapes.

Cores and Peripheries

Despite indigenous development models, inequity between and within countries remains apparent and is often fueled by foreign investment. These landscape differences are better clarified in a **core–periphery** framework. The core–periphery concept expresses the idea that there is a powerful center and a less-powerful edge, a mode of thinking applicable in any context or at any scale. For instance, the world could be portrayed as having a Northern core and a Southern periphery. Or, in terms of Japanese investment in Asia, Japan is a core while its collective recipient countries form a periphery. Geographer Debra Straussfogel (1997) explains that core and periphery bracket a continuum that "represents one horizontal level within the larger global structure." She describes development as a process "whereby one structural configuration is transformed into another." As development alters socioeconomic spaces, restructuring takes place at multiple levels, thereby repositioning the country on the scale of "coreness" or "peripheryness."

An urbanized capital region is usually the core of its own state, with rural areas comprising the periphery. Some countries have several regional cores. For example, China's Beijing, Shanghai, and Guangzhou are all cores with surrounding peripheries. In fact, these and other developing coastal cities can be conceived as a single core with the rest of China as the less-developed periphery. Cities also house core elite groups who hold relatively more power than the peripheral rural population. However, it is important to recognize that although effective power lies with the urban elite, potential power lies with the rural masses.

MULTINATIONAL INVESTMENT

Established core–periphery relationships have global impacts. Multinational corporations and capital investment by the United States, European Union, and Japan place recipient states in a position of dependency. Movement of labor-intensive manufacturing activities to peripheral locations to take advantage of cheap labor has been a hallmark of economic restructuring in the late twentieth century, often with unfavorable consequences. An assembly line might provide hundreds of low-wage jobs, but a downsizing or relocation decision in the core country can create instant joblessness. Many foreign firms in Asia are criticized for their exploitation of labor and resources and their cavalier attitudes concerning environmental quality.

Did you know that most American and British produced cigarettes are now sold overseas?

Smoking Asia

The WHO estimates that the twenty-first century will initiate 10 million tobacco-related deaths a year with more than two-thirds in developing countries. Developing countries already accounted for three-quarters of global tobacco production by 1980, and will consume 70 percent of the world's cigarettes by 2020. Tobacco use and production has declined in the industrialized world, 17 percent in the United States. However, exports have risen by 250 percent and Asia has become a prime marketing target.

Seven companies controlled 45 percent of global sales in 1995. British-American Tobacco (BAT), along with the American companies Philip Morris (PM) and R. J. Reynolds (RJR), are the largest. In 1981, PM, RJR, and Brown and Williamson (a BAT subsidiary in the U.S.) formed the U.S. Cigarette Export Association. Flooding the Latin American market during the 1980s, they have now turned to Asia.

Convincing U.S. officials to claim that foreign restrictions on tobacco product advertising were, in effect, trade barriers, tobacco companies were exempted from national controls in Taiwan, Japan, and Korea. U.S. market shares jumped accordingly. Advertising bans elsewhere are circumvented through sponsorship of sporting and other crowd-attracting events. In Malaysia, tobacco companies held free rock concerts and distributed free cigarettes to attendees.

Promotion is furthered through hats, T-shirts, lighters, and other paraphernalia. Such strategies are aimed at both selling cigarettes to existing smokers and expanding the smoking pool.

Particular target groups are individuals with rising incomes, women seeking greater autonomy, teens, and even children and the poor. Billboards appeal to status, power, and independent thinking. A 1991 study in Japan revealed that 26 percent of teenage girls smoked, yet only eight percent of their mothers did. Cigarettes are often sold individually. This makes Marlboro, Salem, and other brands affordable and ensures that the buyer will not see any warning label. This strategy also allows children to buy cigarettes, and there are increasing numbers of child smokers in the five to nine-year-old age group.

Cigarette smuggling is rampant, with one-third of all internationally traded tobacco landing in the black market. For instance, two out of every three packets sold in Vietnam have been illegally brought from Cambodia, Laos, Thailand, or China. Selling for "introductory" prices makes these more accessible and feeds the new smoker pool. Millions of cartons confiscated by the Vietnamese authorities are subsequently exported.

The tobacco industry can be important economically, contributing significantly, via taxes, to many national budgets as in Sri Lanka and Pakistan. The industry may also be a major employer. For example, in Vietnam, it provides more than 100,000 jobs and generates up to six percent of the national budget. In 1995, China's domestic tobacco industry not only employed around five million people but also ranked as the number one source of tax revenues at 8.6 billion $US. Interestingly, both Indonesia and Korea target a portion of their tobacco revenues for anti-smoking campaigns.

Although they know that cigarette smoking is harmful, tobacco companies produce 1,000 cigarettes every year for each man, woman, and child on earth. Currently, 3 million people die each year from tobacco-related diseases. Costs of medical care and lost productivity are in billions of dollars. Will WHO's prediction of 10 million deaths annually materialize? Should there be curbs on the activities of American or other multinational tobacco companies overseas?

There is a World Tobacco Conference (held in Beijing in 1997) and there is an International Tobacco Control Convention endorsed by a majority of WHO members. The Convention seeks to control smuggling, regulate prices and advertising, and disseminate information. Members are in agreement that tobacco is addictive and that smoking is a type of social disease. Unfortunately, this deadly product is legal, and most limiting or controlling efforts have been thwarted by the tobacco companies.

Anne Platt McGinn (1997), a WWI researcher, believes that our familiarity with tobacco masks its threat. This makes the challenge a cultural one. How can public perceptions of smoking be changed? How can *public* health be made into a paramount value? What are your thoughts on this controversial issue?

The debatable notion of dependency has given rise to **dependency theory.** Dependency theory purports that cores can only grow through exploitation of their dependent peripheries. In order for cores to prosper, spatial imbalances must be maintained. And, to understand a region's low level of development, it must be viewed in context of both local and global core–periphery relationships. From this perspective, trade is not mutually beneficial.

People in the periphery compete to work for very low wages, which drives down the price of goods. However, price reductions are not passed on to the consumer because producers in the core have a common desire to keep prices high. As cores and peripheries become increasingly integrated on international scales, regional economies are structured into situations of dependency on the demands of the cores and are unable to develop their own economies. But, geographies are constantly changing.

In global cores, financial, investment, and assistance rationales and applications are in question. In peripheries, economic independence and local development progress have become paramount concerns. Non-governmental organizations and privatization have taken on stronger roles in development. Such changes mean that dependency relationships are constantly shifting and increasingly complex.

Figure 4-2 shows **Foreign Direct Investment (FDI)** for 1990 and 1995. Generally, investment in Asia has increased with quantum leaps, notably in the case of China. Investment strategies are founded both in the needs of the investor and in the economic and political climate of the recipient country. Change in one area propels change in another. For example, India saw its foreign investment soar after its economic reforms in 1991. In North Korea, investment flows stopped after a 1993 crisis over its nuclear weapons program. Huge investments in China by ASEAN countries and the United States have been stimulated by the revision of China's tax code and its reassessment of incentives for foreign investment.

Any sign of instability can induce investors to retreat. War and political chaos have kept many away from Myanmar, Laos, and Cambodia. Vietnam's mercurial investment policies and disregard for venture profitability have turned off many potential investors. Indonesia's 1994 deregulation improved the investment environment for small- and medium-sized enter-

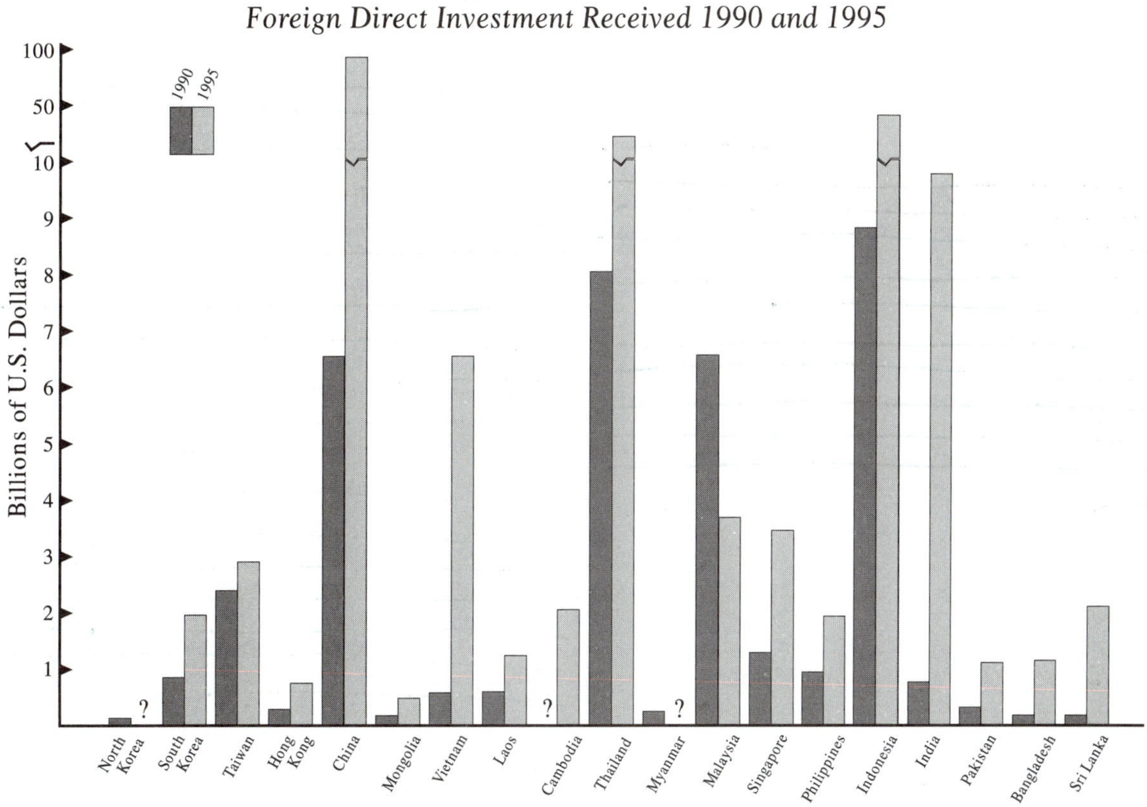

Figure 4-2

Foreign direct investment. Note the increased foreign involvement. Japan is the largest Asian investor in Asia.

prises, and FDI rose significantly. However, political crises in 1996 and 1998 saw dramatic pullbacks. Nuclear testing by India and Pakistan in 1998 ricocheted across the investment world.

Financial crises in Japan, Thailand, and elsewhere in Asia at the end of the twentieth century will have far-reaching ramifications. Reductions in FDI are expected to be countered by mergers and acquisitions, with a gain in assets owned by foreigners. While reduced, FDI is expected to remain China-centered because investors remain attracted to the size of its market potential. Still, the presence of a cheap labor force near Singapore keeps Malaysia attractive for investment in the electronics industry. And, with its skilled work force, an improved infrastructure around Manila, and a new government committed to economic liberalization, the Philippines has also become a positive FDI environment.

FDI can bring benefits and spatial transformation. Governments can use incentives to generate investment in less-developed regions away from the core. Thailand, for example, encourages investment in areas remote from Bangkok. Moreover, Asia is no longer simply a cheap production realm; it is the fastest-growing consumer market in the world. Thus, there is a changing emphasis from fleeting use of a low-cost labor force to long-term commitment, including transfer of knowledge, skills, and technology.

FDI also creates a paradox. While providing employment and raising levels of disposable income, localization of production and management, creation of subsidiaries, and linkages with established firms simultaneously strengthen the multinational's web of influence.

Motorola, based in the United States, entered South Korea in 1967 to make transistors. An initial investment of $6.7 million has generated $260 million in exports, and the industry has spawned a semiconductor and wireless communication enterprise employing nearly 2,000 workers. As the company continues to expand in South Korea, it has earned the title of "Motorola Korea Group."

Swiss-based Nestlé's operational style is to integrate itself in local economies. It has sent experts to virtually every Asian nation (except North Korea and Vietnam) to encourage coffee cultivation and the dairy industry. Large investment in manufacturing and marketing complements local product development. In 1992, Nestle gained U.S.$4 billion from Asian sales, one-third of this amount from Japan.

Multinationals are powerful entities of global proportions. Ford Motors has yearly sales in excess of the GDP of the Philippines. Toyota's sales rival that of Singapore. Interacting with nation-states, such conglomerates are instrumental in the creation of new economic landscapes. According to geographer Brian Berry (1989), "There are different outcomes in different world regions because they play their games on the differentiating checkerboard of culture."

PRIVATE INVESTMENT

This postmodern era is witnessing a dramatic increase in the amount of private capital flowing into Asia. According to World Bank estimates, North–South transfers of money multiplied five times from 1990 to 1996. Deregulation of capital markets in Japan, the United States, and Europe, along with telecommunication networks permitting instant cash transfers, have facilitated this trend. International bank lending rose to its highest levels in 1996, with more loans to Asia than any other region. And in that same year, EU banks surpassed Japanese banks as the paramount lenders in Asia. The two largest borrowers were South Korea and Thailand, followed by Indonesia, Malaysia, China, and the Philippines.

In recent decades, macro-financial flows have enabled developing nations to reverse capital flows in their own favor and fostered the rearing of Economic Tigers. On the other hand, in several countries such as South Korea and Thailand, banks floated excessive loans for too many poorly evaluated development schemes. Hundreds of projects now stand incomplete or unoccupied, their loans outstanding. Furthermore, critics of such large-scale development argue that it serves to widen the rich–poor gap.

But what about the so-called middle class? There is a growing segment of Asian society demanding quality consumer goods, sophisticated leisure activities, better educational opportunities, and access to information. Political input and stability is important to this group, whose skills and purchasing power contribute to national well-being. According to Australian, Asia scholars Richard Robison and David Goodman (1996), "The middle classes represent a new set of social interests that regimes must take into account."

OVERSEAS CHINESE

Who are the major investors in Asia? While Japan, the United States, and the EU are the current leaders, Asians more frequently are investing in themselves. For instance, most of China's FDI comes from Asia-Pacific nations. One pivotal and powerful group are the **Overseas Chinese**, whom China scholar Sterling Seagrave (1995) refers to as *Offshore China*.

With enormous wealth and tightly knit family, clan, club, organizational, and territorial loyalties, Offshore China intricately involves more than 55 million individuals and countless corporations. In Southeast Asia alone, more than a hundred conglomerates have recently appeared, all owned by Chinese living in Indonesia, Malaysia, Thailand, and elsewhere in Asia outside of China. However, most have long-standing connections with regions in southern China. At least 2,000 years ago, their ancestors left the southern provinces to colonize Southeast Asia, and thousands have followed since.

Focusing on business and trade, they maintained, expanded, and intensified their connections with various power structures to the point of becoming gatekeepers for any form of economic interchange. After each of a series of upheavals in China, more would flee or at least channel their capital elsewhere. With the Communist victory in 1949, billions of dollars left the mainland with the outflux of Nationalist Chinese and their supporters. By prudent saving and investment, timely information, and fast action via closely guarded networks, the Overseas Chinese have come to dominate the economies of East and Southeast Asia, with the exceptions of Japan and Korea.

FOREIGN AID

The **spatial duality** of core–periphery development is frequently further entrenched with **bilateral** and **multilateral aid** programs. Most aid is bilateral—government to government transfers of cash, low-interest loans, material goods, food, or medical supplies. Technology and training are also transferred. Technical training, while improving skill levels, lays the groundwork for further capital input and even greater reliance on outside assistance. Aid might also take the form of preferential tariffs, whereby the donor nation agrees to buy goods produced in the recipient nation. Multilateral aid derives from international organizations such as the Organization for Economic Cooperation and Development (OECD), the World Bank (WB), or agencies of the United Nations.

If aid is concentrated in a core area, it strengthens the core and distances it further from peripheral realities. For instance, a full-service hospital is most likely to be located in the core, while poorly supplied clinics are placed in peripheral regions. Transportation development provides another example. Modern road construction typically radiates outward from the capital city. Starting as a paved highway, it eventually degenerates into an unpaved road, and ultimately into a dirt track, often impassable in the wet season. Transportation and other infrastructural conditions decline as distance from a core increases (Figure 4-3).

Tied aid, binding the source and destination in a not-always mutually beneficial agreement, is a common practice. Much official development assistance (ODA) is tied to structural adjustment programs (SAPs). SAPs require privatization, increases in exports, and adherence to free market principles. ODA is frequently siphoned off to pet projects for political reasons or to mega-projects tied to purchases from donor nations. A country might be required to import capital-intensive equipment—road-grading machinery, for instance. Such equipment gets the job done faster but also replaces labor in circumstances of already high unemployment. Moreover, less than a quarter of ODA ever reaches the poor people it is intended to help.

In northern Malaysia, the Pergau hydroelectric dam was financed by Britain with soft loans—subsidized, low-interest money. In 1993, an audit revealed that the British government financed the project even though aid officials considered it a waste of financial resources. Part of the deal was that British companies in Malaysia would be granted additional development contracts. Indeed, British companies

Figure 4-3

This crew is repairing a road in northern China. Note that there are more workers with shovels than machines. Such dirt roads can become impassable in the wet monsoon.

have been major recipients of Malaysian contracts since 1988. Aid is rarely simple giving.

DUAL ECONOMIES

Another spatial duality, notably expressed in urban landscapes, is the **dual economy.** Developing nations are characterized by two levels of economic activity: a *formal sector,* functioning at a large scale in the realm of national and international trade, and an *informal sector,* functioning at a smaller scale at local levels. The formal economy exhibits permanent employment, sets hours of work and pay, and regulates other benefits. The informal sector is characterized by family labor, small enterprise, and self-employment (Figure 4-4). It includes the hawking of goods and begging. These two sectors are not necessarily mutually exclusive and should not imply relative superiority and inferiority.

Figure 4-4

Paper and wood umbrellas are made at this family enterprise in Myanmar. Three generations work in this small factory. Even the waterproofed paper is made on the premises.

Figure 4-5

Here the cook is adding soy sauce to her specialty dish at this Thai food stall.

A significant element of the informal sector in Asian cities concerns the selling of street foods (Figure 4-5). Professor Irene Tinker, one of the foremost authorities on women and world development, has written a book titled *Street Foods* (1997), much of which is devoted to food vendors in Asia. She refers to street vending as *microenterprise* and illustrates its vitality with case studies.

In Bogor, Indonesia, a city of 250,000, there is one food vendor for every 14 inhabitants. Street foods are vital to workers, families, and schoolchildren, and they often absorb secondary agricultural products such as peanuts or corn. Some governments and planners view food stalls and mobile *becak* (pedicab) vendors as detrimental to traffic flow and hazardous to health. However, eradication efforts are rarely effective.

So vital are street foods in local economies that an array of local and external support systems has arisen to protect and enhance this informal service industry. Observers note the value of informal activities such as itinerant trash collecting, knife-sharpening, barbering, and the like for personal survival, sense of self-worth, social networking, and community service. But protection and enhancement ideals conflict with larger reality. Internationalization and newly affluent consumer demand have stimulated construction of multiple retail and service, Western-style malls that also have food courts, fast-food chains, and packaged snacks. These and other modernization phenomena are displacing street vendors who cannot afford to meet higher costs and new regulation requirements (Figure 4-6).

THE SHADOW ECONOMY

Increasing globalization masks a growing **shadow economy** of illegal or at least shady activities. Millions of people are involved in black market—unregulated, untaxed, often unseen business. Smuggling of "protected" wildlife, contraband, arms, drugs, and people is widespread, especially in Southeast Asia. Crossing the border from Thailand to Laos, I was accompanied by a crowd of women, each with a few bars of soap, cigarettes, and other Thai and imported products, which they hid on their person or beneath other bundles under the vehicle's benches. Unavailable in Laos, these smuggled items were to be sold at roadside stands or local market stalls.

Increasing permeability of borders, official collusion, and poverty fosters these and other survival strategies. According to Ed Ayres, editor of *World Watch* (1996), "The growth of the shadow economy can be seen as pervasive evidence of human needs not being met by traditional institutions."

Figure 4-6

Lot 10 is a popular mall in downtown Kuala Lumpur, Malaysia. Inside, escalators take you to several floors of shops and restaurants. Many stores are the same as those in your malls at home. In the basement is a fast-food court and a grocery store.

Growth of unregulated and illegal activities is also related to accountability. Not permanently based at a particular place, they are difficult, if not virtually impossible, to trace. Unregistered settlements, occupations, anarchic groups, and migratory populations transcend and escape the capabilities of extant legal strictures. Uncontrolled movement, especially rural to urban migration, is a major contributor to this situation. Every day, tens of thousands of individuals and families leave the hopelessness of their rural origins seeking opportunities perceived to exist in urban destinations. Lacking requisite skills, they swell the populations of squatter settlements and slums and bolster the ranks of the unemployed and the shadow economy.

Did you know that many products we buy are made by children, and that child labor is often necessary for survival?

Working Children

Tens of millions of children work in Asia. At least 15 percent of all 10- to 14-year-olds labor in fields and factories. Millions more, older and even younger, work 10 or 12 hour days in dangerous, medieval conditions in glass, match, and carpet factories, gravel pits, and brick kilns. They tote bricks for construction, carry baskets of raw sewage out of drains, and pick through garbage for anything salable. For the "boss" or factory owner, child labor for a pittance is better than machinery for a fortune. Long hours and lack of nutrition, combined with virtually no safety standards, mean that accidents are commonplace. Dust, fibers, and chemicals cause respiratory damage and blindness. Severe injury can be fatal, because a lost job can lead to death.

Most people condemn these practices and decry the cruelty of child labor no matter what the circumstances. Sanctions and boycotts might be successful in removing significant numbers of children from deplorable work environments, but the story is more complex.

Realities often make child labor essential for family survival. Many children of poverty have been sold into servitude by desperate parents. Many have crippled or diseased parents, or none at all. Without income they will literally starve to death. Working children are facts of life in many parts of Asia.

Enforcing child labor laws is problematic given disguised factories, hidden children, corrupt officials, and dire poverty in certain regions. Furthermore, not all work environments are terrible. Significant numbers of children work as contributors to the family farm or business in relatively decent conditions. Many go to school as well.

Education works against child labor. In fact, where schooling is compulsory, child labor rates decline. Still, families in poverty must make a decision as to whether long-term gain, via education, is worth the wait in terms of their immediate needs. Education calls for expenditures for books, uniforms, and food. This money and time spent must be balanced against lost wages.

Official and other organizations, while making efforts to end enslavement, brutality, and unsafe and unhealthy work environments, try not to further impoverish families or children by removing them from the work force altogether. Instead, they attempt to get them and keep them in school and improve working conditions. Half-day classes are common in such countries as Myanmar, and children work before or after school. In Bangladesh, a garment worker's union actually runs schools.

Child labor may be unacceptable in societies flush with palliative programs for the less fortunate, but in societies characterized by large numbers of desperate poor and limited or no safety nets, a single working child might be the only solution to survival for many.

Urbanization

Asian urbanization has skyrocketed in the twentieth century (Figure 4-7). In 1950, only 6 Asian cities were among the world's 30 largest. This number increased to 11 by 1970, and to 14 by 1990. Population and spatial extent of cities grow by rural–urban migration; natural increase; and accretion of adjacent settlements. Some urban populations have risen dramatically only because of administrative boundary expansion. For example, Tokyo grew by 9 million people in the 1990s. But 6 million of these were added by a boundary expansion that incorporated 87 surrounding towns and cities. Moreover, forecasts of urban growth concentrating only in existing giant cities have not always materialized. In the case of China and India, it is medium-sized cities that are experiencing higher growth rates. Whatever the case, variation in city definition makes comparative urban data indicative rather than predictive.

Although cities grow significantly with incoming migration streams, growth from NI is even greater. These two factors are related as rural migrants have more traditional ideas about preferred family size. Consequently, NI rates in slums and squatter settlements are significantly higher than those in middle-class and elite districts. It is these poor city dwellers that many elite and middle-class groups consider to be the "population problem."

Today, most of the world's largest cities are in DCs, with increasing numbers of these in Asia. More than half of Asia's largest cities are in India and China. With 38 cities of at least a million inhabitants, China has more large cities than any country on earth. Some countries show a marked degree of **urban primacy,** whereby the population of the largest city is greater than the combined populations of the next three largest cities (Figure 4-8). The largest city is then referred to as a **primate city.** Countries exhibit varying degrees of primacy depending on the population differential between the largest and next largest cities. Primate cities house relatively large proportions of national populations. Greater Seoul, for instance, houses 35 percent of South Korea's urban population, while more than a quarter of city-dwelling Japanese live in the Tokyo metropolitan region. Expansion and accretion absorbs increasing shares of national populations into the urban world.

Primacy is not tied to a country's degree of urbanization, nor does it relate to the number of cities. Both South Korea and Japan are highly urbanized countries. Neither Thailand nor Indonesia are highly urbanized, yet Bangkok and Jakarta are primate cities, dominating their respective economies. Shanghai is not a primate city at the national level because there are several simi-

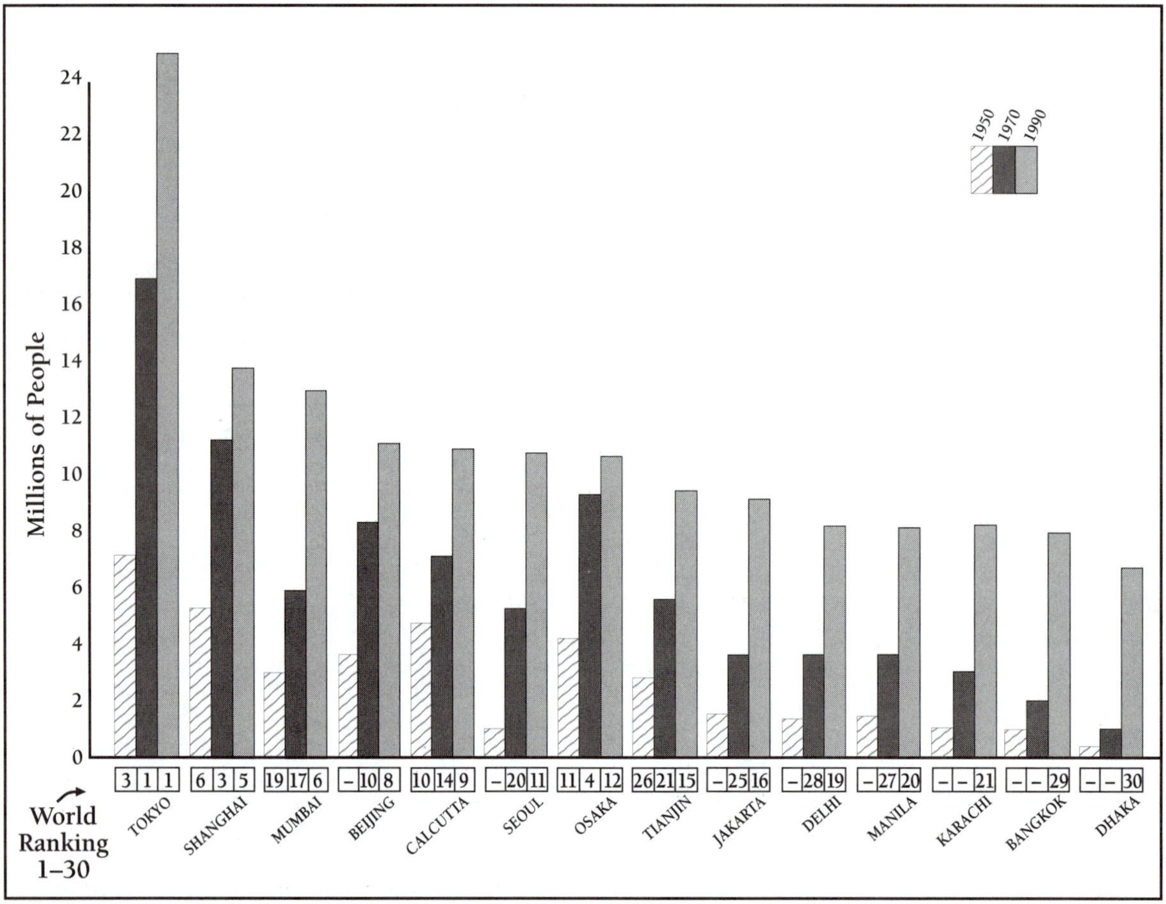

Figure 4-7

Urban growth. Cities grow in three ways: natural increase; in-migration; and expansion of boundaries. Source: United Nations 1994. World Population Prospects. *New York: U.N.*

larly sized cities in China. However, Shanghai might be seen as a primate city in a regional context, as most of China's economic development is concentrated in the eastern provinces. Primacy indicates spatial dominance at varying scales of observation.

MEGA-URBAN LANDSCAPES

Confucius, in the fifth century B.C., postulated 3,000 persons as an ideal city size. In the twenty-first century, the Beijing-Tianjin metropolitan region is expected to house 37 million persons. In the early twentieth century, a mere 26 million lived in India's cities. At the turn of the twenty-first century, India's urban dwellers total more than 220 million, and India has 23 cities of more than a million people. Such dense accumulations of humanity have reached unparalleled proportions with entanglements of human and environmental problems beyond the scope of simple solutions.

Polarization of development has created **mega-cities.** Clusters of cities, or **urban agglomerations,** have evolved into **mega-urban regions** that extend beyond metropolitan boundaries incorporating towns, villages, and transportation corridors. These are com-

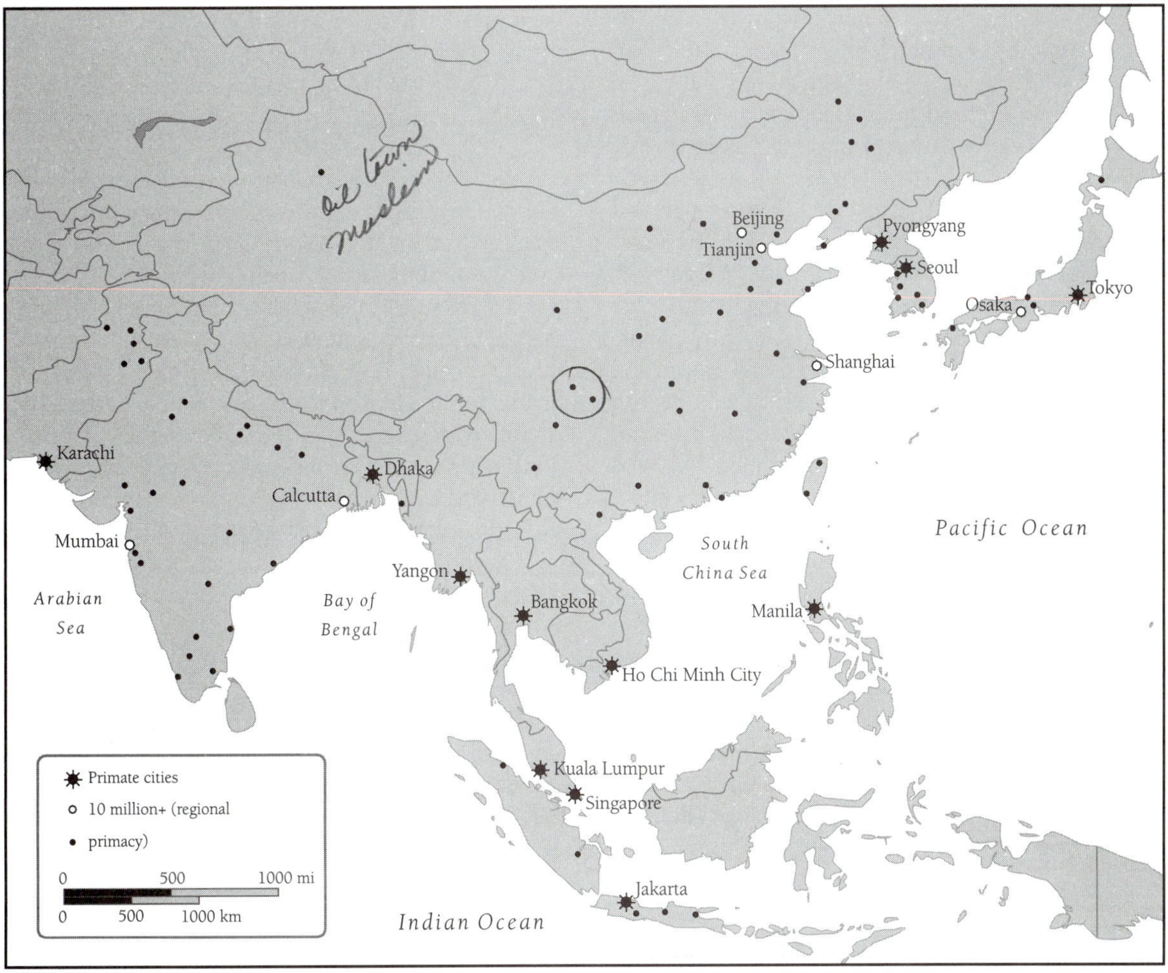

Figure 4-8

Cities of 1 million plus. India and China both have several cities of a million or more population, but most of their people are farmers.

plex fields of urban and rural interaction reaching 62 miles (100 km) beyond urban nodes.

Geographer T. G. McGee (1989) has coined the term ***kotadesasi,*** joining the Bahasa Indonesian words *kota* (town) and *desa* (village), to represent the coincidence of urban and rural activity in the same spatial territory. This concept varies from *urbanization,* which implies a central urban system dominated by urban phenomena, processes, and values, and harboring only remnant rural elements and traditions.

Kotadesasi zones are characterized by an increase of diverse nonagricultural activities in predominantly agricultural areas. Members of the same household participate in sundry occupations in both realms, making economic linkages within the *kotadesasi* zone as important as those with the dominant city. Larger numbers of women work in nonagricultural activities. Proliferation of cheap transport such as bicycles, motorbikes, and scooters, along with public buses and trucks, allows for great mobility of people and goods within and between these and other zones (Figure 4-9).

These mega-urban, *kotadesasi* landscapes are kaleidoscopes of activity, including rice paddies, water buffalo, bullock and horse-carts, factories, golf courses, and traffic-jammed multilane highways, among other things. Industries locate in and subsequently pollute surrounding agricultural land. Agricultural production shifts from monocrop cultivation (principally rice) to diverse production of vegetables, livestock, and fruit for regional and international consumption.

In this ambiguously rural and urban environment, governed by overlapping jurisdictions, management of infrastructure can only be a nightmare. However, since the definitions of "urban" and "rural" are highly variable and frequently change, *kotadesasi* zones are sort of gray territories where urban and industrial regulations may not apply. Small-scale and informal sector operators along with squatters, who find it difficult if not impossible to conform to city regulations, see these as spaces of opportunity. The result, according to McGee (1989), is that "there is no clear-cut division between rural and urban relations; rather, activities in the two sectors are fused and complementary."

URBANIZATION AND EMPLOYMENT

Cities and megacities are part of an all-encompassing international urban hierarchy topped by **world cities** such as London, New York, and Tokyo. The entire structure—from world cities, to powerful international economic centers such as Singapore, to regional market-oriented capitals such as Kuala Lumpur, through smaller regional and local urban centers—is interconnected by political and economic transactions conducted across earthspace and cyberspace. Information and its technology disseminate from **technopoles** such as The Netherlands' "Randstad" region, England's "M-4 Corridor," or Taiwan's "silicon valley," centered on Hsinchu. From such world cities, and technopoles, global corporations manage their global networks and subsets of urban and rural production and processing sites.

Research has shown that higher levels of urbanization are frequently related to degrees of dependency as measured by FDI, which has traditionally focused in cities. Urban job creation is a gravitational pull on the unemployed in the rural hinterland. Rural to urban

Figure 4-9

Agricultural fields and commercial establishments show mixed land use outside of Hanoi in Vietnam. This town specializes in ceramics, and factories exist amid the rice paddies. The main service road to the port of Haiphong is to the right.

migration draws uneducated and unskilled people into the urban employment realm. More people require additional infrastructure, which generates more jobs, which stimulates more in-migration. This ongoing process results in a permanent excess of workers over jobs.

Underemployment is also boosted with urban growth in DCs. Underemployment is a Western notion referring to the underutilization of labor. It might take the form of jobs available only on a cyclical basis such as agriculture, construction, and tourism that operate on daily, weekly, or seasonal schedules. Another type of underemployment is the hiring of several individuals to do the job ordinarily held by one person. Also, there is *hidden underemployment,* based on group or family solidarity, where everyone participates in the work whether they are needed or not.

While unemployment denies means of livelihood, underemployment provides both income and sense of self worth (Figure 4-10). Homeless women in Mumbai chop gravel, carry bricks, mix cement, and do other arduous tasks. Meanwhile, they can live within the project they are building. When it is finished, they will move on to another construction site. Children work alongside their parents. Economically, at the national level, this is not an ideal situation because the cost of services for surplus labor takes away from potentially broader expenditures. But, individually, at the local level, underemployment is preferable to no employment.

With disproportionate investment and capital concentration, cities exhibit dramatically disparate landscapes. Landscapes of power and privilege contrast with those of powerlessness and denial. Yet, many poor are employed by their rich counterparts as house servants, gardeners, drivers, dog-walkers, and the like (Figure 4-11).

Another vital urban phenomenon is the **trans-border development region.** Trans-border development regions include economic development zones in adjacent political units. A region may include several countries, as in the case of the "development triangle" involving Malaysia, Singapore, and Indonesia. Or, it might involve several provinces, as in the case of southern China. Development zones are integrated across borders by transportation corridors and telecommunication links into the global system of production, commerce, finance, distribution, and consumption.

The pace of urbanization and landscape change in general is exacerbated with global economic interchange that, in essence, ignores political boundaries, human costs, and environmental concerns. Fortunately, there is growing awareness that we are running

Figure 4-10

I took this picture at about 14,000 feet (4,200 m) in the Indian Himalaya. No machines were available for the repair of this main road. Instead, three women operated a shovel. In this system, one digs the shovel into the dirt, then the two others pull the rope that pulls the shovel to move the dirt. Some shovel-rope teams comprise only two people. This may not seem very efficient, but it provides more employment than a machine. Moreover, machines are not very practical on such narrow roadways.

Figure 4-11

This beautiful house in India shows how the upper classes live. Note the gardener watering the plants. I took this picture on Christmas day, and the star above the doorway indicates that this is a Christian household. The star is lit at night.

a race on a deteriorating track. Development, whether indigenous or international in scope, is consuming the social and natural capital upon which it depends.

An important dictum is that "everything is related to everything else." Have you thought about the connections between information in this chapter and that of previous chapters?

Tale of the Tiger Prawn

Prawn fever in Asia has been triggered by consumer demand in such countries as Japan, Singapore, Canada, and the United States. Tiger prawn farms made their initial appearance along the mangrove coasts of China, Taiwan, and South Korea in the 1970s and have since diffused into South and Southeast Asia.

Investors convince rice farmers or fishers to sell their land or take up prawn farming for quick profits. Using their property as collateral, farmers take out high-interest-rate loans. However, what is promoted as a boon to the local economy favors only the "aquabusiness" investors, who also own cold-storage facilities and sell their own feed, chemicals, and pumping equipment. Companies are both national and transnational in scope, with an array of business interests. Mitsubishi of Japan and Aqua Star of the United States are two of these.

Coastal mangroves are cleared to make space for the prawn ponds. Fresh seawater must be brought in regularly, and the ponds must be pumped clean of shrimp excrement and chemical additives. This wastewater further destroys coastal ecology, and ground water salinization ruins fields for rice. Finally, pollutants poison surrounding seawater, rendering it unsuitable to flush ponds. With increased contamination, prawn production declines, and ponds are ultimately abandoned.

Big business opens ponds elsewhere, but small entrepreneurs are doomed. No longer able to make loan payments, they become landless and unemployed. Developers take over the once productive rice fields for industrial, housing, and tourism projects. Those displaced provide a ready-made source of construction labor.

Thailand's eastern coast, a prawn production haven in the 1980s, has reaped this destruction. With mangroves gone and rice paddies abandoned, coastal communities have only tourism in their sights. Prawn companies have since moved to Thailand's western coast. The Thai government is loath to end the practice, not only because various officials are involved in the business, but also because it generates a significant portion of export earnings.

In Bangladesh, large aquabusiness investors made millions on prawns while ruining farmlands of thousands of already poor rice farmers. As prawn ponds were abandoned, investors purchased the land for next to nothing, and thousands of families migrated to nearby towns and cities.

The tale of the tiger prawn illustrates the interrelationships among numerous factors: big business; transnationals; foreign investment; foreign exchange; international trade; ecosystem destruction; environmental pollution; cropland abandonment; family ruination, displacement and unemployment; rural to urban migration; growth of cities; and expansion of urban-powered development into the countryside.

Postmodern Thinking

Geographers and others now recognize that *development is multidimensional* and that *there is no single appropriate development framework*. Geographer John Agnew (1987) emphasizes the return of cultural context to the concept of development, noting that every part of the world has its unique relationship with other places individually and collectively. New concerns, such as *meeting basic needs* and *economic reform*, have emerged. Fresh categorizations have been created as well.

The United Nations Development Programme (UNDP) published its first *Human Development Report* (HDP) in 1990. Although it argues the necessity for growth in national production, it also recognizes that some states have achieved high levels of human development with only modest per capita incomes. The HDP devised the Human Development Index (HDI), which ranks 174 countries according to a composite measure based on income, health, life expectancy, and education information. For example, in 1995, Canada ranked number 1 and China ranked 108. That year also witnessed the first compilation of a Gender Development Index (GDI), based on similar data, which indicates discrepancies between males and females (Figure 4-12).

Intermediate technology has become fashionable. Beyond the digging stick and behind the tractor, such technology uses local human and material resources to effect positive change. For example, it uses local building materials rather than relying on foreign imports. Other examples include using solar energy or biogas for power. Such "appropriate" technology has become significant in many places in Asia and is tied largely to private organizations.

Non-Government Organizations (NGOs) reflect contemporary culture-based perceptions of the development process. All have an agenda, some political, others religious. Nevertheless, NGOs tend to focus in peripheral settings such as urban slums and disadvantaged rural areas. Laos, with 13 NGOs in 1985, had 63 operating in 1997. Over half of these were involved in health and human resource assistance programs. Ideally, culturally derived development models and programs can be articulated with larger, international frameworks of change. Unfortunately, disparity between the ideal and the real seems unavoidable.

Because they address people's grievances and could become rallying points for political protest, NGOs are seen as potentially threatening by some governments. Indonesia, Malaysia, and others attempt to reign in NGO activities with tight controls and bureaucratic minutiae. Other countries (e.g., South Korea and Taiwan) have more comfortable relationships. Here, more than 20,000 NGOs maintain projects in areas ranging from environmental protection to political freedom and women's rights.

PRIVATIZATION

As development proceeds, demand for public services, infrastructure, and shelter rises. Not only do sizable debts and shortage of revenues constrain governments,

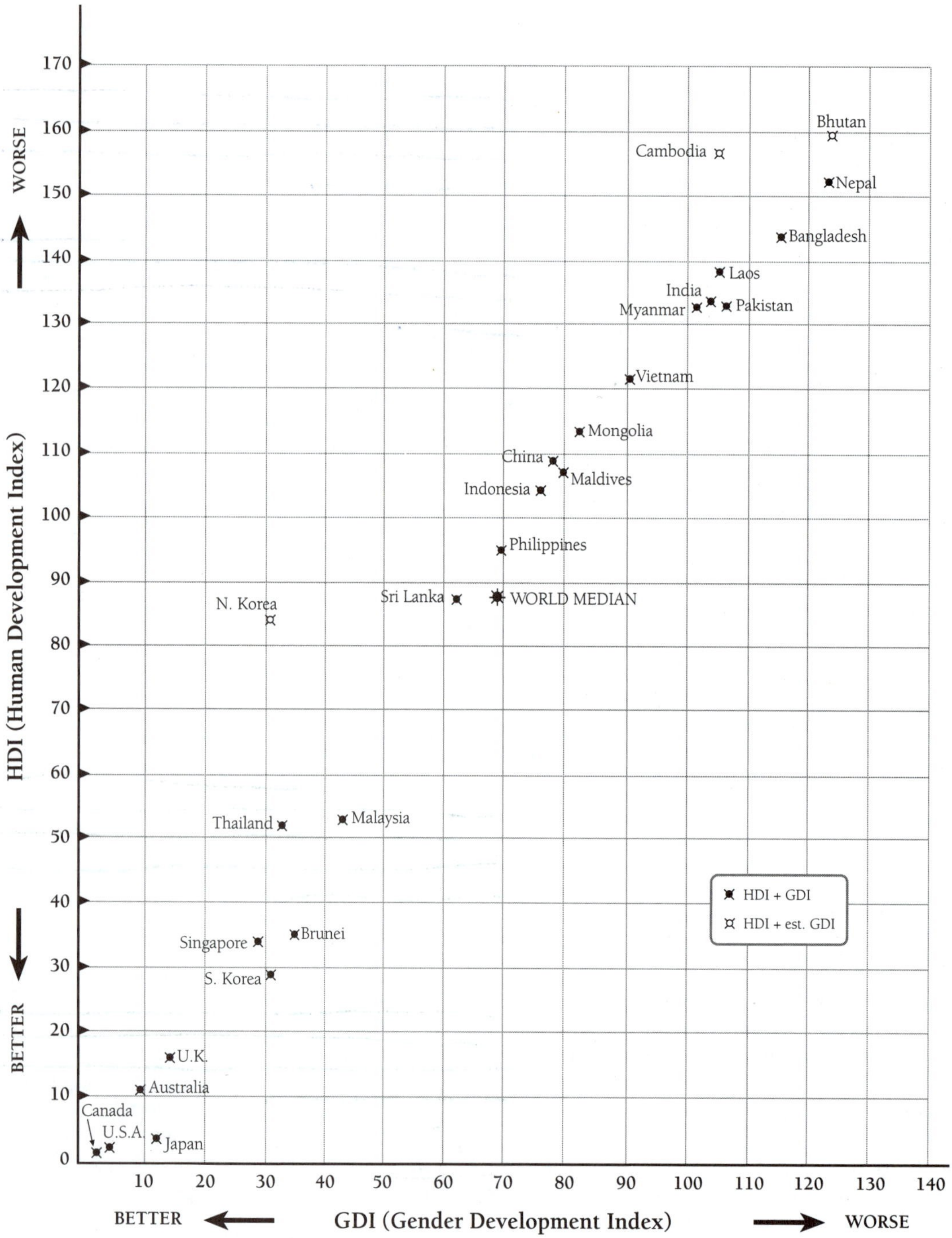

Figure 4-12

The human condition. Combining the Human Development Index with the Gender Development Index gives us a better idea of living conditions than such measures as per capita income or number of TVs per person. Source: *United Nations. 1998.* Human Development Report.

but also corruption, patronage, and mismanagement frequently render state-owned enterprises money losers. Efficiency is precluded by labyrinthine regulations and bureaucracies. For example, it can take as long as 10 years to get a phone installed in Pakistan. Community phones are the rule in most Asian countries, and call completion rates are generally low.

Growing dissatisfaction with central planning and government management of services has promoted *privatization* in various forms. Public assets or enterprises might be sold to private investors, as in Bangladesh. Or as in China, public agencies are *corporatized* and required to cover costs and manage operations more efficiently. More commonly, governments are pursuing policies encouraging private-sector participation in public service provision. Countries such as Thailand, Malaysia, and South Korea are promoting public service delivery by community, religious, or professional NGOs, cooperatives, and small operators. The Catholic church operates schools in the Philippines, and physician's groups have opened clinics in crowded urban areas of Vietnam.

Privatization has affected the airline-passenger industry. State-owned Singapore Airlines and Malaysia Airlines were privatized in 1985, and Philippine Airlines and Thai Airways International were privatized in 1992. Deregulation has encouraged the development of new airlines such as EVA Airways and Asiana. Geographers John Bowen and Thomas Leinbach (1995) note the importance of airlines in the development trajectories of NICs. By increasing access to places, they open up new domestic and international markets. Asian airlines are now playing significant roles in global interchange. Even so, the retention of vested interests by many governments indicates their reluctance to relinquish full control over what they perceive to be a strategic element of infrastructure.

Housing is another sector in which governments have sought private help. Sites and services programs have gained significance since the 1970s. Government agencies install infrastructure and provide essential services on cleared land that is sold with low-interest loans. Poor families contract with private builders or construct their own dwelling with subsidized materials. In India, the government registers and assists housing cooperatives to purchase land and procure financing for low-cost housing. In several countries, private corporations are responsible for the development of complete communities inclusive of residential, commercial, educational, and recreational facilities. Despite these ambitious programs, most shelter for the poor still derives from the informal sector as squatter or temporary housing.

Privatization is not readily accepted everywhere. Individuals and constituencies with vested interests in state controlled enterprises strongly resist privatization efforts. In Indonesia and Malaysia for instance, opposition is voiced by economic nationalists and Muslim fundamentalists, who detest Western consumerism and capitalism. These countries also fear that the already economically dominant Chinese will be the main beneficiaries. In Thailand, opposition comes from the military, which exercises significant control over public corporations to ensure employment for high-ranking retirees. Labor unions and other groups who depend on patronage, paybacks, and similar benefits also oppose privatization. In light of increasing urbanization, the provision of shelter and services will continue to challenge both private and public institutions.

FOREIGN AID RECONSIDERED

As early as 1969, geographer Yi-fu Tuan drew our attention to the fact that Western domination of the world has sparked a "contemporary reaction, one form of which is to elevate cultural difference and cultural pluralism into popular ideologies." The notion of homogenization of landscapes via technological and other imposition from without has retreated in the face of appreciation of differentiation and valuing of local customs. However, Professor Tuan goes on to say that extreme pluralism can take the form of separatism, fueled by resentment rather than pride. Such considerations are pertinent to the purpose and design of foreign aid packages that, to some degree, are being revamped in light of local needs and accomplishments.

The nature of any particular foreign aid package represents first the interests, ambitions, and intentions of donor nations. According to geographers Richard Grant and Jan Nijman (1997), these form the potent content of the relationship between means and ends. Foreign policy in the United States was driven primarily by strategic or security interests prior to the 1990s. In Japan, economic concerns and alliances were the driving forces behind their aid programs.

In 1994, a new foreign assistance act was proposed in the United States. In this, security interests were lowered to a middle priority, while environmental threats and development goals ranked highest. Sustainable development, promotion of democracy and peace, humanitarian concerns, growth through trade and investment, and the advancement of diplomacy were the designated concerns.

Japan has always focused in the Asia-Pacific region, and aid complements direct investment strategies. When the American presence was reduced after the Vietnam War, Japan increased its aid flows to Thailand and the Philippines. In 1992, it announced a foreign aid charter in which it declared the purpose of aid as supporting international interests such as democracy, environmental issues, and the support of women in the development process. It also introduces conditions for recipients based on arms proliferation, human rights, and similar concerns. Countries are targeted according to criteria such as geographic proximity, diplomatic ties, economic importance, and humanitarian needs.

Japan, like the United States, recognizes the significance of environmental and humanitarian considerations, as well as global interdependence. Although the U.S. foreign aid (1994) proposal has not been formally passed, it gives a fresh lining to foreign aid packages and postulates a new mode of thinking about foreign assistance in the twenty-first century.

CHANGING STRUCTURES OF CHANGE

A hallmark of postmodernism is rapid and ongoing change. As views of development have changed, so have tools employed to interpret developmental processes. The **flying geese model** of economic development will illustrate.

The flying geese model, developed by a Japanese economist in the 1930s, was used in the 1960s to describe industrial lifecycles in recently industrialized nations. Industry- and setting-specific at first, the model was advanced to examine changing industrial structures and international shifts in location.

According to the model, an industry initiates with a product import or technology transfer from an industrialized nation. An imitation is domestically produced and substitutes for the original product import. As domestic production exceeds domestic consumption, imports decline and an export surplus is created. Subsequently, exports and production begin to decline. As domestic costs increase, industry looks elsewhere and becomes a net importer again. Industries are thereby cycled in an orderly fashion, akin to the structured pattern of flying geese. Over time, industries rise and fall, with the old being replaced by the new. A country that produced textiles in the 1920s might have focused on steel in the 1940s, chemicals in the 1960s, and automobiles in the 1980s. Each industry is increasingly sophisticated and technologically complex, illustrating the development process.

The "flying geese" analogy can also be applied to a sequence of countries as each moves along its flight path of industrial development. For example, in the 1950s, a labor-intensive manufacturing focus (textiles) characterized Japan as a NIC. This has been replaced by capital-intensive machine industries to the extent that Japan was classified as an industrialized country in the 1980s. From an international perspective, the geese of modernization are seen flying from more developed countries to less developed ones. At the same time, a sophisticated division of labor emerges among Asia-Pacific nations.

The 1990s have witnessed criticism of the flying geese model of development. Perhaps applicable in the past with its vision of Japan leading the Asian pack down the road of progress, the model overemphasizes Japan's role and underplays that of the United States as an export market for Asian goods. Further, it does not account for the fact that imports might be from one's own international subsidiaries or joint ventures, or exports might be dominated by primary (nonmanufactured) products.

Another criticism rests on the fact that industries do not pass *in toto* from one country to another in a linear pattern. Rather, change is manifested through subsector specialization. Each time a new technology is commercialized, a new pattern of clustered economies and division of labor will appear. Furthermore, postmodern era production is interdependent, at once concentrated and dispersed across hierarchies transcending multiple borders. According to economists Bernard and Ravenhill (1995), firms no longer simply seek comparative advantage in location relative to raw materials or labor force for an entire production process. They assign location in terms of "the comparative and technological advantage of *particular* territorial sites for *particular* parts of the production process." Parts of the same product may be produced within the same industry coincidentally in different countries.

As you will see in the following regional chapters, the meaning, practice, patterns, and consequences of development are in constant flux. What does *development* mean to you? How has it changed your community or city? Think about these things, because according to the Chinese philosopher Hsun-tzu (298–238 B.C.), "Through what is near, one understands what is far away."

Main Point Summary

1. The concept of *development* is difficult to define; its meaning varies spatially and temporally. Designations of *development* and measures of *progress* have been derived from Western notions and models. Western ideas have been superimposed on other landscapes, often with undesirable consequences.
2. PCI, a common indicator of level of development, has many shortcomings. The rich–poor gap and other variables are not revealed by this measure.
3. Asians are attempting to define their own culturally appropriate means and modes of development. This is no simple task in light of involvement in global technological and financial networks.
4. FDI has furthered core–periphery development patterns from local to regional and global scales. Multinational corporations have benefited at the expense of environmental and human health in NICs. As cores and peripheries become more integrated internationally, regional economies are structured into dependency relationships. Spatial patterns and networks of power relationships are in constant flux.
5. FDI in Asia has increased dramatically from the 1980s to the 1990s. In the late 1990s, political crises in Indonesia and monetary problems in Japan, Thailand, and elsewhere witnessed falling FDIs. China is of special interest to investors because of the size of its market potential.
6. FDI generates increased incomes and job opportunities but further entrenches local economies in multinational webs of subsidiaries and alien control mechanisms.
7. As a growing middle class of consumers demand quality goods and services, private investment is being sought to counter the hold of foreign-owned enterprises. Japan, Korea, Taiwan, Singapore, and other Asian countries are investing more in Asia. Overseas Chinese are a powerful source of investment in East and Southeast Asia.
8. Most Asian countries have dual economies with formal and informal sectors. Informal activities such as food vending are essential for the survival of millions. Many rely on the hidden, illegal activities of the shadow economy. Drug and cigarette smuggling, child labor, and prostitution are means of profiteering for some and survival for others.
9. Urbanization has skyrocketed, with almost half of the world's 30 largest cities in Asia. Asian cities are part of a hierarchical urban system topped by world cities such as Tokyo and London. This global, urban network is linked through technopoles via cyberspace.
10. Ongoing rural–urban migration and further population increase through NI place demands on city infrastructures and services that cannot be met. FDI in urban projects further attracts migrants and boosts unemployment and underemployment.

11. Megacities and mega-urban regions have emerged to include a disorganized mixture of urban and rural activities. *Kotadesasi* describes the coincidence of urban and rural land use in the same territory. Economic inefficiencies and jurisdictional confusion are problematic.
12. Recent development strategies include trans-border development regions, NGO involvement with the urban and rural poor, government-supported self-help programs, especially in the realm of housing and services, and the implementation of intermediate technologies.
13. Some governments are releasing control over corrupt and inefficient enterprises to private groups. Services such as public transportation and housing provision have benefited significantly from privatization efforts.
14. Western domination has triggered a reaction to homogenization of landscapes and cultures. Greater emphasis is being placed on regional and cultural differentiation, which has the potential to backfire in factionalism. Consumerism and capitalism are being reassessed in context of both traditional and contemporary values and belief systems.
15. In this postmodern era, change is fast-paced and far-reaching. Commonly accepted models of development such as "flying geese" are no longer applicable, given the uncertain complexities of development from local to international scales.

Terms to Know

rich–poor gap
newly industrializing country (NIC)
developing country (DC)
First, Second, and Third World
North–South dichotomy
core–periphery
Foreign Direct Investment (FDI)
spatial duality
bilateral and multilateral aid
tied aid
dual economy
two parts of dual economy
shadow economy
underemployment
megacities
urban agglomerations
mega-urban regions
Kotadesasi
world cities
technopoles
trans-border development region
intermediate technology
Non-Government Organization (NGO)
"flying geese" model

Bibliography

Agnew, John. 1987. Bringing Culture Back In: Overcoming The Economic-Cultural Split In Development Studies. *Journal of Geography* 86: 276–81.

Ayers, Ed. 1996. The Expanding Shadow Economy. *World Watch* 9: 10–23.

Bell, Morag. 1994. Images, Myths and Alternative Geographies of the Third World. In *Human Geography: Society, Space,* and Social Science, ed. Derek Gregory et al., pp. 174–99. Minneapolis: University of Minnesota.

Bernard, Mitchell, and John Ravenhill. 1995. Beyond Product Cycles and Flying Geese: Rationalization, Hierarchy, and the Industrialization of East Asia. *World Politics* 47: 171–209.

Berry, Brian. 1989. Comparative Geographies of the Global Economy: Cultures, Corporations, and the Nation-State. *Economic Geography* 65: 1–18.

Bowen, John T. Jr., and Thomas R. Leinbach. 1995. The State and Liberalization: The Airline Industry in the East Asian NICs. *Annals of the Association of American Geographers* 85: 468–93.

Cho, George. 1995. *Trade, Aid and Global Interdependence.* New York: London.

Douglass, Mike. 1995. Global Interdependence and Urbanization: Planning for the Bangkok Mega-Urban Region. In *The Mega-Urban Regions of Southeast Asia,* ed. T. G. McGee and I. M. Robinson, pp. 45–77. Vancouver: University of British Columbia.

French, Hilary. 1997. When Foreign Investors Pay for Development. *World Watch* 10: 8–17.

Grant, Richard, and Jan Nijman. 1997. Historical Changes in U.S. and Japanese Foreign Aid to the Asia-Pacific Region. *Annals of the Association of American Geographers* 87: 32–51.

Lipton, Michael. 1977. *Why Poor People Stay Poor: Urban Bias in World Development.* Cambridge: Harvard University.

McGee, T. G. 1989. *Urbanisasi* or *Kotadesasi*? Evolving Patterns of Urbanization in Asia. In *Urbanization in Asia: Spatial Dimensions and Policy Issues,* ed. Frank J. Costa, Ashok K. Dutt, Laurence J. Ma, and Allen G. Noble, pp. 93–108. Honolulu: University of Hawaii.

McGinn, Anne, Platt. 1987. Cigarette Traffickers Go Global: The Nicotine Cartel. *World Watch* 10: 18–27.

Quarto, Alfredo. 1993. The Vanishing Mangroves. *Third World Resurgence* 39: 4–6.

Ransom, David. 1996. The Poverty of Aid. *New Internationalist* 285: 7–10.

Robison, Richard, and David Goodman. 1996. The New Rich in Asia: Economic Development, Social Status and Political Consciousness. In *The New Rich in Asia: Mobile Phones, McDonalds and Middle-Class Revolution,* ed. R. Robison and D. Goodman, pp. 1–16. New York: Routledge.

Rondinelli, Dennis A., and John D. Kasarda. 1993. Privatization of Urban Services and Infrastructure in Developing Countries. In *Third World Cities: Problems, Policies, and Prospects,* ed. John D. Kasarda and Allen M. Parnell, pp. 134–60. Newbury Park: Sage.

Seagrave, Sterling. 1995. *Lords of the Rim.* London: Bantam.

Straussfogel, Debra. 1997. Redefining Development as Humane and Sustainable. *Annals of the Association of American Geographers* 87: 280–305.

Tuan, Yi-fu. 1969. Cultural Pluralism and Technology. *The Geographical Review* 79: 269–79.

Tinker, Irene. 1997. *Street Foods: Urban Food and Employment in Developing Countries.* New York: Oxford.

United Nations. 1993. *World Urbanization Prospects: The 1992 Revision.* New York: United Nations.

Weissman, Robert. 1997. Move to Make U.S. Tobacco TNCs Accountable Internationally. *Third World Resurgence* 92: 7–10.

Chapter 5

South Asia: Creating Dilemmas of Diversity

"What is here is nowhere else;
what is not here, is nowhere."

THE MAHABHARATA (C. 1000 B.C.)

This chapter initiates the regional analysis that makes up the rest of this text. In it I will set the historical-geographical framework for subsequent discussions of specific countries in South Asia. For an understanding and appreciation of conditions, events, and swirling controversies in South Asia, historical knowledge is indispensable. Therefore, after a brief review of the physical setting, I will focus on South Asia's history up to and including the creation of India and Pakistan in 1947.

The Physical Setting

South Asia is a realm of spectacular highlands, vast deserts, coarse plateaus, fertile lowlands, and magnificent coastlines. Here mountains and coastlines combine to create one of the world's most sharply defined physiographic regions (Figure 5-1). Splaying southward from the Himalayan wall to the north, the Indian subcontinent is differentiated in the west by Punjab, a region through which flow the Indus River and its four major tributaries: the Jhelum, Chenab, Ravi, and Sutlej. Punjab means "five waters." South of Punjab is the Indus River plain. This entire region is guarded by the Hindu Kush and other mountain ranges to the north and west and the Great Indian or Thar Desert to the east. Eastward, the North Indian Plain extends to include the sacred Ganges River and the amazing Ganges and Brahmaputra delta. Southward, the Central Indian Plateau is separated from the Deccan plateau by mountains and waterways such as the Vindhya range and the Narmada River. These physical features have presented barriers to north–south movements throughout Indian history. Further south is the highly dissected Deccan bordered by the Eastern and Western Ghats. The Deccan is cut by several rivers, most notably the Godavari, Krishna, and Cauvery. The island of Sri Lanka rests at the tip of peninsular India, it too a perplexing and troubled region. You will learn more about these regions in this and the following three chapters.

Figure 5-1

Landforms and waterways of South Asia. Compare this map with Figure 7-2 and note how topography influences settlement patterns.

Monsoon and Life Cycles

As I noted in Chapter 2, a large part of life including harvesting and planting, coordinates with the dry and wet monsoons. This is especially the case in South Asia, where the landscape evolves in a never-ending cycle of brown to green, sparse to lush, and lethargic to energetic. For millions, these cyclical events imitate the greater, overarching cycle of birth, decay, and rebirth. For countless others, they reveal the divine will of Allah or hand of God. Forgotten or improper behaviors and rituals might trigger drought or flood with disastrous consequences.

The great festival of Rathajatra at Puri celebrates northeastern India's transition from the season of drought to the season of rain. A two-story-high vehicle, known as a juggernaut, is constructed to carry a likeness of Jagannath "Lord of the World," a form of Vishnu. To the clashing of cymbals and with a covey of chanting priests, the behemoth is pulled toward the excited throng. As Jagannath nears, people joyously leap and sing and chant "Hari! Hari!" (God! God!). Each creaking and turning of the great wheels speaks of impending change and assures the coming of the wet monsoon. Participation in this and other seasonal events reaffirms rebirth and reasserts one's union with nature.

Archives of Time and Place

Essentially separated from the Sino-Tibetan cultural realm by the Himalaya, the Indo-European culture realm has experienced its major cultural intrusions and contacts via mountain passes through the Pamirs and Hindu Kush from central Asia and via Arab and other sea lanes in the Indian Ocean. Conquerors and adventurers; scholars and writers; soldiers and sailors; merchants and traders; men and women; rich and poor, all were captivated by the richly diverse and fascinating landscapes of what ultimately became the pinnacle of the British Empire—India, Queen Victoria's "Jewel in the Crown."

When the British Empire reached its greatest extent in the nineteenth century, the map of South Asia was quite different. Pakistan and Bangladesh, for instance, did not exist. Actually, when ancient Indian empires were stunning the known world with their spectacular achievements, the British Empire was unheard of because it did not exist. A venture into South Asia's intriguing past is surely worth pursuing.

Geography plays a vital role in history; space and time are twisted together like a two-colored rope. Mountains such as the Himalaya and the Hindu Kush, and deserts such as the Thar (Great Indian), marked defensive frontiers. In the days when territory was traversed on foot or by animal, terrain conditions could serve as barriers or at least make distances impractical to pursue. Control over strategic sites such as river fords and mountain passes could make or break empires. For instance, a dozen or so rulers chose Delhi as their capital because of its strategic situation. Delhi guards the western gateway into the **Doab** (two rivers), the level and fertile land between the Ganges and its main tributary, the Yamuna (Figure 5-2). Much of India's history concerns mountain passes, especially those in the Hindu Kush. The Khyber is one of the most fought over passes in the world (Figure 5-3).

With slow and difficult travel, overextension could easily lead to territorial losses. The Deccan, for example, was a difficult region to conquer and keep, and it typically was subdivided among several rulers. Many an empire or state bordered the Vindhya Mountains, which form the northern flank of the Narmada River, which runs westward into the Gulf of Cambay. Accessibility, or lack of it, played a key role in the relatively separate development of southern India.

Lack of consideration of the surrounding Indian Ocean blinded Indian rulers to the seriousness of maritime encroachment by European powers. Mere trading posts along the shores were not read as clues of impending conquest, as no one had ever conquered India by sea. The sea was seen as the *black water,* and its monsoons would certainly deter even the most daring intruders. Geographic perceptions along with physical

Figure 5-2

This is the Red Fort in Agra, India. Built by the Mughal Emperor Akbar between 1565 and 1573, its red sandstone exhibits Indo-Islamic architecture. It stands guard over the Yamuna River.

Figure 5-3

Here is the Pakistan side of the Khyber Pass. The road was built by the American Corps of Engineers. Note the hairpin turns, switchbacks, and tunnels. I remember traveling this route in a Russian car with a Pakistani driver who careened the road, usually on the wrong side, with his hand constantly on the horn. We had several near misses with huge trucks headed for Afghanistan.

realities were clearly integral to historic developments in South Asia.

Many historians divide the subcontinent's history into Hindu, Islamic, and British periods. However, as landscape expressions of these transcend both time and space, I will use *ancient, medieval,* and *modern* to designate critical shifts in historic evolution. History is a progression through time, which makes necessary the inclusion of dates. Think of these as markers of time, not as something to be memorized. Reference to text and atlas maps will help you envision the upcoming story.

ANCIENT SOUTH ASIA

The discovery of crude stone tools in Punjab shows that human life in India can be dated to the period following the Second Ice Age (400,000–200,000 B.C.). Cultivation of grain commenced around 7000 B.C. in South Asia, but major settlement did not occur until around 3000 B.C., when a notable increase in precipitation and water-control technology allowed large-scale, settled agriculture.

The **Harappan** or **Indus Civilization** emerged as the world's third major civilization after Mesopotamia and Egypt. It is thought that the people were dark-skinned, proto-Dravidians who spoke a precursor of Tamil, now an important language in southern India. With more than 100 settlements spread over a 1,000-mile (1609 km) stretch of the Indus plain, the region also supported three major cities: Harappa, Ganweriwala, and Mohenjo-Daro (Figure 5-4).

When unearthed by archaeologists in the 1920s, the ruins of Harappa and Mohenjo-Daro revealed a phenomenal level of sophistication. Both cities conformed to a general model of ancient city planning found elsewhere. Constructed of baked brick, each walled city had a fortified citadel and a large granary. Oriented to the cardinal directions, streets were grid-patterned. Courtyards shielded two-story, elite dwellings from the smaller quarters of common laborers.

Amazingly, these cities had garbage collection, an elaborate system of running water, and sewage disposal. Elite homes had baths and "toilets," water channels to carry waste to underground sewers and septic tanks. Water was of the essence to the prosperity of these cities and apparently took on ritual/religious significance. The "Great Bath" at Mohenjo-Daro might have been used for ritual ablutions, suggesting the existence of a powerful priesthood. To this day, Hindu temples have rectangular water tanks and ritual washing is important for purification purposes.

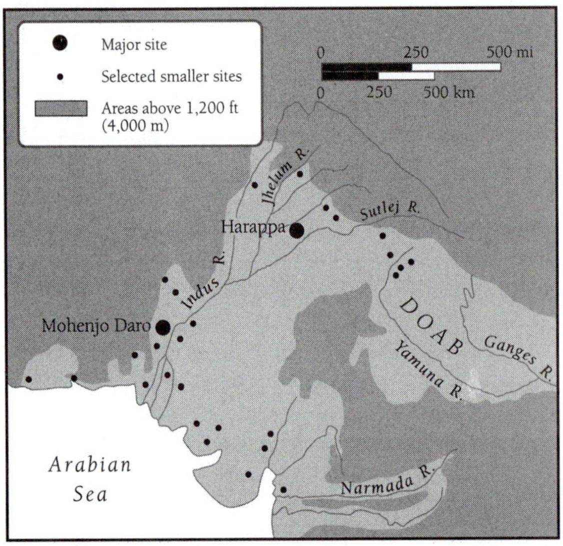

Figure 5-4

Indus Civilization, c. 5,000 B.C.

Religious cults abounded, but the embryos of current beliefs were present. The Bo tree (a type of banyan), under which the Buddha would attain enlightenment centuries later, was sacred then as now. There were solar symbols, such as the wheel of light rays called *su-asti,* "well-being" in Sanskrit. In the twentieth century, this was corrupted and misused by the German Nazis as Swastika, their emblem of Aryan purity. The *su-asti* remains a popular symbol in both Buddhism and Hinduism.

Achievements in urban development imply a sizable agricultural surplus and a well-organized, authoritarian control structure. Harappa alone supported around 40,000 people. Wheat, barley, and rice, along with domesticated sheep, goats, and fowl, plus game provided food surpluses that allowed people to indulge in other pursuits. Some settlements were specialized in certain activities such as mining or pottery. Also, there was trade by land and sea with Mesopotamia. Indian crews navigated the Persian Gulf with the assistance of a "compass-bird"—a crow that would fly directly to the nearest point of land. There is evidence that Indic cotton was a valued trade item. To this day, cotton is India's most important export.

But the Harappans could not control all natural forces. Mohenjo-Daro was rebuilt at least ten times and apparently was abandoned in panic during a severe flood. Recent research shows that despite their mastery of water control, it was climatically changed water balances that were instrumental in the fluctuating fortunes of the Harappan civilization. Steep decline took place from 1800 to 1700 B.C., corresponding with a decrease in rainfall. Water problems and food shortages combined with other crises to drive the **Dravidians** southward.

INVADERS FROM THE NORTHWEST

The Aryans

Around 1500 B.C., lighter-skinned, nomadic herders called **Aryans** appeared out of Central Asia and Afghanistan. The Aryans thrust through the Khyber and other passes of the Hindu Kush onto the plains of the northwest, where another wetter period nurtured pasturage suitable for their kin and kine. It is important to note that while Aryans would have far-reaching influences in South Asian history, from the start they were largely incorporated into existing cultures. In other words, India was not "Aryanized." Rather, it became even more diverse and complex.

These Sanskrit-speaking invaders possessed "Books of Knowledge" called the *Vedas,* which included four categories of texts. Produced around 1200–800 B.C., the mantra provides holy words, and the Brahmana comments on sacrificial rituals. The mantra texts include four collections, the most well known of which is the *Rig-Veda,* the oldest surviving religious literature in the world. The *Upanishad,* philosophical treatises, the *Sutras,* instructions for rituals, and *Shastras,* legal textbooks, were written later, between 800 and 600 B.C.

The *Rig-Veda,* a ten-book collection of 1,028 hymns and 10,600 verses, is rich in detail about Aryan daily life, trials and tribulations, and religious and philosophical views. Almost a quarter of the *Rig-Veda's* hymns are devoted to the rain god Indra, who was thought to be responsible for destruction of the Harappans. Next in importance is the fire god, Agni. There are many other deities, including Varuna, prime mover of the universe.

The Hindu triad of Brahma, Vishnu, and Shiva did not appear until the Vedic pantheon absorbed local popular cults. For example, Brahma, the creator, is, in part, an extraction from Varuna. Accommodation of multiple gods in myriad forms laid the groundwork for faith in reincarnation and transmigration. Since evidence of phallic worship was found at Mohenjo-Daro, it is thought that Shiva, of whom the linga (phallus) is a symbol, also emerged from Harappan cults.

The Vedic god Krishna is interesting because he is dark-skinned (often blue). A divine herdsman and lover of milk-maids, he was a hero of the dark-skinned Dravidians. His earliest cult was in south and central India. The *Mahabarata,* the world's longest poem, chronicles Krishna's battle defeat of Indra, champion of the fair-skinned Aryan invaders.

Vedic literature traces the southern and eastern Aryan movements. Other war stories take place in the Doab of the Yamuna and the Ganges, indicating that primeval forests were giving way to sedentary agriculture. The first capital structure at Delhi (called Indraprastha for Lord Indra), and the earliest foundations of Varanasi (Benares) are examples of late-Vedic and post-Vedic developments.

In a later work, the *Ramayana* (the Story of Rama), the action takes place at Ayodhya, 300 miles (483 km) east of Delhi. Rama's faithful wife Sita is carried off by Ravana, the demon-king of Lanka (Sri-Lanka). Assisted by Hanuman, the monkey-general, Rama defeats Ravana and rescues Sita. However, Rama refused to take his wife back before she proved her chastity by ordeal of fire (look ahead to the box entitled "Women: Subjugation and Sati").

As a woman, Sita was suspect. The fire god Indra took her to Rama but years later, doubt again was cast on her virtue, and once more she was to face the fire. This time her Mother, Earth Goddess, took her away from these foolish, suspicious men. Virtuous and subservient Sita has long stood as a model for the ideal Indian woman: loyal, dutiful, and compliant. Indian mothers often say "Remember Sita" when goading their daughters into "proper" behavior.

Historian Stanley Wolpert (1991) believes that the *Ramayana* can be read as an allegory of how Aryans perceived their conquest of the "uncivilized demons" who inhabited southern forested regions. These stories remain popular, inscribed in monumental bas reliefs, recounted in plays and movies, and dramatized on television (Figure 5-5).

The Aryans brought cows and horses to India. In fact, cows were an early form of wealth and currency. Eventually these would become sacred in Hinduism. Apparently, horses were employed to pull chariots—the basis of cavalry warfare, which spread quickly throughout the continent. Horses made possible the expansion of empires, fostered the militarization of states, and enabled more expedient collection of taxes in kind. As horse-breeding in India was difficult, they were brought from Arabia and Persia at great expense.

While copper and bronze were already in use, the discovery of iron near Patna, combined with Aryan knowledge of iron-working, were instrumental in territorial conquests. Used for weapons, tools, and plows, iron facilitated both forest removal and agricultural settlement.

Figure 5-5

This wall painting in Udaipur, India, is of Hanuman, the monkey hero of the Ramayana. A devotee of Lord Rama, he brings success to armies.

Settlement along the Ganges had been limited to the Doab and the northern foothills and tributary valley, possibly because rivers were more easily crossed there and there were dense jungles in the lowlands. Eventually another dry period narrowed the rivers and the jungle receded. Iron implements furthered settlement on the Ganges plain, and forest dwellers were either conquered or absorbed. In the *Mahabarata,* these non-Aryan forest dwellers were called *nagas* or "serpent people."

Unlike the Harappans, the Aryans did not have a script for their spoken word. For centuries, *Ramayana* and other stories were memorized and passed on verbally. A script was not developed until the fourth century B.C., but religious taboo prevented the writing of religious tomes. Nevertheless, Sanskrit is rich in pejorative terms, and these were willingly used to denigrate non-Aryan peoples. Many were described as physical freaks, and these images became part of Medieval European perceptions of the Far East.

Seeds of Caste Division

The Aryans, whose own name means "noble" or "first born," used the term *dasa* for people with darker complexions. Initially, the term meant "enemy" and later, "subject." *Dasa* always implied dark-skinned people, and the preference for lighter skin persists. Even today, marriage ads often list "wheaten complexion" as a desirable characteristic.

Social norms including social hierarchies were clarified in the *Shastra* legal texts. These were the "Laws of Manu," Manu being the first man. Vedic societies became organized around four classes of people, or **varnas,** which means "color." The four classes were ***brahmins*** (priests); ***kshatriyas*** (warriors); ***vaishyas*** (peasant farmers); and ***sudras*** (serfs). Subjugated people, including the darker Dravidians, could only enter the system at the lowest level. Those who had closest contact with earth, fire, or water, such as artisans, smiths, and millers, were despised. The most skilled artisans were Dravidian. Sudras were excluded from ritual affairs because they were seen as impure, and ritual purity was of the essence. This was the precursor of the ***caste system,*** which still endures. The English word *caste* comes from the Portuguese word *casta,* used to designate people of different colors in Brazil.

By the sixth century B.C., regional rivalries had produced a string of states across the northern plains. A sophisticated commercial sector evolved, with all the necessary trappings such as coinage, taxes, and bureaucrats. Prosperity fostered economic specialization and *varna* subdivisions, known as *jatis,* multiplied.

A ***jati*** is typically defined and named by occupation. It is also an extended kinship group that practices group **endogamy.** This practice of marriage within the jati ensures continuity of its principles and practices. Caste and *jati* are at once divisive and unifying. While fragmenting society at large, they provide structured mutual exchange and support systems at communal levels.

Within a growing mercantile class, the *varnas* had to be further modified. Commercial dealers were taken into the *vaishya* class and farmers were demoted to the *sudra* class. Serfs and other low-order *sudras* became casteless (i.e., outcasts whose designation ultimately hardened into **untouchables**). Today, many of these *jatis* prefer the term ***dalits,*** meaning the "oppressed."

After the establishment of Aryan supremacy in the Indus region and across the northern plains, an Indo-Aryan culture emerged. Large numbers of Dravidians were driven southward across the Deccan. There, they developed a very distinct civilization with their own languages. In fact, Dravidian languages are fundamentally different from Indo-Aryan languages. This difference has induced linguistic strife even into the twentieth century.

Persians and Greeks

India's cultural complexity was further enhanced by a continuing stream of intruders from Central Asia. In the sixth century B.C., the armies of the Persian king Cyrus the Great crossed the Hindu Kush into the western fringes of the Indus valley. Later, Alexander the Great of Greece battled his way from the Khyber Pass into Punjab and crossed the Jhelum in 326 B.C. He pushed on as far as the Beas, where his exhausted troops mutinied and forced his return to Persia. When the Greeks occupied this northwestern part of the subcontinent, many were absorbed into the local population.

BUDDHA AND MAHAVIRA

Two major religious events occurred in the fourth and fifth centuries. First, Gautama Buddha preached the paths and precepts of Buddhism across northern India. Buddha's "Middle Way" was appealing as it lay between the extremes of worldliness and asceticism. He founded *sanghas* (communities for monks) and *viharas* (wet-season refuges). Buddha himself was not worshipped for another hundred years, and the first sculptures of him did not appear until the Christian era.

At first, his followers carved simple cells and halls out of mountain faces, and Indian stone carvers ultimately excelled in their craft. Many carvings and wall paintings still survive in the wake of Buddhism's diffusion throughout Asia. Later, lands were provided for monasteries, and a monastic code was devised. This code and the Buddhist scriptures were recorded in a script called Pali. Even today, ancient Pali prayer scrolls can be seen in Buddhist monasteries. After his death in 483 B.C., Buddha's relics were distributed and encased in structures called *stupas* (Figure 5-6).

Figure 5-6

This awesome stupa is Swayambunath in Kathmandu, Nepal. Sacred relics are confined inside the stupa, from which the eyes of Buddha see all things. Devotees move around the stupa in a clockwise direction, chanting and twirling their prayer wheels on the way.

The second event was the founding of **Jainism** by **Mahavira.** Both Buddha and Mahavira were *kshatriyas,* both disdained social class distinctions, and both addressed the fears of never-ending rebirths. Mahavira was from the region of Bihar. He became a wandering ascetic at the age of 30. He preached in the Ganges valley until he died c. 612 B.C. In Jainism, the Buddhist injunction against killing became all inclusive: people, animals, insects, and even certain plants. Jainism was nonmaterialistic, but since at first it was only property holding that was forbidden, the religion became popular with the mercantile class. Cultivators and soldiers obviously failed to meet the nonviolence standards. Concentrated mainly in the state of Gujarat, Jains have managed to retain an unbroken tradition in India.

THE MAURYAN EMPIRE

Out of a period of uncertainty that followed the Greek withdrawal arose a new leader named Chandragupta Maurya. By the end of the fourth century B.C., he had acquired an empire stretching from the Hindu Kush to the Bay of Bengal. This was the first time the political gap between northern India and southern India had been bridged. While not controlling the entire subcontinent, the *Mauryas* did control regions of the Deccan, important trade routes, and coastal regions. Great mercantile wealth was accumulated (Figure 5-7).

The Mauryan Empire is regarded as the first expression of the political notion of a state on the subcontinent. It was highly centralized, and many economic activities such as mining and land clearance were state monopolies. The fact that death was imposed for defrauding revenues or killing an elephant (a capital asset) suggests that justice was harsh.

Cities had been built from Harappan times onward, many as capitals of various kingdoms and republics in the Ganges valley. The legacy of these remains in the fact that most of this region's largest cities originated as ancient capitals. Exemplifying the power of the Mauryan empire was the city of Pataliputra (Patna in Bihar). The city wall had 64 gates and 570 defensive towers. The brick-structured city was moated and protected on two flanks by the Ganges and a tributary. A tree-lined Royal Road, posted with distances and directions, and dotted with wells and inns, linked the Mauryan frontiers with the capital.

The most memorable Mauryan ruler was **Asoka,** who ruled from c. 268–233 B.C., the first fairly well-documented period in Indian history. Asoka is remembered not only because of the size of his empire, but also because he adopted Buddhism as the state religion. He sent Buddhist missionaries, including his own son and daughter, to Sri Lanka. Asoka became a vegetarian. However, meat, including beef, was still available in the markets. The cow was not yet considered sacred. Asoka established hospitals, animal shelters, wells, and water tanks. His ruling principles and moral doctrines were inscribed on rocks and sandstone pillars around the country. The capital of the 40-foot (12 m) pillar of Samath, Gujarat is a magnificent carving of four back-to-back lions. This official seal of Asoka was adopted as India's national emblem.

THE GUPTAS AND INDIA'S GOLDEN AGE

When Asoka died, political conditions became chaotic. New Aryan migrants such as the Kushans, arrived from the northwest. The Kushans put their capital at Peshawar, gateway to the Khyber Pass. Around 320 A.D., northern India became united under the Guptas, under whom art and Sanskrit literature flowered. Derived from ancient sources, the *Puranas* were shaped and polished by the **Guptas.** These valued historic documents include Hindu myths, philosophical dialogues, ritual prescriptions, and genealogies. Art, science, and mathematics flourished, as well. The Gupta period marked **India's Golden Age** (Figure 5-8).

Under the Guptas, great strides were taken in mathematics. Numerals had been carved on some of Asoka's pillars, and these were further developed with the concepts of zero and nine numbers. The Arabs, who introduced them to the West, called them Indisa (Indian numbers). Europeans called them Arabic numerals. Aryabhata was the first to solve some basic problems of astronomy when he calculated pi to 3.1416 and the length of the solar year as 365.358605 days. He knew that the earth was round and that

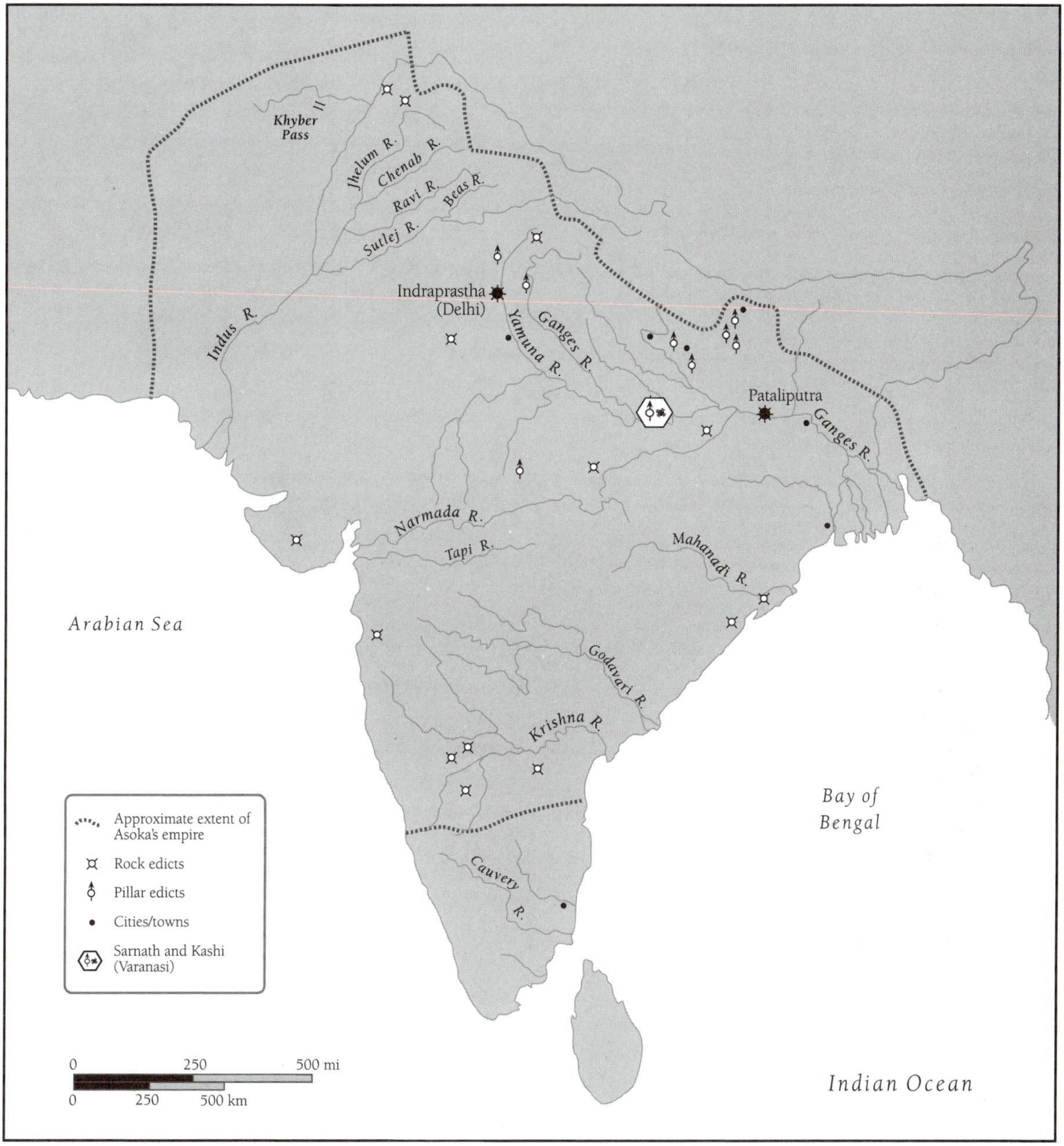

Figure 5-7

South Asia in the third century B.C. *The significance of the Indo-Gangetic Plain in Asoka's empire is unquestionable.*

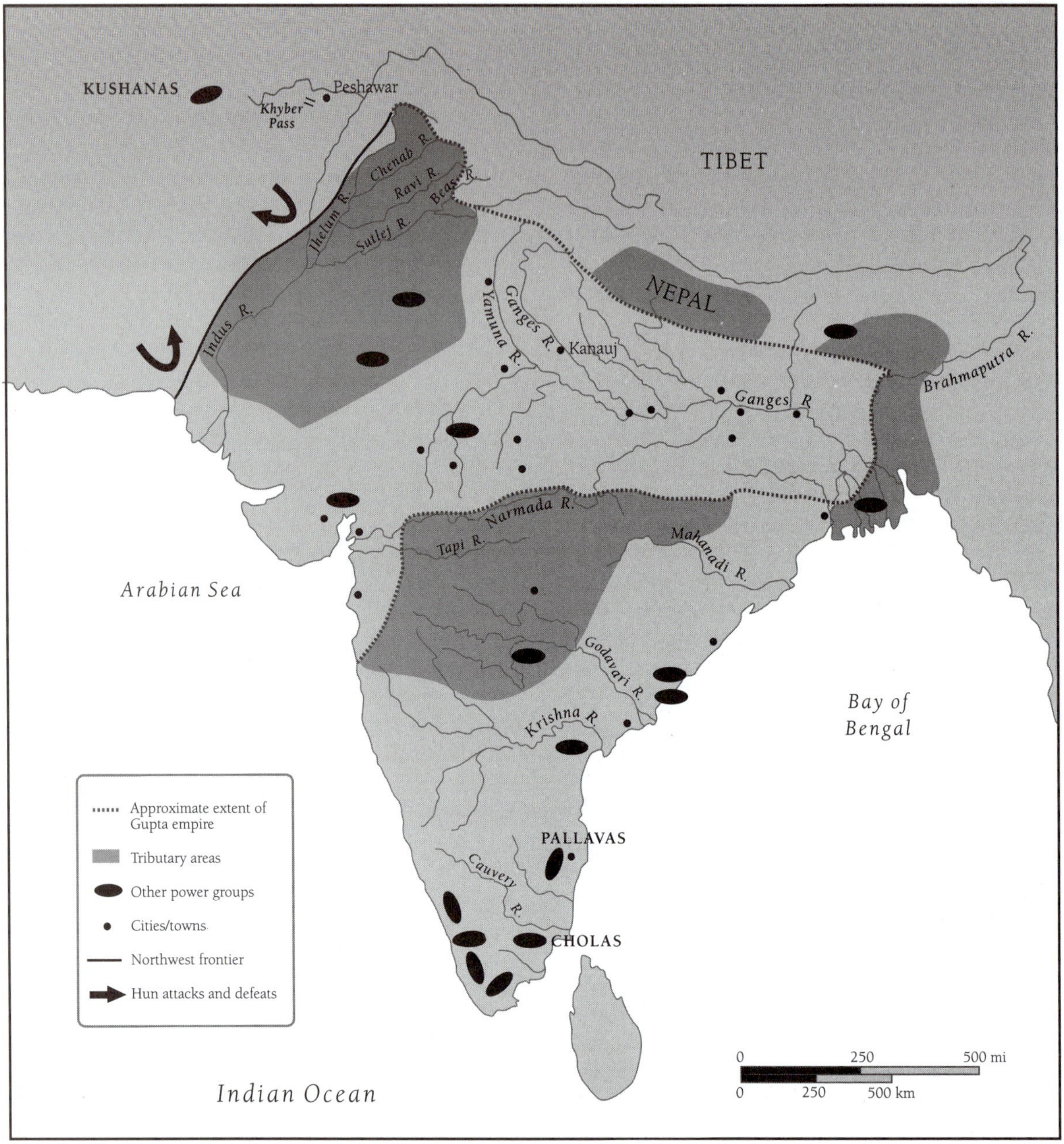

Figure 5-8

South Asia, c. 550 A.D. While the Indo-Gangetic Plain remained important in the north, river and coastal regions fostered several powerful communities in the south.

it rotated on its axis. He knew that an eclipse was caused by the moon's shadow blacking out the earth. And all this was determined in India at least 1,500 years ago!

Fa-hsien, the first of three great Chinese Buddhist pilgrims to visit India, journeyed from China to India and Sri Lanka (399–412). He recorded that although the Gupta kings were Hindus, Buddhism flourished in the countryside. He commented on the well-being of the people and the fairness of the justice system. "Even those who plot treason only have their right hands cut off." He also noted that when certain people appeared, pieces of wood were struck to ensure avoidance of their evil presence. Fa-hsien had witnessed the caste system in action.

SCOURGE OF THE HUNS

The Gupta empire faded with the onslaught of Atilla's Huns via the Hindu Kush. This disrupted important international trade routes in the northwest, thereby diminishing one of the Gupta's key financial sources. A subsequent attack gave the Huns dominance over a region extending into Kashmir, Rajasthan, and into Uttar Pradesh. After the fall of the Guptas, the center of activity moved from Patna to the Delhi-Agra region. Northwest India was once again in the purview of Central Asia.

In the middle of the sixth century A.D., the Huns were defeated by Indian troops. And when they were crushed by the Turks in Central Asia, their day was done. However, there were significant consequences of their short-lived rule. City razing meant that north Indian urban culture was destroyed. Buddhists (and others) were slaughtered and their monasteries sacked. Buddhism in northern India never completely recovered from this terrible scourge.

After the fall of the Guptas, Bengal lasted for a time as a center for Buddhist expression and learning. Benares (Varanasi) was the key center of worship and scholarship was pursued by more than 10,000 monks at the Nalanda monastery. But with successive Muslim purges, Buddhism ultimately faded from its land of birth. Survivors fled to Nepal and Tibet.

India's Middle Ages

North India was the focal point of Indian history until the beginning of the sixth century A.D. Then, regional kingdoms became dominant, only to be reabsorbed after being toppled by more powerful ones. During this period, central and southern India rose into prominence, and events occurred that would alter the geography of the entire Asian region. You will learn about Indian exploits in Southeast Asia in later chapters.

The Chinese pilgrim **Hsuan-tsang** spent 13 years in India (630–643) and reported to the domain of **King Harsha,** whose empire spread from the Punjab to Orissa and as far south as the Narmada River. Harsha was the last north Indian ruler to encroach south of the Narmada for 600 years. When Hsuan-tsang returned to China, he brought 20 horses carrying 657 Buddhist texts and 150 relics. He personally translated 74 of these texts, furthering the spread of Buddhism in China.

Harsha had shifted the capital westward to Kanauj in the Doab. This meant that eastern India was then more able to pursue its own path. History shows that the eastern regions of Bengal and Orissa frequently enjoyed freedom from larger empires to the west. Buddhism had taken a new turn in Bengal, incorporating magical and Tantric practices. This form of the faith diffused from Bengal to Tibet, where it fused with Mongolian adaptation of Mahayana Buddhism.

Harsha's western presence brought attention to influential western tribes such as the **Rajputs** of Rajasthan, who had settled around Jodpur. As *kshatriyas,* the Rajputs were highly militarized and exalted chivalry (Figure 5-9). They were also noted for their patronage of the arts, including the famous temples of Khajuraho.

The Rajputs honored and protected women and even educated some. At the same time they promoted child marriage and female infanticide. Being married to a military chief was a dangerous affair because wives were expected to die with their husbands on or off the battlefield. **Sati,** being burned alive on a husband's funeral pyre, was enforced. A low-class wife had the option of having her head shaved by a male untouchable, eating one meal a day, and enduring the degradation of widowhood. These practices diffused throughout India.

Figure 5-9

In Jaipur, India, these Rajputs are sword makers. This occupation has been passed down through the family for generations.

"Funeral flames dispel suspicion, honour lives when woman dies!" The Ramayana

Woman: Subjugation and Sati

Born into a male-dominated world, girls, if they survived the common practice of female infanticide, were worthy only as breeders of sons. No matter their caste, women were linked philosophically with Sudras, the lowest caste group. Morally and intellectually, they were seen as weak. Chastity, loyalty, submission, and subservience were the watchwords of a woman's life. The essence of a woman's being was her relationship with her husband and his family. Without a husband, there was no point to her existence.

Child marriages were common. In the last census of the nineteenth century, there were 10,000 widows under the age of four, and over 50,000 between the ages of five and nine in and around Calcutta alone. In marriage, a girl "took away" wealth from her own family and was "gifted away" to the groom's household, where she was expected to be selfless and sacrificial (literally). A properly dutiful wife was expected to climb willingly onto the funeral pyre of her husband and burn alive to purge the sins of the entire family. A pregnant woman was expected to burn on a separate pyre after giving birth. With this act of courage and devotion, a wife would earn the title of sati and be venerated and honored forever. "Suttee" is the English version of sati.

Sati comes from the word *sat* meaning "truth." A sati was a woman "true to her ideals." Since a woman's only value was to serve her husband, his death eliminated justification for her living. Those who got away with widowhood suffered countless indignities and lived the remainder of their lives in depredation and disgrace. Remarriage was forbidden to higher castes.

The origins of sati are unclear, although references to such an act are made in some of the post-Vedic scriptures such as the *Mahabarata*. Increasingly prevalent after 400 A.D., actual dictates appeared around 700. By that time the life of a woman apart from her husband was declared sinful.

Sati soon became a "religious duty" for *kshatriyas* like princes and warriors, but soon spread to other castes. Although a wife was supposed to self-immolate voluntarily, most had to be forced onto the pyre. In situations where there was a harem, all the wives would burn, commit mass suicide, or be killed. Some were buried alive.

Initially, brahmin women were forbidden from seeking this sort of salvation. A Brahmin sati would be guilty of suicide and would therefore disgrace herself and her husband. However, eventually they too adopted the practice.

In the eighth century, Muslim rulers attempted to check sati but were unsuccessful. Islam had already introduced the customs of veiling and **purdah**—the complete seclusion of women. By the 1200s, Muslim male and female social lives were completely separate. While the Hindu Laws of Manu instructed women to worship their husbands as gods, the Koran sanctioned similar behavior, and added the practice of polygamy. Hindu and Muslim customs regarding women became intertwined, especially in the north.

The British, who came to India in 1608, did nothing about sati for 200 years. From 1815 to 1828, there were close to 8,000 (reported) cases in Bengal alone. The British finally banned it in 1829. However, the practice continued in most regions, including its stronghold in Rajputana (Rajasthan).

Wife-suicide or killing was not unique to India. It was practiced in China, Egypt, Russia, and elsewhere. Look to Chapter 6 for more on this topic.

The caste system was well entrenched in this era. Caste taboos meant that scavengers and butchers were confined to mud quarters beyond the city wall. Moreover, they were compelled to walk only on the left side of the road.

Southern India was not entirely isolated from activities in the north. Cultural influences diffused from the northwest southward via the western highlands. Many Sanskrit words were incorporated into Dravidian languages. Contact with the Mauryas brought new ideas of statehood. Brahmin families were instrumental in the spread of Hinduism. Translation of sacred texts and orthodox practices strengthened the hierarchical structuring of society. Oppression of low castes and untouchables was frequently intense.

INVASIONS AND EMPIRES

The history of southern India is a fractious one, with various groups vying for control of the Godavari and Krishna River lowlands as well as prominent upland locations. The geographer Ptolemy, who visited around 150 A.D., mentioned capital cities existing in both regions. Maritime trade was very important. Such items as pearls, precious stones, and pepper were traded with the Roman Empire from the Malabar Coast. Coromandel ports, which served as entrepots for oil from the Ganges plain and silk from China, eventually eclipsed the Malabar. Later, many of its ports such as Poduka (Pondicherry) and Supatma (Madras/Chenai) fell into European hands.

In the south, the first group to extend political control beyond their core area were the **Pallavas.** Although Hindus, they supported Buddhism. They were noted for carving sculptures out of solid rock, and their steep-pyramid stone temple carving style eventually found its way to Java at Borobudur.

The Cholas took over from the Pallavas in south India near the end of the tenth century. Their state expanded from their core in the Kaveri delta and was based on power over people rather than territory. Agricultural settlements were linked to commercial towns. These served as redistribution centers for both local and external products. Most important were the merchants who formed an elite group dominating national and international trade. Such functional Hindu social and economic hierarchies promoted increased specialization, greater numbers of noncultivators, and expansion of empires.

Maritime power was of utmost importance to the Cholas, who captured India's southwest coast and extended their hegemony around the Bay of Bengal to the mouth of the Ganges. They conquered Sri Lanka and took the Maldives. The Andamans were taken in 1017 by the powerful King Rajendra.

In 1020 Rajendra launched an expedition to Sumatra and Malaya. There his army defeated the mighty Shrivijayan empire and gained control over the Sunda and Malacca (Melaka) Straits. The Chinese had sent envoys to the "Countries in the South" in the tenth century, expressing their interest in increased trade. Rajendra sent several delegations and embassies and in 1033, China recognized the Chola kingdom as one of its "tributary states" (a definite honor). Consequently, other Southeast Asian states sought connections with the Cholas.

Eighty years after the death of Muhammad in 632, an Arab empire extended from the Pyrenees in Spain to the Indus valley. The Arabs initiated their territorial gains in India when they established themselves in Sind at the mouth of the Indus in 712. Gujarati rulers tried to fend off conquest, in part by welcoming Zoroastrians who fled Persia and formed India's Parsi community, now concentrated in Mumbai. Parsis, for whom fire is the essence of purity, would later make significant contributions in the finance, trade, industry, and the arts. It is thought that the Persian water wheel was introduced at this time (Figure 5-10). The Arabs also introduced paper and gunpowder. By 871, they had established independent dynasties in the Sind and the Punjab and in 1000, a holocaust began.

Figure 5-10

This animal-powered Persian water wheel in Rajasthan, India, is evidence of Persian influences.

Mahmud (971–1030) of Ghazni (in Afghanistan), son of a Turkish slave, plundered India with 17 invasions over the next quarter century. Called "The Sword of Islam," Mahmud and his armies swarmed out of the Khyber with the dry monsoon. After devastating everything in their path, they sped back to Afghanistan before the wet monsoon could make the Punjab Rivers impossible to ford. In 1023, they sacked the temple at Sarnath and slaughtered 50,000 Hindus. Interestingly, Mahmud was never interested in acquiring territory. For him, India was merely a treasure trove.

Why were these invasions so successful? First, fragmentation in terms of social and political divisions precluded any united front against them. The Cholas were far from the fray. The Rajputs with their independent group structure were never able to coordinate their efforts, despite their strong fighting abilities. The caste system was another hindrance to correlative action. In contrast, anyone could join Muslim armies. With ability and loyalty, even slaves could rise to top positions. Second, Indian armies fought with relatively cumbersome elephants. Muslim horse-archers could easily outrun them. With a rapid charge, they would suddenly swerve their horses sideways, release a hail of arrows, and race away.

Did you know that the game of chess was probably derived from elephant warfare?

Elephants and Chess

Around 500 B.C., the elephant arrived on the Indian battlefield. Elephant warfare strategy is faithfully represented in the game of chess, probably invented in India in the sixth and seventh century A.D.. The king directs the battle from atop his elephant. His moves are restricted to protect him because if he is killed, the army is vanquished. Generals, cavalry, and runners conduct the battle. Infantrymen (pawns), primitively armed, slow, and untrained, are valued only for their numbers. Flanks are guarded by elephants, who take front-line positions in the last stages of the battle.

The game diffused via Persia to Europe. *Chess* is thought to be derived from *shah,* the Persian word for king. *Shah mat* approximates "checkmate," meaning "the king is dead."

This military strategy remained in place for 2,000 years, reaching its pinnacle under the Mughuls in the sixteenth century. By that time, firearms and artillery, brought from Europe by the Turks, had been introduced into India. The combination of raining arrows from mobile horse-archers, bullets from musket-men sharpshooters, and explosive barrages from ranks of artillery brought the era of elephant warfare to a close.

In the twelfth century, Arab motives changed. Muhammad the Ghur began a series of conquests for territory. By 1202 he had reached Bengal. Although he did not get Kashmir nor was he successful in annexing Assam, he did occupy northern India. In 1206, he was murdered near the Indus.

A few years later in 1221, Chingis Khan and his Mongol hordes reached the Indus. He left troops in Punjab who harassed the Delhi Sultanate through the thirteenth century.

After 1000 A.D., people fraught with fear in the wake of terror turned increasingly to faith for solace and reprieve. In the more peaceful south, the *bhakti* (devotion) movement arose. Stressing love of God and rejection of sectarian causes, it spread to the north. By the fourteenth century, it had amassed a huge following of both Hindus and Muslims.

THE DELHI SULTANATES

The Arab conquerors inaugurated Delhi as their capital in 1229 and for nearly 90 years struggled to maintain control of northern India. The second sultan, Raziyyat, only lasted for three years. A chronicler of the time described her as being a wise ruler and competent military leader. "But of what use were these qualities to her as fate had denied her the favour of being born as a man?" Raziyyat was ultimately deposed and executed.

After defeating the Mongols in 1279, the notoriously cruel Balban had thousands of Mongol prisoners trampled to death by elephants as entertainment for the Delhi court. Afterward, the victims' heads were piled as a pyramid outside the city gate. Balban had copied the standard Mongol style of revenge.

In 1309, Balban's son Alauddin invaded southern India. The loot, supposedly including the famous Koh-i-noor diamond, was hauled back by 1,000 camels. Another profitable campaign required 612 elephants to carry plundered treasure.

Alauddin, who ruled until his death in 1316, understood the principle of distance decay. He knew that his policies would be closely followed only within a hundred-mile radius from Delhi; beyond that, his control would increasingly diminish. Nevertheless, he instituted strict reforms to be carried out with the aid of spies. He confiscated officials' landed property, assessed farmers of half their standing crop, and taxed pasture land. He fixed prices and forbade alcohol. He built large grain storage facilities and forced transport workers to settle at specified distances along the Yamuna to ensure the safe passage of grain to Delhi. All this plus loot and tribute payments from subject kings supported a very large army. Alauddin's efforts were the only measured attempt by a medieval ruler to centralize authority and interfere with market forces.

During Alauddin's rule, society became even more partitioned along religious lines. Cities and commerce were predominantly Muslim. Entire castes of artisans converted to Islam to escape their low- or no-caste status. By and large, farmers were Hindus, represented by their village headmen. They could not accumulate property as a consequence of Alauddin's belief that "the Hindus will never become submissive and obedient till they are reduced to poverty."

In traumatic times of cruelty and suffering following Alauddin's death, the Delhi Sultanate began to crumble. Madurai declared independence in 1334 and Bengal in 1338. The Sultanate ceased to exist in 1398, when Timur (Tamerlane) exterminated the entire Hindu population en route and in Delhi. Various clans rose to power, and Agra was built as the capital for Sikander Lodi, ruler of the central region. Meanwhile, Timur ravaged the Himalayan foothills.

During the tenure of the Delhi Sultanates and thereafter, many Arab Muslims married Indian women. The ruling class accumulated harems of inordinate size. Furthermore, they promoted the practice of purdah. In many countryside villages in northern India, even Hindu women adopted this practice of veiling and remaining secluded from nonfamiliar males and strangers in general.

Other changes occurred as well. The Delhi court favored the sophistications and fineries of Persian culture. Persian was the official language, influencing Sanskrit and maintaining its presence even in some parts of the south into the nineteenth century.

Persianization

For an 800-year period, Persian culture dominated elite culture in north and northwestern India as well as in other Muslim-controlled areas. The Persian connection was an ancient one. However, the process of Persianization was initiated with the Turkish invasions of the eleventh century and furthered by subsequent immigration of large numbers of Persian soldiers, administrators, scholars, and others into India. Urdu, cultivated with Islamic orientations and written in a Persianized, Arabic script, evolved as the national language of Indian Muslims.

Even the Mughals, the last Indian court of empire, retained Persian dress, literary, artistic, and architectural styles, and the Persian language. Without the later introduction of English by the British, Persian could well have remained the language of political and cultural elites in South Asia.

Elsewhere, Hindu kingdoms such as Orissa in the east and Vijayanagar (City of Victory) in the south preserved Hindu institutions and customs (Figure 5-11). The Gajapatis (Lords of the Elephants) of Orissa controlled the east coast from the mouth of the Ganges to the Godaveri from the thirteenth century. In a flurry of temple building, they created a complex of more than 500 lakeside shrines at Bubaneshwar, the fabulous temple to the sun god Surya at Konarak, and the famous Jagannath temple at Puri. In Vijayanagar, Telugu-speaking Brahmins commanded fortresses, veritable strongholds in the highlands. Patronage and prestige fostered loyalty. Moreover, armies now had firearms to enhance their arsenal. These were highly militarized realms reliant on unswerving loyalties of local *rajas* (rulers) and religious personnel.

Two significant events took place in the sixteenth century. First, the Portuguese appeared in Vijayanagar's empire, where they destroyed Hindu temples at Goa and Madras. Ultimately a peace treaty was signed and Vijayanagar agreed to purchase horses shipped from the Persian Gulf region. Goa fell to Portuguese control in 1510. Second, the Mughals rose to power and ushered in India's modern era.

India's medieval period ended with the arrival of the Mughals, who introduced the modern territorial state. Until this time, India was more a collection of structures than of territories. Cities, towns, villages, and temples; rulers, soldiers, and priests; merchants and farmers; all were ensconced in not always well-coordinated networks. There was a large degree of local autonomy. All this was to change in the era of Mughal authority and British hegemony.

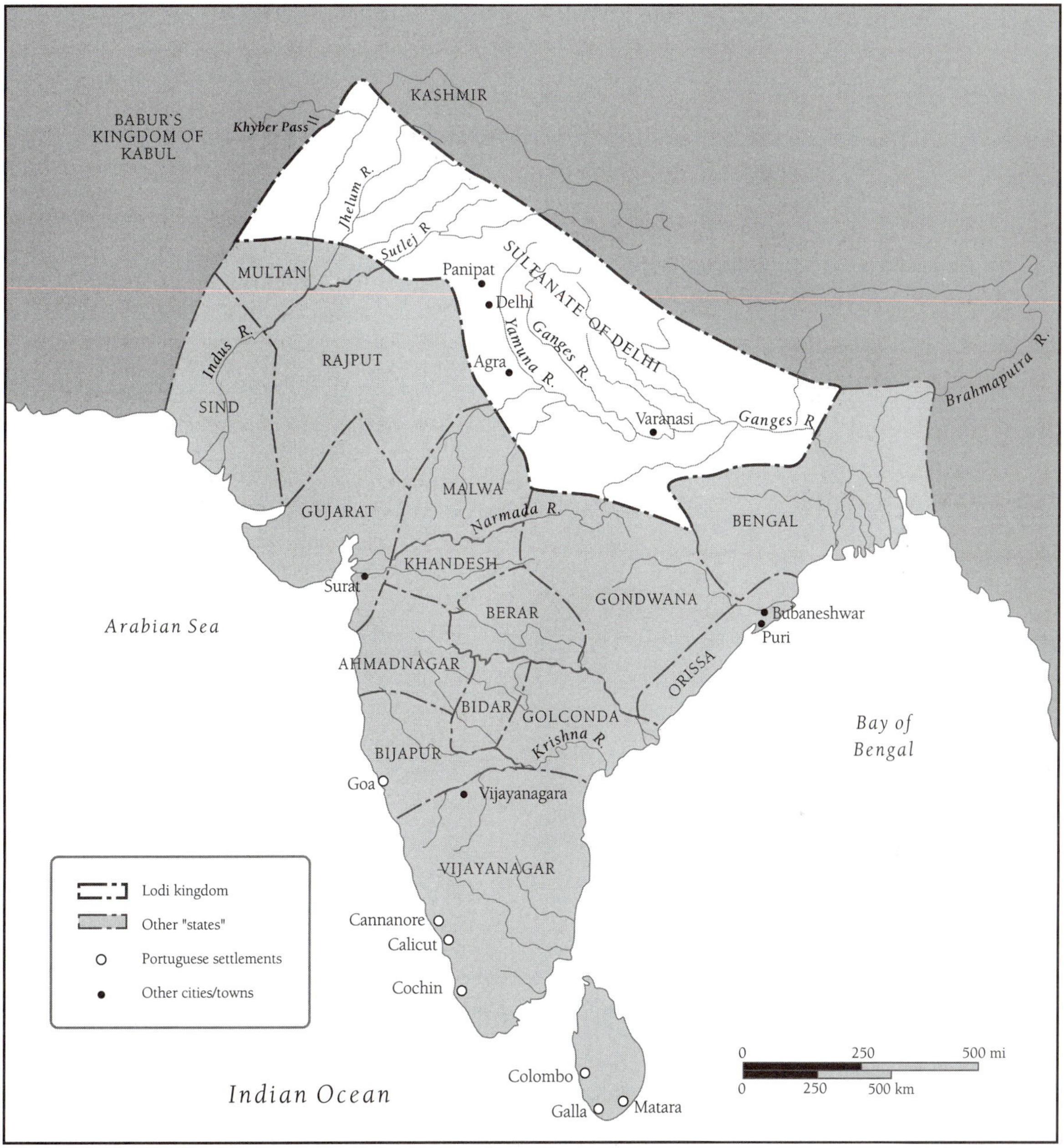

Figure 5-11

South Asia, c. 1525. While the north was unified under the Delhi Sultanate, the west and south were fragmented among various rulers. The Portuguese were the first Europeans to set up trading posts, most of which were on the Arabian Sea.

India's Modern Era

In 1526, on the traditional battlefield of Panipat, 50 miles (80 km) from Delhi, thousands of elephants, horsemen, and footmen of the Sultan of Delhi, Ibrahim Lodi, amassed to face the impending terror Babur (1483–1530). In only a few hours Lodi was dead, his massive army crushed by an onslaught of artillery, a cavalry attack from the rear, and elephants run amok. Babur's mobile artillery overwhelmed army after army, including the Rajputs. After sieging Delhi and other fortified sites, he swept victoriously into Bengal. Babur, "The Tiger," direct descendent of Timur and Chinghis Khan, thus became the first Great Mughal. *Mughal* derived from the Persian word for Mongol.

THE MUGHAL EMPIRE

Babur, while establishing order, treated the Indian people as subjects rather than prey and forbade his soldiers to maraud them. He and his son and successor, Humayun, initiated functional divisions in administration. Humayun died prematurely in a fall, leaving his young son Akbar to rule.

Akbar (1542–1605) quickly consolidated his empire in further military campaigns, defeating the Rajputs and conquering Gujarat and Bengal (Figure 5-12). Akbar's belief in one God and his insistence in the greeting *Allahu Akbar* (God is Great) ritualized and legitimized his role as an Indian king. This idea of rule by divine grace added an enduring charismatic element to the position. Akbar promoted religious tolerance and, despite being illiterate, was eager for knowledge. He had every known book read to him and entertained the ideas of scholars and celebrities from around the world. His fifteen provinces were each headed by a governor and had its own team of tax collectors. He had land measured and assessed for its productive value and brought a hierarchical structure to military and civil officials. These and other structural changes made administrative functions largely automatic. A consummate politician and soldier, Akbar was also an effective administrator and a great patron of the arts and literature.

Akbar was poisoned by his son, Jahangir the "World Seizer." Jahangir's beautiful Persian wife, Nur Jahan (Light of the World), became very powerful and was instrumental in furthering Persian culture and building in northern India. Politically astute, she arranged the marriage of her niece, Mumtaz, to the next truly Great Mughal: Shah Jahan (Ruler of the World).

Shah Jahan (r. 1627–1657) extended Mughal sway into the south. Setting up his capital on the Yamuna at Shahjahanabad (Shah Jahan's city), he sponsored sites of architectural magnificence including Jama Masjid, India's largest mosque, and the Red Fort (in today's Old Delhi). Living in sumptuous splendor, Shah Jahan had 5,000 women in his harem. Nevertheless, when his beloved Mumtaz died, he had what has become India's most famous landmark—the Taj Mahal—erected in her honor (Figure 5-13).

For a while, the empire had enjoyed relative peace and prosperity. Vassals of earlier regimes, tribal leaders, petty rulers and the like were absorbed into the larger regime, making it increasingly top heavy. Recognition as *zamindars* (landlords) bought their loyalty but encouraged peasant exploitation. Trade fostered urban growth supported by agrarian masses. Mughal culture was essentially urban and increasingly differentiated from the majority rural population. The Urdu language expanded, with contributions from Persian, favored by the elite; from Arabic, employed by religious scholars; and from Hindi, spoken by the general populace. Linguistic differences exacerbated other divisions.

After the reign of Akbar, interest in science and technology waned in India. Court brilliance, architectural display, and military prowess took precedent. Lavish expenditure, constant warfare, and court intrigue peaked under Shah Jahan's son, Aurangzeb.

The Mughal empire reached its greatest extent under Aurangzeb (r. 1658–1707). But Aurangzeb's cruel, anti-Hindu policies and uncontrolled expansionism promoted his empire's downfall. Non-Muslims—Hindus, Jains, Buddhists, Sikhs—were brutalized mercilessly. Non-Islamic structures were razed and conversions forced upon pain of torture and death.

Tax burdens in general became increasingly oppressive to finance court excesses and military ventures. Newly conquered areas of the Deccan yielded proportionately less revenue than the fertile northern

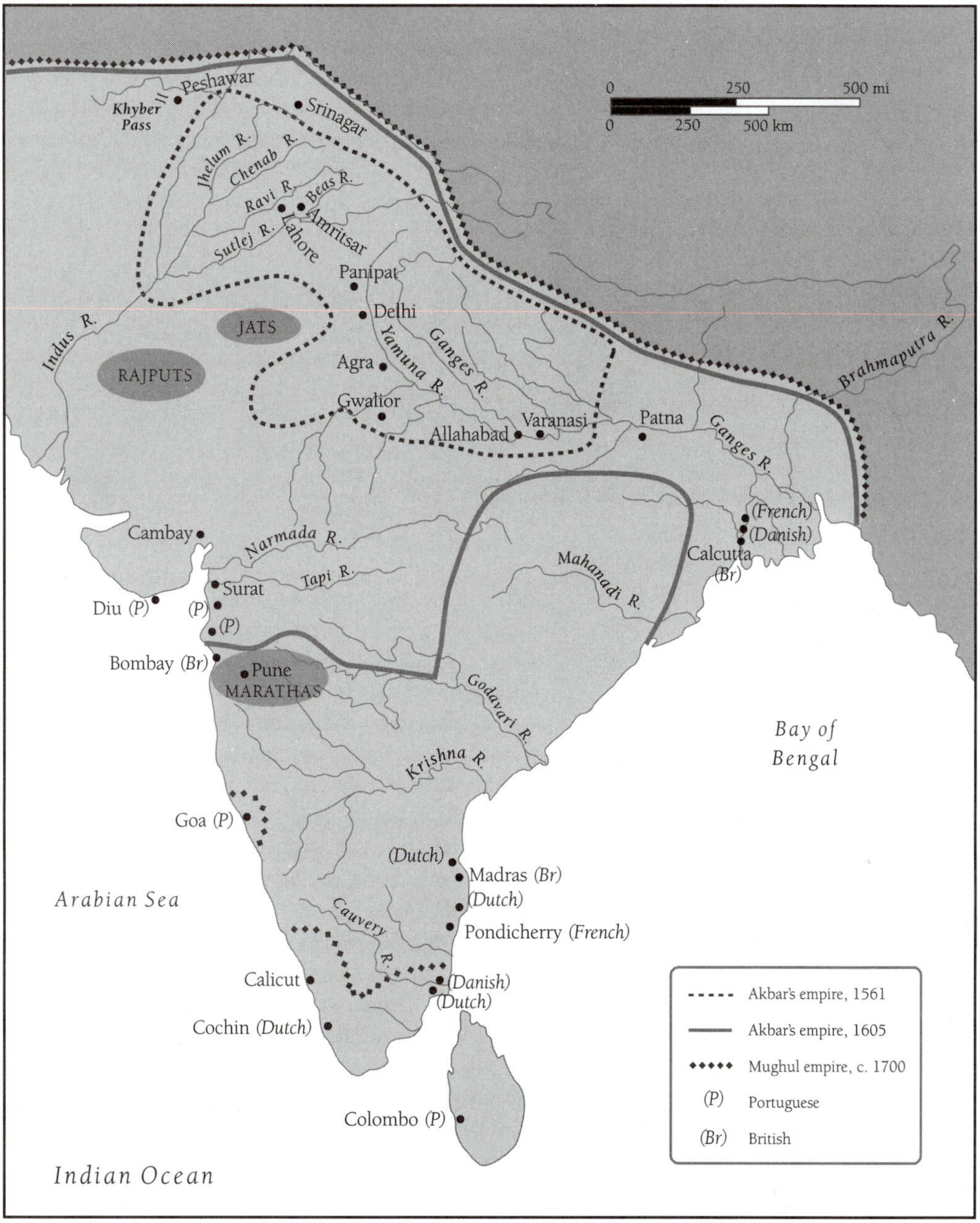

Figure 5-12

South Asia, c. 1700. Again the focus is the Indo-Gangetic Plain from which the Mughal Empire expanded southward. More Europeans including the British, Dutch, French, and Danes are on the scene. Note the concentration of activity near Bombay (Mumbai).

Figure 5-13

This is the magnificent Taj Mahal in Agra, India. It was built from 1632 to 1653 by the Mughal Emperor Shah Jahan as a tomb for his wife Mumtaz who died bearing her fourteenth child. Thousands of workers and designers were brought from Europe and Asia to erect this monument of marble and semi-precious stone. As the costly edifice progressed, one of India's worst famines ravished the countryside.

plains. Revenue collection was stepped up and special taxes placed on Hindus. Discontent was rife, but when protesters gathered outside the Red Fort, Aurangzeb ordered them crushed by the Imperial elephants.

THE SIKHS

Sikhism emerged from the bhakti movement in the fifteenth century. Its core area was the Punjab, a region of incessant Hindu–Muslim conflict. Its founder, **Guru Nanak** (1469–1538), taught that "Man will be saved by his works alone . . . There are no Hindus and no Muslims. All are children of God." He opposed all forms of discrimination, including caste. His teachings, compiled by his disciple and successor Guru Angad, became the **Sikh** holy book, the ***Adi Granth.*** Guru Angad met with Akbar, who gave him a temple site at Amritsar near Lahore.

In the sixteenth century, some Sikhs became involved in political intrigue, and their leader was executed by the Mughul ruler Jahangir. Subsequently, the Sikhs began to take on a militaristic character and by 1650, perceived Punjab as Sikh territory. Persecutions by Aurangzeb intensified their militarism. Under Guru Govind Singh, they vowed not to cut their hair, which is still worn knotted under a turban. They swore to avoid tobacco and alcohol and renounced purdah and sati. Sikh women were accorded almost equal status with men. Sikhs also adopted the surname of Singh (Lion). There is a saying that "All Sikhs are Singhs but all Singhs are not Sikhs."

Sikhism grew as a force to be reckoned with, eventually becoming the last bastion of resistance to British takeover in India. Only in 1849, after two British campaigns, were the Sikh armies defeated. Eventually Sikh regiments were incorporated into the British-Indian army. See the next chapter for more on the Sikhs and Punjab.

In order to alleviate the friction of distance that made central control impossible, Aurangzeb moved his capital to Aurangabad. Even so, encumbered with an army of hundreds of thousands of elephants, guns, artillery, cavalry, and at least half a million camp followers (a moving camp 30 miles (48 km) in circumference), Aurangzeb was no match for the guerrilla-style tactics of Shivaji, "The Mountain Rat."

Shivaji (1627–1680), the zealous Hindu leader of the Marathas, had erected a complex of fortifications at defensible sites in the fractured western Deccan. Headquartered at Poona (Pune), behind Bombay, Shivaji was able to lash out, then quickly retreat to his mountain strongholds. The Marathas turned out to be the Mughal Empire's most formidable opponent. In 1644, he sacked the main Mughal port at Surat. In ensuing confrontations, Aurangzeb the Muslim and Shivaji the Hindu stood as protagonists of their respective faiths, deepening the division between believers.

As disenchantment with the Mughals grew, common gripes and shared concerns bonded individuals in competing groups. The Sikhs—bound by faith—the Jats—farmers bound by kinship—and the Marathas—bound by nationalistic aspirations—challenged the Mughal hegemony. Many revolts were led by the Zamindars. Mughal retaliation was unforgiving. The situation worsened in 1702–1703 when famine and plague eradicated more than 2 million people from the Deccan.

The empire disintegrated in chaotic circumstances, with the strongest and cleverest manipulators and power-grabbers staking out their territories. And so arose the powerful **maharajas** (Great Rulers) of states such as Baroda, Gwalior, and Hyderabad. During this fray, the Persians invaded and sacked Delhi in 1739, carting off the jewel-encrusted Peacock Throne and other treasures. As Delhi wasted, cities such as Lucknow and Hyderabad became centers of Mughal culture.

EUROPEAN SEA POWER

When Babur raged into India, the Portuguese already controlled the Indian Ocean with a monopoly on the pepper trade. Their fortified outposts at Goa and elsewhere served as customs stations where Asian merchants had to obtain letters of protection in order to avoid Portuguese attacks on the high sea. The land-based Mughals were interested in international trade insofar as it was a source of silver and gold. They remained unconcerned about the intricacies of its functioning as long as the precious metals kept coming.

With the Portuguese monopoly, European competitors were welcomed by the Indians who could play them off against each other. In London, an East India Company was founded in 1600. Two years later, the Dutch East India Company was founded in Amsterdam. Indian textiles were desired products, and factories (storehouses) were erected in the foreign concessions. But, as the textile trade boomed, factories became involved in ordering, directing, and financing the industry. Soon they became instruments of foreign control.

By 1740, the British had eclipsed the Dutch, who then focused their attention on Indonesia. Starting with a collection of trading posts, the British fought the Mughals, forged alliances with local rulers, and eventually ousted their European competitors. By exploiting local rivalries, jealousies, Hindu–Muslim and other tensions, they were able to expand their influence. Three key commercial centers emerged.

On the Coromandel coast, Fort St. George developed as Madras. The British East India Company had edged Asian traders out of the interregional trade known as the "country trade" with China and elsewhere, and Madras flourished as an entrepôt.

In western India, Gujarat was the main seat of international trade. With the dissolution of the Mughal empire, however, Surat shriveled in importance. Many merchant ships moved to Bombay. Blocked from the interior by the Western Ghats, Bombay was less accessible. However, its natural harbor was safer, protected by the British fleet. As the empire further dissolved, Indian shipping networks floundered and interregional trade with Bengal and the Malabar Coast dissipated. Bombay had its sights on trade with Europe.

At a swampy, unhealthy, deep-water anchorage 80 miles (129 km) up the Hooghly River, Britain's Fort William expanded to become Calcutta, the empire's capital until 1931. British Bengal trade boomed in the eighteenth century, and further inroads were made into the interior. There, factory agents established contact with the spinners and weavers, and British artisans came to train them in the ways of British fashion. However, little investment was made in the means of production, and the workers remained in poverty. Farmers were forced to grow industrial, cash crops in order to meet their tax obligations. This exacerbated food shortages, increased indebtedness, added millions to the ranks of landless and unemployed, and enriched the *zamindari* land owners, tax collectors, and money lenders. Indian industry declined, and India became an importer of European manufactures and an exporter of raw materials such as raw silk, cotton, indigo, sugar, and opium. The British shipped the opium to China to satisfy their imperialist ambitions there.

By 1750, distrust of the British was widespread, and Calcutta had grown to 100,000 people. Even then, British racism marred the landscape. The city was divided into "Blacktown" and "Whitetown." Attempts to get rid of the British were immediately thwarted. With Colonel Robert Clive's connived victory over a vastly larger force at the Battle of Plassey in 1757, they achieved control over Bengal. British sea power was transformed into British land power.

For the next hundred years, the British struggled to expand their territorial control and suppress opposition on the subcontinent. Repeated wars were fought with various groups such as the Marathas and the Sikhs. In 1819, the British gained control of Rajasthan and became the largest land power on the subcontinent. By 1846, they had taken the Sikh capital at Lahore, the fertile lands of the Punjab, and the exquisite mountain vale of Kashmir. Then, in what seemed an insignificant move at the time, the British sold Kashmir to the highest bidder, a Hindu chieftain, Gulab Singh. By 1850, they controlled an empire in ruins, extending from the Indus to Bengal and from Kashmir to Ceylon (Figure 5-14).

Colonial transformation was unstoppable. As the landed aristocracy saw their lands appropriated, Western influences were instilling new ideas about development and missionaries were challenging Hindu beliefs and practices, especially the caste system. English replaced Persian as the official language of government and education. According to historian Lord Macaulay (1800–1859), "A single shelf of a good European library is worth the whole native literature of India and Arabia." Macaulay's ideas led the way in developing India. The British were to bring up a "class of persons Indian in blood and colour, but English in taste, in opinions, in morals, and in intellect." Schools and colleges were built to do the job.

Lord Dalhousie (1848–1856) launched an ambitious program of modernization. Expanding on the previous work of British engineers, he constructed the first 450 miles (724 km) of the Ganges Canal, which today provides hydroelectric power for a large region and irrigates 2 million acres of farmland. Under the Dalhousie administration, electric telegraph services were installed, an efficient postal system developed, and the first 300 miles (483 km) of railroad track laid. While it is true that India ended up with Asia's best rail system, there were negative consequences.

On the one hand, railways permitted the relief of famine with grain shipments, while on the other hand they created famine in areas from which the grain was sent. Grain shortages drove prices beyond the reach of the impoverished masses, and millions starved. For example, 10 million in Bengal and Bihar starved in 1769–1770. At least 23 famines took their toll between the eighteenth and twentieth centuries. The last severe food crisis was in 1943, when 1.5 to 3 million died as a result of grain shortages, usurious prices, profiteering, and government negligence and incompetence.

Railroads also opened up India as a vast consumer market. But, instead of stimulating interior development, they favored coastal cities through new cotton milling and shipping facilities. Local textile mills closed. Towns and cities bypassed by the rail line declined. By 1879, three-quarters of India's textile mills were in and around Bombay.

In the minds of Indians, infrastructural change did not have the impact of social change. The British had outlawed the practice of sati in 1829. In 1856, another law permitted the remarriage of widows, anathema to Hindus. By British law, Hindu states could not be inherited by any other than a blood heir. Such "heirless" properties reverted to the Crown. States, even those loyal to the British, were annexed under the guise of mismanagement. Such attacks on Indian culture and Hinduism itself were cause for alarm, and in 1857 there was a mutiny.

The army of the East India Company comprised mainly British-trained Indian troops called **sepoys.** Intrigue sparked rumors that cartridges, which had to be bitten off to load the new Lee-Enfield rifles, were greased with pig and cow fat. With the pig being unclean to Muslims and the cow sacred to Hindus, a mutiny erupted. The Sepoy Rebellion of 1857 triggered nationwide violence. Whole communities of Europeans or Indians were slaughtered, and cities were reduced to rubble. Some states such as Hyderabad, Baroda, Rajasthan, and Punjab remained loyal to the British and the rebellion was ultimately crushed.

In 1857, the British government, horrified at the extent of atrocities, abolished the East India Company

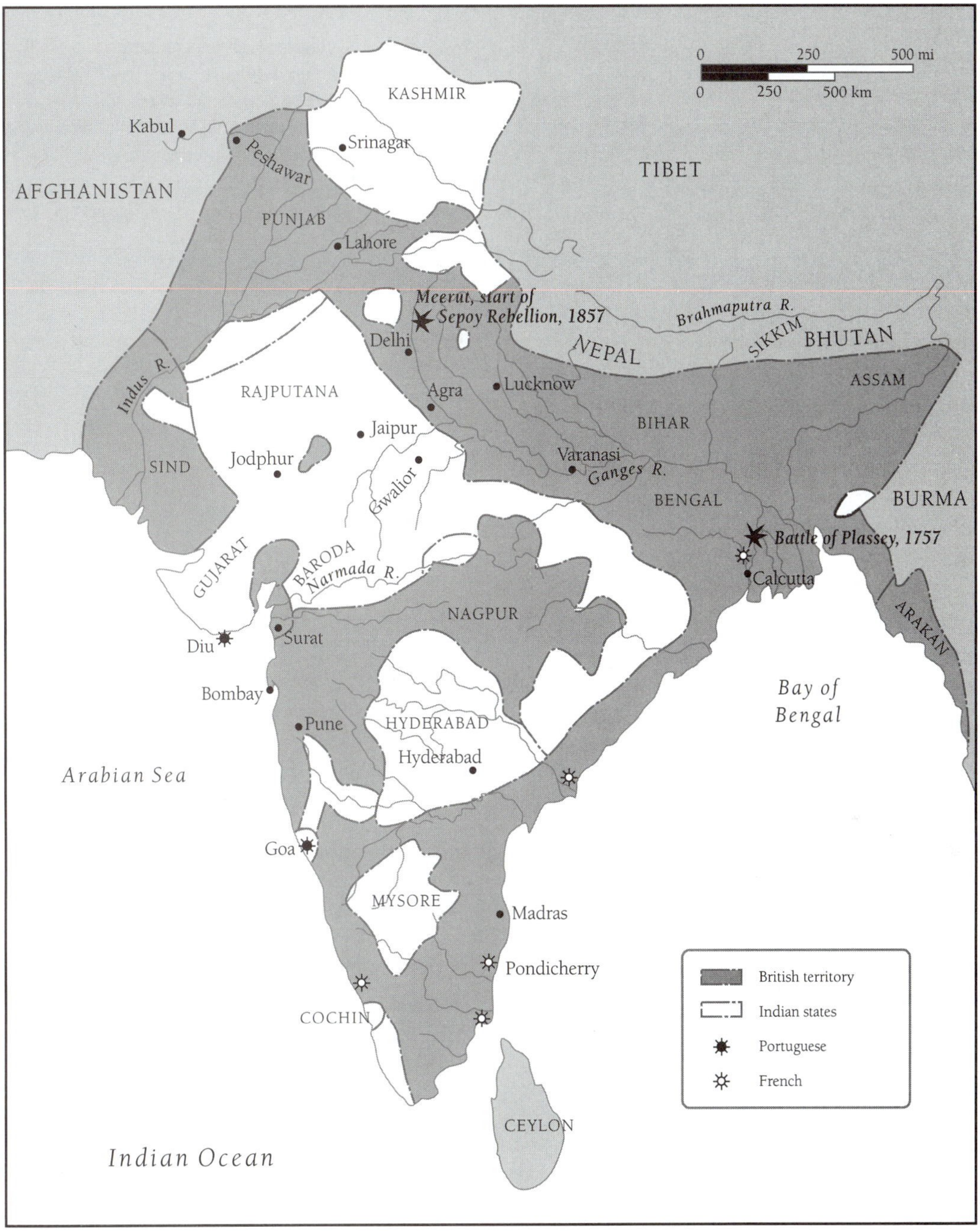

Figure 5-14

South Asia, 1856. By this time Britain had control over most of the region. Other colonial possessions were reduced to small coastal colonies.

Figure 5-15

Clinging to the mountainside is the hill station of Simla founded by the British in 1819. Simla served as the summer capital from 1865 to 1939. It was much cooler than either Calcutta or Delhi. All government materials were hauled by train and caravan from Calcutta (Delhi after 1912) each year.

and formally took over India as a colony, a status it would retain for nearly a century. The era known as **the Raj,** meaning the "rule," had begun.

THE RAJ

Under British government control, social reforms were abandoned and attention was given to maintaining the status quo. The army was reorganized under British officers. The British legal system was introduced, and the Indian Civil Service was formed. This initiated the cumbersome bureaucracy that, like the road and railway system, would tie the country together. In all of this, Indians were deliberately excluded from positions of power. The country was measured, cataloged, mapped, organized, and centralized. The Grand Trunk Road stretched more than a thousand miles between Calcutta in Bengal and Peshawar near the Afghan border.

In 1877, Queen Victoria was declared Empress of India, and her representative in India became the viceroy. New treaties offering varying degrees of autonomy were made with the various princely states. There were more than 500 of these *native states,* occupying nearly 40 percent of the landmass. Larger states such Hyderabad, Punjab, Bengal, and Assam had governors or commissioners who reported directly to the viceroy. Smaller states had advisors. But these political alliances with their incumbent privileges contributed to even greater division of this already very divided region.

The British avoided contact with their subjects. New towns, suburbs, and military stations were built across India. In cities, military cantonments and civilian districts known as **civil lines** were spatially and socially separate entities, complete with their own accommodations and services. These British mini-worlds in the lowlands had their counterparts in the highlands: **hill stations** (Figure 5-15).

Hill Stations

Hill stations are mountain-amenity landscapes offering opportunities for rest and recreation at relatively cool elevations of 5000 to 7000 feet (1500–2200 m). They existed throughout the colonial world and remain important centers of commerce and tourism in many parts of Asia.

In India, hill stations were of great significance, and not just as social, economic, and political centers for Europeans. They symbolized the paramountcy of the Raj and revealed prevailing attitudes of nineteenth-century Europe: racism based on beliefs grounded in environmental determinism.

Then, it was believed that white people could not withstand the rigors of torrid, foreign lowlands for extended periods. The combined onslaught of heat, dust, and "natives" would surely damage European sensibilities and possibly cause offspring to be less than normal. It was essential that Europeans be restored in more familiar, temperate climes. As one Englishman remarked, "like meat, we keep better there."

Hill stations have been described as "island(s) of British atmosphere hung above the Indian plains," and "comforting little piece(s) of England." At the highest location, a protective nest of European architecture, trees, plants, and parks implanted British ideals of restorative havens, salubrious and picturesque. In contrast, the "eyesores and nuisance" of the Indian quarters and bazaars on the slopes below were discreetly hidden behind "attractive" buildings and hedges.

Hill stations thrived as social centers. They glittered with banquets, pageants, and balls and catered to European tastes with afternoon teas, activities at the Club, and the frivolities of games and sport. Indian servants graced only the background, ears cocked for their next command. In later years, educated Indians were permitted to join the fun but remained fundamentally detested by the British.

The British perceived India's climate to have three seasons: the cold, the hot, and "the rains." They adjusted their schedules to suit regional variations. In the north, for example, the cold lasted from October to March. Then, woolens and warming fires were needed, and British ladies remained on the plains with their husbands. By April or May, the plains were roasting and **amenity-migration** was made to the cooler hill stations, where thick-walled bungalows reduced sensible temperatures even further. These small, one-story houses and the name **bungalow** diffused from India throughout the Western world.

Some hill stations became so popular that they were used as summer capitals by government functionaries. For instance, the governments of Bombay repaired to Poona (Pune); Madras to Ootacamund known by the British as "Ooty," and Bengal to Darjeeling. In 1864, the Viceroy of India proclaimed Simla in the Himalaya as the empire's summer capital, and the entire government and all its files and documents were moved there from Calcutta each year from 1865 to 1939.

Hill stations did not boom until the opening of the Suez Canal in 1869 and railroad construction in the colonies. A narrow-gauge rail line carried its first passengers to Simla in 1903. Prior to that time, people and all their trappings were literally carried or carted up the mountain track by their Indian servants. As late as 1932, nearly a third of Simla's summer residents were porters or rickshaw pullers, most working to pay debts at home.

Increased population stimulated commerce. In 1830, Simla had only 30 houses. By 1881, there were over a thousand. The opening of the railway saw the population spiral to more than 30,000 by 1898. Market gardening contributed to increased commerce between hills and plains, and the town's commercial sector grew rapidly.

Indian areas frequently were targeted for improvements, which meant replacing their "unhealthy" warrens with environments suitable for European pursuits. Indians were forced by displacement to live in even more crowded conditions.

Simla, as the Raj capital for at least eight months a year, hosted salient political events. Here in 1913–1914, the British, Chinese, and Tibetans met to fix their common boundaries. It was in Simla that the Indian National Congress was conceived and shaped and that the Muslim League had its beginnings. Here both organizations participated in discussions leading to India's independence and partition in 1947. And here, too, Prime Minister Indira Gandhi granted statehood to Himachal Pradesh in 1971. Simla, with its Viceroy's mansion and British architecture crowning the teeming bazaars below, is now a popular destination for Indian upper- and middle-class tourists and is still a popular site for national and international negotiations.

India had been Britain's major trading partner since 1840. With the opening of the Suez Canal in 1869 and the expansion of railways, the route to Britain was significantly shorter and India was drawn increasingly into the international economy. British capital was invested in rail works, and coffee and tea estates (plantations). India exported these, along with raw cotton, wheat, and jute (fiber). They imported machinery and textile manufactures (made in England from Indian cotton). Eventually an industrial base was established.

India's industry rose and fell with British policies. Her economy was used to make up trading deficits with other industrialized nations. Once-renowned manufactures such as cotton, brocades, and silks were eliminated. However, raw material-based industries such as textile mills and jute, cement, and sugar factories were eventually built. Old trading or religious centers evolved as industrial centers, luring in landless peasants as a cheap labor supply. By 1900, Calcutta was awash with slum dwellers and squatters. British India's most famous author, Rudyard Kipling, described it as the "city of Dreadful Night."

Indians were left to their own resources to develop other types of industry. J. N. Tata (1839–1904) was one of several Indian industrialists. Tata, for instance, founded the country's steel industry, with the Tata Iron and Steel plant opening in Jamshedpur, Bihar, in 1907. The Tata family continues to reign as India's largest industrial dynasty.

Grand architecture of what became known as the Victorian-Muhal style housed train stations, post offices, and other important buildings. Bombay's Taj Mahal hotel was one of the grandest of all (Figure 5-16).

In 1911, Queen Victoria's grandson, King-Emperor George V, and Queen-Empress Mary arrived in Bombay, the first British royal visit in history. At a key moment in the spectacular durbar (assembly) in Delhi, the monarchs announced from Shah Jahan's Red Fort that the imperial capital was to be transferred from Calcutta to Delhi, where a new city would be constructed.

Figure 5-16

This Victorian-Mughal hotel was built by the great Indian industrialist Tata. Opening in 1903, it had granite staircases, electric fans, and 400 beds. It was the only luxury hotel in Bombay that Indians were permitted to stay at.

Designed by British architects and finally inaugurated in 1931, New Delhi exhibited broad, straight avenues, well-suited for parades; prominent Victorian-Mughal government buildings, stately palaces for the Viceroy and other notables; and spacious bungalows for the elite. European in plan, yet Indian in style, New Delhi was a contradictory counterpart to the traditional maze of Old Delhi (Figure 5-17).

THE ROAD TO INDEPENDENCE

The notion of self-rule had its roots in the nineteenth century, when there was a renewed interest in Indian history, culture, and religion. European scholars pursued translations of ancient works, and these were avidly read by growing numbers of educated Indians. At the same time, Indian religious scholars had reinvig-

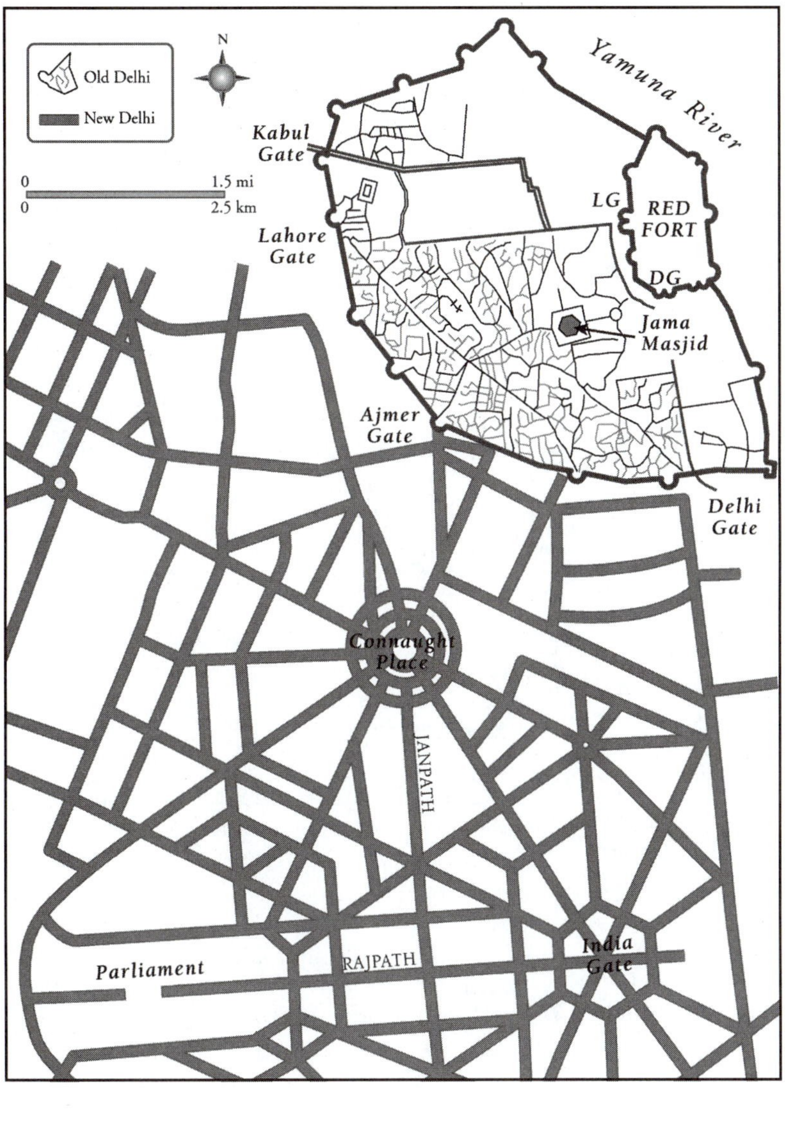

Figure 5-17

Old and New Delhi. Fascination with geometry and perspective, concern for order and accessibility, and desire for grand vistas and impressive layouts led to New Delhi's design. This plan stands in stark contrast to the narrow streets and winding passageways of the fortified old city. While New Delhi was planned, Old Delhi grew organically.

orated Hinduism. Some even suggested modifications more appropriate for a modern society—the abolition of caste, for instance.

One of the most notable religious reformers was Vivekananda (1863–1902), a disciple of the mystic Ramakrishna. Vivekananda spent a year lecturing in England and the United States where he founded a mission in Ojai, California. His focus was twofold: uplifting the downtrodden masses and restoring Indian history and Hindu culture to their past elevated standing. Gandhi was deeply influenced by Vivekananda.

The Indian National Congress, formed in 1865, provided a forum for nationalist sentiments, and became the main instrument for the independence movement. The **Congress Party** continues to be one of the most influential in India. Expressions of desire for greater autonomy were disdained or ignored by the British.

When the state of Bengal was halved into Hindu-dominated and Muslim-dominated parts in 1905, without consultation with any Indian leaders, Bengal erupted with huge demonstrations. Picketers were beaten and leaders were whisked away without trial. Although four years later, the division of Bengal was revoked, nationalistic mind-sets had hardened and moved underground, from where a campaign of terror against the British was directed.

During the same period, Muslims were actively looking out for their own interests. The Muslim League was created in 1906. While Muslims made up more than 20 percent of the population, they were scattered and not well represented in positions of power. Certainly the Congress Party was run by Brahmin Hindus. In 1916 it was agreed that Muslims and Hindus should have separate elections and representation.

World War I (1914–1918) was a significant event in that 1.2 million Indian troops served the Empire. In the aftermath, British colonies such as Australia and Canada were granted significant control over their own affairs. Surely India would be rewarded in a similar manner. It was not.

Imperialist ways would not prevail. Anything short of independence was unacceptable. After all, the Chinese Revolution of 1911 and the Russian Revolution of 1918 demonstrated that popular uprisings could overthrow despotic regimes.

British opposition reached its lowest point in 1919 with a massacre of unarmed Indian demonstrators at Amritsar in Punjab. The toll was 400 dead and 1,000 injured. The British Raj lost the loyalty and support of millions. **Mohandas Gandhi** (1869–1948), considered a "saintly fanatic" by the viceroy, was profoundly affected by this event.

Did you know that Martin Luther King, who led the civil rights movement for African Americans in the United States, was strongly influenced by the philosophy of Gandhi?

Gandhi: Great Soul

Mohandas Gandhi (1869–1948), born in Gujarat, had experienced discrimination as an Indian practicing law in South Africa. There, on behalf of the Indian community, he developed the strategy of nonviolent protest against injustice. Upon his return to India in 1914, he abandoned his European dress and adopted the lifestyle of a saddhu. Traveling far and wide, he used traditional Hindu values such as *ahimsa* to spread the concept of peaceful, mass protest against British oppression (Figure 5-18). He employed traditional symbols such as the spinning wheel to rebuild an Indian sense of nationalism.

Jawaharlal Nehru, in his autobiography (1942), said that Gandhi was "like a beam of light that pierced the darkness." Moreover, rather than direct from the top down, "He seemed to emerge from the millions in India . . . and the essence of his teaching was fearlessness and truth." Gandhi's notion of nonviolence was not simply the absence of violence; compassion and responsibility for the welfare of others were vital components of the concept.

Gandhi spoke out against Hindu–Muslim animosities as well as violence against women. He pointed to male chauvinism as being responsible for the genesis of sati. He abhorred the in-

Figure 5-18

This Delhi sign shows Gandhi promoting vegetarianism. Note that the sign mentions ahimsa*–the sanctity of all forms of life. Note also the use of both Hindi and English.*

equities of the caste system, saying that untouchables should be called *harijans:* children of God. Beloved by the masses and many political leaders, he was given the title *Mahatma,* meaning Great Soul. Mahatma Gandhi was instrumental in bringing about India's independence from Britain. He is often referred to as the "Father of Independent India."

Gandhi, who came to lead the Congress movement, launched three mass campaigns of civil disobedience in 1920, 1930, and 1942. These marches, sitdowns, boycotts, and blockades resulted in retaliatory violence and imprisonment for thousands. But the British did make concessions. In 1932, they began recruiting Indian military officers. In 1935, the Government of India Act and a new constitution put the power of eleven states in the hands of Indian representatives. Indians moved into higher ranks of the civil service. In sports, science, and literature, Indians were receiving world recognition—Olympic gold medals and Nobel prizes.

In 1929, Gandhi had nominated **Jawaharlal Nehru** (1889–1964) to succeed him as leader of the Congress movement. The dynamic, Western-educated Nehru, who could eloquently express his empathy for the Hindu masses, would eventually become India's first prime minister.

The only serious rival to Congress was the Muslim League, founded in the Bengali city of Dhaka. It was led by European-educated **Muhammed Ali Jinnah** (1876–1949), who believed that Muslims and Hindus had evolved as "two nations." As Nehru's Hindus gained power, thousands of Muslims joined the League. With Nehru's refusal to recognize Jinnah as the spokesperson for all Muslims, an attempt at power sharing failed. Jinnah then turned to convincing Muslims that a Congress election victory would be a Hindu victory.

World War II (1939–1945) was the final catalyst in bringing about India's independence from Britain. As Britain had declared war on Germany without consultation, the Indian provincial representatives resigned. Being called upon to "Quit India," in 1942 the British arrested the entire Congress leadership and 60,000 activists. Then, they invited Jinnah into the government. With Congress power diminished, Jinnah was able to put the Muslim League on an equal footing. The British had agreed to Indian independence at the end of the war, but now it was too late. Muhammed Ali Jinnah demanded a separate Muslim state: Pakistan "Land of the Pure." Unable to speak Urdu, he remarked in English, "If not a divided India, then a destroyed India."

Hindu–Muslim violence was widespread, and on August 16, 1946, the "Great Calcutta Killing" event cost the lives of 5,000 Muslims and Hindus. Bloodshed swept across northern India, with violence and atrocities sparing no community. Faced with the prospect of complete anarchy, Nehru and other Hindu leaders became convinced that a Muslim state was indeed necessary.

A new viceroy, Lord Louis Mountbatten, was sent to India in 1946 to negotiate an independence agreement. He quickly saw that the division of India into Muslim and Hindu entities was inevitable. Pakistan was to encompass heavy Muslim concentrations in the northwest and northeast, meaning that the new nation would consist of two separate entities over 860 miles (1400 km) apart. Moreover, to give Muhammed Ali Jinnah his Pakistan, both Punjab and Bengal would have to be divided. The consequences were terrible.

THE PARTITION OF INDIA

As both Punjab and Bengal had mixed populations, no boundary line would be satisfactory to all. Punjab, in particular, was communally mixed with Hindus, Muslims, and Sikhs interspersed. The decision was made to award 62 percent of the land and 55 percent of the population to Pakistan. As of midnight January 14, 1947, millions found themselves trapped in a country hostile to their faith.

The partition of India in 1947 spurred the largest migration in human history. Around 12 million people moved between West and East Pakistan and India seeking refuge in their own religious communities. The ensuing savagery was unparalleled. Entire villages were massacred, and trainloads of people were butchered and torched. At least a million people were killed in the wake of India's partition.

These events branded Hindu–Muslim animosities deeper than ever. A horrified Gandhi took up residence in the Muslim quarter of Old Delhi and vowed to fast to the death if the communal violence did not cease. Although calm returned several days later, many Hindus were enraged by his seeming support for the Muslims. On January 30, 1948, Gandhi was assassinated by a Hindu fanatic. A distraught Prime Minister Nehru spoke to his nation: "The light has gone out of our lives and there is darkness everywhere."

In that same year, Muhammed Ali Jinnah, Pakistan's first political leader, died as well. Spurning Muslim prohibitions against pork, alcohol, and smoking, he died of lung cancer.

The nearly 600 princely states were diverse geographically, socially, and politically, precluding any uniform policy regarding their status. With the creation of India and Pakistan, each state was given the option of joining either country. Most acquiesced to India but some, such as Hyderabad, had to be coerced by the Indian army. Kashmir presented a special case, its complexities defying easy solution.

The Maharaja of Kashmir was a Hindu, but the majority of his subjects were Muslim. While he wavered on his decision, hoping to remain independent, Pakistan tribesmen rebelled. The Maharaja then acceded to India and asked for military help. Indian troops were unable to oust the Pakistanis. With stalemate in 1949, a "line of control" became the de facto boundary. This resolution was not a solution, and Kashmir remains a region of conflict to this day.

Nationalism and Regionalism

Complex, diverse, and paradoxical—this aptly describes the cultural mosaic of South Asia today. With ancient roots and an amalgam of influences from within and without, empires have risen and fallen, leaving indelible legacies on the landscape. Despite continual religious, linguistic, and political conflict and restructuring, India and the newer nations of South Asia remain intact. The interminable dynamics of nationalism and factionalism continue to support the adage of "unity in diversity."

The idea of the territorial state was introduced by the Mughals, who developed an ordered bureaucracy. This merged into the British civil service, an all-encompassing, multileveled network of control and regional division. Yet, persistent national agitation did reap change. Constitutional reform ended with a British, parliamentary-style, democratic system. And the federalism that the British employed in their devolution of power was retained after independence. Highly centralized power of the Mughals and British fed nationalistic tendencies and was furthered by the partition of India. Yet federalism, which allows for regional differences and varying degrees of local autonomy, has been fundamental to the stamina and resilience of the Indian and other governments in the region. At the same time, it has facilitated regionalism, localism, and communalism. National and regional ambitions and inequities underlay countless examples of instability and strife marring the subcontinent throughout the twentieth century.

Main Point Summary

1. Physical geography has impacted Indian history. Mountain passes and river valleys invited settlement yet channeled invasions. Mountain ranges and rugged plateaus afforded protection and insulated from intrusion. Monsoon cycles gave rhythm to invasion strategies, seasonal migrations, and shipping schedules. Agricultural, religious, and social activities continue to be geared to the wet and the dry seasons.
2. Favorable site factors, exemplified by the fertile Indus valley, the Doab, and the larger Ganges plain, were conducive to the support of large populations. Natural harbors provided both shelter and accessibility, fostering the development of cities such as Bombay (Mumbai), Calcutta, and Madras (Chenai).
3. The Indus valley produced one of the world's greatest civilizations. Those who built Harappa and Mohenjo-daro were predecessors of the Dravidian people who now inhabit southern India. Pressed southward by the various Central Asian invaders, their languages remain distinct from those spoken in the north.
4. Around 1500 B.C., Aryans invaded via the Khyber Pass, introducing Sanskrit and the Vedas to northwestern India. This and subsequent writings became the foundation for all modes of thought and behavior.
5. The Aryans brought horses and cows to northern India. Cavalry warfare allowed effective dispatch of traditional elephant forces. Cows, initially significant as beasts of burden and sources of sustenance, eventually became sacred.
6. The division of society into caste—*varnas* and *jatis*—provided group cohesion and standardized intergroup relations. It also engendered social divisions that became even more splintered and rigid with time.
7. In the fourth and fifth centuries, Buddhism and Jainism emerged. Buddhism spread across Asia and engendered different schools of thought such as Mahayana, Theravada, and Tantrism in the process. Overrun by Islam, Buddhism ultimately vanished from the land of its birth. In South Asia, only Sri Lanka has remained Buddhist. Jainism, with its strong emphasis on ahimsa, retained its foothold in Gujarat.
8. The Mauryan Empire of the third century was the first expression of a political state on the subcontinent. But India's Golden Age was in the fourth century, when under the Guptas, great advances were made in literature, art, science, and mathematics.
9. Subsequent invasions from Central Asia destroyed the landscape and brought an end to Buddhism in India. The following centuries saw the rise and fall of numerous states and empires. The most extensive was that of King Harsha, who was the last Indian ruler to reach south of the Narmada River.
10. By this time, caste was firmly entrenched, as were attitudes to women. Sati was widespread among

upper castes, especially the warrior Rajputs. With further Islamic invasions bringing the customs of veiling and purdah, restrictions for women intensified.

11. During the eleventh century, the north was in turmoil but new empires arose in the south where maritime trade was flourishing. The powerful Cholas established footholds in Southeast Asia and, through commercial interchange, even gained recognition from China.
12. Focus turned north again when Delhi was inaugurated as capital of the Delhi Sultanate in 1229. Under successive rulers, heavy taxation for military ventures forced millions into poverty. And, Hindus were widely persecuted. Meanwhile, the court became increasingly Persianized, adding Persian refinements to upper class landscapes and lifestyles. The excesses of the Delhi Sultanates were halted by Mughal forces at the Battle of Panipat in 1526.
13. In the sixteenth century, the Portuguese and other Europeans established trading posts along the Malabar and Coromandel coasts. Ultimately, the British would dominate and change the Indian landscape forever.
14. The sixteenth century also saw Babur take power as the First Great Mughal. Under his successors, the empire was consolidated and controlled through a hierarchical structure of government. Taxation increased to provide the lavish expenditures required to construct architectural wonders and fund military exploits. Exploitation of an already overburdened and indebted peasantry allowed the accumulation of fortunes of urban-based landlords and money lenders. Rulers and ruled became distanced economically, politically, and socially. Finally, the Mughals were denied their invincibility by the Marathas and the empire collapsed into a mosaic of competing states.
15. European powers took advantage of internal dissension to advance their territorial ambitions. The British, with established settlements at Bombay, Calcutta, and Madras, consolidated their power through conquest, intrigue, and alliances with local potentates. They advanced their control with the construction of roads and railroads. Economic activities shifted to coastal locations and interior cities declined spurring labor-migration to foreign-dominated port cities.
16. Attempts to oust the British were thwarted. With the Battle of Plassey in 1757, British control of India was essentially complete. Racist policies, cultural interference, and economic exploitation by the British East India Company fomented rebellion. Strife and famine impelled the British Government to take over India in 1857.
17. Under the Raj, the subcontinent was cataloged and mapped in detail and an elaborate bureaucracy put into place. However, Indians were excluded from positions of consequence. British settlement was encouraged by both the opening of the Suez Canal in 1869 and railroad expansion. Myths of environmental determinism convinced the British of their racial superiority and "necessary" social separation was epitomized by spatially divided cantonments and hill stations. The British designed capital at New Delhi, which differed sharply from traditional Old Delhi, fixed perceived cultural differences on the landscape.
18. The nineteenth century witnessed a resurgence of Hindu and Muslim nationalism and anti-British sentiment. Western-educated leaders Jahawarlal Nehru, Muhammed Ali Jinnah, and "Mahatma" Gandhi led their followers in an independence movement that resulted in the Partition of India in 1947. This division into Hindu India and Muslim Pakistan unleashed terrible violence and triggered the largest migration in history.
19. While unified by British infrastructure and economic and political systems, the region was divided by religion, language, and caste. Moreover, Pakistan comprised two pieces, spatially and culturally distant. South Asia of 1947 was a complex and paradoxical mix of unifying and disintegrative forces that would wreak havoc throughout the twentieth century.

Terms to Know

Harappan (Indus) Civilization	endogamy	Rajputs	The Raj
Dravidian	untouchable	sati	civil lines
Aryan	*dalit*	purdah	hill station
varna	Jainism	Sikhs	amenity-migration
brahmin	Mahavira	Sikhism	bungalow
kshatriya	Asoka	Guru Nanak	Congress Party
vaishya	India's Golden Age	Adi Granth	Mohandas Gandhi
sudra	Fa-hsien	Maharaja	Jahawarlal Nehru
caste system	Hsuan-tsang	sepoy	Muhammed Ali Jinnah
jati	King Harsha		

Bibliography

Allchin, F. R. ed. 1995. *The Archaeology of Early Historic South Asia: The Emergence of Cities and States.* Cambridge: Cambridge University Press.

Basham, A. L. 1968. *The Wonder That Was India,* 3rd ed. London: Macmillan.

Brown, J. 1977. *Gandhi and Civil Disobedience.* Cambridge: Cambridge University Press.

Choi, Dong Sull. 1997. Origins and Realities of Suttee in Hinduism. *Comparative Civilizations Review* No. 36: 38–53.

Davies, C. Collin. 1963. *An Historical Atlas of the Indian Peninsula.* London: Oxford.

Dickason, David G. 1975. The Indian Hill Station. Geographical Record. *The Geographical Review* 65: 115–116.

Dutt, Ashok K., and M. Margaret Geib. 1987. *Atlas of South Asia.* Boulder: Westview.

Farmer, B. H. 1993. *An Introduction to South Asia.* New York: Routledge.

Kenny, Judith T. 1995. Climate, Race, and Imperial Authority: The Symbolic Landscape of the British Hill Station in India. *Annals of the Association of American Geographers* 85: 694–714.

Kulke, Hermann, and Kietman Rothermund. 1990. *A History of India* (revised edition). New York: Routledge.

Lapierre, Dominique, and Larry Collins. 1975. *Freedom At Midnight.* New York: Simon and Schuster.

MacMillan, Margaret. 1988. *Women of the Raj.* New York: Thames and Hudson.

Murphey, Rhoads. 1996. *A History of Asia.* New York: Harper Collins.

Narasimhan, Sakuntala. 1990. *Sati: A Study of Widow Burning in India.* New York: Penguin.

Nehru, Jaharwalal. 1942. *An Autobiography.* London: Chatto and Windus.

Pandian, Jacob. 1995. *The Making of India and Indian Traditions.* Englewood Cliffs: Prentice Hall.

Pubby, Vipin. 1996. *Shimla Then and Now* (2nd edition). New Delhi: Indus.

Schmidt, Car. 1995. *An Atlas and Survey of South Asian History. Armonk:* M.E. Sharpe.

Smith, B. 1995. *Classifying the Universe: The Ancient Indian Varna System and the Origin of Caste.* New York: Oxford.

Sopher, David. 1964. Man and Nature in India: Landscapes and Seasons. *Landscape* 13: 14–19.

Spear, Percival. 1963. *A Study of the Social Life of the English in Eighteenth Century India.* London: Thames and Hudson.

Watson, Francis. 1979. *A Concise History of India.* London: Thames and Hudson.

Wolpert, Stanley. 1991. *India.* Berkeley: University of California.

Wolpert, Stanley. 1984. *Jinnah of Pakistan.* London: Oxford.

Chapter 6

South Asia: Pakistan and the Himalayan States

"From your past emerges the present, and from the present is born the future."

MUHAMMAD IQBAL (1873–1938)

In this chapter, I will focus on the Indian subcontinent's northern regions and tell you about Pakistan; the disputed region of Kashmir; Nepal; and Bhutan (Figure 6-1). These mountain zones are both physically and culturally fractured. High altitudes and deserts have limited accessibility and foster cultural differentiation, communalism, and regionalism.

Within the formidable Himalaya and Karakorum are the tenuous and oft-disputed boundaries between Pakistan and Afghanistan; Pakistan and India; and Pakistan, India and China. Political instability has marked the evolution of virtually every state in the region. I will start with Pakistan at the time of India's partition in 1947 and explain the factors leading to the transformation of East Pakistan into Bangladesh in 1971. Then, I will describe current conditions and problems across the region, beginning with modern Pakistan.

Nationalism and Regionalism: The Devolution of Pakistan

In 1948, British geographer O. H. K. Spate wrote that he saw the creation of East and West Pakistan as "the expression of a new economic nationalism that has inevitably taken into its hands the immensely powerful weapon of immemorial religious and social differentiation." In 1947, Pakistan had two "wings," the eastern wing being only one-sixth the size of the western wing yet holding eight million more people. This was only one of a litany of differences between the two parts of this new nation—centrifugal forces that would ultimately result in its political demise.

For the split nation of Pakistan, unity could not be maintained with diversity. Contrasts were too numerous, and conflicts were too intense. West Pakistan is dry; access to water and drought is of critical concern. East Pakistan is humid; flooding is a yearly event. Aside from differences in size and population, the west

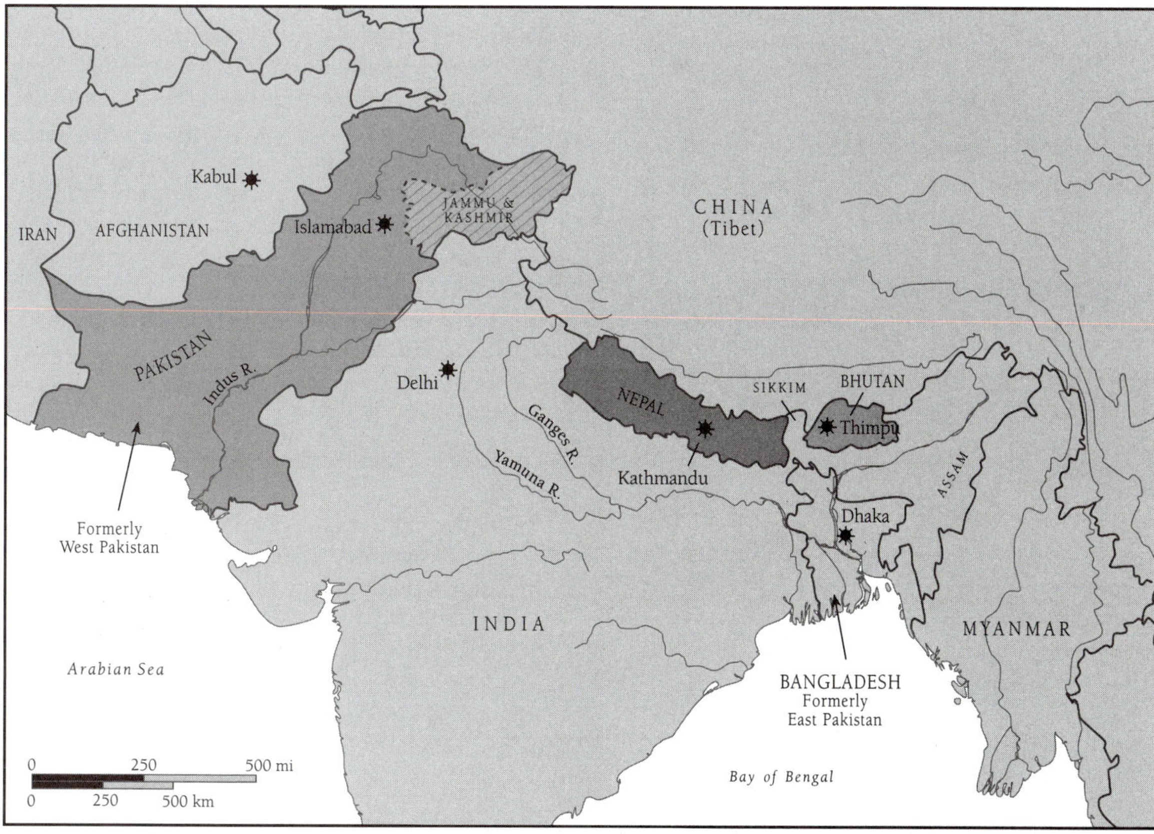

Figure 6-1

Northern South Asia. The political complexity of this region is compounded by its rugged physical geography.

possessed 80 percent of the cultivable land, and its share per farmer was more than double that of the east. West Pakistan had the country's capital and only port at Karachi. East Pakistan had far less infrastructure and was much poorer economically.

Cultural differences are exemplified in linguistic differences (Figure 6-2). There are several languages spoken in the west. These are from different language families and possess numerous dialects. Bengali, a branch of Sanskrit, is spoken in the former eastern wing of the country. In 1948, the Karachi government declared Urdu as the national language of Pakistan, enraging Bengalis and deepening the chasm between east and west. Linguistic division would prove problematic in the years ahead.

Language and Nationalism

Pakistan (formerly West Pakistan) has reaped the consequences of lacking linguistic cohesion. About two-thirds of the population speak Punjabi. Punjab is the most economically productive province and contains the nation's capital, Islamabad. Sindhi is spoken by about 12 percent, mostly in Pakistan's second industrial prov-

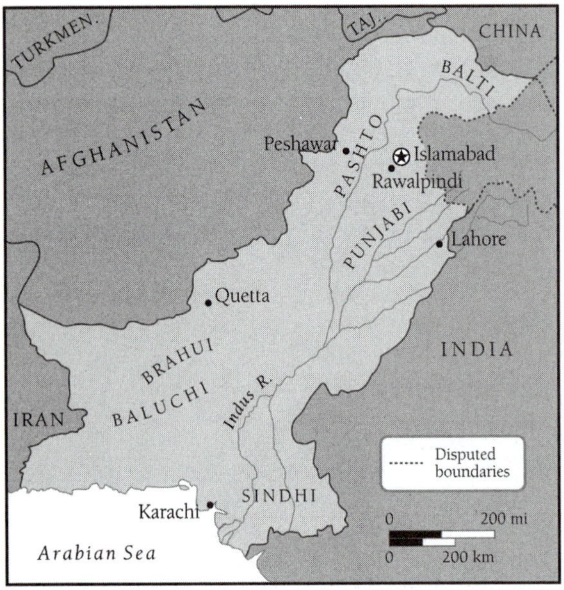

Figure 6-2

Major languages of Pakistan.

ince of Sind. Urdu is spoken by only 8 percent, yet it was proclaimed the national language. Along with English, it is the language of the elite and government. Urdu speakers are concentrated in cities, especially Karachi. Language riots broke out in Sind in the 1950s when university students were forbidden to write their essay questions in Sindhi. More riots occurred in 1965 when Urdu signs and radio programs replaced Sindhi ones in Hyderabad. In the 1970s, some of Karachi's Urdu-speaking Muhajirs, post-partition Muslim migrants from India, demanded equal representation in government even though they were a minority population. Others agitated for a separate state.

These incidents epitomized the sense of deprivation felt by Sindhi-speaking leaders and the emerging middle class. A Sindhi Language Authority was established in 1990. Language still fuels violence in Sind and elsewhere.

Several other languages are spoken, with the greatest diversity in the mountainous regions. Pashto and Baluchi belong to the Iranian branch of the Indo-European language family. Sindhi, Urdu, and Punjabi belong to the Indic branch. Brahui is a Dravidian language. Balti is a Sino-Tibetan language.

Since language is the context for thought and action, and conceptualizations in one language do not necessarily exist in another, differences can act as centrifugal or divisive forces in any centripetal drive for national cohesion.

Divisions sharpened with the economic and political policies pursued by the government in West Pakistan. There were arguments over exploitation of natural resources; industrial development; acquisition and allocation of monetary resources; communal and regional representation; and the role of Islam in government.

Although both parts of Pakistan were essentially Muslim, there were differences in practice. A history of communal strife marred the relationship between Muslims, Hindus, and Sikhs in the west. Moreover, some ethnic groups (e.g., the Pathans) were extremely fervent in their application of the mores and laws of Islam in their daily lives. For Bengalis, religious interaction with Hindus had become routine. They attended each other's festivals and reciprocated in temple offerings.

The question of Islam's role in government remains a critical aspect of political instability in Pakistan. In 1956, under pressure from conservative Islamists, the government declared Pakistan an Islamic Republic. Government was to operate according to the Quran and sunna, the words of Muhammed. Religious moderates, who had pushed for modernization and democracy, were sidelined.

Other problems centered on the rich–poor gap and other real and perceived inequities. After partition, many business-class Punjabis moved to Karachi's Sind Province, where they took over former Hindu businesses and came to dominate the economy. Added to the fray were the **Muhajirs**, Muslims who had previously lived in India. The urbane, Urdu-speaking Muhajirs, many of whom were educated and experienced in business and civil service, came to dominate administration. These Punjabi and Muhajir urban elites were resented by the Sindhis. As the government struggled to establish some degree of equity under the weight of vested interests and expected patronage, the rich–poor and urban–rural gaps widened. Urban riots and peasant revolts led to the imposition of martial law in 1958.

With ensuing economic and political instability, East Pakistan looked to a new political party—the Awami League led by **Mujibur Rahman (Mujib).** Mujib would fight for greater political autonomy for the east. The following years of hectic east–west relations served to strengthen the resolve of the Awami League.

In 1965, Pakistan went to war with India over border demarcation in the Rann of Kutch as well as in Kashmir. The year-long war was costly. Since both border disputes were with West Pakistan, East Pakistan felt unduly burdened. More disturbances again drew military rule, which lasted until 1970 when elections were to be held.

In Karachi, the Pakistan's People's Party (PPP), under the leadership of Zulfikar Ali Bhutto, ran on the slogan "Islam our Faith, Democracy our Polity, Socialism our Economy." Bhutto also declared a "thousand years war with India."

Elections were delayed because of a crippling cyclone in East Pakistan. Then, entirely dissatisfied with West Pakistan's relief efforts, the Bengalis swept the Awami League to victory, a pivotal event because the proposed system of "one person one vote" would give the more populous east a majority in the National Assembly. As it became clear that Mujib rather than Bhutto might rule Pakistan, the National Assembly was "postponed" by the military.

Withholding power from Mujibur Rahman and his Awami League was a fatal error. East Pakistan erupted with strikes, demonstrations, and insurgencies. The Bengalis refused to pay taxes. West Pakistan countered with brutal military action—complete with rampant murder, pillage, and rape. Immediately, 250,000 refugees poured into India. Ultimately 10 million would seek refuge across the border.

Faced with hordes of penniless refugees in the already poor state of West Bengal, India seized the opportunity to retaliate against its archenemy in Karachi. Indian intervention brought the east–west war to an end, and in 1971, East Pakistan became the independent nation of Bangladesh, with Mujib as its prime minister. Bhutto became prime minister of West Pakistan, now called Pakistan.

Pakistan: Land of the Pure

DIVERSE LANDSCAPES

Pakistan of 1971 was politically divided into four highly differentiated and unequal parts: Northwest Frontier Province (NWFP); Punjab; Sindh; and Baluchistan (Figure 6-3). To understand contemporary trends and problems we must look at the different characteristics of each region.

Northwest Frontier Province (NWFP)

The ragged hills of the NWFP are capped by a dramatic knot where the Karakorum, Pamir, and Hindu Kush have buckled together. Afghanistan's Wakhan Corridor, a 16,000- to 24,000-foot (4,880–7,315 m) buffer zone between Tajikistan (part of the former Soviet Union) and the former Indian Raj, lies in this region, touching China in the east. Below glaciers and barren, rocky slopes are forests of pine, deodar, and evergreen oak. At lower and dryer elevations, these give way to acacia, dwarf palm, and coarse grasses. Wood and stone houses cling to steeply walled ravines, and finely constructed rock terraces support irrigated agriculture (Figures 6-4 and 6-5).

This dramatic and broken terrain is, in fact, a **shatter zone,** a kaleidoscope of independent and feuding tribes. Many transcend the border with Afghanistan. Most are Pathan, a general term for tribes such as the Afridis or Waziris. Most speak Pashto, an Iranian dialect. While some Pathans had accepted British rule and laws, others followed *pukhtunwali,* (local Pathan law). Other groups such as the Chitralis are not Pathan.

Cutting through the mountains are the Kabul and Swat Rivers. These converge in the Vale of Peshawar, a

Figure 6-3

Physical geography and provinces of Pakistan. Compare this map with Figure 6-11. What geographic relationships do you see?

Figure 6-4

This scene shows Mount Trich Mir, at 25,260 feet (7,578 m). I took the picture overlooking the Chitral River in northern Pakistan. The alluvial terraces are irrigated by gravity flow of water. Rice, fruit, such as apricots, and vegetables are grown here.

Figure 6-5

Here in Bahrain, Pakistan houses cling to the sides of the rocky slopes. The structures are made of wood and stone with no mortar. The building in the foreground is a grain grinding house where water power turns the huge stone grinding wheels.

lowland separated from the Indus River by a sandstone ridge. Under irrigation, the deep alluvial soils support wheat, corn, sugar beets, and tobacco (Figure 6-6). In Peshawar and other towns of the NWFP, Punjabis became prominent in business. This was also a budding industrial center.

Punjab

In 1947, Pakistan's portion of the Punjab region (formerly India) held 60 percent of the population. It also dominated the armed forces and produced more than half of the GNP. Punjab long had been the breadbasket of the subcontinent. Here flow the Indus, Jhelum, Chenab, Sutlej, and other vital rivers. Out of the mountains and foothills, streams and their doabs (between the rivers) slope downward with masses of alluvium fanning onto the Indus plain. Wheat, cotton, rice, oil seeds, and sugar cane thrived under irrigation. This region had great potential for development.

Sind

Sind, more than any other region, is wedded to the Indus. To the north and west of Sukkur lie the Kirthar

Mountains. Alluvial fans permit cropping, but this Western Valley is underlain with clay and therefore subject to waterlogging and *salinization*—accumulation of salts. Northward, the land turns to clay-desert, called ***pat.***

Centuries of seasonal flooding below Sukkur have produced a doab between the Indus and the Nava Rivers. Each part of this eastern valley at some time has been traversed by an Indus stream. This fertile and productive region has invited water control works through the centuries. The doab, actually the ancient Indus delta, fades eastward into the Thar Desert (Figure 6-7).

Further south, small but significant limestone outcrops support the Sukkur water barrage and the city of Hyderabad. The delta of the Indus merges with the mud and salt flats of the Rann of Kutch. Since the river distributaries have long been in flux, the coast is dotted with dead ports. When Indus floods converge with high ocean tides, the coast floods inland for 20 miles (32 km). Mangrove and tamarisk thickets fringe old and new beach-ridges and bands of clay-like silts.

Sind epitomizes heat and aridity. Summer temperatures can be well over 100°F (37.8°C). At midnight! It rarely rains, but it does pour. Karachi, with an annual average rainfall of about 8 inches, has recorded 12 inches (305 mm) of rain in 24 hours.

Second to Punjab in size and wealth, Sind had great agricultural potential and in 1947 was in the process of being colonized mainly by immigrants from Punjab. However, rural Sind was controlled by zamindari such as Ali Bhutto. Zamindar-peasant tensions were only compounded by Punjabi and Muhajir intrusions.

Baluchistan

Although twice the size of Punjab, rugged and arid Baluchistan possessed the smallest population and was the poorest province. Sedentary farming, pastoral nomadism, and related crafts and trade were ways of life among the various tribes of the craggy outcrops, plateaus, and mesas of the Suleiman and other ranges sweeping southward and westward to the Iranian border. The southern Makran coast has been described as "bizarre," comprising gigantic pillars and pedestals, layered and etched like old teeth. Northward to Afghanistan lie volcanoes, expanses of black pebbles called ***dasht,*** shifting red dunes, and salt-encrusted ***playas*** (dry lakes) called ***hamun.*** A local proverb refers to Baluchistan as "the dump where Allah shot the rubbish of creation."

Baluchistan's only notable urban center was Quetta, a hill station strategically situated between two passes into Afghanistan. In the hills above Quetta at 5,490 feet (1650 m), juniper trees were cultivated. Oil of juniper berries gives gin (e.g., Bombay Gin) its distinct flavor. Coal reserves offered some economic potential but these have since proved to be of poor quality. This region was also known for its ***karez,*** underground irrigation systems using hand-dug, gravity-flow tunnels to divert water from aquifers to surface channels and ponds. *Karez* systems still exist in arid lands from Iran to western China.

ROLLER COASTER DEVELOPMENT

Pakistan was saddled with an infrastructure purposefully laid to connect cantonments with strategic frontiers and tuned to economic and military needs of the former Raj. Moreover, the bulk of the road and rail network remained in India. As goods historically moved overseas via Bombay, Karachi was a rather pathetic port. Pakistan gained only 5 percent of the Raj's industry. Although the British had constructed canal projects, new water power and irrigation development schemes were essential. In essence, Pakistan received only the leftovers of the vast British system.

Physical geography, cultural variation, and impractical infrastructure did not foster the economic or political integration of Pakistan. In fact, its economic history has been turbulent. Wracked with political instability, policy shifts, and changing focus, Pakistan muddled its way from one crisis to the next into the twenty-first century.

Although Pakistan was an agricultural country, early stage planners, many of whom were urbanites from India, turned to industrialization. Cotton and jute mills were built for fiber processing previously done in Bombay and Calcutta. Agriculture received

Figure 6-6

Here in Pakistan's Northwest Frontier, water buffalo pull carts of tobacco to the market at Mardan. The water buffalo are also suppliers of milk with a high fat content.

little attention, farmers were poorly paid for their crops, and lack of incentive hindered productivity. By 1950, this decline, combined with rapid population growth, made Pakistan a net food importer for the first time in its history.

The sluggish economy was revived in the 1950s when landlords returned to the political arena. With a decade of support from the *zamindari,* and benefits of the Green Revolution, Pakistan's agricultural yields rose to be among the world's highest.

Under another military government, a major geographic shift took place. Karachi, with its immigrant population, squatter settlements, and other unseemly aspects, was perceived as unsuitable as the capital of

Figure 6-7

This aerial photo shows irrigated lands along the edge of the Thar Desert in India.

a new Islamic state. For this and more practical reasons, a new capital, Islamabad, was to be built adjacent to Rawalpindi (Pindi) in Punjab, the true heart of Pakistan. Aside from its governmental capacity, it would function as a ***growth pole.*** With the evolution of Islamabad, industry and commerce would be decentralized from Karachi, Punjab would become developed, and a presence would be maintained close to the unstable regions of NWFP and Kashmir. Islamabad, meaning "City of Islam," also exemplifies the concept of a **forward-thrust capital.**

Islamabad took on its function as capital in 1963. As there are restrictive laws as to who can live in the new capital, most workers actually live in Pindi and are bussed back and forth to their government jobs.

International relations also affected Pakistan's roller coaster development. In 1949, the long-established economic interchange between Pakistan and India ground to a halt in a disagreement over monetary policy. But the Korean War (1950–1953) generated a market for leather, wool, cotton, and jute products, perfect timing to support farmers and the newly constructed processing mills. In 1954, Pakistan signed a Mutual Defense Agreement with the United States, and this brought substantial amounts of American investment. But this economic boom ended in 1965 with the outbreak of war with India over Kashmir (see the box on page 147).

The United States stopped military assistance to Pakistan and reduced its level of monetary assistance. With the breakaway of East Pakistan in 1973 and the perception of India as a "friend of the United States and enemy of Pakistan," Pakistan turned away from South Asia and toward the Islamic countries of the Middle East.

Beginning in 1974, hundreds of thousands of workers migrated to the oil-exporting countries of the Middle East to work in construction, oil, and other industries, and in services. Large numbers of these came from poor, ***barani*** (dry/not irrigated) regions. Pakistan continues to receive large amounts of remittances from the Persian Gulf states and has recently incorporated Iran and adjacent Islamic states into its sphere of interest.

Relationships with China were also a factor. Border disputes had been settled in the early 1960s and both countries pursued closer ties, taking a stand against India. In 1965, a joint-venture project was begun to link Pakistan and China by a highway through the Karakorum. Islamabad would be linked to Beijing via the NWFP and Kashgar, gateway to China's ancient Silk Route. Built largely with human labor (500 died), the 500-mile (805 km) road through the 14,928 feet (4478 m) Khunjerab Pass and Hunza was officially opened in 1986. Environmental problems continue to plague this cold and wind-swept region. Not only is the pass closed in the winter months but also landslides frequently render the highway impassable.

The Karakorum Highway Brings Change to Hunza

The new highway enabled change. Migrations from relatively crowded lowlands brought new cultural features, including building styles. Lowland cement block structures now mingle with highland stone and mud brick ones. Consumer goods appeared and merchants gravitated to the highway to take advantage of increased trade with plains cities and China. Mechanization, use of chemicals, and other Green Revolution techniques brought improvements in agriculture, increasing crop yields and reducing animal losses.

As the Monsoon does not reach this far inland, irrigation is still tied to mountain snow melt. Cropping is tied to water availability and timing. Winter-sown wheat is a priority in spring. After wheat harvest, barley and vegetables are planted. Apricot, walnut, and other fruit and nut orchards are tended as well. Working near their homes, women collect fodder and fruit, plant and care for vegetable gardens, and tend poultry and ungulates. Men look after irrigation systems, plant and harvest grain, and milk goats. Viewed as symbols of masculinity and procreation, goats can only be milked by men. Unfortunately, larger, free-ranging animal populations limit cropping area, eat seedlings, and accelerate soil erosion.

Another important event occurred in 1979 when the Soviet Union invaded Afghanistan. Suddenly, for the United States, Pakistan became a **frontline state,** a strategic barricade against the spread of communism and Soviet access to the Arabian Sea and Persian Gulf. Aid was restored, and Pakistan became the third-largest recipient of American aid after Israel and Egypt.

The war in Afghanistan brought millions of refugees into Pakistan, where they joined with their ethnic kin, not only in refugee camps but also in economic pursuits. In cities, Afghans replaced workers who had migrated to the Middle East. Afghans who migrated to the United States sent remittances to their relatives in Pakistan. In addition, the war stimulated commerce in guns and drugs that supported new economic investments. However, vital structural problems were still not addressed.

Zulfikar Ali Bhutto had been ousted in a military coup in 1977 and was later executed for conspiring to murder a political opponent. In 1988, when **Benazir Bhutto,** his daughter, became the youngest and first female leader of a Muslim state, Pakistan was essentially bankrupt. When Iraq invaded Kuwait in 1990, triggering the Gulf War, a sudden leap in oil prices, loss of remittances, and reduction of exports put Pakistan in a critical position. Assistance was provided by Saudi Arabia in return for the service of 6000 Pakistani soldiers deployed as part of Operation Desert Shield, sponsored by the United States. Even with additional support from the World Bank (WB), Pakistan has nowhere near the resources needed to escape its economic problems. Moreover, these are confounded and exacerbated by an increasing rich–poor gap, rapid population growth (2.8 percent in 1998), increased religious conservatism, and political instability.

Twenty-four years of military dictatorship, interspersed with short periods of unstable civilian government, have been characterized by mismanagement, patronage, nepotism, corruption, and violence. Leaders have been deposed, executed, or have died under mysterious circumstances. Benazir Bhutto has held office twice (1988–1990 and 1993–1997), only to fall each time under corruption charges. In recent years economic growth has faltered while communal strife and international disagreements intensify.

Azad Kashmir and Jammu and Kashmir

Muslims first arrived in Jammu and Kashmir in 711, and by the sixteenth century it was known as an "ornament of the Mughal Empire." After a period of Afghan and Sikh rule, the British, having defeated the Sikhs, sold the regions of Gilgit, Ladakh (now in India), and Kashmir to Gulab Singh, a Hindu, Dogra chieftain.

Muslim agitation began in the 1930s. With partition in 1947, Gilgit, Hunza, and other predominantly Muslim mountain states acceded to Pakistan. The British "encouraged" Kashmir's Hindu ruler to join India. That this mainly Muslim region would become part of an independent, mainly Hindu India was anathema to Pakistan. Religion was only one factor. Jammu and Kashmir contain the five headwaters of the Indus River. Control of these rivers means control of Pakistan's fate (Figure 6-8).

With Muslims agitating to join Pakistan, NWFP tribesmen entered the scene to "liberate

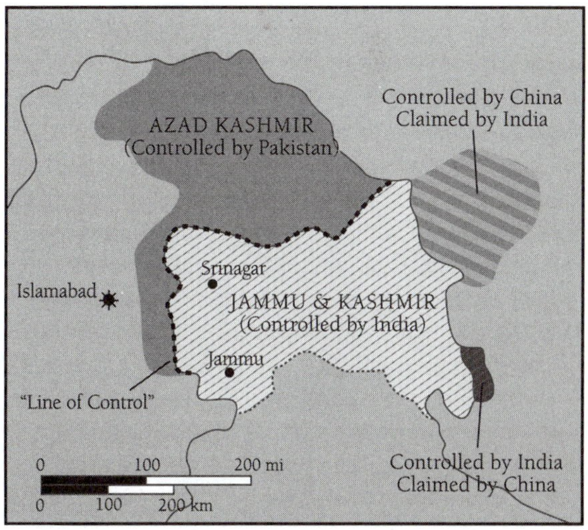

Figure 6-8

Territorial disputes. The Siachen Glacier at 22,000 feet (6,600 m) is the highest glacier and battlefield in the world.

Jammu and Kashmir" from Hindu India's grip. A revolutionary government was formed in northern Kashmir, which became known as **Azad Kashmir.** This stimulated a significant migration of Hindu Kashmiris to India.

Fearing for the lives of his fellow Hindus, the Maharajah fled from Srinagar to Jammu and signed an agreement to join India. The Indian army took control and promised a plebicite. In 1949, a United Nations' Cease Fire Line dividing Azad Kashmir from Jammu and Kashmir was accepted by both India and Pakistan. In 1956, Jammu and Kashmir formally entered the Indian Union. The promised plebiscite was never held.

In 1965, with accusations of infiltration, spying, and sabotage, another war broke out and lasted six months. The following year, in a meeting at Tashkent in the Soviet Union, Pakistan and India agreed to settle their disputes peacefully.

Ideals of peace were shattered in 1971 with the East–West Pakistan war and Indian intervention on East Pakistan's behalf. Fighting in Kashmir continued until 1972, when a new "Line of Control" was drawn. A Kashmiri Accord was signed by India and Pakistan in 1974.

Since that time, military fighting has been sporadic. Skirmishes take place mostly on the Siachen Glacier. This ice mass is strategically important because it commands a China–India-China triangle formed by China's 1962 annexation of India's Aksai Chin region.

Civilian conflict increased in the 1980s and 1990s. Muslim uprisings, election boycotting, and anti-Hindu terrorism have intensified. In 1993, the Jammu and Kashmir Liberation Front and Hizbul Mujahideen united for the liberation of Kashmir. The Indian military has been ruthless in its suppression of insurgents. One result is that, tourism, once a mainstay of Kashmir's economy, has picked up.

Srinagar and the Vale of Kashmir are open for business even as their vehicles, shops, and facilities remain damaged or closed for repairs. Unlike in the past, tourists are coming mainly from India and not from America and Europe. Many travellers see the potential for more violence. After all, this is still disputed territory.

PEOPLE WORKING THE LAND

Supplying Food and Water

In the 1990s, almost half of Pakistan's labor force worked in agriculture. Only about 40 percent of Pakistan is suitable for cropping, and 90 percent of this land is on the Indus River plain in Punjab and Sind. Traditionally, crops fall into two seasonal categories: ***rabi*** and ***kharif*** (Figure 6-9). It is estimated that women do around 60 percent of the farm work in fields close to the house. Since strict purdah is the norm in the countryside, women work outside only out of necessity and they rarely venture far from the family compound.

Agriculture plays an important role in the overall economy. Cotton and related products accounted for 37 percent of exports in 1992. Apparel manufactures added another 16 percent. Cotton acreage has been increased, and with new seed varieties, yields are above world averages. Maize, wheat, and rice are the major grains, and Pakistan specializes in the production of basmati rice. Most wheat and rice is consumed domestically—increased yields are absorbed by a growing population. Even with Green Revolution and other improvements, grain and other crop yields continue to fall short of world averages. Pakistan is fifth in the world in terms of sugar cane acreage but only sixty-ninth in productivity. Inefficiencies and inequities are systemic.

Since independence in 1947, Pakistan has increased its cultivated land by one-third, largely through expansion of irrigation. Acquisition of water has always been a challenge—waterworks date back 4000 years. Both the Mughals and the British constructed irrigation canals. In 1960, the **Indus Waters Treaty** gave Chenab, Jhelum, and Indus water rights to Pakistan and gave Ravi, Beas, and Sutlej water rights

KHARIF and RABI CROPS

KHARIF CROPS
- sown after the onset of the wet monsoon (June–July)
- harvested in autumn
- dominate southern Pakistan in Sind and Baluchistan
- floodplains and tank irrigation

RABI CROPS
- sown at the end of the wet monsoon
- harvested in spring
- dominate northern Pakistan in Punjab and NWFP
- doabs and uplands

TYPE OF CROP		MAIN PRODUCTION REGION (B = Baluch; P = Punjab; S = Sind; NWFP = Northwest Frontier Province; IRV = throughout Indus River Valley)
KHARIF	RICE	P & S
	JOWAR (MILLET VARIETY)	P & IRV
	BAJRA (MILLET VARIETY)	P & IRV
	SESAMUM (SESAME)	S & P
	MAIZE (PULSE: LEGUMINOUS GRAIN)	NWFP
	GROUNDNUTS (PEANUTS)	P
	COTTON	IRV
	VEGETABLES	EVERYWHERE
	PEARS	B & NWFP
	PLUMS	B & NWFP
	GRAPES	B
	DATES	S
RABI	WHEAT	P
	JOWAR	P & IRV
	BARLEY	P & NWFP
	GRAM (PULSES—GRAM = CHICKPEAS	P & S
	—MASOOR = LENTILS	P
	—MATTAR = PEAS	S
	—MUNG = KIDNEY BEANS)	P
	UNSEED (OIL SEED)	P
	RAPE (OIL SEED)	IRV
	MUSTARD (OIL SEED)	IRV
	TOBACCO	NWFP
	VEGETABLES	EVERYWHERE
	CHILIES	S
	APPLES	B & NWFP
	APRICOTS	NWFP
	BANANAS	S
	CITRUS	P
	GUAVA	P
	MANGO	S
	PEACHES	NWFP

NOTE: SUGAR CANE IS GROWN IN RABI AND HARVESTED IN KHARIF IN PUNJAB.

Figure 6-9

Karif and Rabi crops.

to India. Then, the WB-sponsored Indus Basin Development Fund was established to build new waterworks and link western rivers with the fertile soils of eastern Punjab. The giant Mangla Dam on the Jhelum and the Tarbela Dam on the Indus provided flood control, hydro-power, and irrigation. The 2-mile (3.2 km) wide Tarbela, completed in the 1970s, is the world's largest earth-filled dam. Today, Pakistan has the world's most extensive, continuous irrigation system.

Increasing water supply to croplands has a downside. Added water raises the water table—the upper level of groundwater—and root systems can be inundated. In Punjab and Sind for instance, the Indus water table used to be 53 feet (16 m) below the surface. With increased irrigation, it rose to within 10 feet (3 m) of the surface. In addition, salts accumulate with rapid evaporation of moisture. To counter the loss of cropland from waterlogging and salinization, thousands of pumps, tube wells, and conduits have been installed (mostly private).

Traditional sociopolitical relationships affect the effectiveness of the system. Although the government controls the main power nodes and channels, intermediary and smaller channels are regulated by longstanding local power structures, relationships, and legal codes. Consequently, water is not necessarily delivered when crops most need it, and those at the system's extremities might not get enough, if any. The whole system was set back in September 1992, when floods destroyed countless irrigation networks and displaced as many as three million people.

Who Owns the Land?

Inequities remain apparent in farm ownership. Pakistan has struggled with land reform in efforts to reduce absentee landlordism, shrink large farms and enlarge small ones, consolidate fragmented holdings, and increase peasant ownership. However, these efforts have met with strong opposition from the landed rich, who also hold political power. In 1947, a mere 1 percent of farmers owned a quarter of the agricultural land, and 65 percent of farmers held only 15 percent of the land. While there has been some size reduction and an increase in the number of medium-sized farms, 34 percent of farms are still under 5 acres (2 ha).

About one-third of Pakistani farmers are tenant farmers required to give 50 percent of their crop to the landowner. And, despite 1992 laws opposing the practice, **indentured servitude** is widespread, especially in Sind and parts of Punjab. Poor farmers, forced by circumstance to borrow from *zamindaris,* become victims of usurious interest charges and can remain eternally indebted to a land-owning family willing to exploit this situation for generations.

Industrialization and the Question of Energy

With government backing, industrialization has progressed at a rapid pace. Even so, heavy industry is limited. Pakistan has only one integrated steel mill, a Soviet Union–aided venture near Karachi. This mill depends on imported iron and coal and still runs at only partial capacity. Moreover, industry is very much tied to agriculture. For example, in 1993, manufactured goods made up 64 percent of exports, mostly cotton products and processed foods. Pakistan's major trading partners are the United States, Japan, and Persian Gulf nations. Virtually no trade is conducted with India.

In recent years Pakistan has pursued trade relationships through supranational organizations. In the mid-1980s, the **South Asian Association for Regional Cooperation (SAARC)** was formed to disassociate political conflicts from economic cooperation. SAARC includes Bangladesh, Bhutan, India, Maldives, Nepal, Pakistan, and Sri Lanka. Many meetings have been held. The **Economic Cooperation Organization (ECO),** established by Pakistan, Iran, and Turkey in 1985, now includes Afghanistan, Azerbaijan, Kyrghizistan, Tajikistan, Turkmenistan, and Uzbekistan. In 1994, construction started on a highway connecting Pakistan with Turkey via Iran.

Energy has been of paramount concern both for industrial production and improvement of quality of life. Water-generated power is of great significance but alternative sources are being sought. Unfortunately, Pakistan's coal is relatively poor quality, but natural gas sources have been developed in Baluchistan and oil is produced in Punjab. Natural gas pipelines link Islamabad, Faisalabad, Peshawar, and Quetta, and another line from Iran via Afghanistan is under construction. Additional oil fields have been discovered in the south. At least 20 international companies are involved in their exploration. Development has been limited by lack of security in both Sind and Baluchistan.

The fact that a third of Pakistan's power is still derived from animal dung, wood from already severely deforested areas, and ***bagasse,*** sugarcane fiber, implies that the country must continue to seek alternative sources for power production.

Nuclear power is one alternative. Pakistan's first heavy-water nuclear reactor was built near Karachi with Canadian aid in 1965. In 1974, when India exploded its first atomic device, Ali Bhutto pledged an equal response: Pakistan would build an "Islamic Bomb" if they had to "eat grass" to do it. In the 1990s, another plant may have opened at Chashma on the Indus about 149 miles (240 km) south of Islamabad. Fuel is provided from a uranium enrichment plant at Kahuta, also near Islamabad.

Although North American and European nations have stopped providing nuclear technology to Pakistan for fear of its military intentions, China continues to do so. And in 1998, Pakistan exploded an atomic device. India responded with its own nuclear event. You can read more about the political intrigue associated with this dangerous situation in Chapter 7.

Electrification of villages has been a high priority. Power in the countryside can light homes, pump water, and grind grain. The government is convinced that improving rural conditions will stem the ongoing rural-to-urban migration streams that have turned Pakistan's cities into unmanageable, crowded behemoths.

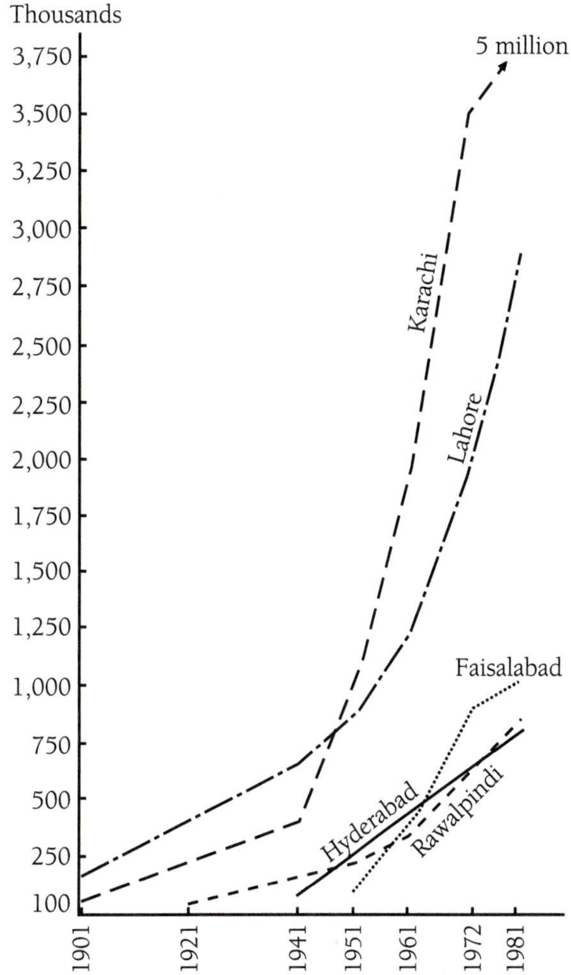

Figure 6-10

Hyperurbanization in Pakistan's five largest cities. Source. Pakistan Statistical Yearbook. 1985. *Karachi: Federal Bureau of Statistics.*

HYPERURBANIZATION

The spectacular growth of cities in Pakistan can be summed up with the example of Karachi, which had a million people in 1950. This rose to 8 million by 1991 and 10 million by 1994. Lahore, the second largest city, had 850,000 in 1950, and now has more than 3 million people. City growth in Pakistan is fast enough to be called **hyperurbanization** (Figure 6-10).

City development has always been tied to the availability of water. Urban centers have the power structure to influence water projects. Understandably, patterns of urban distribution remain associated with rivers and canal systems, and administrative strength and core elites continue to be linked with urban places.

Under colonialism, cities were created and supported to serve colonial purposes. Along the Grand Trunk Road, towns were developed at river fords to function as centers of control, administration, collection, and distribution. With the introduction of railroads, urbanization gravitated to rail stations and lines.

Bypassed, small communities were isolated, and many became deserted. Expanded irrigation projects increased agricultural production, but the Green Revolution and land reform favored medium-sized farms. Very small farms were absorbed. Large-scale, rural–urban migration ensued, and towns became specialized in various agriculture-related processing and manufacturing activities.

Three-quarters of Pakistan's urban places are situated in the Indus Basin (Figure 6-11). Most of the remainder, in the uplands of Baluchistan and NWFP, are more like overgrown towns. In fact, it is the largest cities that are experiencing the greatest growth. For example, Karachi and Lahore grew more than 250 percent from 1951 to 1981. Most of this growth was from in-migration. Afghan refugees have been major contributors to rapid growth in Quetta and Peshawar.

There have been several Five Year Plans and urban plans devised since independence. Political instability, vested interests, and resistance to change have rendered the overall results disappointing. Diffusionary tactics, including growth-pole development (moving the center of power to decentralize activities), have floundered because new industry and ancillary activities tend to gravitate to existing agglomerations and infrastructures. Furthermore, in Pakistan rail rates favor locating as close as possible to the port of Karachi and the newer Port Quasim.

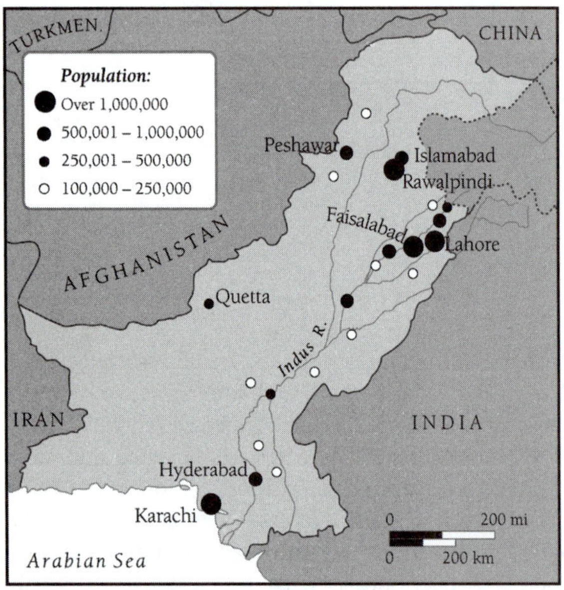

Figure 6-11

Pakistan's cities 1994.

Karachi, the country's primate city, continues to grow by accretion of ***katchi abadis***—squatter settlements on the periphery. Efforts to decentralize into new *metrovilles* outside the city have been largely unsuccessful. Land intended for the poor is purchased by speculators, who leave it vacant. Even some of the low-income participants, after moving into metroville housing, sell their properties to upper-income buyers for immediate monetary gain. The city government now concentrates on improving services and supporting self-help programs in the *katchi abadis* and other poor areas of the city. Meanwhile, the larger infrastructure is crumbling, resulting in contaminated water, accumulation of refuse, and power shortages for everyone. Frustrations arising from these problems are exacerbated by rivalry between the richer, Urdu-speaking Muhajirs and relatively poorer Sindhis, Pathans, and other ethnic groups.

A significant aspect of urban populations in Pakistan is that they are largely rural people. Consequently, urbanization is couched in rural tradition. Most work is in the manufacturing and service sectors and most is labor-intensive. The informal sector of the dual economy is critical to the survival of millions.

Urbanization is obviously tied to population growth. Although Pakistan has had family planning since 1952, programs have not been very effective. The rural fertility rate is 6.6 children per woman; the average age at marriage is 17. With only 11 percent of women using any form of contraception, most are exposed to child bearing for 33 years! Needless to say, infant mortality rates are high. The already-low 1985 sex ratio of 100:91 is decreasing indicating better care and increased life expectancy for males versus female infanticide, neglect, and high infant mortality rates for girls.

Part of the problem is the repression of women and lack of education. Enforced seclusion and traditional attitudes prevent most females from attending school or even learning to read. A 1996 WB report says that Pakistan's literacy rate is around 26 percent, but it is only 16 percent for women. Although schools have been built for girls in several areas, less than 10

Figure 6-12

In Saidu Sharif, these boys are rolling balls of opium to be sold in local shops for processing elsewhere.

percent of girls ever attend. More girls go to school and attend university in urban centers, but life roles of "suitable positions," arranged marriage, and general subservience to men are still expected for most. The status of men among men is vital in South Asian, Muslim society and must be maintained in part through modest and deferential behavior of women, especially those in a man's immediate family.

Muhammed Ali Jinnah had been a strong supporter of rights for women. Under previous governments, gains were made such as women regaining the right to hold property. The election of Benazir Bhutto suggested that progressive strides would be made. But modernization has impacted primarily elite women. Very few women work in formal or informal sectors full-time, if at all. There are scores of women's groups such as the Women's Action Forum (started in Karachi in 1981), but opposition to women's rights has only grown stronger. With Islamization being carried forward in a relatively conservative fashion, women's movements (largely urban and elite) face tough obstacles. Increasing violence against women in recent years has put women's groups into a position of defending the few rights they have rather than acquiring new ones.

Pakistan's rich–poor gap has not narrowed significantly. A 1970 study showed that 80 percent of capital wealth was held by 22 families. In 1991, the poorest 20 percent had 8.4 percent of the nation's income while the top 20 percent controlled 40 percent. With nearly 80 percent of Pakistan's population over 25 having no schooling, this situation is unlikely to change.

Endemic poverty has spurred the informal economy. In recent years the drug and arms trade, strengthened by upper-level collusion and mafia-like operations, has become a critical issue within the international community (Figure 6-12). Not only does this trade generate income but it allows Pakistan a measure of political pull in this and other regions of turmoil.

Today, Pakistan is beset with problems. Regional and structural economic imbalances; rapid population growth and urbanization; and sociopolitical conflicts riddle this turbulent land. When considering Pakistan's volatility, keep in mind its rocky beginnings and increasing complexity.

Himalaya: Abode of Ice and Snow

Now that you know something about the trials and tribulations of Pakistan and the turmoil in still-disputed Kashmir, I trust that you will be interested in learning about the often-called "exotic" countries of Nepal and Bhutan. Both are landlocked, isolated, and very mountainous domains. Both face dilemmas associated with modernization.

The Himalaya extends across eight nations: Afghanistan, Pakistan, India, China, Nepal, Bhutan, Bangladesh, and Myanmar. The relative isolation promoted by the world's highest peaks and deepest gorges has sustained both natural and human diversity amongst the 120 million that live here. Many people speak Sino-Tibetan languages and practice **Tantric** or Vajrayana **Buddhism** (Figure 6-13).

The critical positioning of these various peoples between the political giants of India and China has deeply affected their ways of life and threatened or even eliminated their autonomy. Only Nepal and Bhutan remain as independent states in the Himalaya. Ladakh and Sikkim once independent, are now part of India. Border conflicts and unstable political boundaries are facts of shatter-zone existence.

THE KINGDOM OF NEPAL

Landlocked Nepal is pinched between Tibet (China) and India. Nepal's history has encompassed both survival in context of environmental constraints and maintenance of autonomy and cultural identity within the clutches of competing world powers.

Ten Himalayan peaks mark Nepal's boundary with Tibet (Figure 6-14). Here glaciers and treeless tundra give way to **cloud forests** that coat the slopes from 8,000 to 13,000 feet (2,400–39,000 m). These mountain rainforests are blanketed year-round in fog and mist. Up to 140 inches (356 cm) of rain fall from June to September, but even in the dry season, the forest drips with moisture. Oaks, laurels, maples, and magnolias mix with ancient spruce, fir, and cedar trees. Orchids, morning glory, azaleas, giant ferns, and rhododendron color the gray-green forest mass. Flowering stands of rhododendron, standing 60 feet tall, turn acres of land crimson. This once-pristine forest houses rare and unusual animals such as the red panda, sun bear, clouded leopard, fishing cat, and goat-antelope. Human encroachment has put this habitat in danger.

Southward from the high Himalaya, the land drops precipitously into lower mountains and foothills. This region contains the 12-by-15-mile (19 by 24 km) alluvial Kathmandu Valley. Here are Nepal's main urban centers, including the capital of Kathmandu with about 425,000 people (Figure 6-15). About two-thirds of Nepal's people live in this hill region, terracing the slopes with rice and wheat. Yields are generally poor. Below lies the marshy plain known as the **Terai** at 900 feet (270 m). In recent years large

Figure 6-13

These are only two of a plethora of gods in Tantric Buddhism practiced in the Himalayan regions of South Asia.

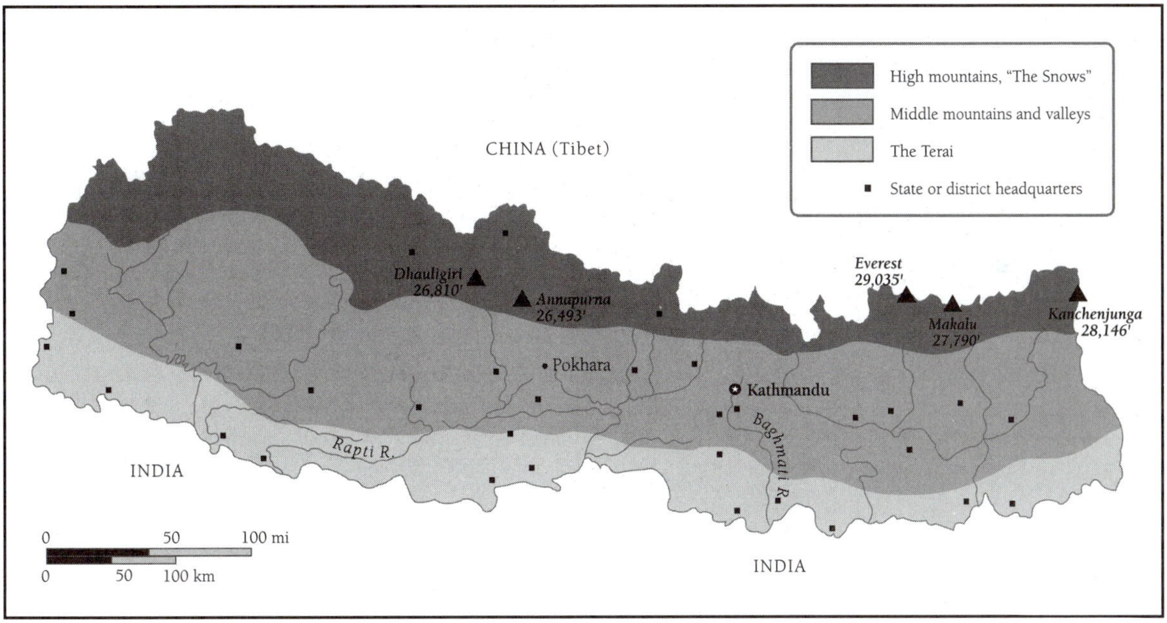

Figure 6-14

Physical geography of Nepal. Nepal serves as a buffer state between China and India.

numbers of migrants from the hills have descended into Terai to find jobs (Figure 6-16).

The Terai is an extension of the Indo-Gangetic plain. Approximately one-third of Nepal's population lives here. Many of these are high-caste Hindus of Rajput extraction from India. The Terai constitutes only 17 percent of the land but generates 60 percent of Nepal's GNP from commercial agriculture and industrial concerns. Competing for space is the Royal Chitwan National Park, with its tropical monsoon forests, giant sal trees, 12-foot (3.6 m) high elephant grass, and diminishing populations of tigers, one-horned rhino, sloth bear, and an array of other animals and birds. Here too flows the sacred Rapti River, a tributary of the Ganges.

One of my most memorable journeys was in a dugout canoe moving silently along a narrow channel of the Rapti. Resting on the narrow banks, in the shadow of the towering elephant grass, were mugger crocodiles. A one-horned rhino stood belly deep in the water munching on river plants. As the sun rose higher, the muggers slithered into the water, under my canoe! I was further startled by a ruckus in the upper branches of the tall sal trees that suddenly came alive with bird cries. Hundreds of parakeets soared overhead, streaking the blue sky with their brilliant green plumage.

Nepal, with its population of 23.7 million (1998), is culturally and socially diverse. Nepali is an Indo-European language related to Hindi but infused with Tibeto-Burman. Another indicator of the country's crossroads situation is religion. Although most people are Hindu, Buddhism and Hinduism intermingle in practice. Life is further complicated by the influence of the Hindu caste system. Here people are differentiated not by occupation but by group customs. The caste system affirms group cohesion and its flexibility adds to the confusion.

Nepal is the world's only Hindu Kingdom. The king is the head of state. However, it is also a parliamentary democracy headed by a prime minister. Historically, the region was fragmented in competing states. In the eighteenth century the western Gorkhas defeated the ruling Newars of Kathmandu, and in

Figure 6-15

This photo of Kathmandu shows a Hindu temple backed by a modern office building. Notice the cloth market held on the temple steps.

1790, the country was unified. Border disputes with British India led to the Anglo-Nepalese war of 1814–1816, a war that was fought with the rhythm of the monsoon. In the end, Nepal had to accept a British resident.

From 1816, Britain accepted Nepal as a buffer state. Nepalese troops—the **Gurkhas**—fought for the British against the Indian Mutiny of 1856 and made their way into the annals of military history for their fighting ability and heroism. Gurkha regiments would fight for Britain in many battles, which contributed to the granting of Nepal's independence in 1924.

Gurkha regiments stationed overseas have contributed significant remittances to Nepal. The return of Hong Kong to China in 1997 and pullbacks from other areas of the world have reduced British Gurkha troops by thousands. While they are stationed in Singapore and elsewhere, most Gurkhas now fight for the Indian army. Remittances have been severely reduced and many older Gurkhas, dismissed without pension, live in poverty.

Nepal is one of the world's least developed countries. Its problems derive from both internal and external causes. Political instability, corruption, and favoritism have characterized the government throughout most of this century. Rule through a tiered *panchayat* (council) system has been largely ineffective at the village level. The established power structure controls funds and reaps the most benefits. Interestingly, a recent political party called itself Panchayat. It became very unpopular; to speak of "village panchayats" now is politically incorrect. Political dissention continues to plague Nepal's government even after the overthrow of the king in 1990.

Nepal's buffer situation places it in a difficult position. Large numbers of refugees have entered from Tibet, especially after China's takeover in 1959. The Chinese want this migration stream to be stopped. Nepal has bought arms from China, and India wants that activity to cease. India wants rights to purchase land in Nepal, first rights for power from Nepal's hydro developments, and an end to border smuggling. Given the physical setting, a strong relationship with China is unlikely. Physical and cultural propinquity position India as the country to be dealt with.

Indians are already prominent in Nepal's industrial and commercial realm. Town development in the southern Terai, dam building on at least five rivers, the beginning of a power grid shared with India, manufacturing enterprises of items such as garments, carpets, cement, sugar, bricks, soap, and matches are largely Indian investments. Interestingly, apparel and carpets are exported primarily to the United States and Germany. Imported manufactures come mainly from India.

Aside from politics, there are numerous and complex economic, social, and environmental problems. Recent research has demonstrated that multiple agen-

Figure 6-16

This is a Tharu house in the Terai region of Nepal. The Tharu migrated here from India many years ago. The house is of wattle and daub (reeds and mud), with few openings, to repel insects. The inside of the house is very dark, but this is not so important because most work is done out of doors. The roof is a convenient place to air bedding and dry clothes.

cies working in specific areas have failed to recognize the interrelationships among all of these. Moreover, integrated and holistic approaches have been difficult to implement in context of unfair power relations. The rural poor with their "overpopulation" and "inappropriate" land practices were usually blamed. However, *blame the victim* philosophy is no longer in vogue.

Researchers are beginning to recognize that the people of Nepal (and elsewhere) managed to survive in their difficult environment for centuries prior to Western impacts; modernization and commercialization have severely damaged both fragile environments and cultural integrity. Many "solutions" have backfired, primarily because they are couched in the industrialized world's "scientific expertise" and "development" (read economic growth and consumerism) mentality.

While Nepal has one of the highest rates of foreign aid, it ranks one of the lowest in human development. Nearly half its population is below the age of 15. Prenatal care is free but only 24 percent of newborns receive it. Contraception use doubled from 1987 to 1997 to 29 percent. Overall agricultural production is declining, and about half of the labor force works only part time. Much aid has been siphoned off into the coffers of bureaucrats and politicos. Unlike some other Himalayan regions, lack of road development keeps its hill regions isolated and in chronic poverty. Road building, by creating access to the more developed lowlands, could bring health care, education, and basic consumer goods. But lack of roads also prevents commercial forestry and other environmental impacts.

Because change is slow, attitudes persist even in light of new approaches to solutions. Even with grassroots participation and co-management programs, according to geographer Barbara Brower (1993), "the old ideology, though explicitly repudiated, lives on in the assumptions, goals, plans, and tactics of Western-trained professionals seeking to involve local people in programs of resource management."

Nepal relies heavily on tourism, although this benefits primarily the hill and mountain regions. Strangely, Nepal, under the guidance of Western advisors, has invested little in this sector. Tourist activities are largely uncontrolled. According to geographer Nigel Allan (1990), "Nepal has become a petting zoo for upscale, middle-class foreign tourists who are anxious for a holiday experience in the exotic and mysterious 'East'."

DRUKYUL: LAND OF THE THUNDER DRAGON

Drukyul, more commonly known by the Indian assignation of Bhutan, is a Lamaistic Buddhist kingdom struggling to keep its culture and identity in the face of outside influences and modernization. Landlocked and isolated, it was virtually closed from the outside world until the 1960s. Like Nepal, Bhutan is one of the world's least developed countries.

This Switzerland-size kingdom once comprised monastic power bases centered on ***dzongs***—fortified monasteries. Its deep valleys were linked by cultural and trade contacts facilitated by iron-chain suspension bridges constructed in the fourteenth century. Given the highly differentiated terrain and natural communities ranging from glaciers and tundra in the north to tropical rainforests and malarial swamps in the south, cultural variety is to be expected (Figure 6-17).

The Bhote tribes of northern Bhutan are originally from Tibet and adhere to the Drukpa Kagyudpa sect of Tibetan Buddhism. Commonly called **Drukpas,** they revere the king as their spiritual leader. The Nepalese of the south are primarily Hindu. Most people speak Dzongkha, a Tibetan dialect. Others speak Sharchop or Nepali or one of at least 18 different languages or dialects.

In 1731, Tibet imposed suzerainty over Bhutan. Forty years later, in light of constant regional struggles and raids into Assam and Bengal, the British became involved. After a lengthy civil war one family triumphed and in 1907, the first hereditary king was enthroned with British support. Four years later, Bhutan signed a treaty with the British to counter Chinese designs on the region and later agreed to rely on the advice of India (Great Britain) for action in foreign affairs. This agreement was formalized by India in 1949.

While Bhutan's current foreign policy leans toward isolationism, China's 1950 claims to Bhutan and its 1959 occupation of Tibet cemented stronger ties with India. The first road from India to the capital of Thimpu was opened in 1962. Now, several roads from India are furthering cultural intrusion and the trappings of development. For India, Bhutan is a convenient buffer state between it and China. It is a strategic training and staging ground for troops that could be quickly employed against China or separatist groups in Assam. Bhutan is also a source of hydroelectric power, convenient to India's industrial region west of Calcutta. India is active in dam construction and has negotiated power-sharing agreements. In fact, most of Bhutan's generated power is used by India. As most foreign aid and development assistance is also from India, Bhutan finds itself in an ever-tightening dependency relationship.

Modernization began in the 1950s. However, the country is too small to develop any economies of scale. For example, Druk Air now has two planes. Even the

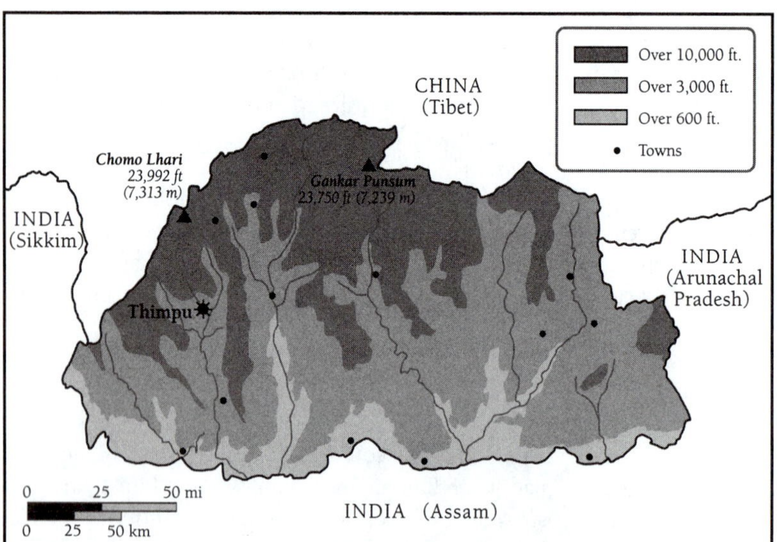

Figure 6-17

The Kingdom of Bhutan.

capital of Thimpu has only 20,000 people. But, the king once pointed out that for his people, "gross national happiness" supercedes "gross national product." Although there is some small-scale industry in the south, including chemicals, cement, and wood products, the majority of Bhutan's inhabitants work directly with the land.

Subsistence farmers cultivate grains and vegetables, and herders raise yaks and cattle and practice *transhumance* (moving livestock between the highlands and the lowlands). A mere 6 percent of the land is cultivated. Most cropping is confined to the flatter alluvial valley floors and narrow terraces. However, Bhutan is no longer self-sufficient in cereals. Efforts are being made to diversify from traditional maize and rice to fruit and vegetables, especially for export to India. One long-standing export crop is **cardamom,** a spice commonly used in South Asian cuisine. There is evidence that some commercial crops are culturally and environmentally inappropriate.

The Land of the Thunder Dragon is in trouble. Nepalese have resided in Bhutan since the nineteenth century. In the mid-twentieth century, many of Nepal's soaring population have illegally migrated across the border, settling in the southern lowlands of Bhutan alongside the existing Nepalese community. When Bhutan conducted a census in 1988, it discovered that ethnic Nepalese made up an unspecified but significant proportion of the total population. The former kingdom of Sikkim had already been overrun by Nepalese and, for protection, elected to join India. Bhutan's king did not want a repeat performance in his realm.

In an effort to salvage Bhutanese culture, the king launched ***Drignam Namzha,*** a "traditional values and etiquette" policy that promoted Dzongkha in the schools and required all people to wear traditional Bhutanese dress for certain occasions. Many Nepalese objected. But when the king declared 1958 as the cutoff date for legal immigration and asked for proof of citizenship, they were outraged. The Nepalese demanded political representation and were ultimately expelled from Bhutan. Currently an estimated 80,000 to 100,000 Nepalese live in refugee camps in southeastern Nepal. Recently, Bhutan has softened its border policies and set up screening agencies for immigrants. The influx has slowed dramatically. But, the refugee problem remains unsolved, and violent acts in Bhutan are increasingly common.

Bhutan's ruler and his Drukpa subjects are very concerned about the precarious state of their Buddhist culture and the endangered state of their physical environment in light of modernization. Consequently, tourists are limited to 3000 a year and must be on guided tours. Sixty percent of the land is designated as forested, and sustainable forestry is national policy. Nevertheless, reforestation programs are not keeping up with deforestation. Large tracts have been set aside as floral and faunal preserves. With its array of biomes, Bhutan houses the world's only population of golden langur and some of the few remaining populations of one-horned rhino, tiger, and Asian elephants. A greenbelt and elephant corridor have been created along the Indian border.

Even with sustainable development policies, Bhutan has numerous social and environmental problems. Poverty, disease, illiteracy, high infant mortality, and low life expectancy are universal. However, Bhutanese women have more freedom and power than in most other South Asian societies. Environmental problems are increasingly widespread and are similar, if not as intense, as in other Himalayan communities.

MOUNTAIN DEVELOPMENT AND ECO-DESTRUCTION

Magnificent, awesome, challenging—such descriptors belie the reality of the Himalaya's geological instability and eco-vulnerability. In this oft-perceived "barren wasteland," millions struggle daily to eke out a living. Long subject to earthquakes, landslides, and flash floods, Himalayan populations more and more are confronted by problems related to environmental fragility and degradation.

There are three geoecological zones. The Outer Himalaya has an average elevation of between 3,280 and 6,560 feet (1,000–2,000 m). This lowest region extends almost unbroken from the Indus River to the Brahmaputra and includes numerous long, flat-bottomed valleys called duns. These are filled with gravelly alluvium (look again at Figure 2-1).

In between this and the Great Himalaya is the Middle Himalaya. About 129 miles (80 km) wide and between 9,840 and 16,400 feet (3,000–5,000 m) in altitude, this zone comprises the Himalaya foothills. Ranges run obliquely from the Great Range or stand disconnected. Most people live here, concentrated in such valleys as the Kathmandu Valley of Nepal and the Vale of Kashmir. This is the region of most critical environmental and human damage.

The Great Himalaya, the highest zone, is sparsely populated. But, in recent years people from the Middle zone have come into the region, placing additional stress on limited resources. This is also the environment affected most by tourism.

Tibetan-influenced Buddhism predominates in the region of the Himalaya, although Hinduism and Islam are important in places closer to India. Monasteries, chortens, mani walls, prayer flags, and pilgrimage tracks are Buddhism's impress upon the Himalayan landscape. Dwarfed by Chomolungma, "Mother Goddess of the World" (Everest), Kanchenjunga, Dhaulagiri, and other pillars of the gods, it is little wonder that millions believe in the sanctity of Himalayan peaks.

Mount Kailas

Not the highest, but most significant in human terms, is 22,028 feet (6608 m) Mt. Kailas. Believed by both Hindus and Buddhists to be the sacred center of the universe, Mt. Kailas is central to religious conceptions of the world known as *mandalas.* From this sacred pinnacle flow the four great rivers: the Sutlej, Indus, Ganges, and Tsang Po (Brahmaputra). In reality, the sources of these important rivers lie within a 60-mile (197 km) radius of Mt. Kailas.

The mountain also figures prominently in pilgrim circulation. Thousands of pilgrims converge on this site each year to complete the 34-mile (55 km) circumnambulation upward to 18,600 feet (5580 m). In 1900, Kawaguchi Ekai, a young, Zen monk from Japan, walked across Nepal and traversed an east–west trail across the Tibetan Plateau. As recorded in Scott Berry's book *A Stranger in Tibet* (1989), the monk wrote of his first vision of Mt. Kailas and its sacred lakes. "Verily, verily it was a natural mandala. The hunger and thirst, the perils of dashing stream and freezing blizzard, the pain of writhing under heavy burdens, the anxiety of wandering over trackless wilds, the exhaustions and lacerations, all the troubles and sufferings I had just come through, seemed like dust, which was washed away. . . ."

Although the Himalaya were apparently remote and inaccessible, isolation was not complete. Traders plied precipitous routes, transferring goods such as salt and wool from Tibet and silk from China to eager merchants in India. Through trade and religious pilgrimage, connectivity was maintained within and without the region. Although contact with Tibet is now limited, yak caravans continue to trek remote areas, linking them with newly built truck routes. Aside from salt and other raw materials, the caravans and trucks now carry cans of cooking oil, tanks of propane, and other manufactures. Pilgrimage continues to be a means of information and economic interchange. But the pilgrims now talk of how much their lives have changed, perhaps about the new satellite dish atop the local monastery.

Roads and Dams

The colonial era brought British explorers to the Himalaya. Thereafter, the region was open to commerce with the densely peopled plains of India. Then imported mass-produced goods were introduced to mountain communities, and the region was exploited for resources such as minerals, timber, and cheap labor supply by foreign power structures. Himalayan peoples soon became enmeshed in political and economic dependency relationships outside their control.

Development schemes such as roads and dams are typically based on the interests of outsiders. India has been very active in the region, understandably in light of its strategic interests regarding Pakistan, China, and Bangladesh.

After a border war between India and China in 1962, India built numerous road networks as high as

17,000 feet (5100 m). Road construction—especially dynamiting—destabilized the hills and soil, and the roads gave lumber companies access to virgin forests. Forests and wildlife quickly vanished. The threat of tigers was replaced by the threat of landslides.

Road construction has also delivered sweeping cultural change. Truck routes funnel in new products and new ideas. They are also conductors for increased prostitution and the diffusion of AIDS. Commercialization, resource extraction, and mass tourism have effectively devastated natural and human landscapes to the point that the integrity of traditional human–environmental relationships has been destroyed.

The Himalaya functions as the water tank of Asia. It has the highest runoff of any mountain range, providing water for one-tenth of the world's population. However, torrential rivers invite dam construction to generate hydropower for electricity and irrigation. Since the 1950s, dams have proliferated, particularly in the upper Ganges and Indus tributary areas. Dam construction destroys forests, degrades land, and displaces mountain and hill people, forcing them to move to other already stressed regions. Pakistan's Mangla Dam, completed in 1967, receives so much silt and debris from the Jhelum that its operational life has been cut from 100 to 50 years.

Soil erosion is accelerated by development projects and general overuse of natural resources. Rivers run brown with silt, precious topsoil required to sustain crop yields. Some 300 cubic meters of topsoil annually wash from the Nepal Himalaya to the plains of the Ganges and Brahmaputra below. The stain of Himalayan silt can be seen as far as 400 miles (645 km) from shore in the Bay of Bengal.

Logging

Forestry has exacted the greatest toll from the ecosystem. Large-scale cutting commenced in the latter nineteenth century, when the British hauled off 6.5 million railway ties from one of the world's finest deodar forests. Since 1950, logging, agriculture, and urbanization have destroyed nearly half of the forest cover in the Indian Himalaya alone.

Throughout the Himalaya, original village forests are so degraded that people (mostly women) must spend hours every few days to find and gather firewood. Imagine carrying at least 50 pounds of wood on your back, secured by a single wood-fiber strap around your forehead, trudging barefoot to get home at sunrise to cook a meager breakfast for the family, do household chores, feed the animals, and do farm work. Every year, a Nepalese family spends at least two months' worth of time acquiring firewood.

In 1970, a huge storm raged over the great mountain Nanda Devi (25,650 feet, or 7,695 m). Tons of earth, boulders, and trees crashed down slopes. A tributary of the Ganges rose more than 60 feet (18 m). Six road bridges, 24 buses, and some 600 homes were swept away, and nearly 200 people were killed. Mountain dwellers were stunned at the magnitude of this disaster.

Realizing that vast forest areas had been cleared by commercial companies, they were determined to fight back and save their environment. In the north of India's Uttar Pradesh state, where leopards and tigers once prowled the forests, a group of women literally held on to trees as loggers wielded their axes. This grass roots movement became known as *Chipko Andolan,* meaning "the movement to embrace the trees." As women in other areas engaged in "tree-hugging" protests, the **Chipko Movement** drew international attention and became instrumental in the formulation of India's conservation policy. Although village preservation groups have become ubiquitous, population and other pressures limit their effectiveness.

Tourism

Tourism and other forms of commercialization have had far-reaching impacts. The tourist count on the slopes of Mt. Everest alone is 18,000 a year. Tourism is a double-edged sword. It does generate revenue and employ significant numbers of locals. On the other hand, tourist facilities and activities trash and pollute land and water, trample delicate vegetation, and place even greater demand on wood supplies than local populations. Villagers strip forests to provide hot water for "needy" tourists. Tourists deface rocks and religious sites, and some even go so far as to steal the inscribed stones of sacred mani walls. Tourism also compromises cultures.

One people to be drastically affected by tourism are the **Sherpas** of Nepal. When Sir Edmund Hillary became the first Westerner to ascend to the top of Mt. Everest from Nepal in 1953, he was lauded throughout the world. Standing in the shadow of his glory was a Sherpa named Tenzing Norgay. The Hillary expedition is now known as the Hillary–Norgay expedition.

Although Sherpas are famed for their mountain-climbing abilities, they are not traditionally mountain climbers. Seeking new pastures for their yak herds, their ancestors migrated into the Everest region from Tibet around 500 years ago. They began trading with people in lowland Nepal and India. When the Tibetan border was closed by China in the 1950s, Nepal was opened to foreign climbing expeditions.

Sherpas had already served as servants and porters for British climbers attempting to surmount Everest from Tibet. However, when Hillary and Norgay achieved the summit from Nepal, the word *Sherpa* was heard around the world.

Sherpas are now dependent on tourism. Although some lead expeditions to conquer peaks, others guide trekking groups at lower elevations. Some Western climbers admit that they pay Sherpas to take risks they are unwilling to take. On some occasions, injured Sherpas have been abandoned by climbers too anxious to reach a summit. There are other negative consequences. Having earned money to be there in the first place and then dirtying the land is offensive enough to the gods. A terrible death in the snows even furthers the detriment to one's karma and diminishes chances of a good rebirth.

Master–servant relationships are now transforming into client–guide relationships. Today, many climbers go lightweight and even without oxygen. Sherpas are employed to maintain base camps. However, media attention is usually given to the large-scale expeditions, and those guides who do climb seldom receive recognition equivalent to that of their clients. Have you ever heard of Sundare Sherpa, who has ascended Mt. Everest five times without oxygen? Or Kaji Sherpa, who in 1998 climbed the peak in 20 hours and 24 minutes, two hours faster than the previous record?

About Yaks

Imagine a one-ton, hairy ox standing 6 feet high at its shoulder. Imagine a beast that is able to carry as much as 300 pounds for 20 or more days at an elevation of 18,000 feet. This is a yak, the most useful animal for residents of the high Himalaya. Aside from their packing abilities, they provide meat, milk, butter, cheese, wool, and leather. Yak products are used in traditional medicine and their tails make fine dusters for the sacred scrolls of Buddhist temples.

First domesticated in Tibet, yaks cannot survive well below 10,500 feet (3,500 m). A blood hemoglobin is thought to be responsible for its ability to exist in thin air at high altitudes. It begins to lose its condition around 7,000 feet (2,100 m). Females, called *dimos,* are smaller than males and give about one quart of milk a day. Dimos produce a calf once every two years; populations do not reproduce quickly. Yaks can live 25 years.

There are around 12 million domesticated, black, brown, and gray yaks. Wild yaks are much larger and are always black. Wild yaks are very rare; their populations diminished by illegal hunting and loss of habitat.

Yaks can be crossed with cattle to produce animals better-suited to lower elevations. This animal is called a *dzho* and looks like a cow with long hair.

Tourism is one of the fastest-growing industries in the Himalaya region (Figure 6-18). Nepal now earns one-quarter of its foreign exchange in this manner, but only 10 percent of this ever reaches the rural poor whose environment is the main attraction. Sadly, mountain dwellers can now claim to live in the world's "loftiest garbage dump."

Who is to blame for this disaster? Some argue that the mountain people have ruined their own land through population increase, overfarming, overgrazing, and careless cutting of trees. Others target outsiders—Japan, India, the West—as harboring the guilt.

Figure 6-18

This is your friendly "rent-a-yak." Tourism is slowly entering the remote regions of the Himalaya.

This dichotomy can be clarified in light of the paradigm that everything is interrelated and that a holistic approach is essential.

PEOPLE AND THE MOUNTAINS

For thousands of years, Himalayan villagers have endured by subsistence farming and herding. For many, transhumance was a way of life. But all this has changed as the region becomes more entangled with the forces of modernity.

Populations increased with immigration of Indians from the twelveth century and again with eighteenth- and nineteenth-century food surpluses. However, road construction in the 1950s allowed the advent of modern, although limited, health care. This resulted in a drastic decline in death rates, an increase in life expectancy, and, inevitably, a population explosion. Land use intensified on already limited farming and grazing areas, resulting in reduced crop yields and ruined pastures. Greater wood requirements meant that woodlands were rapidly depleted. Productive land use systems, that sustained smaller Himalayan populations for centuries were unable to support the added numbers.

The average population growth for the entire Himalayan region is 2 percent. This, combined with a decrease in the area of cultivated land, and a mere 1 percent increase in farm yields means that local food production can no longer keep pace with human requirements. Because living conditions are deteriorating, women feel the need to have more children to help them.

With self-sufficiency destroyed, ecological migration began in earnest. People sought new land in higher and even more fragile environments, where too-steep slopes and thin soils cannot support intensive cropping or grazing. A second, mostly male ecological migration to lower elevations is an ongoing process. Cities and towns of the hills and lowlands swell with unemployed males, while women and children assume added burdens in their mountain villages.

Many outsider "experts," agencies, and governments have held the view that the people were "destroying their own environment" with their primitive and harmful methods. But proposed solutions have been drawn from experiences elsewhere. These are not appreciative of the intricacies of varied niche environments of mountain settings. Consequently, modernization efforts and palliative programs have worsened the situation and increased dependency relationships with outside political and economic forces.

Commercialization has transformed a barter economy to a cash economy. This, along with public intervention, improved accessibility, media exposure, and tourism have combined to raise expectations. Simultaneously, social and economic inequities have intensified, with little opportunity for improvement.

Widespread input substitution has upset the traditional balance. For example, following the decline of mountain pastures, sheep and goats were substituted for cattle. This has further destroyed habitat and exacerbated erosion. Sheep and goats, lacking upper front teeth, tend to rip out plants, unlike larger ruminants that are able to cut vegetation. This ruins pasture and encourages soil erosion.

Shortage of firewood has increased the use of animal manure (mixed with straw) for cooking fuel. Therefore, chemical fertilizers must be substituted for the lost manure. Farmers, now dependent on alien manufacturers, can be totally ruined if they lack the purchase price or improperly use the product.

With declining grain yields, farmers have been encouraged to switch to cash crops, meaning that people who traditionally produced their own barley, rice, corn, and other crops are compelled to buy imported grains. Some commercial crops are environmentally and culturally inappropriate. New technologies, such as hand tractors, have opened nutrient-poor submarginal lands, quickly making them unfit for production. Irrigation water shortages have been created by diversion to industrial and other uses.

For development, local governments must promote links with external markets. Horticulture, hydropower, irrigation and other resource extraction, and tourism are rarely sensitive to local needs. Moreover, mountain regions tend to interact with mainstream markets on unequal terms. External markets perceive mountains as cornucopias of resources ready for the taking, and very little revenue is ever plowed back.

Government efforts in land reform are strapped by delays and muddied by vested interests. Resource-centered programs such as aforestation or pasture development focus on techniques rather than user perspectives. They are generally area-specific, not taking into account the potential impacts on related areas.

Research in the region of the Himalaya has made clear at least three important facts. First, a society's inability to live within the usable limits of a biophysical resource base generates responses that inevitably cause environmental alteration and degradation. Second, market forces can be singled out as the most significant factors underlying those responses. Third, policy makers and farmers perceive neither problems nor solutions in the same way. This derives from differing degrees of closeness to the problem and differing stakes in outcomes of either problem or solution.

While the situation is critical in many areas, especially environmental, for many people general well being has improved. Land reform, health care, income subsidies, increased education, self-help programs, and so forth have exerted some positive impacts. However, most of these have bonded mountain communities to outside power structures more tightly than ever.

Clearly this northern region of South Asia is rife with complex problems. Some of these arise from environmental constraints. Others emerge from transitional political settings. At the turn of this twenty-first century, however, two trends are apparent. On the one hand, international and global economic and strategic concerns are increasingly overriding the impacts of local geographies. On the other hand, global frameworks are countervailed by cultural and territorial identity-consciousness. This is an era of rampant and proliferating nationalism. Unfortunately, in many regions environmental integrity and social well being have become lost in the shuffle.

Main Point Summary

1. Political instability has marked the evolution of practically every state in this region. After the partition of India, a split Pakistan could not withstand the centrifugal forces that ultimately brought about its political demise. Differences in size and population were compounded by eco-

nomic imbalances and cultural contrasts. Language and religion were particularly potent forces in communal strife and the question of Islam's role in government remains a critical question in Pakistan.

2. After partition, many business-class Pujabis moved into Sind and came to dominate the economy. Others, known as Muhajirs, came to dominate administration. These groups are resented by the Sindhis, and internecine strife persists.
3. Pakistan has fought India over border demarcations. Kashmir remains an unsolved issue.
4. When East Pakistan protested the draining of its resources for these military forays and other expenditures in the West, West Pakistani troops countered with brutal military action. This stimulated a massive refugee movement into India's West Bengal. Indian intervention ensured the creation of an independent Bangladesh in 1971.
5. Pakistan is characterized by dramatic regional variation. Its rugged Northwest Frontier is a cultural shatter zone, difficult to control, and with ties to tribal groups in Afghanistan and Kashmir. Punjab, the land of five rivers, is the breadbasket of Pakistan. It is also important industrially and politically. Sind is tied to the Indus River and is second to Punjab in size and wealth. Pakistan's largest city and port—Karachi—is here. Baluchistan is a rugged and remote tribal area with little development.
6. Pakistan's infrastructure was tuned to the economic and military needs of the British Raj. This fact has combined with physical and cultural divisions, and political instability, to foil efforts at balanced economic development.
7. Pakistan's more recent roller coaster development is tied to its relationships with India, China, and the United States. China is viewed as a supporter. India is seen as a protagonist (especially in the case of Kashmir), and the United States is perceived as a friend of India.
8. The Karakorum highway has permitted greater interchange between Pakistan and China. Increased traffic, especially of tourists, has brought about dramatic change in formerly isolated tribal areas.
9. Pakistan is a troubled country suffering decades of military rule and political instability. The question of jurisdiction over Kashmir remains unsolved, and hostilities there continue.
10. Almost half of Pakistan's labor force works in agriculture. Maize, wheat, and rice are the major crops. Irrigation is critical to production, and giant dams have been constructed on the Jhelum and Indus for flood control, hydro-power development, and irrigation expansion. Additional surface water can induce waterlogging and subsequent salinization. Thousands of tube wells have been installed to avoid this problem. However, unfair practices regarding control over various parts of the system mean that many farmers do not get water when they need it.
11. Inequities remain apparent regarding farm ownership. About one-third are tenant farmers. Many are landless and spend their lives in servitude to land owners.
12. Industrialization is limited by resource availability, internal conflict, and corruption. Development of oil and gas fields by international companies is strapped by lack of security in Baluchistan and Sind. Proposed pipelines from the Middle East are hampered by political instability in Afghanistan.
13. Pakistan is a nuclear power.
14. Hyperurbanized, three-quarters of Pakistan's cities are in the Indus Basin and the largest cities continue to grow even larger. Rural-to-urban migration is a serious problem, and squatter settlements accrete to city fringes.
15. Urbanization is tied to population growth. Fertility rates are very high, and women have little say in their lives. The preference for males is reflected in a highly imbalanced sex ratio. Gender inequity is only one of a host of problems faced by Pakistan, where the majority live in poverty.
16. The Himalaya extends across eight nations. Relative isolation among the world's highest peaks has fostered cultural diversity. Nepal and Bhutan are the only independent states within the Himalaya. Both must balance their lives between the powerful influences of China and India.

17. Fantastic forests coat mountain slopes below the treeless expanses of glaciers and tundra. Populated zones, including Kathmandu, are in alluvial river valleys and foothills. Slopes are terraced with rice and wheat but yields are generally poor. About one-third of Nepal's population lives in the Terai, an extension of the Indo-Gangetic plain.
18. Nepal is culturally diverse, the world's only Hindu Kingdom, and one of the world's least-developed countries. Political instability and corruption have led to a widened rich–poor gap and environmental exploitation. India continues to play an influential role in Nepal's "development."
19. Nepal has one of the world's highest rates of foreign aid, yet it remains one of the world's poorest in terms of human development. Lack of infrastructure prevents certain forms of development but also deters environmental exploitation.
20. Nepal relies heavily on tourism, but this impacts mainly the hill and mountain regions. Both environmental and cultural integrity are severely threatened.
21. Bhutan is a Lamaistic Buddhist kingdom struggling to keep its identity under the pressure of tourism and development.
22. Bhutan, like Nepal, serves as a buffer state between India and China. Bhutan leans toward isolationism but is strongly impacted by India.
23. Large numbers of Nepalese live in Bhutan, many illegally. This is a source of tension. Bhutan does not want to go down the road of Sikkim and lose its independence to India. The Bhutanese government has launched a "traditional values" policy and program to ensure the maintenance of cultural integrity.
24. The Himalaya has three geological zones: the Outer; the Middle; and the Great. Environmental damage is most severe in the Middle zone.
25. Tibetan influences are apparent on the landscape and Mount Kailas is an important center of pilgrimage. Increased tourism is despoiling this and other sacred places.
26. Development schemes are based on the interests of outsiders. India is especially active in road and dam construction. Although the Himalaya has long been accessible by yak caravan, roads have opened the region to trucking—and pollution, prostitution, and AIDS. Human–environmental integrity has been compromised in many areas. Sherpas are examples of this predicament.
27. Overpopulation and development have led to ecological migration, which has incurred even greater environmental damage. Outside agencies blamed the people for their troubles and imposed unsuitable "solutions." Commercialization, input substitution, environmental exploitation, external trade requirements, population increase, and other factors have combined to bring ruination to many parts of mountain regions. On the other hand, modernization has improved the standard of living for countless individuals and families.

Terms to Know

Muhajirs
Mujibur Rahman (Mujib)
shatter zone
pat
dasht
playas
hamun
karez
growth pole
forward-thrust capital
barani
frontline state
Benazir Bhutto
Azad Kashmir
rabi
kharif
Indus Waters Treaty
indentured servitude
South Asian Association for Regional Cooperation (SAARC)
Economic Cooperation Organization (ECO)
bagasse
hyperurbanization
katchi abadis
Tantric Buddhism
cloud forests
Terai
Gurkhas
Drukyul
dzongs
Drukpas
cardamom
Drignam Namzha
Chipko Movement
Sherpas

Bibliography

Agrawal, Arun. 1996. Forest Management in the Indian Himalayas: An Institutional Perspective. *Human Dimensions.* 1: 6–10.

Alexander, Brian. 1987. A Beast For All Seasons. *International Wildlife.* 17: 45–50.

Alexander, David. 1984. Saving the Trees, Saving Ourselves. *International Wildlife* 14: 18–22.

Allan, Nigel. J. R. 1990. Household Food Supply in Hunza Valley, Pakistan. *The Geographical Review,* 80: 399–414.

Allan, Nigel, J. R. 1998. Book Review of P. R. Karan and H. Ishii. 1996. Nepal: A Himalayan Kingdom in Transition. Tokyo: United Nations. *The Geographical Review,* 87: 421–22.

Berry, Scott. 1989. *A Stranger in Tibet: The Adventures of a Wandering Zen Monk.* Tokyo: Kodansha.

Bray, John. 1993. Bhutan: The Dilemma of a Small State. *The World Today,* 49: 213–16.

Brook, Elaine. 1988. Through Sherpa Eyes. *Geographical Magazine,* 60: 28–34.

Brower, Barbara. 1993. Co-Management vs Co-Option: Reconciling Scientific Management with Local Needs, Values, and Expertise. *Himalayan Research Bulletin,* 13: 39–49.

Burki, Shahid Javed. 1991. *Pakistan: Continuing Search For Nationhood.* 2nd edition. Boulder: Westview.

Chopra, Pran. 1986. The View From Bhutan: The Temptations and Traps of Development. *Ceres,* 19: 24–27.

Denniston, Derek. 1993. Saving the Himalaya. *World Watch,* 6: 10–21.

Fisher, James. 1991. Has Success Spoiled the Sherpas? *Natural History,* 2: 38–45.

Flavin, Christopher. 1990. Last Road to Shangri-La. *World Watch,* 3–4: 18–26.

Gerrard, A. John. 1990. *Mountain Environments: An Examination of the Physical Geography of Mountains.* Cambridge: MIT.

Hausler, Sabine. 1993. Community Forestry: A Cultural Assessment: The Case of Nepal. *The Ecologist,* 23: 84–90.

Jodha, N. S. 1995. The Nepal Middle Mountains. In *Regions At Risk: Comparisons of Threatened Environments,* eds. Jeanne X. Kasperson, Roger E. Kasperson, and B. L. Johnson, B. L. C., pp. 140–85. London: Heineman.

Karan, P. P. and Cotton Mather. 1985. Tourism and Environment in the Mount Everest Region. *The Geographical Review,* 75: 93–95.

Karan, P. P., and Shigeru Iijima. 1985. Environmental Stress in the Himalaya. *The Geographical Review,* 75: 71–92.

Mandelbaum, David G. 1993. *Women's Seclusion and Men's Honor: Sex Roles in North India, Bangladesh, and Pakistan.* Tucson: University of Arizona.

McColl, Robert W. 1987. House and Field in the Karakorums: The Interaction of Environment and Culture. *Focus,* 37: 15–19.

Meier, Richard. L. 1985. Islamabad Is Already Twenty-Five. *Ekistics,* 52: 212–16.

Metz, John, J. 1990. Forest Product Use in Upland Nepal. *The Geographical Review,* 80: 279–87.

Rahman, Mushtaqur. 1996. *Divided Kashmir.* Boulder: Lynne Reinner.

Rahman, Tariq. 1995. Language and Politics in a Pakistan Province: The Sindhi Language Movement. *Asian Survey,* 35: 1005–16.

Rauniyar, Durga, S. 1997. Spotlight: Nepal. *Population Today,* 25:7.

Scarborough, Vernon, L. 1988. Pakistani Water: 4,500 Years of Manipulation. *Focus,* 38: 12–17.

Searight, Sarah. 1987. Priming the Pump. *The Geographical Magazine,* 64: 19–24.

Siddiqi, Akhtar Husain. 1989. Urban Development in Pakistan. In *Urbanization in Asia: Spatial Dimensions and Policy Issues,* eds. Ashok K. Dutt, Laurence J. C. Ma, and Allen G. Noble, pp. 47–74. Honolulu: University of Hawaii.

Sopher, David. 1964. Man and Nature in India: Landscapes and Seasons. *Landscape,* 13: 14–19.

Spate, O. H. K., 1948. The Partition of India and the Prospects of Pakistan. *The Geographical Review,* 38: 5–29.

Spate, O. H. K., and A. T. A. Learmonth. 1967. *India and Pakistan: A General Regional Geography,* 3rd edition. Bungay, Suffolk: Richard Clay (Chaucer).

Stephenson, Bryan. 1994. Bhutan: The Development Dilemma. *Geographical Magazine,* 66: 51–55.

Taylor, George C. Jr. 1965. Water, History, and the Indus Plain. *Natural History,* 74: 40–49.

Tayyeb, Ali. 1966. *Pakistan: A Political Geography.* London: Oxford.

Thapa, Gopal, B., and Karl Weber. 1991. Deforestation in the Upper Pokhara, Nepal. *Singapore Journal of Tropical Geography,* 12: 53–67.

United Nations. 1988. *Population Growth in Mega-Cities. Karachi.* Population Policy Paper No. 13. New York: United Nations.

Weiss, Anita. M. 1998. The Slow Yet Steady Path to Women's Empowerment in Pakistan. In *Islam, Gender, and Social Change,* eds. Yvonne Haddad and John Esposito, pp. 124–43. New York: Oxford.

Young, L. J. 1991. Agriculture Changes in Bhutan: Some Environmental Questions. *Geographical Journal,* 157: 172–78.

Chapter 7

South Asia: India, Giant of the Subcontinent

"We wear the dust of history on our foreheads, and the mud of the future on our feet."

SHASHI THAROOR

Scholar and writer Shashi Tharoor (1997) believes that "The central challenge of India as we enter the twenty-first century is the challenge of accommodating the aspirations of different groups in the Indian dream." The reality of a billion individuals, belonging to countless ethnic, linguistic, religious, caste, and other defined groups, makes India's continued existence as a federation amazing (Figure 7-1). *Federalism* is a governmental system whereby central and regional authorities are voluntarily tied in a mutually interdependent interrelationship. In this light, India is a classic paradox; it is a unity of disunity. It is multitextured, a cauldron of competing interests, a collection of centers and peripheries (Figure 7-2).

Independent India was conceived with diversity in mind. Its ability to change, redrawing boundaries along linguistic lines for instance, is a large part of its *raison d'etre*. Federalism thrives on the tension between cores and peripheries, as long as the links are not severed. With this in mind, India's Constitution grants extraordinary powers to federal authorities when unity and integrity of the nation are threatened. Thus far, separatist movements have been quashed. However, after 50 years of independence, the voices of dissention are more frequent, much louder, and more powerful. These arise from social inequity, regional disparity, and political instability. These are India's challenges.

Demographic Challenges

Figure 7-3 shows that India is already densely populated. At its partition, India's population was 344 million. According to the 1991 census, it was 844 million. By 1998, it had reached 988.7 million. Although India's NIR has fallen from 2.1 in 1978 to 1.9 in 1998, overall population growth is phenomenal. India adds at least 18 million a year—2,000 babies are born every hour! Its area is just half that of China, but it will probably outpace it in population in the twenty-first century. Moreover, food production is not keeping pace with population growth.

Fertility behavior is strongly influenced by prevailing social and economic conditions—as well as by

Figure 7-1

Khond dancers in Orissa State. I happened across this group on a road in the Eastern Ghats. They were celebrating the mother goddess, a yearly event. The dancing was accompanied by a huge feast and animal sacrifices. Some of the women danced to the point of going into a trance.

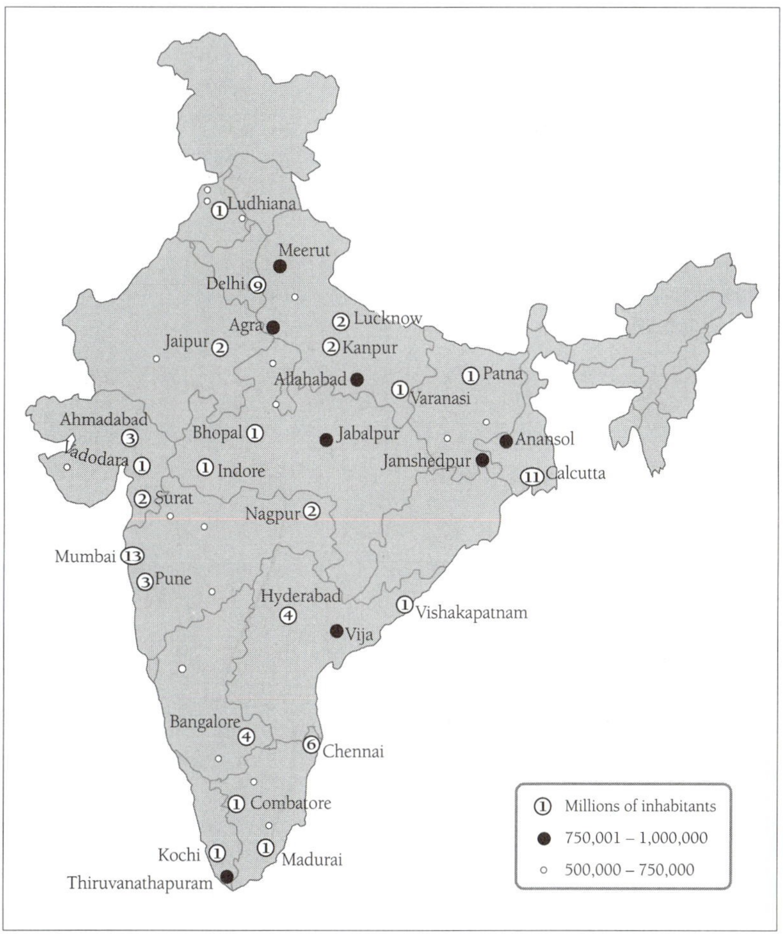

Figure 7-2

States and cities of India.

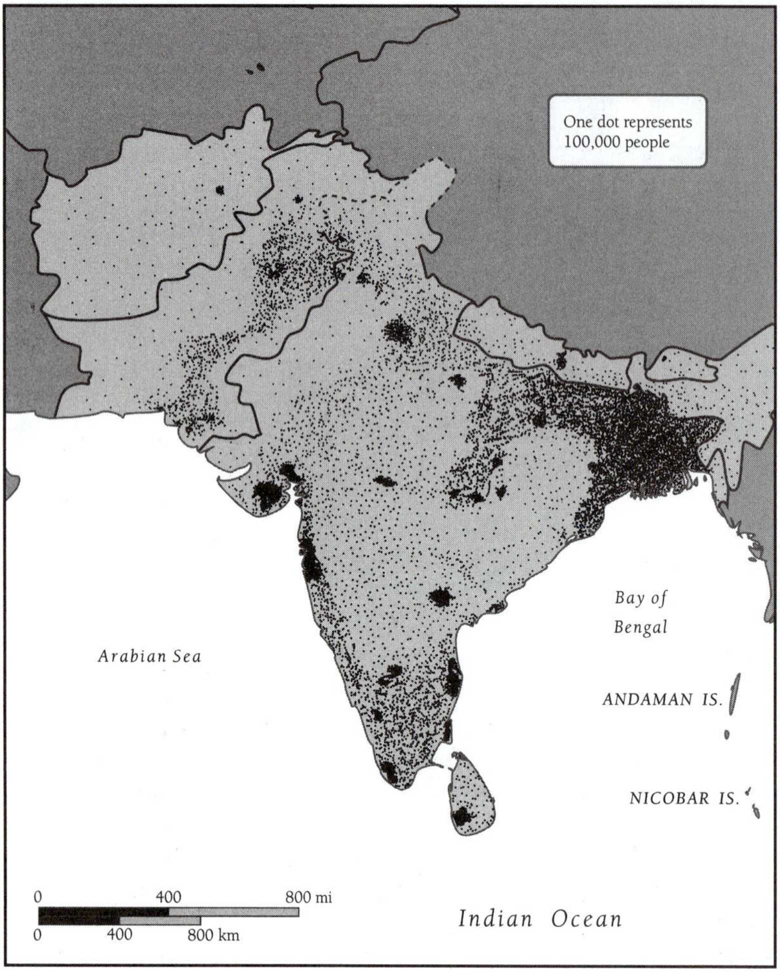

Figure 7-3
Population density.

religious/cultural norms and expectations, such as early marriage. Of the 4.5 million marriages each year, an estimated 3 million brides are between 15 and 19 years old. India has a young population, with 36 percent under the age of 15.

Not all populations are growing at the same rate; regional variations are apparent (Figure 7-4). And, as I noted in Chapter 3, poorer families tend to have more children. Poorer women, who have less control of their bodies or have no access to birth control information or devices, also generate higher NIRs (Figure 7-5).

Less than half of couples of reproductive age use contraception. The most popular form of birth control is surgical sterilization, but this is elected by couples who already have three or four children. India's TFR has fallen from 6.0 children in the 1950s to around 3.4 in the 1990s, but periods of quickly falling fertility have been followed by periods of stable levels. Whether India has reached another plateau is not clear.

In 1996, the Indian government eliminated national targets for acceptance of fertility control methods. The new "Target-Free Approach" puts family planning at the community level, where grassroots

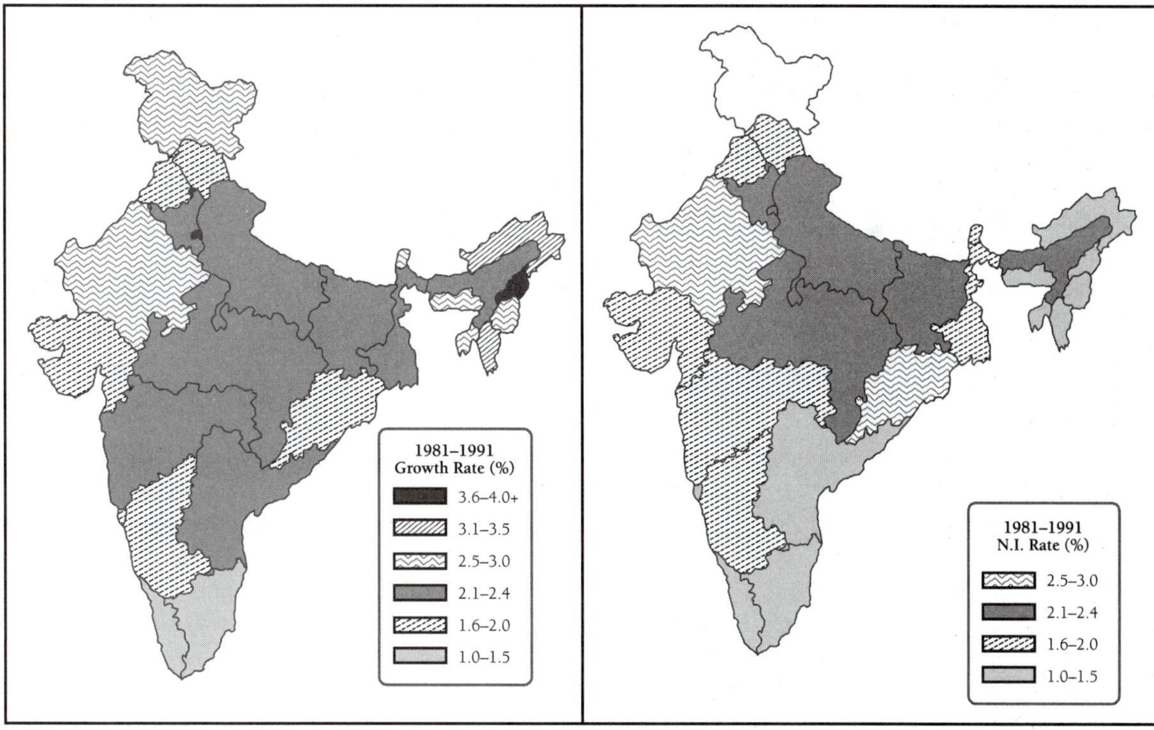

Figure 7-4

Population dynamics. Source: Observer Statistical Handbook 1998. *New Delhi.*

workers set targets according to individual needs. Acceptance of all forms of birth control rose in the 1997–1998 period. The effects are yet to be seen.

For every 1,000 men in India, there are 929 women—a figure on the decrease. This trend of decline has existed unabated since the first census in 1861. The gender gap reflects widespread discrimination against women and is more apparent in the north, where sex ratios are considerably lower than in the south. The states of Rajasthan, Haryana, and Uttar Pradesh, for instance, have much lower ratios than Kerala or Tamil Nadu.

Caste differences play a significant role in these discrepancies. There are more higher caste women in the northern states, especially of the Kshatriya or former military groups. These castes are noted for their sheltering of women from education and outside contacts. Female illiteracy rates are high, and women are expected to produce sons. Female children are not usually killed at birth but rather, neglected in their early years. Fed last and less, their ailments ignored, many girls never reach the age of five.

Another issue is health care. Some 300,000 children, more girls than boys, die yearly of diarrhea. This is attributable to lack of sanitation and neglect. There was even an outbreak of pneumonic plague in 1994. However, this does not mean that health care has not improved overall. Indeed, death rates have been lowered all over the country. Killer diseases such as smallpox have been eradicated, and infant mortality rates have been lowered. The life expectancy of Indians has been raised from age 32 in 1947 to age 59 in 1998. In fact many more Indians are living longer and are capable of having more children.

India has a less than successful history of family planning in terms of programs and their impacts. In

Figure 7-5

This young woman is approximately 15 years old. She is a member of the untouchable, sweeper caste. She lives in a temporary tent on the outskirts of Udaipur in Rajasthan.

1952, it became the first country to establish an official family planning program with the aim of curtailing population growth. Various national and state programs were instituted, including promotion of IUDs and sterilization. Three million men underwent vasectomy in 1970 and 1971.

Economic incentives were common; food rations were withheld from some families with more than three children; others lost a month's pay unless a worker was sterilized. Such programs are believed to have averted more than 85 million births from 1956 to 1986. Even so, progress was problematic.

Ethnic, religious, and caste rivalry fostered fears of persecution and genocide, and violent resistence became frequent. Capitalizing on these fears, politicians have been quick to blame family planning for other social ills. Many organized programs have fallen by the wayside. I recall visiting a village in 1985 where a "two-children" poster had been painted on the side of a school building. As the children thronged around me, I noticed that the poster was in a terrible state of disrepair. It was barely legible. One boy jumped up, pointed at it, and said in English: "Small family very funny." Everyone laughed, the teacher louder than anyone.

The 1991 census results shocked many into realizing that fertility rates must be reduced at all costs and that women must be empowered to control their lives. But women remain last in line for education, job training, credit, medicine, and even food. Constantly changing governments cannot provide the stability needed to sustain long-term empowerment strategies. Yet, experts agree that this is the most effective mode of population control.

India's population growth will continue its spectacular rise, given its current size and youthfulness. Despite the use of family planning, more than 40 percent live in areas where CBRs are in the mid- and upper 30s. WB predictions indicate that even if India achieves a two-child family by 2015, it will still have a population of 1.9 billion!

Kerala: A Woman's State?

Kerala has the highest literacy rates in India, even higher than those for Delhi. Although women still lag behind men, more than 80 percent of both have the ability to read and write. Prior to India's partition, only 11 percent were literate. Clearly, this state has made great strides in educating its population (Figure 7-6). Why is this state so different?

Kerala's government spends 40 percent of its budget to maintain a vast network of schools and colleges. Education has long been valued in society, especially among higher castes. Since

Figure 7-6

A middle-class family in Kerala. This family owns a lumber business. Note the difference in dress between the more traditional parents and the more modern children.

most higher castes were traditionally matrilineal, many women continue to enjoy freedoms unattainable elsewhere. During the Raj, most of Kerala was ruled by independent maharajas who promoted education in the Malayalam language. Literacy rates were much higher there than in states with English-based schooling. About 20 percent of Kerala's citizens are Christian, attaching more importance to equal education for males and females. Access to education at Christian schools was an attractive lure for the conversion of low-caste Hindus. In addition, knowledge of recordkeeping was essential to maintain the status of Kerala's international traders.

In 1990, Ernakulum (Cochin) became the first administrative district to be officially declared fully literate. This was the result of a literacy drive entailing motivational strategies so appealing that the project's founder received commendation from the United Nations. Other communities followed suit with the support of various official agencies. Volunteer teachers were mobilized, two-thirds of them women. Local religious leaders were recruited to back up the media blitz. Interestingly, this was accomplished by a communist state government. Most impressive has been the progress achieved by tradition-bound Muslim women.

Linguistic Conflicts

With 15 national languages and more than 100 tribal tongues in India, it is no wonder that language has been a divisive force. This is especially significant because language is even more important than ethnicity or nationality for individual and group identity. Unlike Europe, where Latin long served as a medium of cultural diffusion, in India, Sanskrit was replaced by Persian as the court language, and this was later replaced by English. Educated elites, versed in one or more of these languages, became divorced from the masses, who remained steeped in their own linguistic worlds.

Four language stocks, each with its own set of language families, produce a complex language map (Figure 7-7). The Munda and Tibeto-Burmese stocks are relegated to relatively isolated areas, including the Eastern Ghats and the Himalaya. The main cleavage is between the Indo-European and the Dravidian lan-

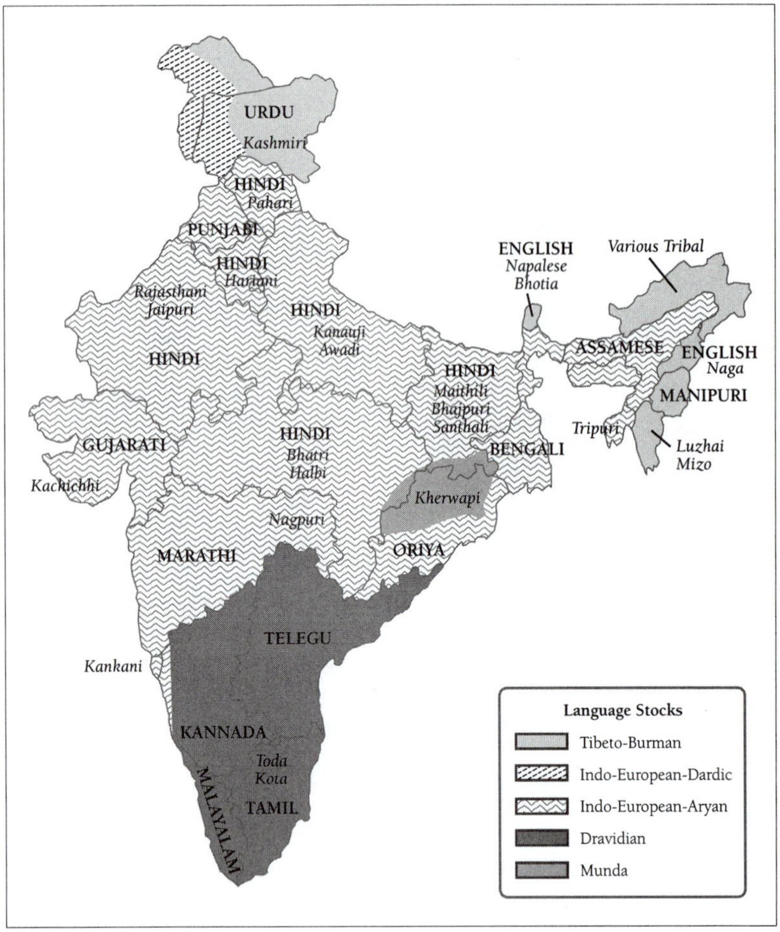

Figure 7-7

Language stocks. Note the division between Aryan and Dravidian linguistic regions.

guage stocks. The most widely spoken language is Hindi, a member of the Indo-Aryan language family. Hindi speakers also occupy India's central states, a core ringed by people speaking unrelated tongues. Dravidian includes 4 main languages and 18 lesser ones. These occupy the southern states, having been pushed out of the Indus valley with historic migrations.

Hindi is India's official language, a source of aggravation to Muslim Urdu speakers. In 1994, riots broke out in several states over the broadcast of a 10-minute news program in Urdu in Bangalore. Hindi derived from Hindustani, a language used in Mughal bazaars and military camps. It incorporates numerous Persian words, as this was the Mughal court language. There are several dialects, but Indian schools teach "Standard Hindi" written in the Devanagari (Sanskrit) script.

Urdu is similar to Hindi in grammar, but from a political perspective it is different. It borrows from Arabic and Persian and is written with a Persian script. Muslims in northern India see Urdu as their "special language."

Transcending this linguistic web is English, which was made the language of high school education in the 1830s. Contrary to Indian advice, the British ignored linguistic boundaries when establishing state boundaries. In 1953, there was a Telugu movement for a linguistic state. In 1956, there were language riots in Bombay between Marathi and Gujarati speakers. Boundaries had to be changed.

Bombay went into Maharashtra. In the south, Kerala and Tamil Nadu were formed to benefit Malayalam and Tamil speakers, respectively. Orissa was defined in the eastern tribal region, and Karnataka was created for Kannada speakers. These are only a few examples of boundary restructuring following partition.

Boundary alterations did not mask the fact that India's elites spoke and read in English—a "foreign tongue." Government efforts to force Hindi were met with resistance. It has spread, nevertheless, via media such as schools and television, and interregional migrations. Hindi is even a *lingua franca* (common language) for mixed migrant worker groups. English remains the language of private schools, colleges, and universities. Moreover, India's modern, urban social class is differentiated by English more than any other cultural force. Many intellectuals argue that an English education should be the means employed to "uplift" the poor masses.

Language issues remain significant. Recent attempts to force Goan English speakers to learn Konkani caused mass demonstrations. Politicians representing Uttar Pradesh and Madhya Pradesh are calling for the eradication of English. Disagreements and conflicts have obvious impacts on political and economic development.

No solution is apparent to India's language woes. In 1983, I left Ernakulum (Cochin) for a couple of days in the Western Ghats. When I got back, there were burned buildings and vehicles along the main road. When I asked what had happened I was told, "Oh it was nothing, madam, just a language riot."

Figure 7-8

Bonda woman and child. Numbering about 6,000, the Bonda live in the Eastern Ghats. They are hill cultivators, growing mangos and jackfruit, but still hunt with bows and arrows. Theirs is a female-dominated society and they speak Koui, an Austro-Asian language. They are geographically isolated and very independent.

Caste Complexities

Even though there is a growing ambivalence toward caste in India's upper social strata, caste rules and regulations apply among virtually all people. Brahmins, Rajputs, Marathas, and Jats are examples of regular castes. But more than two-thirds of Indians belong to *scheduled castes and tribes*. Scheduled tribes are also called ***Adivasis*** meaning "original inhabitants." These include such groups as the Nagas of Nagaland and the Bondas of Orissa (Figure 7-8). Another term employed by the government is *other backward classes* (OBCs) that include the bulk of the population—the "lower middle classes."

The Indian government has published lists including 405 scheduled castes, 255 scheduled tribes, and 3,500 OBCs. These groups are entitled to special social and political benefits. Discrimination in favor of the "backward" classes did not begin with independence. It began in the state of Mysore (Karnataka) in 1921. In this state, 92 percent of the people belong to this group for which university admissions and jobs are reserved. In contrast to predictions at the time, the state did not collapse, and the capital of Bangalore has become the center of India's computer industry.

Tribal Troubles

India has 255 scheduled tribes plus numerous unofficial ones. Many of these groups live in the Eastern Ghats and the hills of Assam. In 1944, a little-known but decisive battle was fought in the hills and jungles around Kohima in Nagaland. Here, the British broke the Japanese advance from Burma, ending their plans to capture the Raj. This remote region has remained fractious, with insurgencies against the Indian government mixing with inter-tribal conflict. Seven states cover the upper basin of the Brahmaputra River. Incursions by Christian missionaries and Bengali Muslims add to the fray.

The Nagas, who became Christian in the nineteenth century, have been up in arms since 1946. They were granted statehood in 1963. In 1993, 316 people were killed in ethnic massacres between the Nagas and the Kuki tribes in the Manipur hills. In Meghalaya, the Khasis are rebelling against outsiders, and Assam has witnessed rioting between ethnic Assamese and Bengali Muslim migrants from India and Bangladesh. If this seems very confusing, be assured that this is only part of the story of tribal troubles.

Smuggling is another part of the picture. Teak, aromatic agar wood, jade, gems, and heroin are among the smuggled goods. Eastern India is a key route for the movement of heroin out of Burma. Drug lords cum politicians rule their territories with personal armies. Border towns are hotbeds of crime and prostitution. Burmese girls, retired from the Thai brothels because they are HIV positive, now work here. Addicts number anywhere from 15,000 to 40,000 in Manipur alone. The government claims that the HIV infection rate is more than 50 percent.

Why does the Indian government spend so much money on maintaining control of this region? After all, it is attached by only a narrow corrider of land at Siliguri. New Delhi's concern lies in the fact that the region is of strategic and economic importance. It borders on four nations: China, Bangladesh, Bhutan, and Burma. Further, it is rich in oil, timber, and tea. It also has great hydroelectric potential. Culturally, it is the meeting point between Hindu-based culture groups and the animist and Christian Tibeto-Burman people of the hills. Keep your eye on this region as one of South Asia's hot spots.

Caste as a form of kinship or marriage lineage circumscribes who can marry whom and delineates mutually supportive behaviors and responsibilities among members. In this context, hundreds of thousands of castes exist. As members and nonmembers are sharply defined, favoritism and nepotism to one group are expected by other groups.

Caste is not static. Castes' positionings in relation to one another are constantly being renegotiated on the basis of fluctuating power, status, ritual behavior, political mobilization, education, and geographical location. Caste holds diverse peoples together via a hierarchical structure permitting flexibility within. Upward mobility is potentially possible. Caste may be associated with occupation—*jati*. While one's *jati* is fixed at birth, its position within the hierarchy is not. Moreover, modernization has confounded birth assignation with practical occupation.

Abolition of discrimination has applied only to *untouchability*, not to caste. Those in low level or polluting *jatis*—the Dalits—seek recognition as regular members of Indian society. Whatever their origin, their subordination derives from perceived ritual impurity rooted in antiquity. Some scholars argue that untouchability was exacerbated by the British, who were overly concerned with caste inequities and thereby drew undue attention to them.

Today, Dalits demand their rights, and government programs have facilitated their move into many upper echelons, including politics. Education, in particular, has invested them with expectations for equal treatment. Indeed, "untouchables" now have many opportunities historically denied them. Nevertheless, stepping outside the bounds of untouchability is no easy task; deeply rooted social taboos are difficult to expunge.

Affirmative action policies have created privileged sections within underprivileged groups. An interesting point is that pro-action policies are part of the official Constitution, and they were written by a former untouchable. The situation that "backwards" go forward by virtue of their "backwardness" is resented in many circles.

The progression of these formerly subordinated classes has political impacts. Many politicians avidly seek the loyalty of entire socially categorized populations. Appointments are made with an eye toward potential mobilization of caste followings. Just as significant is the fact that 85 of 545 seats in Parliament and 22.5 percent of government jobs are reserved for scheduled castes and tribes.

Caste then, in terms of establishing hierarchical relationships and interactions, is not disappearing, although it is exhibiting greater flexibility. The poor and lower-middle-class masses are seeking economic improvement, while the privileged are trying to retain the status quo. Brahmins find themselves taking orders from or eating food prepared by Dalits. Increasing numbers of marriage ads state "caste no bar." Although quotas and special privileges for untouchables have impelled many into upper echelons, these have sometimes triggered violence. Most untouchables continue to exist at the very bottom of Indian society.

Religious Conflicts

In December 1992, a rampaging mob demolished an unused mosque in Ayodhya. Uttar Pradesh claimed that it had been built on the birthplace of Lord Rama. This inflamed the sensibilities of India's largest cultural majority: 115 million Muslims. Communal riots broke out nationwide, and hundreds were killed and injured. This incident was merely a spot on the tapestry of religious conflict in India's officially secular democracy. The overwhelming majority of Indians are Hindus. In fact, in some languages such as French and Persian, the word for "Indian" is *Hindu*. It once meant "people beyond the Indus," but history has rendered it, correctly or incorrectly, a term designating a nation and a religion. However, like India itself, Hinduism is a polyglot of practices and lifestyles. Unfortunately, in the twentieth century, this kaleidoscope has become blurred under the forces of socioeconomic inequities and political ambitions.

The unlikely notion of *Hindu chauvinism* has emerged in the political arena with the Bharatiya Janata Party (BJP). This fundamentalism is not simply religious. It incorporates the resentments and animosities of caste, belief, and regional sectarianisms. It cannot be simply religious, because Hinduism is not a singular faith divorced from all other faiths. Hinduism is not a monolith; it is inclusivistic in nature. But it is convenient to appeal to differences in practice or detail of belief for those with a particular political or economic agenda.

In the early 1990s, the BJP called for a ban on English, the promotion of Sanskrit, and a ban on proselytizing by other religions. By 1999, the BJP had achieved national power and Hindus were killing Christians in Gujarat, Orissa, and elsewhere. The BJP fell from power in 1999, but religious insurgencies continue to plague this vast and complex country.

Quest of the Sikhs

Most Sikhs live in the Indian province of Punjab (Figure 7-9). Beginning in earnest in the 1980s, militant Sikhs pushed for an independent state called **Khalistan.** Their campaign of violence is directed not only against the Indian government but also against Hindus, and even Sikhs who reject their policies. Sikh terrorism has taken several thousand lives and indelibly impacted the political landscape.

Punjab is India's richest province, in part due to its natural resources, and in part due to the skills and ingenuity of its Sikh Jat inhabitants. Punjab is India's breadbasket, and the Jats dominate the nation's trucking industry. Nevertheless, it was here that fundamentalists, under the leadership of Sant Jarnail Singh Bhindranwale, a political–religious leader, directed an insurrectionist movement heard around the world. In the minds of his followers, the Indian government's secularism contradicted the Sikh

Figure 7-9

Sikh gurdwara *or temple in the Punjab. All faiths are welcome to a* gurdwara. *It is a refuge and place of peace. There is no set architectural style for the building of a* gurdwara.

belief in the inseparability of sacred and secular spheres—a doctrine dating to the sixteenth century. The quest for an independent Khalistan was funded largely by overseas Sikhs, especially those residing in North America.

Terrorism and insurrection peaked in 1984 when the Indian army stormed the Golden Temple complex in Amritsar. This was *Operation Bluestar,* directed by Prime Minister Indira Gandhi (daughter of Jahawarlal Nehru), who had been denounced as the "acting Mughal King of Delhi" for her official declaration of India as a secular state. For Mrs. Gandhi to give in to Sikh demands would be to give in to communalism, and that would see the disintegration of India. Bhindranwale and his companions were eliminated, but on October 31, 1984, Indira Gandhi was murdered by her Sikh bodyguards. For three days, Hindu mobs raged in Delhi, causing more than 4,000 deaths. This level of violence was quelled, but the idea of Khalistan has not died.

The Khalistan movement is not entirely religious. The Sikhs have a proud history within and against the British Raj. Modernization has brought much economic change to Punjab. Industrialization requires workers, but this type of work is not acceptable to many Sikhs who find pride in farming and militaristic pursuits. Factory workers have been brought in from poor provinces such as Orissa and Bihar. Their presence is resented, and they are often victims of violence.

Language is another bone of contention. Many Sikhs resent the imposition of Hindi as the national language. Their opposition to Hindi dates back to the 1920s when they campaigned against its spread in education and civil service.

When Punjab was divided in 1947, Sikhs lost that part of their identity tied to their dislike of Muslims (Muslims went to Pakistan). Years of upheaval, often centered on anti-Hindu issues, culminated in further division of Punjab in 1966. The area with the majority of Hindi speakers was carved into two states: Himachal Pradesh and Haryana. Punjab, at least 60 percent Sikh, retains the essence of Sikh nationalism. It also exhibits a strong non-Sikh military presence.

Agricultural Landscapes

Figure 7-10 illustrates the wide variety of crops cultivated in India. While crop yields have risen significantly in the latter half of the twentieth century, they do not equal those achieved elsewhere nor, because of inadequate distribution facilities, are they sufficient to meet population requirements (Figure 7-11). Human and animals continue to dominate labor even as tractors and other machines become more apparent (Figure 7-12).

Water availability has increased via tube wells, storage tanks, and canal irrigation. More than 14 million pumps have been installed, and wells now are the major source of irrigation water. Farmers have been able to put more acreage under production and use high-yielding seeds. Some well owners have started water markets, selling their water to others to open up even more farm land.

Timely application of water can increase yields markedly, but supplies are overexploited in many areas. Other areas have never had enough. Management

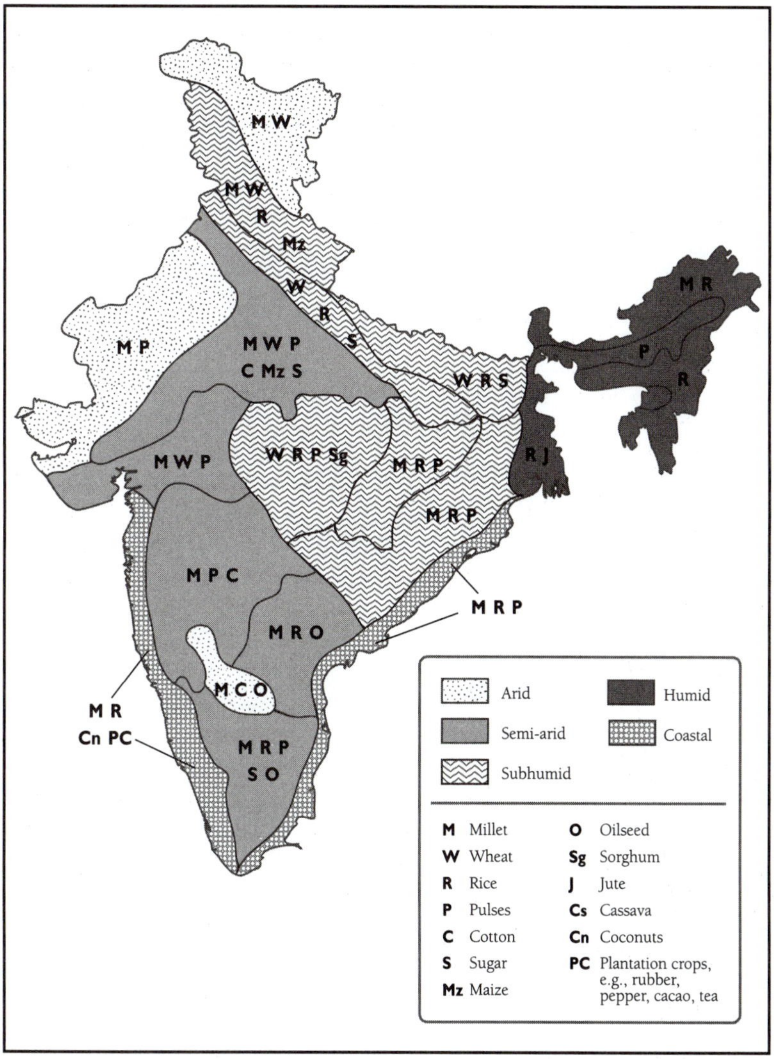

Figure 7-10

Patterns of agriculture. What is the relationship between crop distribution and water availability? Note that millet can be grown in many different environments.

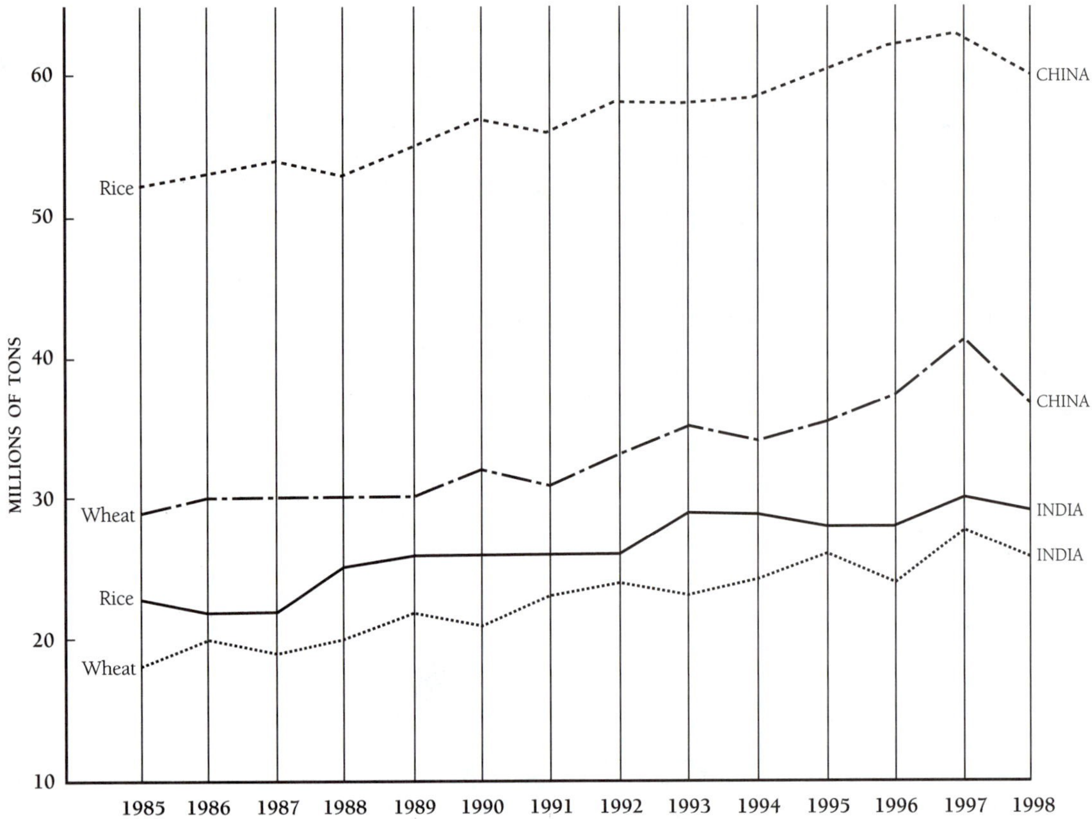

Figure 7-11

China and India: Comparative crop yields. Why do you think China's crop yields are so much higher than India's?

agencies lack coordination, and the entire issue of water, like land, is very political. According to Robert Repetto, writing for the World Resources Institute (1994), "In India, the rights to an immensely valuable resource [water] were distributed gratis in a pattern even more unequal than that of land distribution, reinforcing rural inequalities in income and wealth."

Outright famine has not plagued India for decades. However, food insufficiencies and malnutrition remain troublesome in many areas, especially when distribution networks fail or costs are prohibitive. A critical contributor to improved food supply in India was the 1966 Green Revolution, which began in Punjab with high-yielding varieties (HYV) of wheat and Tamil Nadu with HYV rice. But scientific improvements have not been equitably distributed, and they benefit different crops in different ways. In the case of wheat, for instance, Punjab, Haryana, and western Uttar Pradesh have benefited most because of suitable irrigation water supply. In the case of rice, Tamil Nadu, Kerala, Andra Pradesh, and Jammu and Kashmir have profited most. Other rice-growing states or regions with variable rainfall and inadequate water storage and distribution systems use a diffferent approach. Here, farmers plant a rabi (dry season, winter) HYV wheat crop.

Figure 7-12

Man and water buffalo harrowing a rice paddy in Orissa. This action helps to break up the lumps of clay-like soil. Rice cultivation relies heavily on human and animal labor.

Uttar Pradesh, where one in six Indians reside, has seen variable consequences of the Green Revolution. Uttar Pradesh is India's largest grain producer, yet yields are only half those in Punjab. One government- and WB-supported countermeasure is the reclamation of **sodic land**—land where crusts of salt or other alkalis prevent nutrients and water from going below the surface. Application of gypsum and flushing with water can bring such soils into production. Farmers are given free gypsum, loans to buy water pumps, and counseling on crop and farm management.

Agricultural improvements have not erased food import requirements, which are projected at 6.9 million tons in 2000 and 14.1 tons in 2010, with a further 2 percent increase each subsequent year. These requirements relate to population needs; deceleration of crop yield improvements; economic growth and increased consumption; and greater consumption of nonvegetarian foods. These factors pose challenges to Indian agriculture that does not make the best use of its land in terms of potential productivity.

Seventy percent of India's land is rain-fed and 30 percent is dry cropped. With irrigation, these lands could become relatively affluent. In Maharashtra, experiments with water conservation, tree planting, and grazing control have led to agricultural improvements and the creation of a viable market system (Figure 7-13).

Many Indian analysts believe that India has the potential to be self-sufficient in food as well as an international exporter of food. However, there are obstacles. How can India sustain increased yields given international specifications regarding the use of chemicals? Will there be an endless supply of ever-better HYV seeds? Policies regarding the ownership and use of HYV seeds have become a global issue. How far can biotechnology go? Nearly 500 **transgenic plants** (human-made) were developed in 1994–1995. Will these be available to Indian farmers? Would Indians even want to eat these products? (Think of the beautiful tomato developed to be machine picked; it looks great but tastes like nothing.) Who would buy India's exported food?

Technical improvements may not be appropriate for all farmers. For example, large tractors or harvesters are unsuitable for small fields. Moreover, who can afford such machines? How will fuel be acquired and paid for? Are different religious or caste groups willing to cooperate in the acquisition and sharing of such equipment? Storage, packaging, and shipping to avoid losses and disease are further issues to be grappled with.

Figure 7-13

Rice paddies and coconut estates line the coast of Kerala State. These are the first in a vertically zoned series of crops in the Western Ghats. Rubber, pepper, cardemom, and other crops grow at higher elevations. The coconuts provide copra*—meat—and* coir*—fiber. The man is pedaling a bicycle rigged with paddles to move the water from one field to another. This is another example of the importance of human labor in agriculture.*

Why Waste Waste?

Judicious use of animal and even human waste can improve the lives of millions. Cow dung, when fermented or digested in an enclosed, cement container, produces methane gas, carbon dioxide, and a nutrient-rich slurry. From this *biogas* plant, methane can be piped into homes and used as cooking gas, or it can fire a diesel engine to generate electricity or pump water. The slurry is an excellent fertilizer. Generating biogas is culturally appropriate as well, because all products of the cow are considered sacred.

Biogas was produced as early as the sixteenth century in Persia and has been used in India for almost 100 years. Now, more than two million small "digesters" and a thousand community-size plants have been put in place in rural, agrarian regions. Biogas provides light to millions who are not linked to the larger electrical grid and is a boon to the 80 percent of rural India who lack access to sufficient cooking fuel. The Indian government hopes to construct digesters for the 12 million households that have enough cattle to maintain a sufficient supply of dung.

Human waste can also be put to good use. Sewage-fed lagoons produce 10 percent of the fish consumed by Calcutta. A natural wetland east of the city has been turned into a 12,000-hectare aquaculture operation. While the sewage provides nutrients for the algae that carp and other fish eat, the process dilutes the high concentration of fecal coliform bacteria in the sewage. The resulting effluent can be safely used for irrigation in urban gardens. City gardens in the rich, composted soils of old garbage dumps produce 150 to 300 tons of vegetables a day and employ around 20,000 people in East Calcutta.

Deceleration in crop yields is now occurring in key areas such as Kerala and Punjab. This can be attributed to several factors. Even with the widespread installation of tube wells, water management is ineffective. Often, water control projects have been installed in regions where crops do not require irrigation. Fertilizers and pesticides are often adulterated. Seed multiplication farms are not operating efficiently. Roads have fallen into disrepair, disrupting distribution of both inputs and outputs. Rural electrification, expanded from 10,000 villages in 1947 to 500,000 (out of 580,000) in 1998, plays a significant role in pumping and other aspects of water control. Unfortunately, mismanagement has resulted in intermittent power supply in some areas and weeks of blackouts in others.

Changes in agriculture also have social consequences. I will give you two case studies, but of course there are many more. The first involves caste relationships and the second centers on the changing role of women.

DEMISE OF THE *JAJMANI* SYSTEM

In Punjab the custom was for Harijan laborers to work for their Jat landlords in exchange for grain and fodder. This was called the ***jajmani* system.** Harijan laborers received one-twentieth of the wheat harvest. This was much more than an economic system. It was a centuries-old social contract rooted in caste relationships and practiced among Sikhs and non-Sikhs alike. It required a certain decorum and specific exchange relationships including water, tea, and food.

With the advent of HYV, the one-twentieth share became greater from the HYV fields than the traditional variety fields. Furthermore, some farmers could afford tractors and no longer needed Harijan labor on their large plots. With concomitant economic growth in towns and cities, and extended bus routes, many Harijans sought urban employment. New-found wealth meant that Harijans no longer had to accede to Jat requirements or even requests. Jats purchased motor scooters; Harijans bought bicycles. Activity spaces were expanded, and Harijan–Jat social distance was extended. The intergroup ceremony and camaraderie surrounding the harvest disappeared.

More tube wells were installed, and underground pipes reduced evaporation. Rice was grown in addition to wheat. With quick-maturing seeds (e.g., 60-day lentil), triple-cropping became possible. When labor was needed, migrant workers from the Gangetic plain were hired until that source dried up. Now many Jats do their own labor and must work longer hours to support all their essential equipment. There is a rising level of debt—some sell their land in fear of bankruptcy. Rich land owners buy these parcels, circumventing property holding limitations by registering titles among family members. Small farmers cannot compete and ultimately sell out. Large land holders cannot function efficiently without tractors, and many become overburdened with debt. Technological change alters social circumstances. Clearly, the pros and cons of these changes are in the eyes of the beholders.

WOMEN AND THE WHITE REVOLUTION

India is experiencing a **White Revolution** as its dairy industry undergoes transformation. Western planners introduced new techniques to increase milk yields as they had improved crop yields. The development plan, introduced in 1970, was called Operation Flood (OP). India has more than 240 million bovines including cattle and buffalo. Here 18 percent of the world's cattle stock produce only 6 percent of the world's milk. OP determined that India had too many low-milk-yielding zebu cattle and that new breeds and dairy processing plants should be introduced. The results are mixed.

Urban dwellers now consume more milk, but rural people often cannot afford higher prices and therefore drink less. New cattle breeds require grain, and this means less grain is available for human consumption. Moreover, women who frequently tend and milk zebu were deprived of this central role. Women traditionally collect milk and make *ghee,* curds, buttermilk, and cottage cheese, which they can sell. Dairy processing plants deny women these activities. Farmers take their milk to the factory, where it is reduced to products for purchase by the urban elite. Thus, nutritional sources are diverted from rural areas.

New breeds and cross-breeds are not always suited to tropical or even subtropical environmental conditions, and an investment can be easily lost to disease and death. Dairy cattle require high-quality feed and clean water—items not available to millions of people. They more easily succumb to parasites. Further, milk cows are not well suited as draft animals.

The sacred zebu is well suited to India's environmental conditions. It does give some milk, but more importantly, it provides dung for fertilizer and fuel, and it is India's main beast of burden. Further, it devours anything from straw to newspapers.

The goal of getting more milk from an animal fails to recognize the whole animal and its integral role in traditional cultural complexes. Many scholars attack this reductionist approach to development. While OP claims to have "emancipated" millions of women, in fact it has eliminated an important role and source of income. Very few women are directly involved in the modern dairy industry. Modern dairy enterprises provide better care for dairy animals than many farmers can for their children. Kanti George (1985) notes that in certain parts of India, "it must be more comfortable to be a largeholder's crossbred cow than to be a small farmer's child."

Urban Landscapes

Indian cities have their roots in Harappan civilization. Later, they centered on the Indian princely states. Such indigenous cities were nodes in functional regions, offering protection in return for food. With British intrusion, powerful economic systems shifted emphasis to coastal cities. Interior cities became peripheral in importance to the larger international economic system in which spatially peripheral (coastal) cities became central to economic interchange. Furthermore, two separate administrative systems operated: that of the Indian princes and that of the British.

Similarly when British institutions and infrastructure were superimposed on traditional landscapes, two urban structures emerged. English cantonments and civil lines operated differently than the rest of the city, and straight avenues brought order to the perceived chaos of the Indian areas. Calcutta, Madras, Bombay, and later New Delhi and others emerged as *dual cities* (Figure 7-14).

In 1947, there were 582 princely states and estates of which 115 were **gun-salute states.** The number of gun salutes—21, 15, 10—indicated the importance of the prince, not the economic system tied to the princely city. When states were reorganized in 1950 and 1956 along linguistic lines, princely states lost primacy, prestige, and power. Only five of these are still state capitals: Bhopal; Hyderabad; Jaipur; Srinagar; and Thiruvananthapuram (Trivandrum).

According to geographers Brian Berry and Howard Spodek (1982), in the residential geography of the traditional Indian city, socioeconomic status was the dominant theme. However, modernization is transforming communal caste status to communal class status. High-status areas are central by location; low-status areas are peripheral by location. With rapid growth and change after independence, by the 1960s, distinct familial areas and male migrant areas had emerged around factories and port facilities. And there was some suburbanization.

In the 1960s, planning direction shifted from controlling metropolitan growth to diffusing it. Greenbelts

Figure 7-14

India Gate on one of the broad avenues of New Delhi. Holding the names of victims of the Afghan wars and World War I, the arch is of Roman style. Such arches were commonly built in European planned cities in the eighteenth and ninteenth centuries.

Figure 7-15

Marine Drive in Mumbai. Lined with crowded apartments, Marine Drive, along the Arabian Sea, houses middle and upper class people as well as squatters and the homeless. Overcrowding and the tropical climate combine to give the buildings a run-down appearance.

and satellite towns were installed but were largely ineffective in stemming urban growth (Figure 7-15).

Mumbai is located on a series of small joined islands and reclaimed land. Traffic congestion is exacerbated with north–south commuting. The government decided to develop a New Mumbai (New Bombay) on the mainland to decentralize the ever-choking stream of commuters. Success was deemed likely because similar plans had been effective in Japan, China, Korea, and Singapore. With jobs and housing available in New Mumbai, congestion would be relieved and more industry would be attracted away from the central city. Similar schemes were planned for all of India's major cities. Unfortunately, success at decentralization has been limited, at best.

India is a place where conflicting interests and powerful lobbies can distort any plan. Industry and business, real estate and construction, middle class and poor: all have their goals. Besides, the Indian government has generally stood as champion of the poor, so how could it legitimately support projects benefiting the other groups? Maharashtra state government showed little interest in the plan. Moreover, it permitted further land reclamation and building development at the southern end of Bombay Island, adding to the existing congestion.

In New Mumbai, 50 percent of the housing is for the poor. Tiny apartments house extended families in large apartment blocks. Hurried building has resulted in poor-quality structures, which now appear no better than inner city ***chawls*** (slums). But middle-class housing shows sensitivity to both space and privacy needs. This reveals the class bias on the part of the designers.

New Mumbai housed 156,000 in 1971. By 1991, it had grown 126 percent, with a population of over half a million. Migrants count for at least 50 percent of this growth. Half the residents live in slums, and the city is ringed with squatter settlements.

Indian cities are notoriously crowded, dilapidated, and polluted. Calcutta is regarded by many as a lost cause (Figure 7-16). New Delhi is one of the most congested and polluted cities in the world. With 25 million, Mumbai adds 300 migrant families every day. Many end up as rag pickers in Dharavi, Asia's largest slum. Others join the city's throng of 120,000 prostitutes.

Government efforts to decentralize with developments in interior cities and rural areas have not stemmed this tide of migrants to larger urban centers. Thousands sleep in the streets, yet millions—including the wealthy, the middle class, and the poverty-ridden masses—do manage to survive in often appalling conditions (Figure 7-17).

Figure 7-16

In this Calcutta street scene, you can see the crowded and broken sidewalks. The man to the right is bathing at a public standpipe. These have been installed all over the city to accommodate those with no running water, or the homeless. Note the confusion of wires at the top of the picture. Electricity was added after the buildings were constructed. Moreover, people simply tap into the electrical system illegally, and there is little official control over this situation.

Hunting the Economic Tiger

Figure 7-18 shows India's raw material and manufacturing regions. The primary cores associated with Mumbai, Chennai, and Calcutta are products of historic inertia. But there are new, secondary regions of activity generated by decentralization policies and exploitation of new raw materials and other resources such as labor supply. Calcutta now anchors an industrial district, which focuses on jute and cotton manufactures, engineering, and chemical industries. Coal mining and iron and steel manufacturing center on Nagpur. The region incorporating Maharashtra and Gujarat specializes in textiles, food processing, chemicals, and engineering. Chennai also specializes in tex-

Figure 7-17

Homeless woman in Mumbai. Thousands of individuals and families live in the streets of India's major cities.

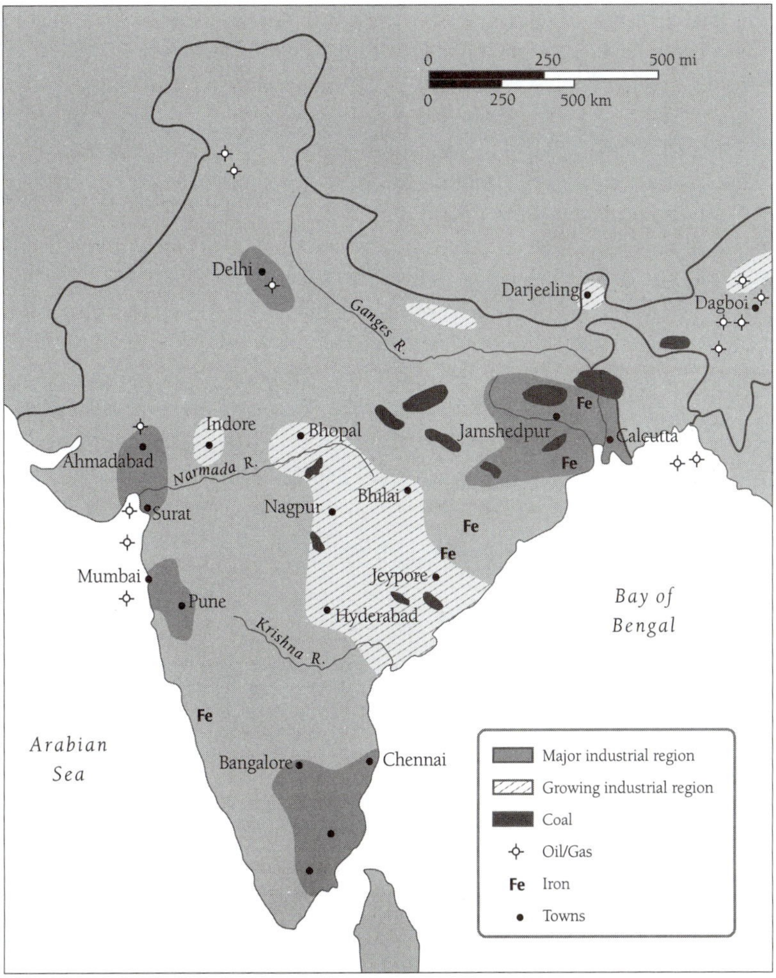

Figure 7-18

Raw material and manufacturing regions. Note the role of former colonial cities as centers of activity.

tiles and light engineering. This region incorporates what has become known as **India's Silicon Valley**, centered on Bangalore.

In 1999 Rakesh Mohan wrote an article in the *Far Eastern Economic Review* that poses this question: "What Ails the Indian Tiger?" It says that India's share of world trade in 1950 was around 2 percent; by 1970 it had declined to less than 1 percent, and it remains at that level. India's exports earn only a quarter the amount of China's exports. Why is India's economy lagging in productivity?

Government policies, founded in socialist ideals, have long served as blockades to economic development in the Western sense. India's export structure has been frozen since the 1970s; textiles, apparel, leather goods, and handicrafts account for almost half of exports for the past three decades. Unlike other nations, where mature, labor-intensive industries move from one leading country to another, less-developed country, India's transformation has been relatively static.

India's import-substitution policies have aimed to promote industrial development with walls of trade barriers. Labor-intensive manufacturing has been reserved for the "small-scale sector." Entrepreneurs have been limited in growth, arguably to protect the jobs of workers in small, nonmechanized enterprises.

Government-operated institutions, state-run enterprises, and entangled regulations have made efficiency unachievable. For example, in 1986, the Steel Authority employed 247,000 Indians to produce 6 million tons of steel. In South Korea, 10,000 people made 14 million tons. It is more efficient for southern India to import coal from Australia than it is to get it from northeastern Indian coal fields. Most ports are not equipped to handle super containers or receive oil and gas. Roads are generally in a dilapidated state, with a national program in the works only recently in 1999.

Recognition of the fact that India is lagging far behind other nations has stimulated new economic policies. As Mohan says, "If there is to be an Indian tiger, it must be freed from the shackles that keep it languishing in its small cage."

In 1991, a package of reforms was delivered. Import duties were slashed, and sectors such as autos, telecommunications, power, and consumer goods were opened to foreign investment. Joint ventures were encouraged. Capital market, banking, and tax reforms were instituted as well. However, many economists say this is not enough to push India out of the economic doldrums. Furthermore, constantly changing governments, shifting political coalitions, corruption, and favoritism make a coherent policy difficult to define and impose. Even so, the economy has been growing at 7 percent yearly since 1984.

In a 1997 issue of *Far Eastern Economic Review,* an advertisement for "International Tech Park: Bangalore" boasts about a $U.S.86 million India–Singapore joint venture infrastructure project. Bangalore's story illustrates some of the problems and trends in modern economic development in India.

Bangalore, an army cantonment during the Raj, later became known as India's *Garden City.* In the 1980s, a spate of colleges and skilled workers along with a pleasant climate drew the attention of foreign corporations. Multinationals like Lipton Tea and Unilever; and American firms such as Texas Instruments, IBM, and Compaq opened plants in the area. Bangalore soon was called India's Silicon Valley.

The 1990s brought more companies, stores, malls, and restaurants, and then an array of woes ground the city to a halt. Population increased from 2.9 million in 1981 to more than 5 million in 1999. There are shortages in supplies of water and electricity. Cars and motor scooters vie with pedestrians in a dangerous rat race; Bangalore is now one of India's most polluted cities. Suburbs are built and occupied before service facilities are installed. Commuter times have doubled. Slums that sprung up to house migrant construction workers have become permanent. The city is collapsing under the weight of its own success.

The India–Singapore joint venture plan was to build a world-class business park, complete with essential infrastructure. Liberalized government policies and continued availability of highly trained workers, a relatively high literacy rate (56%), and the use of English for business combine to make Bangalore's Karnataka State desirable for investment.

Monetary and regulatory changes attract development and improve the lives of millions. However, there are often negative consequences. One infamous case is Bhopal.

Bhopal: The World's Worst Industrial Disaster

Late one evening in December 1984, a heavy gas poured out of a storage tank at the Union Carbide factory. The gas was methyl isocyanate, an ingredient of the pesticide Sevin. The tank was supposed to be cooled, but the refrigeration had been shut down to save money. The gas crept into the workers' slum housing and out into the old city, crammed with people. As their eyes burned and lungs choked, people panicked, but there was no getting away from the deadly gas. More than 2,500 people died that night, and more than 200,000 were affected. The next day, Muslims were buried in mass graves and Hindus were cremated on huge funeral pyres.

Union Carbide admitted "moral responsibility" and promised reparations. The Indian government, under pressure to get money to victims, settled for a $470 million payout. In fact, the money was tied up in legal battles, and most victims have seen little if any payment.

But the consequences of the gas leak are not yet over. Altogether, at least 4,000 people have died, many people are blind, others suffer from respiratory disease or cancer. Birth deformities are another result of this unprecedented disaster. Researchers believe that at least 300,000 individuals will experience long-term damage to their health. Many of the victims believe that because they are poor, they are forgotten. What do you think?

Getting the Energy

THE PETROLEUM BACKBONE

The petroleum industry forms the backbone of India's national economy. One of the oldest oil fields in the world was established in 1889 at Dibrugrh in Assam. Exploration and discovery expanded from the folded mountainous regions to the alluvium-covered shelf in the Brahmaputra Valley. After independence, exploration was expanded to nearly all the sedimentary basins of the country. Several discoveries of oil and gas were made in the Bombay Basin (Gujarat); Assam and Arakan Basin; and the Bombay Offshore Basin. In the 1990s, additional fields were discovered offshore Bombay.

Another recent development has been to commercialize reserves of natural gas. Over the years there has been a significant shift in the domestic pattern of energy consumption. In 1947, more than two-thirds of India's total energy consumption was of noncommercial fuels such as wood, harvest waste, animal dung, and coal. Now, natural gas provides more than half of India's energy, and the private sector has diversified into infrastructure construction, marketing, import facilities, pipeline manufacturing, and the production of an array of byproducts such as naptha, toluene, and lube oil.

India must still import fuel. Multinationals as well as private firms are contracted to seek imports in addition to new sources. One Australian-backed scheme involves a plan to build an undersea pipeline from Iran via Oman and the Arabian Sea Coast to India's west coast. But problems are apparent.

The Arabian Sea is 9,840 feet (3,000 m) deep, and its seabed is a series of rugged mountains and ravines. This means that the pipeline would have to be laid in shallower coastal waters—territorial waters of Pakistan. Pakistan could bar passage of the pipeline through its territorial waters, or demand an inordinate share of the oil. Given the rocky nature of India–Pakistan relations, the future of this project is uncertain.

DAMS AND THE DAMNED

Water control—timing, amount, and distribution—has been a bane of Indian farmers for centuries, and the rapid expansion of irrigation systems has been one of the country's highest priorities since independence. Dam construction has sought to alleviate these stresses, thereby increasing food production. Numerous projects have been installed, mainly large-scale operations involving foreign aid and modeled on America's Tennessee Valley Authority's (TVA) principles of regional hydroelectric planning. Tremendous investments have been made in several areas but the consequences have not always been positive.

The Krishna River, most important for irrigation on the Deccan, was developed with a series of dams, tanks, and canals dating to the Raj. Unfortunately, planners did not take into account farmers' perceptions of their water requirements or desired crops. State boundaries divided stored water supplies from users' lands. Moreover, construction integrity was poor, resulting in silted dams and leaky canals. In fact, dams and canals are rarely able to provide water to even 50 percent of their designated service areas.

Figure 7-19

This poor village in Madya Pradesh has just received a well for potable water. Getting water is a twice-daily chore for women and children.

Another earlier dam project concerns the Damodar River, which drains Bihar and West Bengal, an old industrial core founded on iron and coal supplies. For years, excessive rainfall in the upper valley produced flooding in the lower valley, with devastating consequences. The Damodar Valley Corporation (DVC) was designed with the assistance of the TVA. Floods have been quelled, but the DVC's greatest success is in the area of power generation.

The cheap electricity attracted a number of industries into this once-stagnating region. Urbanization and industrialization increased with clusters of cities focusing on power plants. The area around Chandrapura became a primary growth pole at the expense of some of the other industrial centers. Devised initially to control floods, the DVC has stimulated industrial development in one of India's most heavily populated regions.

South Bihar is one of the richest areas of India in terms of raw materials such as coal, iron, bauxite, and mica. These minerals have been systematically exploited, first by the British, then by Indian industrialists such as Tata, as well as the Indian government with money from the WB. The results have been lucrative for these outside developers but disastrous for the region's inhabitants and the environment.

Mining displaced thousands of poor people. For example, Tata Iron and Steel displaced nearly all the original inhabitants for its massive complex at Jamshedpur. Other consequences are deforestation, soil erosion, and ground water pollution. In addition, the bulk of government expenditure is plowed into industries producing consumer goods for the urban wealthy and upper middle classes.

The industrial region around Jamshedpur straddles the borders of four states: Bihar, West Bengal, Orissa, and Madhya Pradesh. Development has bypassed or negatively affected millions in this otherwise poor tribal region (Figure 7-19). In the 1980s people began to organize politically. Ultimately, a separatist movement emerged. Protesters advocated a new state to be called **Jarkhand.** However, none of the existing states are in any mood to give up their mineral or industrial resources or their associated revenue. Since the upper echelon is in control, blockades, strikes, and bombings by the "left-outs" can only escalate.

Ganga Ma Is Sick

Ganga Ma—Mother Ganges, 1,560 miles (2,510 km) long, is the ancient symbol of creation, preservation and destruction. Its sacred waters emanate the powers of Brahma, Vishnu, and Shiva. But it has become fouled with the profane evils of poison, death, and disease. More than 400 million people depend on its monsoonal waters to flood their fields with nutrient-rich silt. By 2030, there could be as many as a billion people living within its reach. An increasingly sick river could spell catastrophe.

Microorganisms that break down organic wastes require oxygen. **Biological oxygen demand (BOD)** is the measure of this need. Imagine how much BOD is needed to destroy the harmful components of the raw sewage of more than 30 cities, nearly a hundred towns, and thousands of villages. Although sewage comprises most of the waste material, industrial waste is far more dangerous.

Chromium and organic wastes from tanneries; bleaches and dyes from textile factories; thick goop from sugar processing; toxic farm chemicals (many of which are banned in the United States); sludge from oil refineries; human ashes, body parts, and whole bodies: all become part of the sacred river.

An estimated 35,000 bodies a year are brought to Varanasi for cremation. But the required sandalwood is expensive. Many bodies are only partly burned. Others are simply placed in the river by the devout.

Pollution worsens as the Ganges makes its way to the Bay of Bengal. The World Health Organization standards for drinking water is no more than 10 coliform count per 100 milliliters. In Varanasi, coliform counts are as high as 100,000. Typhoid, cholera, and viral hepatitis are common here. In this region, one person dies of diarrhea every minute. Amoebic dysentery, gastro-enteritis, and tapeworm infections are part of the lives of millions. About 150 factories are lined up along the Hooghly at Calcutta. These contribute 30 percent of the waste present in the mouths of the Ganges. Fish put in the water upstream from Calcutta survived only five hours.

Industrial discharge from the Ganges is growing at 8 percent a year. With increased population and industrialization, how can the river possibly possess enough BOD to clean itself?

Consider these facts. Nearly 70 percent of India's available water is polluted, and waterborne diseases such as typhoid and cholera are responsible for 80 percent of all health problems and a third of all deaths. Only 7 percent of India's cities have any kind of sewage treatment services.

In 1985, there was a plan to clean up the river. Corruption, mismanagement, and technological errors rendered the plan useless. "A fundamental reason for the failure was that most of those who have a stake in the river's health were never included in the planning" (Sampat 1996).

India has more than 1,500 dams, making it one of the world's largest dam-building nations. The Narmada River project, the largest ever undertaken, includes two super dams: the Sardar Sarovar in Gujarat and the Narmada Sagar in Madhya Pradesh. In addition, 30 large dams, 130 medium dams, and 3,000 minor ones are planned. The scheme should help feed 20 million, foster an industrial boom, and protect 750,000 downstream residents from floods.

Critics claim that the Narmada scheme is a disaster in the making. Both forests and agricultural lands will be flooded, and up to 1.5 million people will be displaced. The destruction of forests will be particularly devastating because they have housed a diverse community of tribal peoples such as the Bhils and the Gonds, as well as endangered wildlife, including the tiger. Tribal representatives say they have nowhere to go, and when the valley is flooded, they will join India's growing number of **environmental refugees.**

India's Oustees

This is a century of uprooted people: **oustees.** These are people who flee from war and violence; political and/or ideological oppression;

natural and other disasters; economic stagnation; and human-induced environmental displacement. Dams produce the largest segment of environmental oustees, followed by mines. More than 20 million have been displaced by dams alone.

About half of oustees are landless or sharecroppers with no recorded or "legitimate" claims to the land they may have worked for generations. No more than half of oustees are ever resettled. Some are compensated monetarily, but many squander their new-found "wealth" on consumer items and end up landless, jobless, and moneyless.

Many oustees become seasonal laborers in agriculture or construction. In northern India, husbands tend to leave their wives and children with a male relative, perhaps for years at a time. In southern India, husbands tend to take their family with them. These workers join the vast sprawls of squatter settlements on urban fringes. Many become indebted to ruthless bosses and spend the rest of their lives virtually enslaved.

An important point to note is that as India's population increases, every project displaces greater numbers of people. The Narmada project is only one of many. The Sardar Sarovar in Gujarat will flood out 193 villages in Madhya Pradesh. All the associated works will displace an estimated 140,000 people. More than half are scheduled tribes. Thousands of people have marched in protest but the waters continue to rise as the project ensues.

HAVING THE BOMB

Gandhi called the atomic bomb the "most diabolical use of science" ever. Prime Minister Nehru proposed a "standstill agreement" on the nuclear arms race in 1954. These Indian ideas were essentially ignored by the world.

Geopolitical considerations are key to understanding why India "needs" military nuclear capability. In 1962, China invaded disputed territory east of Kashmir. In 1964, China conducted its first nuclear test. India–Pakistan wars ensued from 1965 to 1971. China provided assistance to Pakistan. By 1974, India had exploded its first "peaceful" nuclear device. By 1988 it had tested a missile capable of delivering a nuclear warhead into Pakistan. In 1996 both countries refused to sign the International Test Ban Treaty. Two years later in 1998, Pakistan tested its first nuclear missile. India responded by conducting five underground tests.

Pakistan's missile is named the Ghauri after an Afghan, Muslim king who defeated a twelfth-century Hindu ruler named Prithviraj Chauhari. India's missile is called the Prithvi.

Kashmir has been a flashpoint in stimulating a nuclear competition between India and Pakistan. The Bharatiya Janata Party (BJP) has long advocated a more military India and is committed to making the nation a nuclear power. Pakistan is just as committed to proving its nuclear strength.

Military expenditures on the part of both nations are exorbitant in light of socioeconomic needs. The Indian government spends 3 percent of its GDP on defense—the same as that spent on education and twice that spent on health. In Pakistan, defense spending gobbles up almost 30 percent of the national budget—twice that spent on education. Clearly, military costs are devastating to social and economic needs.

China's role is pivotal in this ongoing arms race. Not only has China supported Pakistan in its anti-India hostilities, but also it has sent it missiles capable of carrying nuclear warheads. Moreover, it has stepped up its military assistance to Burma. China is displeased that New Delhi views Tibet as an "autonomous" region within China. To make things worse, Tibet's Dalai Lama and his anti-communist, expatriate followers are headquartered in India at Dharamsala. You can read more about the "Tibet question" in Chapters 9 and 10.

Paradox: Reality Belies Developed Nation Image

India, with its nuclear capability and its average annual economic growth rate of 7 percent, is the leading economic power of South Asia. Scores of millions comprise a dynamic middle class and urban India, by itself, could be counted as the world's third largest country. Thirty million households are already in the market for various durable consumer products, and millions more are climbing into or aspiring to this category of economic behavior. Fired by the engine of its burgeoning middle class, India is on the move toward becoming a "developed" nation. This is the message delivered by an array of Indian government studies and politicos in the 1990s. These perspectives are at once correct and misleading. The truth is a paradox.

The Indian middle class certainly exists, but it is the middle class of an overall poor country. Only one in five Indians owns a wristwatch; only 48 percent of households own a bicycle; less than 4 percent of households have a telephone. In this country of over a billion, only 1.4 million are regarded as rich. The middle class is, in effect, a poor one unable to generate the kind of demand that can sustain rapid growth, even in elite consumption sectors.

Economic liberalization policies of the 1990s have granted the middle class a role it cannot assume. The credit card industry has mushroomed (Master Card exhibited the highest growth in Asia in 1995). Consumer giants such as Hindustan Lever are spending millions on advertising in the consumer sector (Figure 7-20). But as advertising expenditures have grown 70 percent, sales have risen only 30 percent. Much of the so-called middle class is not able or willing to march to the drum of consumerism.

Satellite TV has exploded, offering multiple channels, foreign programming, and upper middle class imagery: well-off, cosmopolitan, and urban-centric. This has reinforced the notion that a consumptive middle class will lead the masses down the road of development, progress, and prosperity. But the middle class TV owners watch and confirm their lifestyles and ambitions. This does little for the masses. With respect to TV sets, India only has 3.2 sets per 100 people. Malaysia has 14.8. For those convinced that India is on the move, the fact that the country has the largest number of illiterates in the world is glossed over. Glitzy commercials mask harsh realities of urban slums, poor villages, and wasted countrysides.

Deprivation is sanitized by statistics and TV-screen "realities." Even the Indian government admits that 39 percent of the people live below its own definition of poverty. In effect, the small elite and the larger middle class live far removed from the worlds of the masses of destitute Indians. According to one critic, economic liberalization, "gave a flamboyant ideological justification for the creation of two Indias, one aspiring to be globalized, and the other hopelessly, despairingly marginalized" (Varma 1998).

Figure 7-20

With its own satellite, India is now part of the global system of finance. Instant transferral of funds is part of India's growing corporate world.

India's rich–poor gap is being perpetuated. We know that education is an important vehicle for change because through education comes opportunity, empowerment, and self-realization. In Malaysia, the government spends U.S.$128 per person on basic education. India invests U.S.$9 per capita. Furthermore, the bulk of India's huge work force is far below the rest of the world in productivity and quality. For example, the value added per worker in India's manufacturing sector is a tenth that of Japan.

Although multinationals and joint venture enterprises manufacture cars, electronics, machinery, chemicals, and war material, the sad fact is that poverty and ignorance prevent at least three-fifths of Indians from participating in any direct way in bettering their condition in any national context. Many of the middle class "have consigned the poor to being a fixture on a landscape they do not wish to see" (Varma 1998).

Main Point Summary

1. India is the world's largest democracy. Yet, reconciling cultural, economic, political, and social differences is its challenge in the twenty-first century.
2. With a population of one billion and a NIR of just under 2 percent, population numbers and distribution will continue to be at the forefront of many perplexing issues.
3. Regional imbalances in population result from gender and caste inequities as well as migration. Kerala exemplifies accomplishments made possible through the education and empowerment of women.
4. India is characterized by linguistic diversity. Fundamental differences in language families are at the core of divisiveness. Hindi, the most widely spoken language in central and northern India, belongs to the Indo-Aryan language family. The southern states are dominated by Dravidian languages such as Tamil.
5. Although discrimination based on caste is illegal, the rules and regulations of caste are evident in all strata of society. Affirmative action programs to benefit unscheduled castes and tribes are resented by many members of regular castes. Tribal regions such as Assam are of particular concern because they are rich in natural resources. Caste affiliations are becoming increasingly significant in politics as parties jockey for group loyalties.
6. Religious conflicts, which underlay the history of the entire region, continue to stir up trouble on a regular basis. Religious differences have been intensified with the emergence of the Bharatiya Janata Party.
7. The Sikhs, who dominate Punjab, India's richest state, continue to press for an independent state of Khalistan. This idea is anathema to the Indian government because of its threat to Indian federalism.
8. Many improvements have been made in agriculture: expanded irrigation, high-yield seeds, use of chemicals, biogas, land reclamation, and mechanization. Nevertheless, crop yields are below world averages and food imports will continue to rise in light of population growth and environmental limitations.
9. Changes in the structure and organization of agriculture have social impacts, as illustrated by ruptured caste relations in Punjab and curtailment of women's roles in the cattle industry.
10. India has a long history of urbanization. Those cities developed or strongly influenced by colonial powers emerged as *dual cities* with distinct Indian and European areas. Calcutta, Chennai (Madras), Mumbai (Bombay), and New Delhi are examples of this phenomenon.
11. In the traditional city, socioeconomic status was the dominant theme in residential patterns. Modernization is transforming communal caste status into communal class status. High status areas are central; low status areas are peripheral.
12. Indian cities are overcrowded, run-down, and polluted. Efforts to decentralize have mixed re-

sults, as in the case of New Mumbai. Rural to urban migration continues to overpower attempted urban improvements.

13. Patterns of industrialization are uneven with concentrations around former colonial and coastal centers such as Calcutta, Mumbai, and Chennai. Maharashtra is India's most important industrial state. Government efforts to decentralize have spawned new areas in the interior. One important example is Bangalore in the south, which has earned the sobriquet "India's Silicon Valley."
14. Government regulations including limits on foreign investment have long controlled industrial development. The results of loosening these controls in the 1990s remain to be seen.
15. Modern developments have brought catastrophe to thousands of people. The gas leak at Bhopal is noted as the world's worst industrial disaster. Dam construction, forestry, and mining operations have displaced countless individuals and groups powerless to claim their traditional territories. These are the *oustees,* many of whom end up in India's already overburdened cities. The Narmada River dam project is one of the world's most controversial.
16. The Ganges, Hindu India's most sacred river, is dangerously polluted.
17. India, like its rival Pakistan, is a nuclear power. Self-esteem and political disagreements with both Pakistan and China fuel India's participation in the nuclear race.

Terms to Know

adivasis	*jajmani* system	*chawls*	biological oxygen demand (BOD)
Khalistan	White Revolution	India's Silicon Valley	environmental refugees
sodic land	gun-salute states	Jarkhand	oustees
transgenic plants		*Ganga Ma*	

Bibliography

Adhikari, Sudeepta. 1986. Some Aspects of Indian Federalism: A Study in Political Geography. *Singapore Journal of Tropical Geography,* 7: 1–11.

Berry, Brian, and Howard Spodek. 1982. Comparative Ecologies of Large Indian Cities. In *Urban Patterns: Studies in Human Ecology,* pp. 444–53. University Park: Penn State.

Beteille, Andre. 1996. Caste in Contemporary India. In *Caste Today,* ed. C. J. Fuller, pp. 150–79. Delhi: Oxford.

Catliff, Nick. 1990. Killer in the Night. *World,* 47: 76–80.

Critchfield, Richard. 1980. The Greening of Ghungrali. *Asia* (May/June): 38–44.

Duara, Prasenjit. 1991. The New Politics Of Hinduism. *Wilson Quarterly,* 15: 42–50.

Dreze, Jean, and A. Sen. 1996. *India: Economic Development and Social Opportunity.* Delhi: Oxford.

Etienne, Gilbert. 1998. Alarm Bells for South Asia. In *Feeding Asia in the Next Century,* eds. G. Etienne, C. Aubert, and J. L. Maurer, pp. 35–55. Delhi: MacMillan.

Farmer, B. J. 1981. The "Green Revolution" in South Asia. *Geography,* 66: 202–07.

George, Kanti. 1985. *Operation Flood: An Appraisal of Current India Dairy Policy.* Delhi: Oxford University Press.

Gupta, Pranay. 1985. Looking Ahead in India. *Harvard International Review,* 8: 29–32.

Halarnkar, Samar, and Uday Mahurkar. 1999. Burning the Cross. *India Today,* 11 (January): 22–27.

Haub, Carl. 1996. Spotlight: India. *Population Today,* 24: 7.

India Bombs the Ban. 1998. *The Bulletin of the Atomic Scientists* (July/August).

Jacobson, Jodi. 1991. India's Misconceived Family Plan. *World Watch,* 4: 18–25.

Jayaraman, N. et al. 1996. GloomTown. *Far Eastern Economic Review,* 18 (January): 40–2.

Khan, Nizamuddin, and Almatar Ali. 1998. Periodic Markets and Rural Transformation in Gonda District, Uttar Pradesh. *Focus,* 45: 34–73.

Kalam, A. P. J., and Y. S. Rajan. 1998. *India 2020: A Vision for the New Millennium.* New Delhi: Viking.

Karp, Jonathan. 1995. To Have and Have Not. *Far Eastern Economic Review,* 17 (August): 62–5.

Kayastha, S. L., and Prithvish Nag. 1989. The Bhopal Disaster. In *Population and Disaster,* ed. John I. Clarke et al., pp. 206-18. IGU Cambridge: Blackwell.

Kunakrishnan, K. 1991. Handwriting on the Wall. *Far Eastern Economic Review,* 30 (May): 57.

Madeley, John. 1984. Progress Has its Price. *Geographical Magazine,* 56: 226–27.

Maloney, Clarence. 1990/91. *Environmental and Project Displacement of Population in India Part I: Development and Deracination.* Universities Field Staff International: Field Staff Report No. 14.

Marwah, Bhushan. 1995. India Striving for Self Reliance in Oil. *India Perspectives,* 7: 24–6.

Maslen, Geoffrey. 1995. Asia's Aching Hearts. *Asia Inc.* (Sydney Morning Herald Magazine) October : 64–65.

McDonald, Hamish. 1993. Fuel Injection. *Far Eastern Economic Review,* 19 (August): 42–43.

McDonald, Hamish. 1994. Rebel Redoubt: New Delhi Fights Tribal Insurgencies in the Northeast. *Far Eastern Economic Review,* 9 (June): 31–35.

McKibben, Bill. 1996. The Enigma of Kerala. *Utne Reader* (March/April): 102–12.

Mehta, Pratap, B. 1998. India: The Nuclear Politics of Self-Esteem. *Current History,* 97: 403–406.

Mendelsohn, Oliver, and M. Vicziany. 1998. *The Untouchables: Subordination, Poverty and the State in Modern India.* London: Oxford.

Mohan, Rakesh. 1999. What Ails the Indian Tiger? *Far Eastern Economic Review,* 8 (April): 61.

Morehouse, Ward. 1994. Unfinished Business: Bhopal Ten Years After. *The Ecologist,* 24: 164–68.

Nelson, Toni. 1996. Closing the Nutrient Loop. *World Watch* 9: 10–17.

Noble, Allen G., and Asok K. Dutt, eds. 1982. *India: Cultural Patterns and Processes.* Boulder: Westview.

O'Brien, Conor, C. 1988. Holy War Against India. *Atlantic Monthly,* 262: 54–64.

Rain, Nick. 1994. Living With the Landscape. *Geographical,* 66: 24–27.

Rao, Radhakrishna. 1992. On Fertile Ground. *Far Eastern Economic Review,* 13 (August): 70.

Repetto, Robert. 1994. *The "Second India" Revisited: Population, Poverty, and Environmental Stress Over Two Decades.* New York: World Resources Institute.

Saha, Suranjit, K. 1979. River-Basin Planning in the Damodar Valley of India. *The Geographical Review,* 69: 273–87.

Sampat, Payal. 1995. India's Low-Tech Energy Success. *World Watch,* 8: 21–23.

Sampat, Payal. 1996. The River Ganges' Long Decline. *World Watch,* 9: 24–32.

Sarangi, Satinath, and Carol Sherman. 1993. Piparwar: White Industries' Black Hole. *The Ecologist,* 23: 52–56.

Sarkar, Aditi Nath. 1990. Oh, Calcutta. *Natural History,* 5: 74–78.

Sharmer-Smith, Pamela. 1984. The Sikh Identity. *The Geographical Magazine* (September): 442–43.

Tharoor, Shashi. 1997. *India: From Midnight to the Millenium.* New York: Arcade.

The Economist. 1997. A Survey of India: Time to Let Go. *The Economist,* 22 (February): 1–26.

Varma, Pavan, K. 1998. *The Great Indian Middle Class.* New Delhi: Viking.

Visaria, Pravin, and Leela Visaria. 1981. India's Population: Second and Growing. *Population Bulletin of the Population Reference Bureau,* 36: 1–56.

Wallach, Bret. 1984. Irrigation Development in the Krishna Basin Since 1947. *The Geographical Review,* 74: 127–44.

Wallach, Bret. 1996. *Losing Asia.* Baltimore: Johns Hopkins.

Chapter 8

South Asia: Bangladesh, Sri Lanka, and the Indian Ocean

"When unity is evolved out of diversity, then there is a real and abiding national progress."

(MANHAR-UL-HAQUE 1866–1921)

Bangladesh: Nation on the Edge

Imagine 120 million people eking out an existence in a country the size of Wisconsin (population of about 5 million). Think about the facts that two-thirds are illiterate and two-thirds of children under five are malnourished. Add the fact that storms and floods wipe out people and livelihoods by the hundreds or even thousands yearly. Then, consider that this impoverished, primarily agricultural country is the sixth largest (1994) exporter of ready-made garments in the world. What is the story of this relatively new nation of Bangladesh, and what are its future prospects?

Established in 1971, Bangladesh occupies a drainage basin where the Brahmaputra, Padma (Ganges), and Meghna Rivers converge (Figure 8-1). Most of the country is a deltaic, alluvial plain although there are hill districts in the northeast and southeast. Its marshy coastline extends 370 miles (600 km) across the Bay of Bengal.

BENGAL NATION

Bangladesh means "Bengal Nation." This region, once called Bengal, is renowned for its tigers and as South Asia's "rice bowl." Aside from tribal areas in the Chittagong hills, Bangladesh is linguistically homogeneous. Bengali, known locally as ***Bangla,*** serves as a unifying force within this politically fractious nation.

Although not officially a Muslim state, most of its people are Islamic. Bengal, especially in the east, had a tradition of Tantric Buddhism and Hinduism. This was conducive to the introduction of new faiths such as Sufi Islam brought by Arab sea traders in the thirteenth century. The Hindu caste system made Islam and its notions of equality an attractive option. By 1947, more than half of all Bengalis were Muslim. Today, around

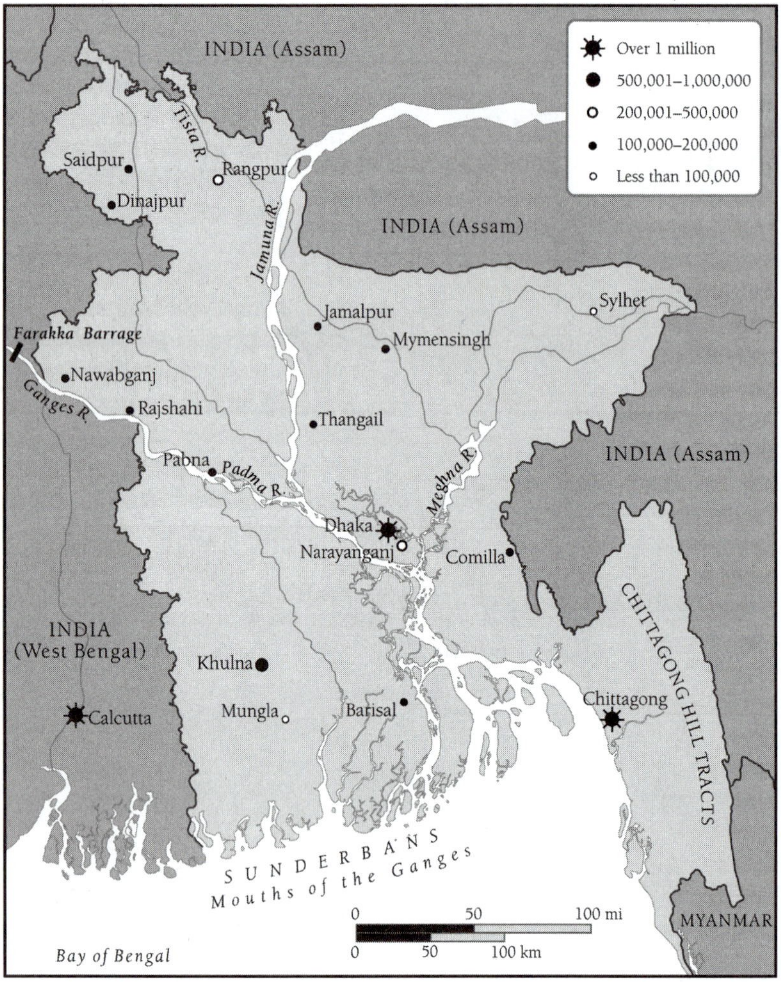

Figure 8-1

Bangladesh and city populations 1991. Source: *1994* Demographic Yearbook.

85 percent are Muslim, 12 percent are Hindu, and the remainder are Buddhist, animist, or Christian.

Christianity was introduced by Portuguese trader's in the sixteenth century. Again, many low-caste Hindus were drawn to Christian ideas of equality and salvation. The Portuguese also introduced cashew nuts, papaya, pineapple, and guava to Bengal.

The region has a history of political subjugation and fragmentation dating back to the sixth century. British *divide and rule* policies only added to this legacy. Calcutta had served as the capital of British Bengal since 1772, but this changed in 1905 when eastern Muslim regions were linked with Assam and were designated as East Bengal and Assam, with Dacca (Dhaka) as the capital. At the same time, majority Hindu districts of West Bengal, Bihar, and parts of Orissa were designated as Bengal.

In 1912, following years of mass resistance, East Bengal was divided into three provinces and West Bengal was reunited with Calcutta. Boundaries were again redrawn with the creation of East Pakistan in 1947. Consequently, Bangladesh's borders are politically, not naturally, defined. Surrounded by India on three sides, Bangladesh also shares a 120-mile (193-km) frontier with Myanmar. Bangladesh has signed agreements with both countries, but territorial control remains controversial among some tribal groups. Marine boundaries are still undefined and problematic because of undersea oil discoveries in the Bay of Bengal.

This history of jurisdictional change has created regional dissention and political factionalism. For example, dissention continues to be problematic in the Chittagong Hill Tracts (CHT) situated in the southeastern corner of Bangladesh, bordering on both India and Myanmar. The indigenous Jumma, incorporating more than a dozen ethnic groups, are being overwhelmed by the Bangladeshis who have continued to move into this territory.

Requests for a degree of sovereignty were denied, and the nationalistic Jana Sanghati Samiti (JSS) was formed. Rebellion ensued, and the Bangladeshi army went on a rampage of killing and destruction in the northwest CHT. Thousands of refugees fled to India's Tripura state. At least 8,000 rebels have died in this two-decade struggle.

A peace accord was signed in 1997, and about 50,000 refugees returned. The Jumma were given some autonomy in the border areas, but the JSS continues to press for further changes. The future of this region remains uncertain.

Regional instability also engendered several core areas focusing on such centers as Calcutta and Dhaka. When East Pakistan was created, the key industrial center and port of Calcutta was lost to India, leaving East Pakistan without a viable industrial base or seaport. Cotton and jute mills were built in Dhaka. Ports were developed at Chittagong in the far south and Mangla in the southwest. Dhaka and Chittagong remain the focus of the modern economy that is only in its initial stages relative to other countries in South Asia.

THE IMPACT OF FLOODS AND CYCLONES

"Bangladesh" is often perceived as synonymous with "disaster." It is true that millions of lives have been lost to natural forces. As recently as August 1998, 10 million people were rendered homeless by floods! As geographers, we want to know why this happens and what can be done about it.

The erratic nature of Bangladesh's climate produces uncertainty. Normally, the monsoon generates 60 to 80 inches of precipitation from April to October, with rainfall typically peaking from June to September. Excessive rains cause rivers to overflow their banks and inundate surrounding regions (Figure 8-2). Flooding frequently derives from heavy snowfall in the Himalaya. With deforestation and rapid runoff, rivers be-

Figure 8-2

These buildings outside of Dhaka are built on stilts to avoid annual flooding. However, they have been washed out on several occasions when floods and storms are especially severe.

come deluges. In 1974, two-thirds of the country was under water; subsequent famine killed 30,000 people. This degree of inundation occurred again in 1998, although with only a few thousand losing their lives.

It is understandable to associate Bangladesh and flood disaster. But Bangladesh and drought? In fact, the country does suffer from drought, especially in the northwest. Since governments are geared to dealing with floods, little if any aid reaches drought-stricken regions. Victims are left to their own devices and must rely on friends and relatives for support.

Bangladesh is also beset by **cyclones.** These tropical storms swirl in a northeasterly direction from the Bay of Bengal and wreak havoc across the nation. In 1970, a cyclone accompanied by a 20-foot **storm surge** took 300,000 lives. A storm surge is a tide that is shaped into astronomical proportions by an increasingly narrowing channel across low, flat terrain. In the case of the Bay of Bengal, as the landform funnel narrows, the wave heightens and wind velocities increase, even to hurricane force.

The 1970 storm surge wiped out fishing communities, killed fishermen, and destroyed boats, to the extent that 65 percent of the nation's fishing capacity was eliminated. This spells human disaster in light of the fact that two-thirds of the country's protein is derived from seafood.

While rain and run-off floods can occur anywhere, the impact of cyclones is most evident in the south. In April 1991, Bangladesh faced the worst cyclone of any since independence. The death toll was 138,000 people, most of whom lived in the south.

The delta comprises 3,720 square miles (6,000 sq km) of ever-shifting mangrove, mudflats, islands, and distributary channels. This region is called the **Sunderbans.** Settlement, although discouraged by the government, is virtually immediate after a flood. Land and water configurations may be different, but any land is space to be seized at the first opportune moment. Even islands barely above sea level are occupied, at least until the next flood. Person-eating tigers are also a threat in the Sunderbans.

WATER, LAND USE, AND SETTLEMENT

India affects the water supply in Bangladesh. In 1974, India installed the Farakka water diversion barrage on the Ganges (Padma in Bangladesh) near the Bangladeshi border, allowing less water to flow to Bangladesh. Bangladesh claimed that reduction in water levels increases salinity in the Padma River basin, thereby preventing some 35 million people from getting irrigation water. After years of deadlock, the two countries signed a treaty in 1996 to guarantee equity and efficacy in water supply from the Farakka Barrage.

Instability of land and water patterns affect settlement and land use patterns. As rivers sway between valley walls, they assume a braided pattern separated by small, sandy islands called ***chars.*** As river waters shift laterally, erosion is exacerbated on the right banks and the whole system exhibits a westerly shift. *Chars* disappear and new channels are formed.

Populations move with *chars,* taking advantage of naturally fertilized land. If a *char* is developed, it can act as a magnet for more people who inhabit the adjacent riverbanks in makeshift shelters. Over half of these displaced people are sharecroppers or rent from large landowners. Others survive in the informal economy. Whatever system is established can be quickly erased in a subsequent flood.

In 1972, it became policy for newly emerged *chars* to revert to the government. The idea was to redistribute these lands to the landless and poor, small farmers. However, the power of landlords, inequities in land ownership, intergroup conflict, and political intrigue hamper the implementation of resettlement. In fact, over half of rural householders are landless. With little chance of participating in the larger economy, they remain as semipermanent clusters in the countryside, sources of urban squatters.

What is being done to mitigate these yearly disasters? In 1967, tidal gauges were installed along the coast. A Flood Action Plan was developed in 1990. A year later, a Cyclone Warning Service was established. However, the system was overused and people became complacent. When warnings of "great danger" failed to materialize on several occasions, many people ignored the warnings.

Today, with help from the World Meteorological Organization, Bangladesh has its own Space Research and Remote Sensing Organization. With links to Indian systems, it can broadcast more accurate warnings ahead of time. In addition, concrete shelters, channels, and embankments have been built across the country (Figure 8-3). Unfortunately, many of these failed to

Figure 8-3

New embankments have been constructed around Dhaka to minimize flooding. However, this one is being repaired from damage in a recent flood.

hold up in the floods of 1998, and even parts of Dhaka were flooded. The nation's main road, connecting Dhaka with Chittagong, elevated more than 6 feet with WB aid, remained above the floodwaters. Unfortunately thousands of miles of roads were washed out and need to be repaired and raised.

Embankments are also magnets for human habitation. Impromptu settlements cling to their sides. In fact, the right bank of the Brahmaputra has been described as "the world's longest linear housing development."

AGRICULTURAL RHYTHMS

As in all of South Asia, the rhythm of life in Bangladesh follows the seasons in which flooding is integral. Flooding also has a spatial pattern and temporal rhythm. For example, the Brahmaputra and Meghna rise in March and April, fed by Himalayan snow melt. At this time, pre-monsoon rains occur in Assam and northeastern Bangladesh. The Ganges rises in May. By June or July, all rivers are swollen with monsoon rains. The Ganges reaches its peak flow anywhere from August to October.

In terms of planting crops, there are three main seasons (Figure 8-4). In addition, cropping practices are adapted to flood conditions. For example, virtually thousands of rice varieties have been planted over the

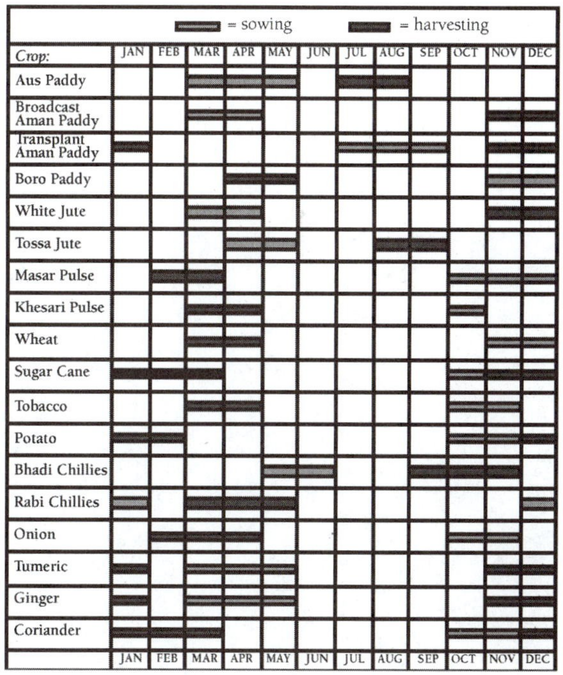

Figure 8-4

Agricultural calendar. Note that June and July are the least busy months. Why do you think this is?

centuries in context of variation in water levels. These include deep-water types that grow in 15 feet (5 m) of water. This rice is harvested from boats.

Bangladeshi farmers produce a wide array of crops, yet rice covers 80 percent of the cultivated area. Unfortunately the high-yielding rice of the Green Revolution is not suited to flood conditions. Consequently, high-yielding rice is limited to rabi or dry season and other varieties are planted for the rain-fed and flooded kharif. Bangladesh must import food to meet the needs of its ever-growing population.

Given climatic factors and the availability of irrigation and fertilizer, most of Bangladesh is limited to one crop a year. About 40 percent of the land can be double-cropped and 8 percent triple-cropped.

Recent efforts to diversify have seen regional specialization. Virginia tobacco is grown in the north, and the Sylhet region in the northeast produces 94 percent of the nation's tea. Harvested nine months a year, exported tea is an important earner of foreign exchange. Rubber and coffee are concentrated in the CHT.

With the installation of tube wells, high-yielding wheat has become prominent in the north, with yields exhibiting a tenfold increase between 1970 and 1983. However, in much of Bengal overuse of tube wells is drawing arsenic from pyrite bedrock into the water system. In Bangladesh, up to 60 million people are exposed to arsenic poisoning. This causes ever-worsening skin lesions and leads to death.

Fiber crops are critical to the Bangladeshi economy. For instance, two-thirds of all the manufacturing jobs in the country are related to the cotton industry. Second to rice in terms of acreage is **jute,** a fiber crop used for sacking, twine, rope, carpets, car interiors, and lining for asphalt roads (Figure 8-5). Green jute is finding increasing use in fabrics. In fact, special jute blankets have been manufactured to protect poor, northern peoples from the winter cold. Moreover, jute stalks are used as fuel, the leaves can be eaten, and any remaining debris can be plowed back into the soil as natural fertilizer.

Sown from March to May and harvested in August, jute can be grown in uplands or lowlands. One variety can withstand 5 feet of flooding. In fact, Bangladesh produces two-thirds of the world's jute. Until recently, synthetic fibers threatened the jute industry and production declined. Fortunately, environmental concerns have made jute desirable once again.

Livestock are of increasing significance in the economy. In fact, livestock rearing—employing about 20 percent of the labor force—is the most important activity for the rural poor. Cattle and water buffalo are not only draft animals, but also provide dung for

Figure 8-5

Jute fiber is one of Bangladesh's most important products. These mats, however, are made of another fiber called coir. *This fiber comes from coconut husks.*

fertilizer, milk and meat for food, and leather, horn, bone, and oil for export. Bovine animals work as much as 80 percent of the agricultural land and are indispensable for threshing grain and crushing oilseeds and sugar cane.

Despite the large numbers of cattle, buffalo, goats, and chickens available, Bangladesh is deficient in meat, milk, and eggs. This can be attributed to genetic deficiencies of livestock, ignorance of breeding, feeding, and maintenance procedures, and lack of veterinary care. As a result, Bangladesh is a net importer of live cattle and milk products.

While Green Revolution technologies have increased yields, beneficiaries are primarily middle and rich peasants who dominate cooperatives and local power structures. It is important to note that more than half the land is owned by 10 percent of households and that 75 percent of rural households are either landless or possess tiny plots around 0.8 ha. Geographer Abu M. S. Ali (1998) found that small farm size is the leading cause of unemployment in agriculture (i.e., frequency of unemployment rises with decreasing farm size). Land reform or any other improvements in agriculture must be viewed in context of population requirements.

POPULATION AND GENDER ISSUES

Bangladesh experienced a population explosion after 1947 with the introduction of health care and improved sanitation. The death rate fell from 3.7 in 1941 to 1.1 in 1982. Increasing numbers moved into marginal lands, where their needs could not be adequately met. This situation was compounded by migration streams between East Pakistan and India.

In Chapter 3, I reported geographer Bimal Kanti Paul's findings that Bangladesh has made great strides in lowering its birth rate. However, a 1997 natural increase rate of 1.74 remains considerable in light of a large base population and a total fertility rate of 3.41. Only about half of Bangladeshi women use contraception and about two-thirds are illiterate.

In this Muslim society, women are traditionally subjugated by men, marriages are arranged, and dowry deaths still occur. The sex ratio is imbalanced with 105.6 males per 100 females. Although there is a tradition of purdah, most women must work outside. Consequently, strict purdah is practiced primarily among the middle classes.

Changes in Female Activity Space

Geographer Bimal Kanti Paul (1992) studied mobility and employment opportunities for rural women in Bangladesh and found that as poverty increased, more women were forced to seek work outside the home. This trend jeopardized traditional spatial contexts defined by village structure, purdah, and patriarchy.

The basic spatial unit is the ***bari***, two to six houses around a square or rectangular courtyard. Within the *bari* are patrilinearly related, nuclear families. Clustered *bari* form a neighborhood, and several of these make up a village. Tradition holds that women's work is confined within or close to the *bari*. However, various factors such as religion and economic status influence the spatial extent of **activity space.** And, the status and position of women change at different times in their life cycle.

Rich women, young wives, and spouses of Islamic religious leaders have the smallest spaces, rarely venturing from the home or *bari*. Other women are confined by more extensive neighborhood boundaries. Elderly, Hindu, and poor women have an even greater range of excursion. Males and tribal women, with freedom of movement, have the largest activity spaces. Of course, this scheme is a model, and there is much variation in reality.

Patterns of women's activities and associated spatial realms are in flux. Certainly programs intended to empower women have encouraged and facilitated the expansion of female activity spaces (Figure 8-6). However, poverty is the basic cause of working outside the home. While paid labor within other households remains the major employer of women, increasing numbers are seeking wage employment beyond

Figure 8-6

Bangladeshi women. Self-help programs are improving the lives of many women in Bangladesh. These women are middle class.

bari and village boundaries. As landlessness and destitution increase, traditional familial support structures fail and the rates of divorce, abandonment, and wife-beating rise. Consequently, more women are working alongside men in the fields, in road and building construction, brickmaking, and numerous manufacturing enterprises. Usually, increased activity space is accompanied by improvement in living standards, but as Dr. Paul observes, "Unfortunately, this relationship has not emerged in Bangladesh."

Recent governments, headed by women, have taken great strides in improving women's lives through education and employment. Widespread media propaganda promote the necessity of empowering women to fully participate in the country's development. Employment in labor-intensive industries and craft and other cooperatives have markedly advanced living standards for some Bangladeshi women.

Bangladesh's population dilemmas derive more from distribution than from sheer numbers. Rural-to-urban migration has overwhelmed cities, especially Dhaka and Chittagong. An urban growth rate of 6 percent (1993) is one of the highest in Asia. In 1961, a mere 5 percent of Bangladeshis lived in cities. In 1991 it was 18 percent. Currently, 26 percent live in urban places, and by 2015 that will increase to 37 percent. In other words, 80 million Bangladeshis will be urban residents by 2020. This is an amazing number in context of infrastructural requirements.

Take power availability as an example. Only 16 percent of Bangladeshis have access to power. The country often faces riots over blackouts. At the opening of the largest generating plant, 30 miles (50 km) northwest of Dhaka, Prime Minister Sheikh Hasina urged people to save electricity at home and at work by shutting off lights. Public and private sector projects are expected to boost the national grid's power to 4,700 megawatts by 2002. But there is another problem: People tap into power lines illegally, and nearly 30 percent of power production is lost to theft.

DILEMMAS OF GROWTH AND DEVELOPMENT

Cities and Industries

Rural push factors drive urban migration streams while growth in production is not expected to meet employment needs. As population experts Professor Abdul Barkat, U. R. Mati, and M. L. Bose (1997) point out, "urbanization in Bangladesh will remain poverty driven." Poor migrants typically end up living in slums, squatter settlements, or as pavement dwellers. There are already about 25 million of these individuals, more males than females.

Dhaka, the country's capital and primate city, attracts most migrants. In 1992, Dhaka had 7.4 million inhabitants. By 2010, it is expected to have 17.6 million, with at least 50 percent in slums or worse. Dhaka's Old City can only be described as a hive of humanity (Figure 8-7). Every crumbling building, room, cubicle, alley, doorway, and indentation reveals people in every imaginable condition. The narrow streets are crammed with people, carts, bicycle rickshaws, trucks and busses, all carrying stupendous loads of oil drums, cotton bales, vegetables, sacks of ice, sheets of metal, thousands of rubber thongs, baskets of poultry—a limitless kaleidoscope of colors, sounds, and smells.

Child labor is common. This is reflected in the fact that only 20 percent of slum children attend school. In addition, at least one-third of slum dwellers are sick at any one time. Slum communities have the highest infant and maternal mortality rates in the nation, in part because health care programs have been geared to rural areas.

Outside Dhaka's Old City are still relatively old areas that are becoming increasingly densified. Here, unchecked settlement outpaces government efforts to develop housing and other infrastructural needs (Figure 8-8). Incentives to decentralize and create communities outside the city have met with limited success.

In the 1960s, industries were given incentives to locate outside Dhaka. However, they located just beyond an established boundary to claim "less-developed area" status and still enjoy the advantages of urban location. These and other efforts at decentralization were disrupted by the 1971 war of independence.

In the 1980s, subdistricts called ***upazilas*** were created in Bangladesh. Each of these 500 subdistricts was to house an average of 186,000 people and have revenue-raising powers and measures of self-government. However, related employment development has not slowed out-migration.

Although foreign investment and new construction are changing Dhaka's skyline, labor strikes called ***hartals*** are frequent, and infrastructural problems remain overwhelming. Power outages are daily events. Inadequate garbage and sewage removal results in clogged canals and drainage backups and overflow. Raw sewage gushes into the Bay of Bengal. Vehicles belch diesel fumes, and industries pour raw effluents into canals and poisonous chemicals into the air. Dhaka is now one of the most polluted cities in the world. Lead poisoning is more than double unsafe

Figure 8-7

Traffic crush in Dhaka. In all my travels, I have never seen anything to equal the traffic jams in Bangladesh.

Figure 8-8

New multistory buildings arise out of the old skyline of Dhaka. Often the money for construction runs out and the buildings are never finished. Of buildings that do get completed, many stand empty for want of occupants.

levels. Thousands of premature deaths and millions of cases of illness result from the city's pollution.

There are more than 175,000 motor vehicles in Dhaka, including more than 40,000 auto-rickshaws that use a deadly combination of engine oil and leaded gas. In 1999, the government was taking steps to reduce lead pollution by banning the import of leaded fuels and calling for catalytic converters on new cars. Auto-rickshaws are to be phased out by 2001. Keep in mind that these are *plans.* Population and urban growth, natural calamities, and political or other instabilities or inefficiencies can easily slow or halt implementation.

Chittagong is about half the size of Dhaka. It has good rail, road, and water connections and is a designated export-processing zone. Its industries are based on primary products including oil from offshore facilities and bamboo harvested from the surrounding hills. Yet, half of Bangladesh's formal employment sector is in Dhaka. Investors prefer Dhaka because of its larger pool of managerial and skilled workers.

Jute and cotton mills and leather industries cluster in and around Dhaka, and these labor-intensive industries remain the country's leading employers. In recent years the government has promoted diversification into new fields such as food processing.

The frozen food industry is founded on fish, prawns, and similar products. More than 11 million individuals are connected either directly or indirectly with the fisheries sector for their livelihood. Oceans, rivers, lakes, canals, and more than a million manmade ponds and tanks are fishery sources. Shrimp exporting began in the early 1970s, earning U.S.$3 million. In 1996–1997 this had increased to U.S.$279 million. Even so, there are now more processing plants than fish. This calls for the development of fish farms and hatcheries. Chittagong and Mungla are the most important processing centers.

Accounting for 86 percent of the country's export earnings in 1997, the garment industry is of most importance. Nike, Ralph Lauren, Tommy Hilfiger, and other well-known brands of apparel are made in Bangladesh. The United States is the largest purchaser of Bangladeshi-made garments.

The garment industry has been a significant employer of women—90 percent of the workers are women. Until recently, children were also employed. In 1993, the U.S. Congress passed a bill to ban Bangladeshi clothing made with child labor. Consequently 50,000 children were dismissed from the industry. Bangladesh has since agreed with international organizations to ban children from the garment sector.

Nongovernment Organizations (NGOs)

Bangladesh is one of the world leaders in NGOs, having more than 20,000 in 1998. According to most reports, NGOs function far more effectively than government organizations such as the Bangladesh Rural Advancement Committee (BRAC). However, BRAC has more than two million members, mostly female. It promotes craft and cottage industries and offers flexible schooling taught by locals.

The most well known NGO in Bangladesh is **Grameen Bank**, which offers small loans to the poor, again mostly women, for shops, sewing machines and other labor-saving devices, livestock, or anything else to empower women. Women are regarded as better managers of household finances and more responsible in paying back loans. Loans are made to small groups. Both peer pressure and group responsibility are critical in ensuring repayment.

World Bank (WB) studies indicate that 50 percent of borrowers get out of poverty in 5 years and that both divorce and birth rates decline. With Grameen as a model, thousands of other organizations have provided micro-credit to the poor worldwide.

Critics point out that the bank is heavily supported by foreign funds and argue that it would collapse otherwise. Some say that Grameen is really controlled by the government. Some say it is anti-Islamic. Others argue that peer pressure drives out the weakest members, leaving them destitute and compelled to a life of bonded labor. Case studies indicate that perhaps two-thirds of Grameen loans are actually controlled by male family members and stringent payback requirements drive some women to usurious moneylenders. Grameen authorities say that such stories are rumors circulated by a jealous BRAC or larger organizations like the WB, which are promoting their own schemes. Whatever the case, micro-loans apparently have helped millions of women in Bangladesh. The full story remains to be seen.

Sri Lanka: Teardrop Weeping

Arabs called it **Serendip**, the Portuguese named it **Cilao**, and the Dutch spoke of **Zeylan.** All were referring to ***Simhala Dvipa***—Island of the Lion Tribe. Known as Ceylon until 1972 when it was renamed Sri Lanka, this tear-shaped island is divided from the Indian subcontinent by the 22-mile- (35-km) wide Palk Strait. While separation fosters cultural distinction, propinquity favors spatial interaction. In the case of Sri Lanka and India, interrelationships have been filled with inherent antagonisms.

THE PHYSICAL LANDSCAPE

The island of Sri Lanka includes both *up-country* and *low-country*. About 30 percent is up-country, where mountains rise above 8,000 feet (2,500 m). From these heights, rivers flow in a radial pattern downward through lush forests to the coastal plains. The longest of these is the Mahaweli Ganga, running in a southeasterly direction for 206 miles (332 km). In the low-country, hills are separated by abrupt escarpments and cut by deep valleys. They descend in a step-like fashion to lowlands, mostly under 1000 feet (300 m). In the north, these lowlands extend as an undulating plain to the Jaffna Peninsula (Figure 8-9).

The island experiences four weather and climate modes in a yearly cycle. From May to August, the southwest monsoon brings orographic rainfall to the southwestern hills. This season culminates with tropical cyclones from the Bay of Bengal, which can affect all areas from October through November. Next comes the northeastern monsoon, which supplies 40 to 80 inches (102–203 cm) of rainfall in the north and northeast through February. From March to April, summer insolation and forested uplands generate **convectional rainfall.** Moisture-sodden warm air rises, cools, and condenses, resulting in late afternoon downpours.

Sri Lanka is tropical with little temperature variation. In fact, nocturnal-diurnal variations are greater than seasonal ones. Although temperatures rarely rise above the low 80s F (25°C), the sun's verticality induces high evaporation rates. This makes irrigation essential, especially in the east. Aridity is a traditional problem. Ancient civilizations fell not simply to invasion but to the destruction of their irrigation systems. Moreover, dry zones were malarial until the end of World War II and the use of DDT. With lack of moisture and the presence of killer mosquitoes, colonization dry zone schemes were dismal failures before

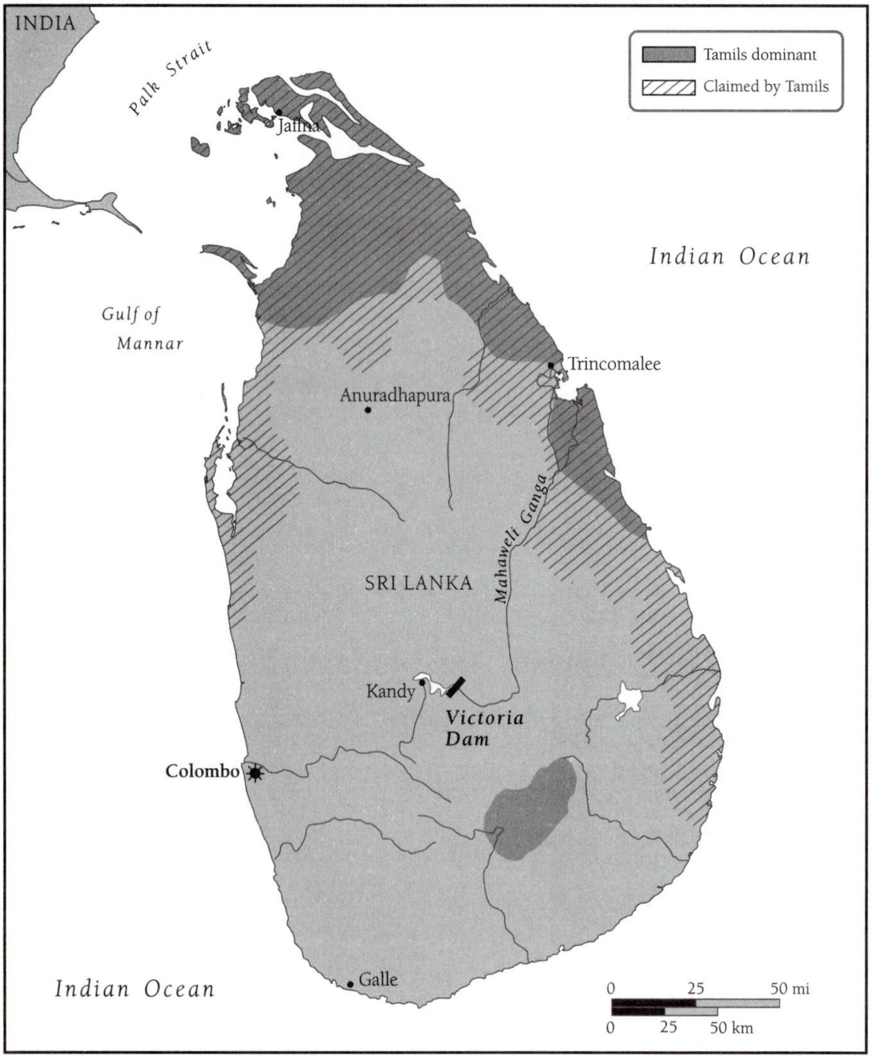

Figure 8-9

Sri Lanka and Tamil territorial claims.

1945. Karst topography in the Jaffna Peninsula allows the drilling of wells into limestone aquifers.

Water availability was pivotal to ancient settlement. Around 500 B.C., Sinhalese settlers developed elaborate water storage and distribution systems. Property ownership was based on the control of water, not land. Canals reached 50 miles (31 km) in length, and the Sinhalese devised water-control valves at dams to accommodate variation in water pressure. These preceded European devices by 1,500 years!

THE INDIAN CONNECTION

Sri Lanka's history is replete with invasions from India. The Sinhalese, descended from the Aryan invaders of Northern India, migrated via the islands and sandbars

Figure 8-10

Picking tea in Sri Lanka. Typically supervisors are male and the pickers are female.

of the Palk Strait from the fifth to third century B.C. They would become Buddhist. In the seventh century there were Hindu, Tamil invasions. In 1017, Ceylon became part of India's Chola Empire. Its history is marred by friction and fragmentation.

The Buddhist Sinhalese complex of Anuradhapura, with its 15-mile walled perimeter, supported more than a million people based on an irrigated rice economy. In an elaborate interlocking system, rivers were channeled to storage tanks, from which diversion channels delivered water to villages. This was one of the most elaborate irrigation complexes ever conceived. However, by the time the Europeans arrived, warfare had reduced the settlement to ruins and the area had become an overgrown wilderness.

EUROPEAN IMPACTS

The sixteenth century witnessed dramatic alterations in Sri Lanka's state of affairs. The Portuguese, striving to capture the Arab sea trade, stopped at Galle and set up spice-trading operations. They also demolished Buddhist and Hindu religious sites and forced conversions to Christianity. Further, they captured the mountain kingdom of Kandy and removed its most sacred relic—an alleged tooth of the Buddha. The tooth eventually found its way back and is currently housed in Kandy's Temple of the Tooth.

A century later the Dutch arrived (1658) and ousted the Portuguese. They pursued commercial development, but as a result of the Napoleonic Wars in Europe, were forced to turn over the island to the British in the nineteenth century. The Dutch did introduce the island's first commercial crop: coffee. But the coffee trade was under challenge from Brazil, and it was ruined by blight in the British period (1869). The British then turned to **cinchona** and tea.

Cinchona is extracted from tree bark and was the only treatment for malaria, widespread at the time. Eventually numerous other plantation/estate crops were grown (e.g., rubber, coconuts, cacao, and pepper). Estates were nationalized in the 1970s but reverted to private management in the 1990s. Sri Lanka is still one of the world's top tea exporters. Tea accounts for about 75 percent of the country's income (Figure 8-10).

Tea: The Global Beverage

More people drink tea than any other beverage (aside from water) in the world. Ancient Chinese texts describe tea as an elixer. Today tea is touted for its antioxidant properties. Tea is widely employed as a social medium. I have enjoyed tea on many occasions around the world: "China" cups of black tea in England; small glasses of mint tea in Morocco, cups of steaming milk-tea in Pakistan; glasses of spicy chai in India; and cups of aromatic green tea in China. Socioeconomic interchange is eased via the ceremony of tea drinking.

Tea diffused from China to Japan, where at first it was used medicinally and in Buddhist monasteries to enhance meditation. Subsequently it spread to Korea. The British introduced tea to India and Sri Lanka in the nineteenth century and the tea trade contributed significantly to the fortunes of the British East India Company.

Tea flourishes best in the cooler air from 3000 to 7000 feet (900–2,150 m). Tea trees are kept small with hand pruning, which stimulates growth. Leaves are hand picked and graded according to size and position on a branch. The bud and top two leaves are the best and earn the label "Orange Pekoe." Tea bags contain small, broken leaves known as "fannings."

The flavor of tea is derived from local growing conditions and processing. For example, Darjeeling tea has a peach aroma and is called the champagne of tea. Earl Grey tea is flavored with oil of bergamot, extracted from an orange-like citrus. Jasmine tea is green tea mixed with jasmine flowers.

There are three basic categories of tea: "black" tea is fermented; "oolong" tea is partly fermented; and "green" tea is unfermented. Fermentation includes bruising, wilting, and drying, followed by rolling. As more surface is exposed to the air, tea juices oxidize. Hot air drying stops the process. In the case of green tea, the freshly picked leaves are dried immediately.

China, India, and Sri Lanka are the world's largest tea producers. However, it is also grown in Bangladesh, Kenya, Indonesia, Japan, Georgia, and Azerbaijan. Are you a tea drinker? Investigate the variety and origin of the types you enjoy most.

The British brought in Tamil laborers from Tamil Nadu in India. These nineteenth-century plantation workers were distinguished from the Sri Lankan Tamil settlers of earlier times and are known as **estate Tamils.** In fact, because of pressure from the powerful Buddhist Sangha, they were disenfranchised in 1948. Since India refused to recognize them as citizens, these Tamils were virtually stateless.

Other minorities are the aboriginal **Veddas** and the **Burghers**, descendants of the Portuguese and Dutch. The Burghers are Christian and speak English. There are also Muslims, a religious minority in this Buddhist country. In 1931, Britain granted Ceylon limited self-government and in 1948, full independence.

A controversial development project was launched with British and other aid in the 1950s, an era when dam building was at the forefront of foreign aid and regional development. Incorporated in the Mahaweli Development Program, the Victoria Dam was the world's largest aid project at the time. Although the dam was to provide electric power and irrigation, it required the displacement of 1.5 million people. Typical of many grandiose schemes, construction costs tripled and the government was forced to raise taxes. Displacees were settled in isolated areas with poor infrastructure and adverse climatic conditions. They also suffered psychologically from being uprooted from their traditional homeland. Eventually they became known as *Victoria's Victims.*

Victoria Dam was part of a larger scheme for agricultural colonies in the eastern zone. Here, the climate was different and the land difficult. More significantly, Sinhalese colonists clashed increasingly with Tamil inhabitants, prestaging the problems to come.

Religious, linguistic, political, and economic differences fomented dissention between Sinhalese and Tamils. Political assassinations and violence continue to revolve around these fundamental differences, although the cleavage is not simply two sided. Anthropologist S. J. Tambiah (1986) makes a salient point: "Sinhalese and Tamil labels are porous sieves through which diverse groups and categories of Indians intermixed with non-Indians . . . have passed through."

Moreover, this already complex situation is exacerbated by international involvement.

The Sinhalese are of Aryan extraction with ancient roots on the island. They are Buddhist and speak **Sinhala,** and Indo-Aryan language with its own script. Sinhalese make up 74 percent of the population and are prominent in administration.

The Tamils, who make up 20 percent of the population, are a Dravidian people. They are Hindu and speak **Tamil,** a Dravidian language with a script different to the Sinhala one. It was not until the 1970s that Tamil was given recognition as a national language.

Since the Sinhalese dominate government, they have tried to inculcate Sinhalese history and values into schools, texts, and the like. There have also been efforts to divorce Sinhalese and Tamils from any common history, such as their relationship with India. These practices have encouraged social distancing and hostility.

English is regarded as a **link-language;** it is supposed to serve as *a lingua franca,* or common language. But English is spoken and understood by a minority elite, and years of anti-English and pro-Sinhala policies mean that it is not a common language. Students have rioted and burned books at Colombo University because 90 percent of the books are in English. For many rural Sinhalese, English is known as ***Kaduwa***—the sword that separated the privileged from the masses.

Violence erupted in the 1950s with a succession of Tamil uprisings against the Sinhalese. After her husband was assassinated, Sirimayo Bandaranaike became the first woman head of state in the world. She nationalized oil fields, companies, and plantations. Communal strife worsened in the following years as Tamil nationalism hardened. In 1972, Ceylon was declared a republic and renamed Sri Lanka. Nevertheless, organized terrorism continued.

In the 1980s, transnational capital entered in the form of an International Monetary Fund (IMF) development program and U.S. aid "to stabilize the region." Subsequently, U.S. warships acquired refueling rights at Trincomalee on the east–west sea route via the Suez Canal. Sri Lanka was to be developed as an "export center" like Hong Kong, and this led to the creation of a free trade zone in Latunyabe. But, interethnic and other discriminatory practices channeled most benefits to the Sinhalese.

The killing of 13 Sinhalese soldiers in Jaffna in 1983 sparked a spiral of violence that has not yet ended. Clashes between Sinhalese and Tamils resulted in 40,000 Tamils seeking refuge in India's state of Tamil Nadu. A series of political assassinations followed, and foreign investment faded.

More fighting led to the intervention of Indian troops. In 1987, the Tamils were granted a degree of autonomy in the north and northeastern provinces, and their language rights were guaranteed. Meanwhile, Tamil militancy had intensified in the form of a strong guerilla force with international ties. Known as the **Tamil Tigers,** the group called for the establishment of a Tamil state to be named **Tamil Eelam.** In 1991, India's Prime Minister Rajiv Gandhi was assassinated by a suicide bomber. The Tamil Tigers are blamed for this act.

The Tamil Tigers

The Liberation Tigers of Tamil Eelam (LTTE) emerged in the 1970s and are responsible for taking at least 50,000 lives. LTTE successes have been sustained by both international support and global access to resources via the worldwide Tamil diaspora. Tamil populations are dispersed throughout the former British Empire and beyond. Western Europe, North America, and Australia are home to more than half a million. Canada alone has 140,000, concentrated mostly in Toronto.

The LTTE has home offices in more than 30 countries and runs a gamut of businesses, including arms dealing and shipping. Norwegian mercenaries instruct guerillas in underwater demolition, Kalashnikov assault rifles are acquired from Pakistan, and Soviet missiles have been sold by corrupt Cambodian generals and smuggled across the Thai border. The LTTE also has drug-running links in Myanmar.

> Arms enter Sri Lanka through Velvettiturai, a historic smuggling/fishing port referred to as VVT. Trading companies are located in such cities as Kuala Lumpur and Dhaka. Money is funneled via Singapore. The LTTE operates a network of businesses in Tamil Nadu and has links to insurgent groups in Assam and Andhra Pradesh. Sri Lankan pressure and international government intervention mean that the operational pattern is always changing. Twenty years of unbending militancy of the Tamil Tigers and protracted terrorism have reduced international support to some degree.

December 1995 witnessed the largest military operation in the island's history. Government forces routed the Tigers and took control of Jaffna. Perhaps 2,000 died, and half a million civilians were displaced in this conflict. A cease fire was broken in 1996 when a bomb exploded in Colombo, killing 91 and injuring 1,400 people. The explosives were determined to be Ukrainian RDX smuggled via Dhaka. Kandy's holy Temple of the Tooth was bombed in 1998. Violence continues.

The Sinhalese–Tamil conflict has not only resulted in the loss of thousands of lives and livelihoods but also has reduced the economy to shambles. The once-lucrative tourist trade is ruined. Refugees from the north crowd urban slums or seek safety in India. With people desperate to survive, countless brothels have appeared on the scene. Colombo alone has an estimated 50,000 prostitutes, including thousands of young boys and girls. The government spends at least 5 percent of its budget on defense. With security of paramount concern, social improvements and economic enterprises are short changed.

Sri Lanka's crisis has international ramifications. Many foreign governments are loathe to support the LTTE but recognize at least some of the Tamils' grievances. Little can be done to bar support for the rebels from within their own Tamil populations. India is concerned about the political implications of a separate Tamil state only a few miles away from its own often contentious state of Tamil Nadu. Telling is the fact that India chose to ignore Sinhalese anti-Tamil actions in the 1990s.

A proposal for political devolution—the granting of a Tamil homeland—remains stalled in Parliament. Sri Lanka would become an "indissoluble union of regions" with each region having considerable autonomy in the areas of education and economic development. Tamils want the north and east but the Sinhalese have sponsored colonization schemes in the east for decades. Aside from this point of contention, Buddhism is recognized as having a "foremost place" in society. Many Tamils, very few of whom are Buddhist, had hoped for a declaration of secularity like India. Many powerful Sinhalese refuse the Tamils any quarter. Consequently, Sri Lanka's future remains uncertain.

More Islands of the Indian Ocean

Four island groups punctuate these waters: the Maldives and Laccadives to the west, and the Andaman and Nicobar to the east. Only Maldives are independent. The latter three island chains belong to India.

MALDIVES

Twenty-seven atolls including 1,200 islands known as Maldives are situated atop a submarine ridge rising from the depths of the Indian Ocean. Each atoll is a ring-shaped, coral reef supporting 20 to 60 islands with 5 to 10 inhabited. Dry during the northeast monsoon, the islands experience heavy precipitation with the southeast monsoon from April to October. Although protective coral reefs serve as breakwaters, in May 1991, a tidal wave caused damage estimated at U.S.$30 million.

Hills and rivers are nonexistent in Maldives. Breadfruit trees and coconut palms shade expanses of dense scrub, shrubs, and flowers. Soils are poor, and only about 10 percent of the land is cultivated.

Agriculture, mainly taro, a starchy root crop, plus bananas and coconuts, is concentrated on the highest island of Fua Mulaku. At an elevation of 9 feet (c. 3 m), its fresh groundwater is less prone to the intrusion of salt water. However, with global warming and rising sea levels, all the islands are experiencing a depletion of fresh groundwater.

Situated on historical maritime routes, Maldives became inhabited by diverse ethnic and linguistic groups including Dravidians from the subcontinent, Sinhalese speakers from Sri Lanka, Arabs, Australians, and Africans. Maldivians speak Dhivehi, derived from an archaic form of Sinhalese. The word *atoll* is Dhivehi.

Maldives was once the major source of cowrie shells, widely used as currency in Asia and East Africa. The renowned explorer Thor Heyerdahl discovered that Maldives was integral to Egyptian, Mesopotamian, and Indus Valley civilizations' trade routes as early as 2000 B.C. He postulates that the first settlers were the Redin, sun-worshipping seafarers. Even today, many mosques in Maldives are oriented to the sun rather than to Mecca.

Islam of the Sunni school was introduced by Arab traders in the twelfth century and quickly replaced Buddhism, established by Sri Lankans in the fourth century. Both civil law and Islamic law or *sharia* are used. Under shariah law, the testimony of one man equals that of two women. However, women do not veil, nor are they rigidly secluded.

In 1558, the Portuguese became the first Europeans to occupy Maldives, administering it from their colony at Goa. The Dutch followed in the seventeenth century but were subsequently forced out by the British. Maldives was a British protectorate from 1887 until it achieved independence in 1965. It became a republic in 1968. In 1988, a mercenary coup was squelched after military intervention by India. This move symbolized India's hegemony over Maldives.

More than a quarter of Maldives' 200,000 people are concentrated in Male, the capital. High birthrates and lack of an official family planning policy means intense population pressure. As rice, the dietary staple, is not grown in the islands, food shortages could be problematic.

Maldives has an active merchant shipping fleet, but fishing employs almost 25 percent of the islanders. Coconut wood ships have been mechanized, leading to depletion of fish stocks. Additional refrigeration and canning facilities are being built with aid from the World Bank and Japan. The Japanese have long participated in Maldives' fishing industry.

In the industrial sector, small-scale cottage industries are dominated by women who weave rope and mats from coconut and palm fiber, and reeds. Women also work in Hong Kong–owned garment-making factories at the former British air base on Gan.

The most important source of foreign exchange is tourism. As most Maldivians are uneducated, Sri Lankans, Indians, and other groups fill all but menial jobs. Tourist resorts, deliberately isolated from settlements, offer gambling, alcohol, and Thai "massage girls." It is government policy to keep Maldivians away from Western influences in order to protect local Islamic values. The majority of tourists are from Germany, Italy, Britain, and Japan. Unfortunately, the seasonal unemployment pattern for the tourism industry coincides with that of the fishing industry: from April to October during the northeast monsoon.

While economic circumstances have improved for some, Maldives remains one of the world's poorest and most isolated nations.

ANDAMAN AND NICOBAR UNION TERRITORY

The Andaman and Nicobar Islands, in southeastern Bay of Bengal, are considered an Indian Union Territory. The Andamans, more than 300 islands, sit on an ancient trade route between Burma and India. The islands served as penal colonies in the ninteenth century. Only a few indigenous hunters and gatherers remain, with the majority population being Indian. In 1991, the Andamans had 240,000 and the Nicobars 39,000 people.

The Andamans comprise a series of highly dissected longitudinal valleys and are densely forested with mangroves to the north. There are few rivers, and fresh water is difficult to get. Only the south Andamans have roads. The main activity is agriculture, focusing on coconut, betel, and other palm nuts, cassava, and spices such as tumeric.

The Andamese language is interesting in that it has no numeric component. There is a word for "one" and another word for "more than one." Andamese is unrelated to any other language.

Main Point Summary

1. Separated from Pakistan in 1971, Bangladesh occupies the largest deltaic region in the world where the Ganges, Brahmaputra, and Meghna Rivers converge and flow into the Bay of Bengal.
2. Bangladesh is linguistically homogeneous. Most of its people are Muslim. Language and religion are unifying forces in a region historically beset by change in both political boundaries and jurisdiction. Regional dissention continues in the southeastern hill tracts.
3. Independence in 1971 left Bangladesh without a major port. Chittagong was developed for this purpose. Dacca and Chittagong remain the focus of the economy.
4. Bangladesh suffers from frequent natural disasters. While floods and cyclones take the greatest toll of lives and property, drought also occurs.
5. Decisions made in India regarding the Ganges (Padma) river impacts water supply in Bangladesh. A case in point is the Farakka Barrage.
6. Shifting water channels create permanent and impermanent land spaces. Sandy islands called *chars*, in braided river channels, are important for settlement, although patterns quickly shift with the next disaster. More than half of rural householders are landless.
7. A system of embankments and improved warning systems have lessened loss of life, but property damage from floods and storms is still severe.
8. Farming activities in Bangladesh are timed with the seasons with field preparation, planting, and harvesting geared to monsoon patterns of wet and dry. Most of Bangladesh is limited to one crop a year. Rice is the most important food crop and jute is the most important commercial crop.
9. Livestock are of increasing importance, yet Bangladesh is deficient in meat and dairy products.
10. Green Revolution technologies have increased yields but beneficiaries are primarily middle and rich peasants who dominate local power structures. Even with improvements in production, Bangladesh must import food.
11. Bangladesh population explosion has slowed with dramatic declines in NIRs. However, with a large base population and a relatively high TFR, the population continues to expand. Education and empowerment of women are seen as means to slowing population growth. Problems with gender inequities are very apparent in the country's imbalanced sex ratio.
12. Female activity space is changing, with poorer women having greater access to larger areas outside the home. This is generated by economic necessity.
13. Cities such as Dhaka and Chittagong are overwhelmed with migrants from the countryside. Infrastructures are unable to handle the pressure of humanity. Areas ringing cities are more crowded than ever, and efforts to decentralize have met with limited success.
14. Bangladesh's urban (and rural) environments are seriously polluted. Increased industrial concentration adds to the problem. High rates of disease are related to poisonous air and water, especially in the cities.
15. Fiber crops are important components of industry. Jute products and textiles dominate industrial production. Food processing is increasing.
16. NGOs such as the Grameen Bank are significant factors in improving the lives of peasants, especially women, by providing monetary and other means to economic startups and maintainance.
17. Sri Lanka, only a few miles across the Palk Strait from India, is a tropical, mountainous island with rivers flowing to coastal plains. Convectional rainfall and the monsoon bring moisture to different regions throughout the year. Irrigation is essential, especially in the east.
18. Dealing with aridity was the bane of ancient civilizations. Even so, the Sinhalese became masters of water control long before Europeans.
19. The European presence began with the Portuguese and continued with the Dutch and the British. Plantation (estate) systems were established for the production of spices, coffee, and later tea and other tree products such as coconuts and cinchona.

20. The British imported Tamils from southern India to work on the estates. These "estate Tamils" are differentiated from Indians already living in Sri Lanka. Moreover, both these Hindu, Dravidian groups are differentiated from the Buddhist, Aryan Sinhalese. Language differences exacerbate the situation. English, supposed to be a link-language, is spoken mainly by a minority elite.
21. Tamils are concentrated in the north and north-east of the island. Since the 1950s, a revolutionary group called the Tamil Tigers has been fighting for an independent state to be named Tamil Eelam. Assassinations and bombings have taken a heavy toll. Government efforts to end this violence have not been sucessful. The Tamil Tigers are largely supported through overseas connections. Years of violence have hurt the economy of Sri Lanka, especially the tourist industry.
22. The Maldives, an independent group of atolls off the southwest coast of India, has little in the way of modern development. Anxious to protect its Muslim way of life, the government is promoting controlled tourism whereby tourists have contact with pristine natural environments in resort settings, and minimal contact with Maldive citizens. The densely forested Andamans, historically used as penal colonies, remain sparsely populated.

Terms to Know

Bangla	*upazila*	*Simhala Dvipa*	Sinhala
cyclone	*hartal*	convectional rainfall	Tamil
storm surge	Grameen Bank	cinchona	link-language
Sunderbans	Serendip	estate Tamils	*Kaduwa*
char	Cilao	Veddas	Tamil Tigers
activity space	Zeylan	Burghers	Tamil Eelam
bari			

Bibliography

Acker, Fabian. 1985. Adam's Tears Turn to Electricity. *New Scientist,* 14 (February): 14–16.

Alam, Mahbubul, ed. 1998. *The Independent Yearbook: Bangladesh 1998.* Dhaka: Media.

Alam, Shamsul, ed. 1997. *Jahan Atlas.* Dhaka: Jahan.

Ali, Abu Muhammad Shajaat Ali. 1998. Toward an Ecological Explanation of Agricultural Unemployment in Bangladesh. *The Professional Geographer,* 50: 176–91.

n.a. 1991. Bangladesh in Troubled Waters. *The Economist,* 11 (May): 4–5.

Barkat, Abul, Mati Raham, and Mark Bose. 1997. Family Planning Choice Behavior in Urban Slums of Bangladesh: An Econometric Approch. *Asia-Pacific Population Journal* 12: 17-32.

Bigelow, Elaine. 1995. *Bangladesh: The Guide.* Dhaka: ABP.

n.a. 1998. Capital and Kindness: Micro-credit is No Miracle. *New Internationalist* 308: 4.

Brammer, H. 1990. Floods in Bangladesh I: Geographical Background to the 1987 and 1988 Floods. *The Geographical Journal,* 156: 12–22.

Daneels, Jenny. 1998. Defying Mother Nature. *Asian Business,* 34: 6–8.

Davis, Anthony. 1996. Tiger International. *Far Eastern Economic Review,* 22: 30–38.

D'Costa, Jashinta. 1994. Changes in Urban Structure in Bangladesh. Urban Geography of Hinduism. *Wilson Quarterly,* 15: 42–50.

Elahi, K. Maudood. 1989. Population Displacement due to Riverbank Erosion of the Jamuna in Bangladesh. In *Population and Disaster,* eds. John I. Clarke et al., pp. 81–97. International Geographical Union. London: Blackwell.

Foster, Gerard. 1969. The Concept of Regional Development in the Indigenous Irrigation Systems of Ceylon. *Association of Pacific Coast Geographer's Yearbook,* 31: 91–100.

Frank, Neil L. 1971. The Deadliest Tropical Cyclone in History? *Weatherwise,* 5: 438–44.

Government of Bangladesh. 1998. *Statistical Pocketbook: Bangladesh* 97. Dhaka: Bureau of Statistics.

Haque, C. E., and MD. Z. Hossain. 1988. Riverbank Erosion in Bangladesh. *The Geographical Review,* 78: 20–31.

Hye, H. A. 1992. *Case Studies on Rural Poverty Alleviation in the Commonwealth of Bangladesh.* London: Commonwealth Secretariat.

Islam, M. Aminul, and Nazrul Islam. 1997. Urban Environmental Issues and Governance in Bangladesh. In *Urban Governance in Bangladesh and Pakistan,* eds. N. Islam, and M. M. Khan, pp. 69–80. Center for Urban Studies Dhaka: University of Dhaka.

Ives, Jack. 1991. Floods in Bangladesh: Who Is to Blame? *New Scientist,* 13 (April): 1–4.

Madeley, John. 1984. Progress Has its Price. *Geographical Magazine,* 56: 226–27.

McDonald, Hamish. 1991. Learning From Disaster. *Far Eastern Economic Review,* 30 (May): 23–4.

Murphey, Rhoads. 1957. The Ruin of Ancient Ceylon. *The Journal of Asian Studies,* 16:181–200.

Paul, Bima Kanti. 1998. Coping Mechanisms Practiced by Drought Victims (1994/95) in North Bengal, Bangladesh. *Applied Geography,* 18: 355–73.

———. 1992. Female Activity Space in Rural Bangladesh. *The Geographical Review,* 82: 1–12.

Potten, David. 1994. The Impact of Flood Control in Bangladesh. *Asian Affairs,* 25 Part II: 156–162.

Rain, Nick. 1994. Living With the Landscape. *Geographical,* 66: 24–27.

Rao, Radhakrishna. 1992. On Fertile Ground. *Far Eastern Economic Review,* 13 (August): 70.

Saenger, P. 1993. Land From the Sea: The Mangrove Afforestation Program of Bangladesh. *Ocean & Coastal Management,* 20: 23–39.

Tambiah, S. 1986. *Sri Lanka: Ethnic Fratricide and the Dismantling of Democracy.* Chicago: University Press.

Taylor, Alice ed. 1968. Ceylon. *Focus,* 18: 1–10.

Chapter 9

East Asia: Center of the World

"To understand is hard. Once one understands, action is easy."

(SUN YAT-SEN 1866–1925)

For some 7,000 years East Asia has been dominated by *Chung Kuo:* the Middle Kingdom. Not only is China the largest country in the region but also its cultural, economic, and political impact has been profound. Japan, Korea, and Taiwan—each owes significant aspects of its heritage to the wonder that was China. In this chapter, I will tell you about the physical characteristics of the region and introduce you to the Chinese worldview relative to those within its sphere of influence. I will then review China's history up to its inception as the People's Republic. Although I will mention Taiwan, Korea, and Japan where appropriate in this chapter, I will save the more salient details of their histories for Chapter 11. Be sure to examine the pronunciation guide before reading further (Figure 9-1).

Topography

Mountains, hills, and plateaus dominate the East Asian landscape. For example, Mainland China is 65 percent and Japan is 75 percent. These physiographic barriers, while permitting the evolution of unique cultures, were penetrable enough to allow outside influences to wend their way from one valley settlement to another. Mountain passes, river arteries, and ocean passages were instrumental in the diffusion of people and ideas across the East Asian landmass to the Korean peninsula and the islands of Japan and Taiwan.

Figure 9-2 shows the topography of East Asia. The mainland comprises rivers, plains, and basins separated by mountains. Geographers demarcate the rugged and arid environment of the west—**China Frontier**—from the gentler and more humid environment of the east—**China Proper.**

CHINA FRONTIER

The region demarcated as China Frontier is dominated by spectacular mountains and plateaus, as well as forbidding deserts and swampy basins. The Altai Mountains and the Tien Shan (Heavenly Mountains) encase the dry Junggar (Dzungarian) Basin. Just to the east of the Tien Shan is the Turpan (Turfan) depression. At 505 feet below sea level (155 m), this is the second lowest place in Asia (after the Dead Sea). The southern slopes of the Tien Shan fade into the Tarim Basin,

PRONOUNCING AND SPELLING CHINESE PLACE NAMES

Since Chinese is written with characters, it is necessary to transliterate these phonetically into the Roman alphabet. The nineteenth century Wade-Giles System of romanization was commonly used before the 1970s. At that time a new system was introduced: Pinyin. Pinyin was made official in 1979.

While Pinyin is found in contemporary books and atlases, Wade-Giles remains in pre-1950 publications. Therefore it is important to be cognizant of both systems. If you learn to pronounce the names of all China's provinces and autonomous regions, you will have a head start on pronouncing other Chinese words.

PROVINCE IN PINYIN	PRONUNCIATION	WADE-GILES SPELLING
ANHUI	AHN-WAY	ANHWEI
ZHEJIANG	JUH-JEEONG	CHKIANG
FUJIAN	FOO-JEEEN	FUKIAN
HEILONGJIANG	HAY-LOONG-JEEONG	HEILUNKIANG
HENAN	HUH-NON	HONAN
HEBEI	HUH-BAY	HUPEH
HUNAN	HOO-NAN	HUNAN
HUBEI	HOO-BAY	HUBEI
NEI MONGGOL	NAY-MUNG-GOO	INNER MONGOLIA
GANSU	GAHN-SOO	KANSU
JIANGXI	JEEONG SHE	KIANGSI
JIANGSU	JEEONG SU	KIANGSU
JILIN	JEE-LYNN	KIRIN
GUANGXI	GWONG-SHE	KWANGSI
GUANGDONG	GWONG-DOONG	KWANGTUNG
GUIZHOU	GWAY-JOE	KWEICHOW
KIAONING	LEEOW-NING	LIAONING
NINGXIA	NING-SHEAH	NINGSIA
SHANXI	SHAHN-SHE	SHANSI
SHANDONG	SHAHN-DOONG	SHANTUNG
SHAANXI	SHUN-SHE	SHENSI
XINJIANG	SHIN-JEEONG	SINKIANG
SICHUAN	SSU-CHWAN	SZECHWAN
XIZANG	SHE-DZONG	TIBET
QINGHAI	CHING-HI	TSINGHAI
YUNNAN	YUOON-NAN	YUNNAN
CITIES		
BEIJING	BAY-JING	PEKING
GUANGZHOU	GWONG-JOE	CANTON
CHONGQING	CHONG-CHING	CHUNGKING
SHANGHAI	SHONG-HI	SHANGHAI
TIANJIN	TEEEN-JIN	TIENTSIN

Figure 9-1

Pronunciation and spelling guide for Chinese place names. Be sure to study this guide. Once you know how to pronounce and spell the province names, pronouncing and spelling city names will be easier.

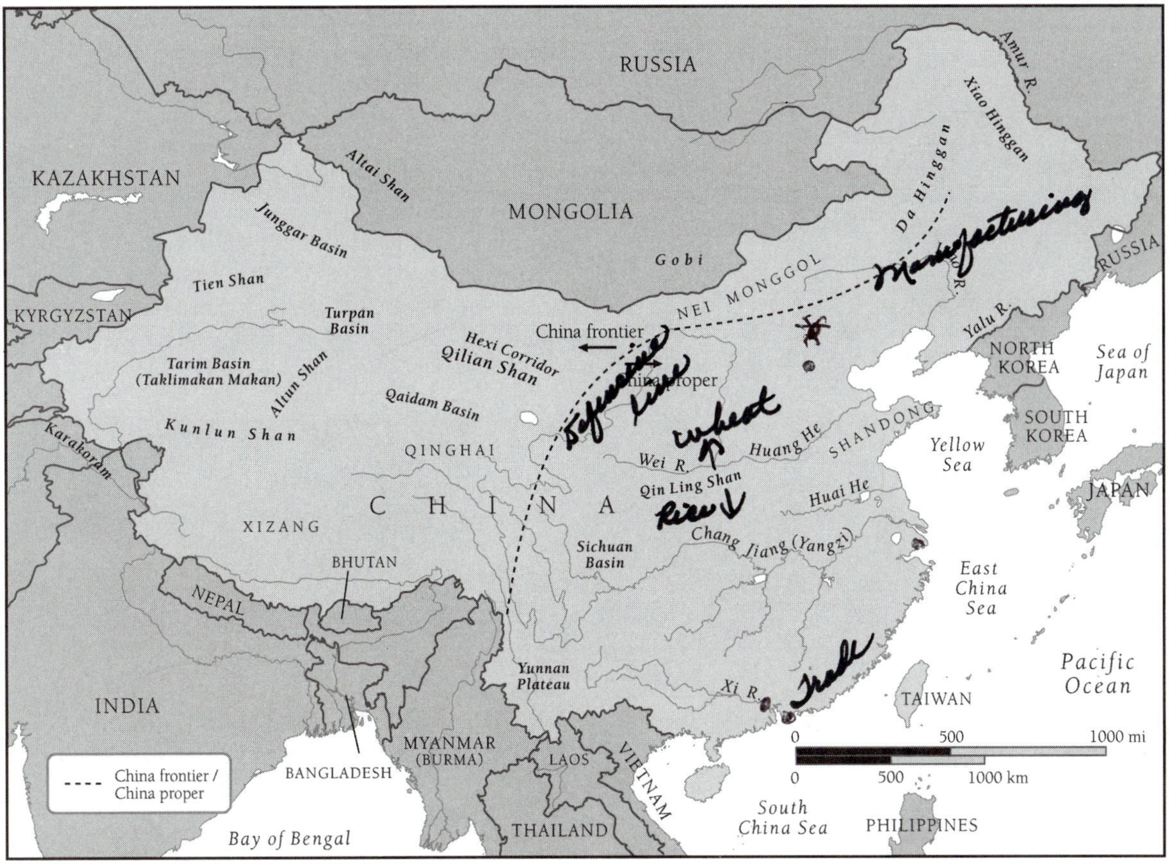

Figure 9-2

Physical geography of East Asia. Study this map carefully and use it in reference to the text. Your atlas map will give more detail. Notice how mountainous China is and how distinct lowlands are. Look at a population distribution map and check the relationship between population distribution and China Frontier and China Proper.

bordered on the west by the Karakorum and on the south by the Kunlun and Altun Shan. Within this basin lies the Taklimakan, one of the world's largest sand deserts. Its name means: "You can go in but you can't come out."

Settlement is possible in desert basins because of the presence of water. The Junggar and Tarim Basins and the Hexi Corridor are **artesian basins** with plentiful water trapped in underground **aquifers** (water-bearing rock), especially within alluvial fans at the base of mountains. With seasonal rainfall, rivers frequently disappear from the surface and flow underground. The use of ***karez*** irrigation systems render mountain-base oases productive (Figures 9-3 and 9-4).

Directly south and southeast of the dry wastes of the Taklimakan lies the Qinghai-Xizang Plateau, with an average elevation of more than 12,000 feet (4,000 m). The Qilian Shan separate the swampy Tsaidam Basin from the steppes of Nei Monggol (Inner Mongolia). These short grasslands fade northward into the Gobi (desert) of Mongolia. Immediately north of the Qilian is a narrow routeway known as the Hexi

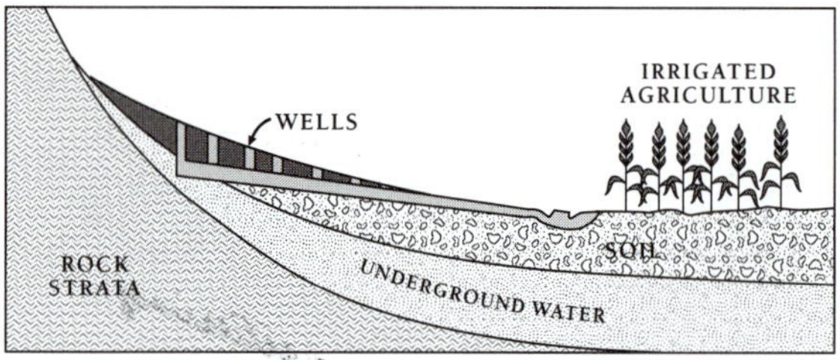

Figure 9-3

Karez wells are vertical shafts connected to an underground horizontal canal that penetrates an aquifer. The water flows by force of gravity to a surface location, where it is diverted to homes and fields. Both vertical and lateral canals must be kept clear of material that could interrupt the water flow. Both digging and clearing wells are dangerous ventures.

or Gansu Corridor, an integral part of China's famous trade route known as the Silk Route.

CHINA PROPER

To the northeast are the Da (Greater) and Xiao (Lesser) Hinggan (Khingan) Mountains, sloping inland to the Manchurian Plain and the Liao River basin. Northward is the Amur River, forming the border between China's northeast and Russia. The Yalu and Tumen Rivers separate China from North Korea.

China Proper is dominated by rivers and plains. To the north is the Huang He, or Yellow River. Flowing 3,395 miles from the Qinghai Plateau, it sweeps northward, wrapping the Ordos Desert in a big loop, cutting southward across the **loess** region of Shaanxi and Shanxi Provinces. Loess is fine, wind-blown soil that is readily eroded. At the confluence of the Wei River, it turns northeast across the North China Plain and empties into the Bo Hai Gulf of the Yellow Sea.

The Huang He is literally yellow, due to the immense amounts of silt it carries in suspension through erosion of the loess plateau. At one time, the Shandong (Eastern Mountain) Peninsula was an island, but eons of alluvial deposition have joined it to the mainland. Its highest peak is sacred Tai Shan at 5,056 feet (1,517 m).

Just south of the Wei River is the Qin Ling Range, an extension of the Kunlun. East of the Qin Ling is the Huai River. As you will discover, both the Qin Ling and the Huai are sentinels of change in both climatic and agricultural patterns in China.

The North China Plain is the largest alluvial landform region in eastern China. It covers 125,000 square miles (323,750 sq km). It blends into the Central Chang Jiang (Yangzi River) Plain and its associated lake basins. The Chang Jiang, the world's third longest river, flows an incredible 3,900 miles (6,279 km) from Tibet to the East China Sea. To the west, surrounded by mountains, is the fertile Sichuan Basin and the Yunnan Plateau. Mountains dominate southern China

Figure 9-4

Road and water channel in Dunhuang. Poplar trees line this dirt road. Straight poplar wood is important for construction. Note the irrigation canal alongside the road. The water comes from a karez and is fundamental to irrigation in this arid region.

until breaking at the Xi River Basin by the South China Sea. Southwestern China is dominated by the longitudinal extensions of the Tibetan-Qinghai ranges.

Climatic Variation

East Asia is characterized by continentality modified by the monsoon. Seasons are accentuated in the land-dominated interior, remote from the moderating effects of water bodies. Mountain patterns add to these contrasts. Winter and summer temperature extremes are experienced in the interior of China, while coastal regions have more comfortable climes. Latitude is another consideration. While rice is being harvested in the Xi Basin, ice sculptures are being created in the Liao Basin.

The summer monsoon (detailed in Chapter 2) brings ever-decreasing amounts of precipitation from south to north (Figure 9-5). Precipitation ranges from around 120 inches (305 cm) on Hainan Island to less than 10 inches (35 cm) in the Tarim Basin.

The Qin Ling, with peaks reaching 12,000 feet (3,600 m), form a topographic barrier to monsoon rainfall from the south and to dust-laden winds from the north. Significantly, the Qin Ling and Huai River approximate both the 20-in. (51-cm) isohyet and the northern limit of double-cropping.

In winter, winds sweep outward from the Siberian high pressure cell near Lake Baikal, bringing relatively dry, cold air to the continent as well as Japan. Sometimes the chilling air is filled with loess, bringing yellow dust as far as Japan's west coast.

Typhoons are important sources of moisture during the summer and fall. These develop in the South and East China Seas, striking China's southern coast around May and reaching Taiwan and Japan in July or later. Typhoons bring a succession of rain squalls and winds that can reach velocities of 150 miles (242 km) per hour, causing severe damage.

Seasons are highly differentiated between both hot and cold and wet and dry. Temperatures contrast greatly with latitude, averaging −13°F (−25°C) in northern Manchuria in January and 68°F (20°C) in

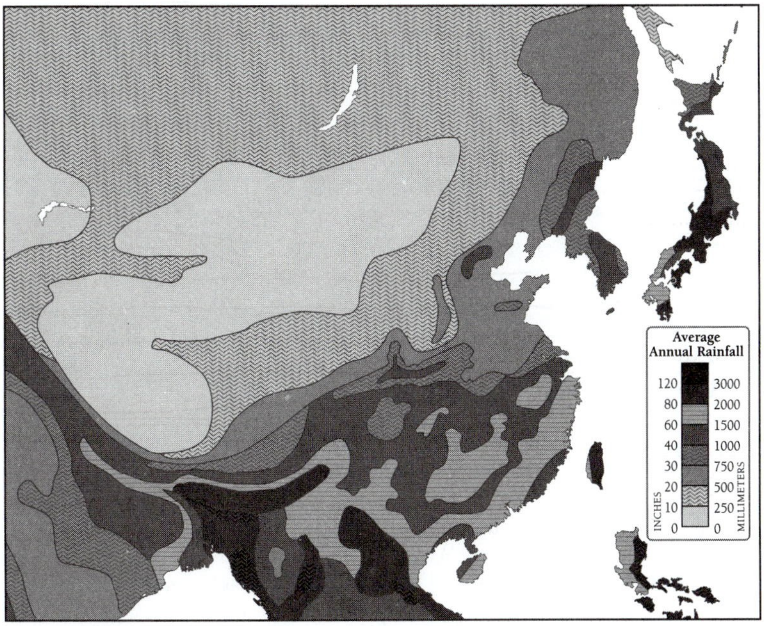

Figure 9-5

Yearly precipitation in East Asia. Note how the amount of precipitation decreases as you move inland. How does this pattern compare with population distribution? Compare this map with Figure 10-10 and see the relationship between precipitation and agricultural patterns.

Hainan in the south. July temperatures average 70°F (21°C) and 84°F (29°C), respectively. Summers are oppressively hot and humid virtually everywhere, including Taiwan, Korea, and Japan. I will provide further details on the physical geography of these countries in Chapter 10.

Peoples of East Asia

It is a common but erroneous perception that all East Asians are of Mongoloid stock. Although this is true of Chinese, Koreans, Mongols, and Tibetans, it is not true of all groups. The Japanese, for example, are thought to be a Mongol–Malay mixture. And, in China Frontier, there are several Turkic populations such as the Uygur. As you will learn in the next chapter, China has many minorities, and **Sinicizing** (making Chinese) these groups is a major policy issue.

A brief look at language patterns reveals the dissimilarities of the various peoples of East Asia. Mandarin and Cantonese are Chinese languages belonging to the Sino-Tibetan language family. Tibetan is included in this family but belongs to the Tibeto-Burmese group. Some of the minority people in southwestern China speak Mon-Khmer of the Austro-Asiatic language family, related to languages in Southeast Asia. The Uygur, Kazak and other Turkic peoples speak Altaic languages of the Turkish group. Mongolian belongs to a different group of Altaic languages. Neither Korean nor Japanese are related to any of these!

The Chinese see themselves as Han Chinese, descendants of the glorious Han Dynasty (202 B.C.–220 A.D.). All others are non-Han people. China's historic perception of its own non-Han people mirrors its perception of other non-Chinese, Asian people. Furthermore, China's perception of itself relative to outsiders has been a major force in shaping its interrelationships with the world.

China's Worldview

The written characters for China translate as *Zhong Guo* (Chung Kuo), meaning central or Middle Kingdom. For much of China's history China has perceived itself as central to the known world. This notion has its roots in antiquity. To appreciate it we must venture back in time 4,000 years.

An ancient creation myth has the confusion of heaven and earth divided into light and bright, and heavy and dark segments, with a primeval being called **Pan-gu** (P'an-ku) growing between them. Pan-gu's body parts transmuted into various parts of the cosmos. Pan-gu's function prepares us for the important role of humans in the transformation of chaos into order and explains the Chinese view of apparently opposing forces as complementary in an organic harmony. These ideas are seeded into the philosophy of Daoism (Taoism) discussed later.

One seminal work regarding the origins of human society is the ***Yi jing (I-ching)***, a work in which hexagrams depict the subtle harmonies of complementary opposites (Figure 9-6). According to the *Yi jing*, hu-

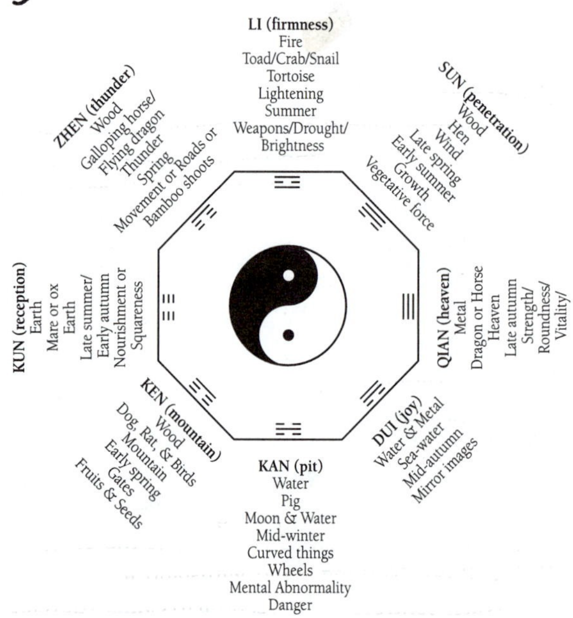

Figure 9-6

This diagram expresses harmony and interrelationships among all things. Understanding these relationships is important in interpreting East Asian landscapes. This diagram is part of the Korean flag attesting to its importance in East Asian culture.

mankind was ruled by a succession of sage kings such as Fu-xi, who invented nets and baskets for fishing. His wife invented feminine arts and patched a rip in the fabric of heaven. Shen-nong, the Divine Farmer, invented agriculture and herbal medicine, and taught people how to hold periodic markets.

> ### Oracle Bones
>
> The rulers of the ancient Shang Dynasty (1799–1122 B.C.) practiced divination by heating certain bones such as animal scapulae (shoulder blades) and interpreting the resultant cracks. Questions and interpretations were often inscribed on the bone. These **oracle bones** were first discovered by scholars in 1899 near the last Shang capital at Anyang in the Wei River valley.

The character Huang Di was of particular importance. Huang Di, the Yellow Lord (emperor), constructed roads and set up a bureaucracy of control. He distributed grain at appropriate times and clarified knowledge of heaven and nature. Educated Chinese believe that humanity reached a peak of order about 2350–2200 B.C. during the reigns of subsequent emperors. This belief in human order is a fundamental theme in Chinese thinking and behavior.

Another fundamental belief is that an ordinary man can achieve great things and that the best government is government by the best men—men of virtue. Shun is held as an example of a man who rose from humble origins. The Confucian thinker Mencius (c. 390–305 B.C.) wrote that when Shun "heard a single good word, witnessed a single good deed, it was like water causing a breach in the dikes of the Yangzi or Yellow River. Nothing could withstand it."

Water control has been central to China's development since the earliest of times. Shun appointed a man named Yu, who assembled a massive labor force and spent 13 years supervising water control and irrigation works. Many of China's unruly rivers, including the Huang He, were rechanneled. Subsequently, agriculture flourished. The concept of water management and agricultural productivity as responsibilities of the ruler is another salient theme in Chinese history.

Yu also surveyed the empire, compiling information on soils, local products, and land revenues. He reported these in a lengthy document called the *Yu Gong* (Tribute of Yu). The reporting of competent ministers to rulers is another important element of proper rule and order. As Shun said to Yu, "The mind of a man is perilous; the mind of the True Way is hard to discern. Concentrate! Be single minded! Hold fast to its center." Years later, Yu's son came to power, and his descendants ruled China by hereditary right for 400 years. Yu is regarded as the founder of China's first dynasty: the Xia (Hsia).

Several centuries later a new dynasty came into being under the Duke of Zhou (Chou). The Zhou Dynasty (1100–256 B.C.) was highly centralized with a vast bureaucracy. Kinship ties were important media of ensuring effective rule, and ministerial families soon amassed great power in their respective states.

Control by diminishing degree has haunted Chinese history and, by association, the history of East Asia. An important component of the model was the tributary zone. People of this zone acknowledged Chinese sovereignty through trade and giving gifts (paying tribute) to the Son of Heaven.

Tribute payments were not always a one-way process. China often paid tribute to barbarians to keep them at bay or to employ their assistance in thwarting other barbarian groups. The northern frontier was of constant concern.

The northern frontier formed a zone where opposing ways of life met and comingled: farming and herding. Sometimes pasturage would give way to sedentary farming. Historian Jacques Gernet (1983) observes, "The problem of defense against incursions from the steppe must be seen in a context which is as much cultural, political and economic as military, thanks to phenomena of assimilation, diplomatic combinations, and commercial exchanges." Here were allied tribes who collaborated in the defense of the empire, outposts, forts and garrisons, military colonies, lands developed by deportees, horse breeding, and nomad encampments.

Two types of troops were located in the frontier: *farmer-soldiers,* known as "soldiers of the irrigation canals" or "soldiers of the granaries," and *soldiers on garrison duty in forward posts.* Each post could contact the other through a system of signals: red and blue flags, smoke by day, and fires by night. Transmission of information was swift, and all was recorded in writing. Precise information about troop movements is given in texts as early as the second century B.C.

The Ways of Writing Chinese

Chinese characters originated as *pictograms,* where a symbol stood for an object. Over time, pictograms evolved more complexly as *ideograms,* where symbols represent ideas. There are thousands of Chinese characters. To be reasonably literate, you need to know at least 10,000.

Chinese is a tonal language. Different dialects have different numbers of tones. Mandarin has four tones. Cantonese has eight. The same sound, pronounced with different inflections, mean different things or ideas. For instance "ma" might imply "mother" or "horse" depending on what tone you give it. However, these two words are written with their own unique characters.

Writing Chinese characters in the Roman alphabet requires "transliteration," a way of transforming one to the other. One system, employed until well after the establishment of communism in China (1949), is the Wade–Giles system. More recently, the communists have introduced a more simplified character and transliteration system called **Pinyin** (see p. 218).

Shan-tung under Wade–Giles becomes Shandong in Pinyin. Ch'ing Dynasty becomes Qing Dynasty, and Mao Tse-tung becomes Mao Zedong.

While most books and atlases now use Pinyin, English publications from Taiwan continue to use Wade–Giles.

Perhaps the most obvious evidence of China's relationships with outsiders is the Great Wall (Figure 9-7). Dates for initial construction and subsequent connections of the wall(s) are more suggestive than real. It is said to have been built during the Qin (Ch'in) Dynasty (221–206 B.C.) when the Yellow Emperor consolidated existing sections with watch towers along the northern frontier. More or less permeable throughout history, it was consolidated again under the Ming Dynasty (1368–1644 A.D.). It is the Ming wall that tourists visit today north of China's capital Beijing (Peking). The western sections have either vanished or are in a state of disrepair. The measured length of the Great Wall also varies, depending on which sections are measured. Estimates range from around 1,500 to 4,000 miles (2,415–6,440 km).

For most of Chinese history, the wall was actually many walls—sections constructed as the need arose in time and space. Their positioning depended on the state of Chinese–barbarian relationships at the time. China spent millions in terms of finances and manpower in warding off such nomadic hordes as the Khitan, Xiongnu (Hsuing-nu), Mongols, and Manchus, not always with success. The Mongols and Manchus, respectively, swarmed Chinese forces and took control as the Yuan Dynasty (1279–1368) and the Qing (Ch'ing) Dynasty (1644–1911). China's influence in East Asia peaked under the Mongols, who sent expeditions against Japan, Burma, Vietnam, and Java using both Chinese and Korean troops and fleets.

Further fascination with proper hierarchical relationships was represented by a system of landholding called the **well-field** (*jing tian*) **system.** Where feasible, land was divided into nine squares. Eight families each farmed one square and kept the entire crop for itself. All worked together to farm the ninth, public field, from which the produce went to the ruler. Young men were granted fields as adults, and upon old age returned them to sovereign power for redistribution. Clearly this would only be practical on level land of uniform quality and it may not have been practiced at all. However, well-field serves as a model of equity, cooperation, and obedience—principles valued to this day.

Figure 9-7

The Great Wall of China was begun in the Qin (Ch'in) dynasty (221–206 B.C.) "China" comes from Ch'in. The wall actually linked several older walls, and in its full length reached nearly 4,000 miles (6,000 km) long. It averages 25 feet (7.8 m) in height and 19 feet (5.8 m) in width. Only parts of the wall survive today. This section of the wall near Beijing was refurbished under the Ming dynasty in the fourteenth century.

The Terra-cotta Army

The 1974 discovery of an entire terra-cotta, life-size army that had been buried underground for more than 2,000 years, astonished the world. Thousands of pottery soldiers, each with unique features, along with war horses and chariots, are regarded as the "eighth wonder of the world." This army is believed to be guarding the tomb of Emperor Qin Shi Huang of the Qin Dynasty (221–206 B.C.). Near the city of Xian (Hsi-an), this site is one of China's top tourist attractions.

CONFUCIUS

K'ung-fu-tsu (Kongzi)—or **Confucius**—was born 551 B.C. in Shandong. After learning and practicing a variety of tasks and skills, he became a teacher of ethics, ritual, and philosophy. His ideals were recorded by his students as ***The Analects of Confucius.*** The *Analects,* in addition to the transmission of his lore over time, imply his adherence to hierarchical structures of loyalty: of ruler to heaven; of subject to ruler; of son to father; of younger brother to older brother; and of women to men. This complex of customs and interrelationships is referred to as **filial piety.**

Filial piety involved "right relationships" that left small room for the individual. Yet, both Confucius and Mencius stressed the vital importance of self-cultivation and education as means to virtuous behavior. Enjoyment of life was essential; emphasis was placed on the pleasure of good food, production of children, and attainment of old age. The greatest sin against filiality was to have no descendants (i.e., male offspring). Even under hierarchical order, Confucius allowed for individual ability through which one could become a sage. No priests were necessary; self-development was the key to sagehood.

Confucius also affirmed the right of the people to rebel against an unjust ruler. Rulers ruled under the **Mandate of Heaven.** If they lapsed from virtue, they forfeited the Mandate of Heaven. However, the apparently contradictory tenet of loyalty to superiors usually won out in practice.

Confucius was also a fervent believer in the importance of culture and learning, as well as hard work. A good and just ruler would surround himself with ministers who achieved their goodness through moral education. He saw the wisdom of the ancients as crucial to solving problems, and he put the Duke of Zhou on a pedestal. Everyone was to play his proper role in the traditional hierarchy as preserved in Zhou moral teachings. Regarding government, Confucius said: "If you set an example by being correct, who would dare to remain incorrect?" Confucius also recognized the need to exert control over distant peoples: "Ensure that those who are near are pleased and those who are far away are attracted." In all these matters, appropriate ceremony was critical to facilitating and maintaining reciprocal relationships.

Confucius created the role of the scholar-minister, a calling of the highest value. He believed that poverty was not an issue, in that any young man could obtain an education and achieve an administrative or even a court position. Confucius died in 479 B.C., but has always been known in China as the "First Teacher." His example underpins the extraordinary respect for education and teachers in China today. Under the Han Dynasty, Confucianism became virtually a state religion, and a system of rigorous examinations were established for would-be administrators.

In the twelfth century, a Confucian philosopher Zhuxi (Chu Hsi) (1130–1200) speculated about the existence of a Supreme Ultimate, a cosmic force. However, Confucius had never instituted priests or religious rituals, leaving any religious aspect in question. Nevertheless, faithful Confucianists did build temples and worship the sage. It was this neo-Confucianism that diffused to Korea, Vietnam, and Japan, where it became the prevailing philosophy, especially for the educated.

In the twentieth century Confucian ideas were disparaged as contributing to China's backwardness relative to the industrialized world. Under communism, the sage was ridiculed and vilified as a symbol of all that was wrong with the Imperial past. But, in recent years of increased freedom from the rigors of communism, Confucius has been returned to good graces and extolled for his teachings of loyalty and obedience to government approval. "Confucian principles" are frequently referred to in context of the economic successes of Singapore, Taiwan, South Korea, and other Asian countries.

DAOISM (TAOISM)

A contemporary of Confucius, called **Laozi (LaoTze)** meaning "The Old One," supposedly debated with Confucius on the right way of being. Daoist philosophy is recorded in the ***Dao de Jing*** (*Tao te Ching*) (*Classic of the Way*) as quite cryptic and even mystical sayings. For example: "The Way is void, yet inexhaustible when tapped;" or "Men should be bland, like melting ice, pure and peaceful like a block of uncarved wood."

Daoism evolved as a religion as it mingled with folk beliefs grounded in nature worship and animism. It stressed the desirable unity of man and nature, and the futility of worldly strivings and government. Man's place in nature's enormous expanse and eternal change was insignificant, "a drop of water in a flowing stream." Water, flowing effortlessly around obstacles, was the ideal Way of the Dao. Daoist priests and monastic orders developed and temples were built. What started as an esoteric way of being became a mass religion, eventually incorporating a pantheon of gods and immortals.

Later Daoists practiced alchemy—the mixing of metals to produce gold. They also searched for the elixers of immortality, roaming hills and mountains in search of appropriate roots and plants. Their search for medicinal herbs contributed significantly to the evolution of Chinese medicine.

Even Confucianists found Daoism attractive. Daoism was particularly suitable when things were not going right, or in old age. Confucian activism and social reformism were counterbalanced with Daoist passivity and "go with the flow" attitude. This dualism was very attractive in the context of even older ideas concerning the integral nature and harmonious balance of all things and the fundamental principles of **yin** and **yang** (Figure 9-8).

Yin and yang are two opposing, yet complementary, primordial forces that govern the universe and symbolize harmony. For instance, without dark there is no light. Without life there is no death. All phenomena contain elements of yin and yang that interact and create cyclical change. Night rolls into day, winter into summer, and so forth. This is the process of the Dao—the universal situation.

FENG-SHUI RELATIONSHIPS

The concepts of the five elements or forces with their Yin and Yang relationships were introduced as early as the Fourth Century B.C. All of the elements are connected: Wood makes fire which burns it to make earth. Earth has metal (ores). Metal has condensation—water, and water nourishes plants and trees—wood.

The relationships of the five elements are in accordance with the laws of nature and heaven. It is important not to interfere with the order of nature or disrupt the natural conditions designed by heaven.

The ten stems and twelve branches were introduced prior to 2000 B.C. Each stem and branch is related to the five elements and Yin-Yang principles. All these forces are significant for proper Feng Shui location and a geomancer must be consulted.

FIVE ELEMENTS	YIN-YANG	TEN STEMS	TWELVE BRANCHES	DIRECTION	ANIMAL
WOOD	YIN	SPROUTING	BUD; BEGINNING	N	RAT
	YIN	SPREAD OF GROWTH	GROWTH	NNE	OX
FIRE	YANG	BLOOMING	SPREAD OF GROWTH	ENE	TIGER
	YIN	MATURITY	ABUNDANCE	E	RABBIT
EARTH	YANG	ABUNDANCE	PROGRESS	ESE	DRAGON
	YANG	ORDER	RENEWAL	SSE	SNAKE
METAL	YANG	FULLNESS	MATURITY	S	HORSE
	YANG	RESTORATION	SMELL OF MATURITY	SSW	SHEEP
WATER	YIN	HEIGHT OF FUNCTION	EXPANDED MATURITY	WSW	MONKEY
	YANG	PREPARATION FOR SPRING	RIPENESS	W	COCK
	YIN		DEATH	WNW	DOG
	YIN		NUCLEUS	NNW	PIG

Figure 9-8

Here the concept of yin and yang are outlined. Yin and yang balance opposite forces to ensure harmony between humans and nature. Yin and yang represent the interrelationship of all phenomena and are critical to feng shui orientation.

Did you know that East Asia has its own "geographic tradition?"

Feng Shui: The Chinese Art of Placement

The Chinese have always stressed relative position in the landscape, the world, and the cosmos. The forces of ***feng shui*** (wind and water) must be dealt with appropriately. For example, Chung-kuo infers the Middle Kingdom, a nation at the hub of the universe. Cities were usually oriented according to the cardinal directions. Numerology was also employed, with the number three and multiples thereof considered auspicious. The Heavenly Altar in Beijing, where the emperor conducted ceremonies for rain and good harvests, is reached by nine steps. The top surface comprises nine rings of stones. Gates to the Forbidden City are covered with nine rows and nine columns of brass studs. Four, representing death, is not a lucky number. So you would not buy a house numbered 444.

Decisions regarding location are made in the context of yin and yang forces in addition to the force of ***Ch' i.*** *Ch' i* is the most important element of feng shui. It is the vital force, the cosmic breath that gives life to all things. Without *ch' i,* there would be nothing. *Ch' i* is a pervasive concept in Chinese traditional medicine and acupuncture, and martial arts such as *Gung fu* (Kung fu).

The land most influenced by *ch' i* is the most habitable. *Ch' i* spirals around the earth, sometimes "exhaling" mountains or volcanoes at the earth's crust. When *ch' i* falls short of the surface, there is no water or flowers, but bad luck. Mountains are viewed as dragons exhaling cosmic breath. Mountains and hills are important as defensive sights and are typically kept to the north in orientation of settlement. "Back to the north; front to the south." The worst location would be on a featureless plain with no protection from anything. Atmospheric *ch' i* molds human *ch' i.* Balance and harmony of yin, yang, and *ch' i* are essential to repel negative forces.

Individuals called **geomancers** are experts in feng shui. They are consulted to determine the proper siting of buildings and even interior design and placement of furniture. Once a house is carefully and correctly designed and oriented, a Daoist priest is called to perform a moving-in or consecration ceremony. The Hong Kong and Shanghai Bank stages dragon dances when opening a new branch, even in overseas locations.

Even colors are subject to feng shui principles. The south is associated with red, representing joy and festivity. A south-facing house is auspicious, and the placement of red door guardians will ward off evil spirits. Green is related to wood and foliage and evokes youthful energy and growth. It represents the dragon and the east. Yellow was a royal color and is worn by Daoist priests for burying the dead and for geomantic blessings. White is the color of autumn. It stands for metal, represents purity, and denotes western orientation. Black is the north and winter. Its element is water. Black denotes the consequence of man, death, mourning, and penance. It is the color of calamity, guilt, and evil.

Feng shui's goal is maximum harmony with natural forces to induce happiness, prosperity, and longevity. Widely practiced among Chinese communities worldwide, it was banned in Communist China as harmful superstition. However, in the new, modernizing China, rules have relaxed, and feng shui and the practice of geomancy have reappeared.

When I visited rural and urban homes in several regions of China, I noticed that many new pieces of furniture were painted either red or green.

BUDDHISM

Buddhism entered China from India in the first century A.D. This new religion spread quickly along the Silk Route (Figure 9-9). By the time of the Sui Dynasty (589–618), it was the dominant faith among the merchants and soldiers plying the trail. Buddhist monks constructed monasteries at oases, and these became inns and even banks for travelers. Dunhuang became a significant staging point, as this is where the choice would be made as to the safest passage westward: the northern or southern route.

Dunhuang housed a large and elaborate Buddhist art treasury. Around 366, monks began to cut caves into cliffs, and these were carved and painted. This labyrinth of caves contains nearly 500,000 square feet of murals and 2,000 statues of various sizes. In the Tang Dynasty (618–907), there were more than 1,000 caves. Today there are 486. These Mogao Caves are one of China's prized tourist attractions.

In the political chaos following the Han Dynasty (see the next section), Buddhism, as a nonsecular belief system, appealed to many people, and Buddhist monks were able to assert their authority. Because there were few texts and many questions, in 399 the pilgrim Fa Xian (Fa Hsien) journeyed to India for manuscripts and advice. His account, *A Record of the Buddhist Countries,* revealed developments in Mahayana Buddhism, as well as China's communications with the outside world.

Fa Xian traveled the Silk Route to Afghanistan and on to India. From the Ganges Delta, he visited Sri Lanka and Sumatra before returning to Shandong in 414. Then, with the support of various rulers who hosted thousands of visiting monks from India, Buddhism made a phenomenal rise in China.

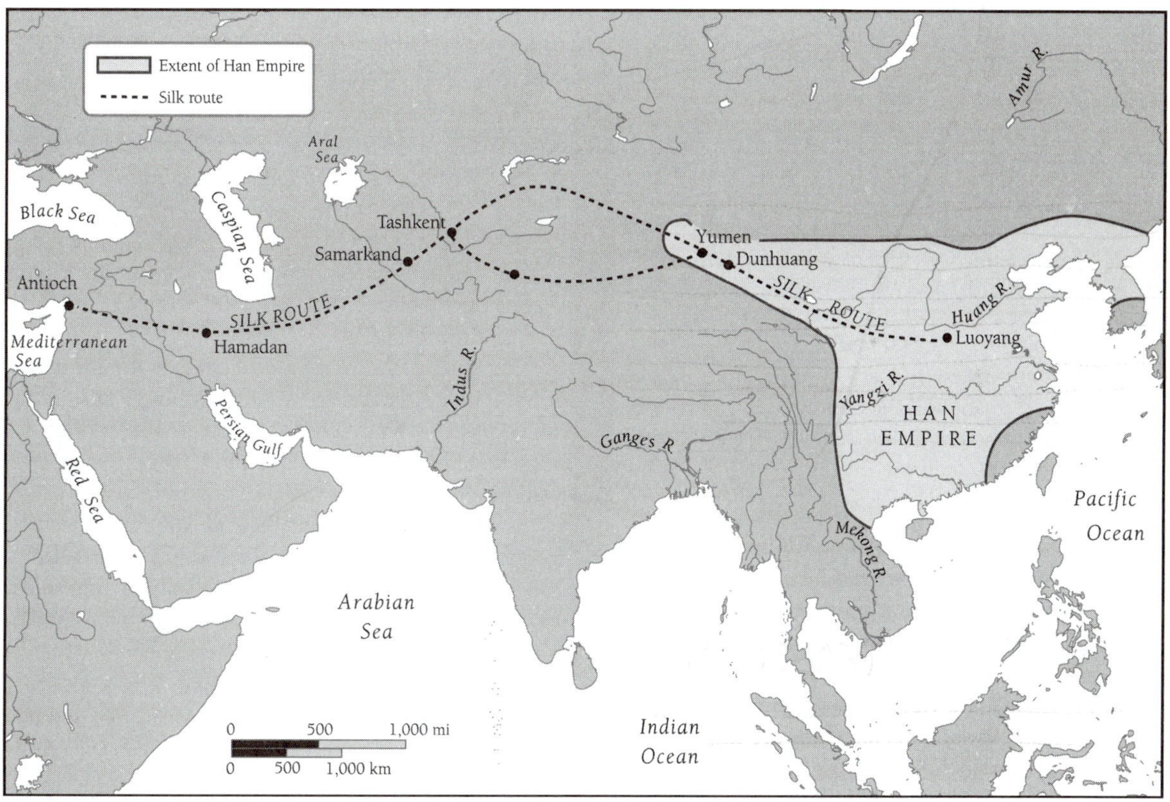

Figure 9-9

Note how the Han Empire focused on major river plains. Compare this map to Figure 9-2 and a physical geography map in your atlas. Trace the arduous journey along the Silk Route where thousands died attempting to complete the trip. Note the preponderance of mountains and deserts.

Dynasties Rule the Middle Kingdom

For more than 3,000 years, the vast region of China was held together—with varying degrees of success—by a series of (206 B.C.–220 A.D.) **dynasties.** Under the dynasty system, emperors ruled for life, under the mandate from Heaven. Upon the emperor's death, an heir became the new emperor.

Natural disasters, disturbances of order, were seen as portents of Heaven's displeasure with the mistakes of rulers and as pretext for rebellion. During the Han Dynasty (206 B.C.–220 A.D.) the emperor presided over special rituals at the imperial capital to intercede with Heaven for appropriate rain, bountiful harvests, and freedom from natural and man-made catastrophes. Even so, many dynasties (periods of rule), having become weak and corrupt, collapsed in coincidence with calamitous events. Floods were often precipitated by circumstantial inability or failure to maintain water control works. Eventually a just leader would come to power and bring order to the land and the **dynastic cycle** would begin again.

A book from the Zhou Dynasty (1100–256 B.C.), called the *Zhou li—The Ceremonies of Zhou,* presented an idealistic model of bureaucracy. According to the model, all major facets of human existence are organized in nested hierarchies of boundaries, duties, and practices. The goal is social equity, orderly goodness, and ceremonies vested in maintaining proper relationships among humans and the cosmic order. The model is geographic, encompassing a symmetrical arrangement of territories centered on the capital of the Son of Heaven (the emperor) and continuing through territories governed in increasingly less direct ways. The region of least control housed the "outer barbarians."

THE HAN DYNASTY (206 BC - 220 AD)

The gathering of territory and coping with barbarians are epitomized by the conquests of the Han Dynasty (206 B.C.–220 A.D.). At first they made agreements with the Xiongnu, mollifying their leader with gifts, and later became powerful enough to wage war against them. Real expansion was made under Emperor Han Wu Di (Han Wu-ti), who ruled from 140 to 87 B.C. Regional control was managed with the establishment of military colonies that were to clear and irrigate the land and defend the area.

The original Han capital was at Ch'ang-an (near present-day Xian) in the Wei River Valley. The capital was later moved to Loyang. By 140 A.D., Loyang had a population of more than a million. Wu-di developed "forward policy" regarding the west, which was occupied by the Xiongnu. By 115 B.C., the Chinese had forced the Xiongnu north of the desert and had occupied the Tarim Basin. The Great Wall was extended, with fortified outposts.

Wu Di was particularly interested in the relatively large horses of the "Western Regions" because they could be used to carry heavily armed soldiers against the Xiongnu and their smaller Mongolian ponies. In 101 B.C., Wu Di's forces conquered the Ferghana Valley and the "Western Regions" were declared a Chinese Protectorate.

Han Wu Di also brought northeastern and southern regions under his control. In Korea, a large military colony was established near present-day Pyongyang. The southern kingdom of Nan Yueh around Canton was also captured by 111 B.C. Kingdoms in the border areas of Sichuan and Tibet also accepted Chinese supremacy. Yunnan, the terminus of an important trade route to India, was subjugated as well. In the years 42–43 A.D. southward expeditions captured Hainan Island, the Tonkin region (North Vietnam), and Annam (central Vietnam).

Expansion toward the south enabled the countries of Southeast Asia and the Indian Ocean to enter into direct relations with China. The development of Indo-Iranian trade brought glass, amber, agate, slaves and jasmine among other products. Indian embassies came to the Han court between 89 and 105 A.D. The first record of contact with the island of Java dates from 132 A.D. One of the great ports of call for this trade was a port in the old Cambodian kingdom of Funan in the Mekong Delta, where ancient Roman coins have been found. Read more about Funan in Chapter 12.

The western border was not always stable, and the barbarians acquired control of the Tarim Basin at the turn of the century. Later in 73 A.D. Emperor Ming attempted to capture Hami, a strategic oasis that was key to the northern trade route along the foothills of the Tien Shan. Hami finally came under Chinese control in

Figure 9-10

This is called "White Horse Pagoda." It was built by a Buddhist monk in honor of his faithful horse who carried him along the Silk Route.

86 A.D. Chinese eventual control (although not permanent) over the entire Tamim Basin facilitated the development of an overland silk trade with the Roman Empire. General Pan Ch'ao is credited with conquest of 15 kingdoms west of Kashi (Kashgar), and he managed to extend Chinese influence as far as the Caspian Sea (Figure 9-9).

Assisted or forced population movement was another policy of the Han. Mass movement of populations to border regions was both a means of control and of evening out the distribution of people. More than 2 million settlers trekked to northern areas under Wu Di's reign alone. Even more migrated to the fertile lands of the Yangzi.

Throughout the Han and later dynasties, borders waxed and waned with the relative strength of Chinese and barbarian. Furthermore, China itself was unified or fragmented depending on internal strengths or weaknesses. Overspent and overburdened, the Han collapsed in 220 A.D.

THE SUI DYNASTY (589–618)

Again, under the Sui, the northern walls were strengthened. In addition, inland waterways were constructed, the most famous of which is the Grand Canal (Figure 9-11). Now cargo could be moved from the Yangzi to the Hwang He, even though the trip might take a year. More importantly, the southern regions were linked with the north.

The Sui introduced new forms of state examinations. Their purpose was to attract the top talent, regardless of social status, into the ranks of China's civil service. Candidates had to score very high in law, mathematics, and Confucian philosophy.

Once more, excessive military exercises, including thwarted attempts to capture Koguryo (Korean) forces across the Yalu River, wrought rebellion and chaos within China. Even so, a battle-weary Koguryo king agreed to accept status as a vassal of China.

THE TANG (T'ANG) DYNASTY (618–907)

Under the Tang, China became the richest and most powerful empire in Asia with frontiers extended even further than those of the Han (Figure 9-12). Two million people lived in and around Ch'ang-an, the capital and largest city in the world. More than 50 million resided in China. This was a period unparalleled in poetry, painting, pottery, printing, and other art forms. The Son of Heaven, as the emperor was known, was perceived as the lord of "all under heaven"

Figure 9-11

The Grand Canal in Suzhou. This is a tangential channel of the Grand Canal. Both the main canal and its tributaries are too shallow and narrow for anything other than low-lying barges and small boats.

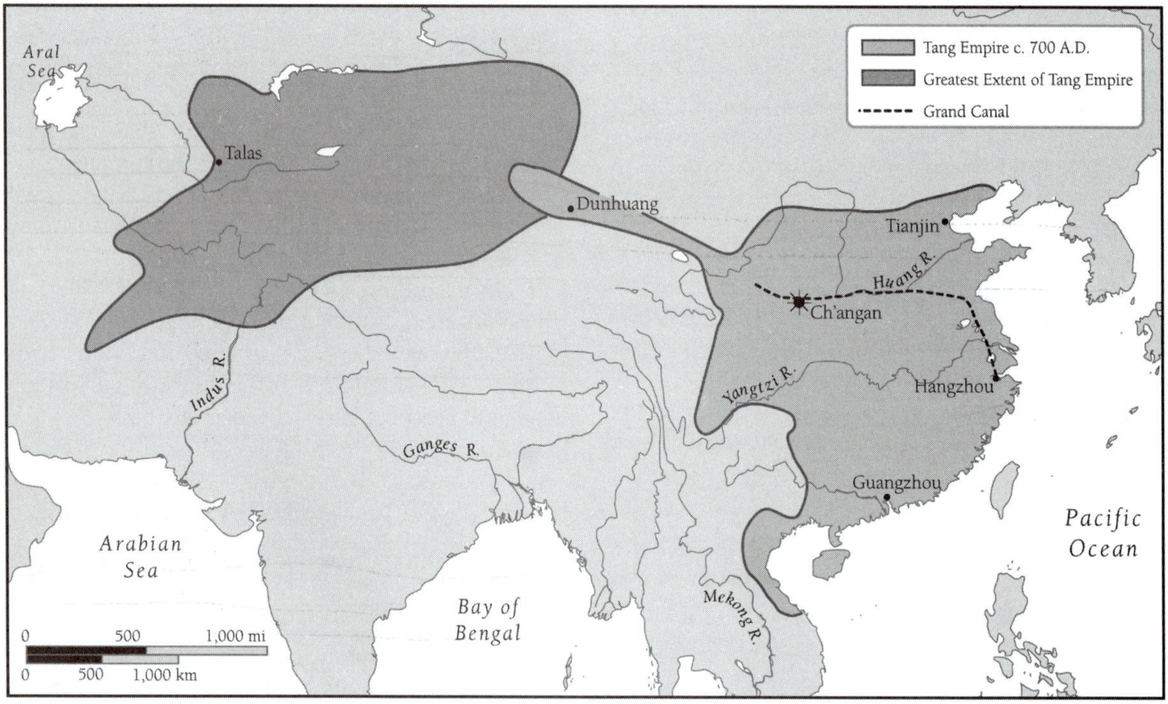

Figure 9-12

Note the expansion of the Tang Empire into Central Asia. The empire, although larger than the Han, was still basically confined to the lowlands and river valleys of the east. Note how the Grand Canal links the region of the Huang He and the Yangzi River basins.

(i.e., the known world). This was acknowledged by all others who offered praise and paid tribute, and imitated everything Chinese.

But again, the dynasty failed to endure. Tang forces were defeated by a coalition of Turks and Arabs at the Talas River near Samarkand in 751. The Silk Route was thereby closed. This battle is significant because certain captured Chinese transmitted the recently developed art of wood-block printing and an older method of paper making to the West. By 1030, the Chinese had developed movable-type printing, long before it spread in fifteenth-century Europe. In fact, China hosted a wide spectrum of inventions long before the West (Figure 9-13, p. 234).

Ch'ang-An was likely the largest planned city ever built. Laid out in a checkerboard pattern, it covered some 30 square miles (78 sq km). Its wide avenues ran north–south and east–west to great gates at the cardinal compass points. The walled city was further divided into quarters and lesser administrative units. The people lived in rectangular wards, each with gates that were closed at night. Markets and entertainment districts served residents and visitors alike. Ch'ang-An was notably cosmopolitan, housing an international set of merchants, traders, artists, pilgrims, students, religious scholars, adventurers, and others. Official tributary emissaries came from Korea and Japan and regions of the south on a regular basis, returning to their homelands with all they could carry from China.

A great drought in the north precipitated a series of rebellions, exacerbating the dissatisfaction already apparent throughout the empire. Rival generals vied for control, but the Tang collapsed in disarray in 907.

THE SONG (SUNG) DYNASTY (960–1279)

The Song eventually saw the maintenance of far-flung frontiers as unprofitable. Vietnam, Korea, and other frontier regions were chronic drains on China's wealth and military manpower. However, both were willing to accept nominal tributary status, which allowed the Chinese to save face. The Song focused their attention south of the Great Wall, by now a region of 100 million people and the largest, most developed, and most productive state in the world.

The Song placed their capital at Kaifeng, which became a commercial and manufacturing center of more than a million inhabitants. Iron and steel production boomed—China produced more metal goods than all of Europe in the eleventh century. But this prosperity was not to last. The nomadic Mongol Jins (Jurchen), pressured by other barbarians north of them, descended to control the region north of the Yangzi River. This initiated the era of the Southern Song with a new capital at Hangzhou (Hangchou or Hangchow).

Foot Binding—The Golden Lotus

Golden Lotus is a Chinese euphemism for the **foot binding** of women. The first reliable reference to this atrocity is from the Southern Song (1127–1278). Supposedly, it began in court circles and spread among the upper classes of society. The idea was to keep women's feet very tiny by folding the toes under the foot and tying the bent foot tightly with cloth. In this manner, the feet would be deformed, but small. This very painful process was started between the ages of five and ten. Needless to say, the victims were incapable of walking normally. Judging by poetic references, the bound foot was an erotic sight to some men. Moreover, foot binding created a gait that would eventually restructure the pelvic bones to enhance men's pleasure during sexual intercourse.

The Mongols of the Yuan Dynasty (1279–1368) did not allow their own women to have bound feet but encouraged it among Chinese women. Foot binding was at its height during the Ming Dynasty (1368–1644). The Manchu or Qing Dynasty (1644–1911) tried to ban the practice but it continued into the early twentieth century. Very few, very old women with bound feet are alive today.

Attention was turned to southern commerce. Cities such as Suzhou (Suchou or Suchow), Fuzhou (Fuchou or Foochow), Hangzhou, and Canton

50 AMAZING CHINESE INVENTIONS/DISCOVERIES

INVENTION	IN CHINA (century)	IN THE WEST (century)
Row cultivation of crops	6th BC	16th AD
Iron Plow	6th BC	16th AD
Multi-tube Seed Drill	2nd BC	16th AD
Horse Collar	3rd BC	7th AD
Quantitative Cartography	2nd AD	15th AD
Mercator Map Projection	10thAD	16th AD
Cast Iron	4th BC	13th AD
Crank Handle	2nd BC	9th AD
Steel from Cast Iron	2nd BC	18th AD
Water Power	1st AD	13th AD
Suspension Bridge	1st AD	19th AD
Deep Drilling for gas	1st BC	20th AD
Belt Drive	1st BC	13th AD
Chain Drive	10th AD	18th AD
Underwater Salvage	11th AD	19th AD
Lacquer (Plastic)	13th BC	19th AD
Petroleum and Gas as Fuel	4th BC	19th AD
Paper	2nd BC	12th AD
Wheelbarrow	1st BC	12th AD
Fishing Reel	3rd AD	17th AD
Horse Stirrup	3rd AD	6th AD
Porcelain	3rd AD	20th AD
Biological Pest Control	3rd AD	19th AD
Mechanical Clock	8th AD	14th AD
Printing Movable Type	11th AD	15th AD
Playing Cards	9th AD	14th AD
Paper Money	9th AD	17th AD
Spinning Wheel	11th AD	13th AD
Circulation of Blood	6th BC	12th AD
Deficiency Diseases	3rd AD	19th AD
Thyroid Hormone	7th AD	19th AD
Immunology for Smallpox	10th AD	18th AD
Compass	4th BC	11th AD
First Law of Motion	4th BC	16th AD
Seismograph	2nd AD	16th AD
Modern Geology	2nd AD	17th AD
Kite	5th BC	15th AD
Relief Map	3rd BC	13th AD
Parachute	2nd BC	18th AD
Ship's Rudder	1st AD	12th AD
Multiple Mast Ships	2nd AD	14th AD
Watertight Compartments	2nd AD	19th AD
Canal Pound-Lock	10th AD	14th AD
Rotor and Propeller	4th AD	19th AD
Paddle-wheel Boat	5th AD	15th AD
Chemical Warfare	4th BC	19th AD
Crossbow	4th BC	2nd BC
Gunpowder	9th AD	12th AD
Fireworks	10th AD	12th AD
Land Mines	13th AD	14th AD

Figure 9-13

How many of these inventions are you familiar with?

emerged as great ports connected by an extensive network of navigable creeks and canals from the Yangzi Delta to the Xi Delta. Each of these cities had well over a million people.

These large urban populations were sustained by a shift to rice cultivation and double-cropping (as opposed to single-cropped millet and wheat of the north). A special drought-resistant variety of rice was introduced from Champa, modern Cambodia. Cities also grew from rural–urban migration—an exodus of peasant-farmers who were indebted to rich land owners. Reforms were controversial and largely unsuccessful because of growing corruption and self-interests of administrators and landed gentry.

Hangzhou remained notable for centuries. Marco Polo, who saw it much later in the thirteenth century, said that even then, it was the greatest city in the world. In the fourteenth century, the Arab geographer Ibn Battuta reported that Hangzhou was three days' journey in length and subdivided into six towns. Hangzhou was a thriving center for commerce, art, drama, literature, and music. It was also a center of landscape painting and gardening, both famous during the Song for their portrayal of human–nature harmony (Figure 9-14).

This, too, was the era of the Neo-Confucianist Zhuxi (Chu Hsi). Like many philosophers, he was interested in "the investigation of things." His journals record that uplifted rock strata above the current sea level contained marine fossils. This was the world's first statement regarding the geomorphological theory of uplift.

Agriculture benefited from the invention of various tools and machines for cultivation, harvesting, and threshing. Water power was used to grind grain and run weaving machines. Water clocks were common as well. China appeared to be on the road to an Industrial Revolution exceeding or at least rivaling that of eighteenth-century Europe. But that did not happen. A barbarian onslaught crushed a Song Dynasty that had become complacent, factionalized, and more concerned with Confucian bureaucracy than military preparedness. This onslaught was triggered by the formidable **Genghis** (Chinghis or Ghenghiz) **Khan** (1155–1227).

THE YUAN DYNASTY (1279– 1368)

Genghis Khan was a military genius who amassed his nomadic hordes into the most deadly cavalry armies the world had ever seen. In 1211, he crossed the Gobi into northern China. His horsemen carrying sparse rations of grain and mare's milk could cover 100 miles a day. They could fire arrows on the run—all in all, a terrifying force. A Chinese army of at least 150,000, stationed behind high walls and moats and armed with

Figure 9-14

Moon Gate in Suzhou. The concepts of yin and yang and feng shui are commonly illustrated in gardens. Here at the Garden of the Nets in Suzhou, a round moon gate symbolizing heaven harmonizes with a rectangular gate representing earth. A carefully orchestrated rock and plant garden is just beyond the gate.

catapults and flame-throwers, was effectively massacred by the mobile Mongols who swarmed into Beijing in 1215 to enjoy "a most glorious slaughter." The death of Genghis Khan in 1227 saved the Southern Sung for half a century.

Genghis Khan's three grandsons ruled the North China Plain, Mongolia, and the Far East. **Kublai Khan** mastered the North China Plain and moved his headquarters to Beijing. He began his invasion of southern China in 1271. Sweeping into Hangzhou, he crushed the last Chinese resistance at Canton in 1279. Kublai Khan thus became the Mongol Emperor of China.

Korea, northern Vietnam, and the previously non-Chinese southwest were also conquered, driving forced migrations southward into Southeast Asia. Southern Vietnam, Siam (Thailand), Burma, and Tibet were then compelled to accept tributary status.

Kublai Khan did have one important defeat. In 1274, he launched an attempt to occupy Japan. Confronting Japanese troops with an advance force, his main force was prevented from landing on the southern island of Kyushu by an impending storm. The Mongols beat a hasty retreat until after their victories in China. In 1281, 150,000 Mongols landed on Kyushu, where they battled for 53 days. Then, a typhoon sunk most of the Mongol transport ships and the invasion failed. The Japanese named this typhoon ***kamikaze:*** the divine wind.

The Mongols were now in control of the largest empire in the world (Figure 9-15). While they filled some top administrative posts with their own people and abolished the civil service exams, they left much unchanged in the traditional methods of Chinese government. They reconstructed defenses, extended the Grand Canal to link Beijing and Hangzhou, and turned battle-scarred Beijing into a beautiful city. From 1280 to 1300 was a period of peace and prosperity known as ***Pax Mongolica.***

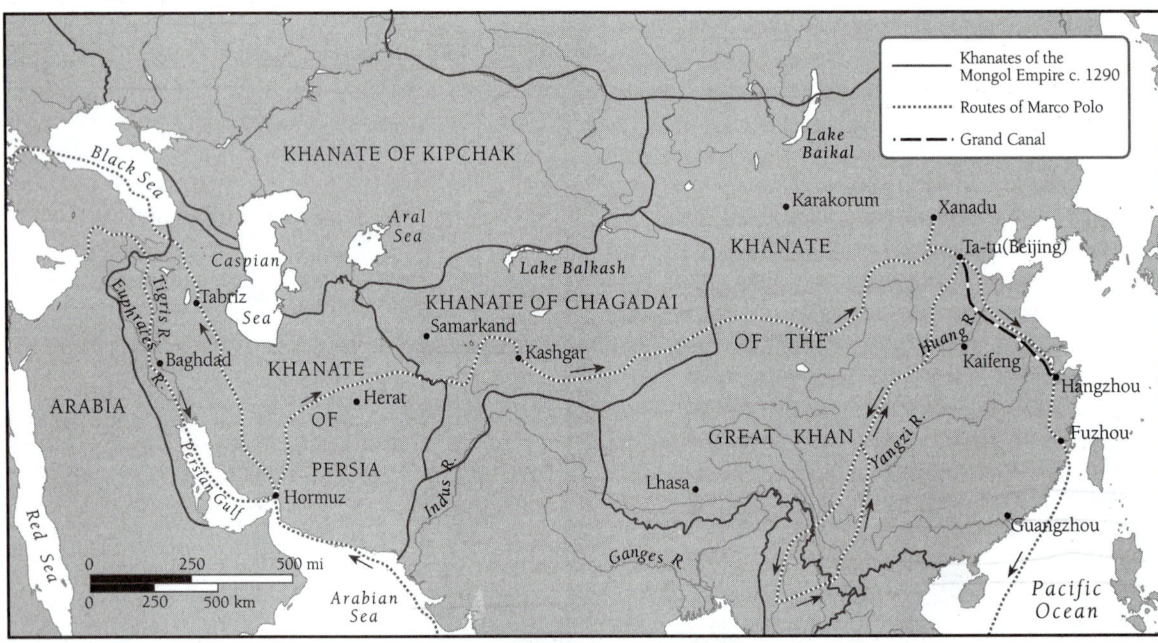

Figure 9-15

China Under the Mongols. At this time China was linked with other parts of Asia ruled by various allied Khans. Some scholars believe that Marco Polo never made his journeys. He failed to mention chopsticks in his diaries which seems strange given the detail of his other observations. Read more about Marco Polo and see what you think.

Merchants, traders, diplomats and myriad others, including Marco Polo, journeyed the Silk Route to the cities of the vast Mongol Empire. Exchange of goods and ideas between East and West proliferated. But while commerce and the arts flourished once more, hatred of these foreign overlords did not wane. Kublai was followed by increasingly inept leaders, leading to widespread revolts by 1330. Famine, earthquakes, and floods took the lives of millions. Populations were further decimated by the Black Death that was ravaging Europe. One peasant-rebel leader welded the Chinese forces together and drove the Mongols back to the steppes and in 1368, announced the formation of the Ming Dynasty.

THE MING DYNASTY (1368– 1644)

What was Chinese culture like at this time? With the Sinification begun under the Han and solidified under subsequent dynasties such as the Tang and Song, Chinese culture had reached great heights. The scholar-bureaucracy was in place, as was a landed gentry. Civil administration included multiple bureaus and a military administration. The countryside was managed largely by the gentry and county magistrates. Most important was the family that, in Confucian form, controlled members' actions and settled disputes.

The Yung-lo Emperor commissioned an encyclopedia of all knowledge, on which 3,000 scholars worked for 5 years. This was followed by a great medical encyclopedia that listed more than 10,000 drugs and prescriptions mostly unknown to the West. This document also recorded the use of inoculation to prevent smallpox, a medical advance not discovered in Europe until the eighteenth century. There was also a handbook of industrial technology describing techniques in weaving, the manufacture of porcelain, ship building, canal locks, paper making, and many other items and pursuits. Ming led the world in science and technology at this time.

It was decided to move the capital back to Beijing because of its proximity to the troublesome northeast and the threat of the Manchus of Manchuria. Moreover, the Grand Canal facilitated the movement of food northward from the Yangzi Basin.

The walls surrounding Ming Beijing were 40 feet (12 m) high and nearly 15 miles (24 km) in circumference. There were nine double gates with watchtowers. Inside the city was the Imperial City with its own walls, 5 miles (8 km) around. Inside the Imperial City

Figure 9-16

Hall of Supreme Harmony viewed from the Gate of Supreme Harmony. This is the largest and most important building in the Forbidden City. Leading to the Hall is a three-tiered "dragon pavement" over which the Emperor was carried. Dragon motifs cover the pavement.

was the moated Forbidden City—2 miles around—containing the palace, throne halls, audience halls, and other buildings along a north–south axis. Buildings were set on terraces of white marble and had gold-colored tiled roofs. The entire complex was designed to awe and impress (Figure 9-16).

Rice was the dominant element in the diet, supplemented by wheat noodles (brought back to Venice by Marco Polo as spaghetti). Food was eaten with chopsticks, a method quickly adopted by Korea and Japan. Bean curd (dofu or tofu) was a major source of protein, as were fish and chicken. Oxen and water buffalo were used as draft animals but eaten only if they died naturally. Whatever was put on the rice or noodles was sliced finely and cooked quickly in a wok. Deforestation and a shortage of fuel fostered the need for stir frying.

The landscape became more and more cultural than natural as irrigated and terraced rice paddies and fish ponds proliferated. New towns and villages emerged with their craft industries and periodic markets. From Tang times, the house became a common center for socializing and negotiating business deals and marriage contracts. It was not necessary to travel far—people rarely ventured beyond the nearest village. The wheel barrow and the flexible bamboo carrying pole were the major means of transporting goods on land, methods still used in China and much of Southeast Asia to this day. Peasants adhered to a folk proverb: "Work when the sun rises, rest when the sun sets. The emperor is far away." In essence, while dynasties rose and fell, the fundamental order of Chinese society persisted.

The first Ming emperor made valiant attempts to reopen the Silk Route, sending repeated expeditions against the Mongol Khans. Further, he repaired the northern frontier and made Korea a vassal state. More important, however, were Ming sea voyages.

Cheng-ho sailed south to Malacca (Melaka) in 1403. Two years later, his junks reached Ceylon, East Africa, and the Arabian Sea. But it was Malacca that would become the main regional entrepot—thousands of Chinese merchants flocked to Malaysia, Sumatra, and Java. And, in 1509, six Portuguese ships dropped anchor at Malacca. Soon they were seeking the land they called Cathay. In 1514, they were in Canton's harbor. Japanese pirates terrorized the coasts from the Shandong Peninsula to Hainan Island.

Between 1400 and 1600, China's population rose from 70 to 140 million. This growth spurt was accompanied by an agricultural revolution, and rice was 70 percent of food production. The 100-day maturing Champa rice coated valleys and hillsides. Then, in the 1550s came the "American" crops: potatoes, maize, and peanuts. Colonists planted formerly undeveloped hills and established new farming communities. This agricultural revolution was not one of mechanical devices, it was one of crops. Unfortunately, clearing hillsides caused so much erosion that the rivers silted up and flooded more than ever before.

Meanwhile the Portuguese pressed their case for trade. In 1557, the Chinese allowed them to build a colony on the isthmus of Macao. Spanish colonists in the Philippines who were trying to break into the lucrative China trade even contemplated attacking China in 1586. The Jesuit Matteo Ricci arrived in Canton in 1582 and expressed interest in Chinese technology. Some years later he was allowed to live in Peking, where he worked with Chinese scientists. This was designed to bring knowledge of European firearms including heavy cannon to a Ming Empire worried about its northeast frontier.

The Japanese had been harassing China's coastal trade for years. Things were so bad that all settlements 30 miles (48 km) inland were removed, and maritime trade was officially forbidden. This edict was largely ignored. And in 1592 an invasion force of 200,000 Japanese attacked Korea and occupied Pusan, Seoul, and Pyongyang. Since China held sway over Korea, in 1593, the Chinese crossed the frozen Yalu River and began years of fighting on the Korean Peninsula. The extraordinarily talented Korean Admiral Yi Sun-sin, who had developed iron-covered ships called *turtle ships*, managed to sink 70 Japanese ships in a single engagement. As the Ming armies contrived to defeat the Japanese on land, they were suddenly faced with an invasion from the northeast. The Chinese defeated the Japanese in Korea in 1593 but the Manchu tribes stood on the horizon to ultimately take over all of China.

THE QING (CH'ING) DYNASTY (1644– 1911)

Internal chaos, peasant rebellions, and Manchu invasions combined to destroy the Ming, China's last native dynasty. The Manchus ruled as the Qing (Ch'ing) Dynasty. *Qing* means "pure." In many ways the Manchus became more Chinese than the Chinese themselves. However, there were looming problems. China population was increasing at an unprecedented rate, and there were food shortages. Foreign powers with their gun boats were nipping at Chinese sovereignty; missionaries were pursuing converts; Western merchants were bargaining for silk, tea, and art treasures; and peasant uprisings threatened the entire fabric of government.

The Manchu attacks had caused chaos, and a stable government was long awaited by a war-weary populace. The Manchus pressed southward, setting up garrisons and "enjoining" the allegiance of the local populations. One determined Chinese opponent was Kok Seng-ya, better known as Coxinga. Coxinga was finally routed at the Battle of Nanjing in 1659, but he fled via the island of Quemoy and the Pescadores to Taiwan, the island known by the Portuguese as Formosa "the beautiful" (Figure 9-17, p. 242).

Thousands of Chinese emigrants were already living on the island, and the Dutch had been shaping a colony there since 1623. Coxinga defeated the Dutch but died shortly thereafter. The Manchu fleets came across the Straits of Formosa and eventually defeated Coxinga's forces. In 1683, Taiwan, for the first time, became part of the rapidly expanding Qing Empire.

The Manchus were cousins of the Mongols, and their appearance, clothing, customs, and writing system were different from those of the Chinese. They insisted that all males adapt the Manchu queue (pigtail)—otherwise they left Chinese culture alone. In fact, they tried to inject more honesty and enthusiasm into government, which had become corrupt and complacent.

The Russians had been penetrating the northeast, and Cossack frontiersmen clashed with Manchu border patrols from 1650 to 1689. Then the Manchu emperor negotiated the Treaty of Nerchinsk, the first of many attempts to stabilize the Chinese–Russian frontier. Emperor Ch'ien Lung settled colonists in Xinjiang and sent troops to Tibet to counter Gurkha incursions from Nepal. He also forced Annam and Burma to recognize Manchu authority.

Meanwhile, China's population was increasing, probably doubling between 1700 and 1800. Ch'ien Lung authorized huge public works to provide peasants employment in times of crisis. He encouraged farmers to grow more wheat and, wherever possible, to plant cash crops such as cotton, sugar, and indigo. Related industries were also established. His foresight provided a long period of peace. Unfortunately, this was not to last. European powers and overseas Chinese were pressing at the gates of commerce.

A New Breed of Invaders

THE CHINA TRADE

The Chinese intelligentsia were not keen on trade as a profession. Nevertheless, thousands of Chinese were involved in it within and without China. Portuguese, British, Dutch, Spaniards, and Danes—all were anxious for footholds on the mainland. By the time the first American ship sailed into Canton in 1784, the emperor had already decreed that all trade must be channeled through Canton and that factors or traders must confine themselves to *factories* (storehouses) leased from local landlords. Foreigners were thus crowded into small compounds the Chinese called *hongs*.

In 1793, a British trade delegation arrived at the court of Ch'ien Lung. The emperor accepted the gifts of clocks, guns, and telescopes but, in a letter to England's King George III, pointed out, "There is nothing we lack. . . . We have never set much store on strange or ingenious objects, nor do we need any more of your country's manufactures." Ordinary Chinese saw things differently, and thousands went abroad to participate in the business of trade. The favorite destination was the island of Luzon in the Philippines, which became the first major center of the China trade. Silks, tea, chinaware, and rhubarb were exchanged for high-grade silver mined in Spanish America.

The emperor would not deign to have contact with the "foreign devils." Instead, civil servants did the dealing and reported their activities via horse relay to the Son of Heaven. Misleading reports and mixed interpretations of information soon wrought trouble between the traders and port officials. Accidental deaths of three Chinese led to the strangulation of a British and an American sailor according to Chinese law.

Meanwhile the British discovered that the most lucrative import of all was opium, which was grown in abundance in Bengal and could be sold in China. When the emperor tried to stem its import because of rapidly increasing addiction, as well as its cultivation in Yunnan, demand increased and the illegal trade boomed. Several efforts were made to end this pernicious trade, but to no avail. At one point, the British were forced to seek refuge on their ships anchored in the Hong Kong harbor. They responded by turning the ships into opium trade headquarters. When Chinese junks tried to intercept transport ships, four were sunk by British gunboats.

Britain responded by attacking the Chinese coast and threatened Nanjing (Nanking). This first Opium War (1839–1842) culminated in the Treaty of Nanking in 1842, whereby Britain acquired the island of Hong Kong, a cash indemnity, and permission to trade at five *treaty ports:* Canton, Xiamen (Amoy), Fuzhou (Fuchow), Ningpo, and Shanghai. Shortly thereafter, Britain was given "most favored nation" status, which meant that it would enjoy the provisions of treaties signed with any other nations. In 1844, the United States signed a trade treaty.

The Doctrine of Extraterritoriality

Extraterritoriality refers to consular jurisdiction. This means that a foreigner is subject to the laws of his own country. For example, a British sailor who committed a crime in China would be punished under British law. This practice was introduced to China by the Russians in the Treaty of Nerchinsk in 1689 in which the privilege was mutual. Extraterritoriality was one of the provisions of the Treaty of Nanjing of 1842, but in this case, it was not a reciprocal arrangement. Europeans were subject to European laws and so were the Chinese. This was one reason why trade treaties with the West were called the Unequal Treaties.

The Nationalist Chinese began to negotiate the removal of extraterritoriality in 1928. Unsuccessful, they abolished it by decree in 1932. New treaties, minus the Doctrine of Extraterritoriality, were signed with Britain and the United States in 1943.

Even with signed treaties, the Chinese interfered with trade whenever possible. The arrest of the crew of a British ship in 1856 triggered the second Opium War. The outgunned Chinese were forced to sign the Treaty of Tientsin (Tianjin) in 1858. The emperor, still refusing to deal directly with the issues at hand, fled to Mongolia. Foreign troops invaded and looted then burned the Summer Palace. In 1860, the Chinese capitulated and signed the Convention of Peking, agreeing to pay another indemnity and add Tientsin to the list of treaty ports. China's long period of isolation was over.

THE MISSIONARIES

Another intrusion involved Christian missionaries. By the end of the eighteenth century, the Jesuits had amassed around 300,000 converts among the peasantry, but largely failed to impress many scholars still steeped in Confucian beliefs. However, historian Jonathan Spence (1999) makes an important point that they introduced the Chinese to new ways of looking at the world, not just through religion, but also by means of their versions of art and perspective in drawing, mathematical and astronomical calculation, and other interesting items such as telescopes. Furthermore, "these same missionaries were themselves the representatives of expansionist Western societies determined to increase their bases of wealth and power by territorial conquests and the development of new markets overseas."

British and American missionaries built their missions, schools, and hospitals within the confines of the treaty ports but soon spread to the countryside with the China Inland Mission. Scottish and Irish missionaries pushed north into Manchuria. French Catholics went into Sichuan, where converts who accepted food handouts were called *rice-Christians*. Missionaries were helpful in times of famine and trouble. Nevertheless, there was widespread violence against them, and the annual rate of conversion numbered only a few hundred people.

The French Sisters of Mercy set up an orphanage in Tianjin (Tientsin) but unwisely offered to pay the Chinese for "orphans" brought under their care. In Chinese eyes, these foreign women were really kidnappers. When the Chinese gathered to protest, the French opened fire. This sparked the 1870 Tientsin Massacre of Christians. Western gunboats steamed into port, and any semblance of good relations between Europe and China was ended.

Heavenly Kingdom of the Great Peace

At the time of the Opium Wars, the Hakka people of the southern provinces of Guangdong and Guangxi had been influenced by Christianity. Led by Hung Hsiu-chuan, a schoolmaster, they formed the beginnings of a revolutionary program. Hung taught that land should belong to the people, crop surpluses should go to areas of shortage, and women should be equal to men.

Resistance to the Manchus, obviously corrupt and ineffective, began in 1850. Hung announced the coming of the Taiping Tien Kuan: "Heavenly Kingdom of the Great Peace." Hung attacked along the Yangzi valley and occupied Nanjing, making it his capital. From there he launched two armies to liberate the rest of China from Manchus and foreigners. By 1860, the "God Worshippers" had brought a third of China under their control. This was the **Taiping Rebellion.**

With the Emperor in Mongolia and the Manchu government in disarray, the Taiping seemed assured of victory. But Hung and his advisors underestimated the reaction of the Western powers, who were concerned that the Taiping might capture all the treaty ports. The British decided to give the emperor just enough support to defeat the Taiping. Nanjing surrendered in 1864, and the rebellion that had ravaged China for 14 years was finished. Hung committed suicide. The Taiping Rebellion cost China 20 million lives! But the real significance in terms of modern history is the fact that it demonstrated the potential power of the peasantry. Properly equipped and led, China's peasant masses could achieve anything.

RESPONSE TO THE WEST

After the Opium Wars, many Chinese emigrated to Australia and the United States. Waves of Chinese *coolies* (laborers) were hired to work in mines and build railroads. Australian natives resented their capacity for long hours and hard labor. Restrictive laws were passed, race riots took place, and widespread anti-Chinese feelings persuaded the government to adopt a "Keep Australia White" policy. In the United States, *Chinatowns* emerged and race riots were frequent. America's first zoning laws were designed to keep the Chinese out of "white" areas. People spoke of the "Yellow Peril," and in 1882, strict immigration laws were passed to stem the Chinese influx. Nevertheless, Chinese communities and business ventures prospered.

Within China a *Self-Strengthening Movement* was taking place. Certain senior scholars recommended that China should learn from the West in order to get rid of the barbarians, including the corrupt Manchus. They pointed to Japan, a small nation that was busy industrializing like the West. Many young Chinese went abroad to Japan, Britain, and the United States to study. However, there were few lasting results in terms of reform.

Work began on China's first railroad out of Shanghai in 1875. Peasants believed that the steel road was disturbing the *feng shui* so they smashed it to bits. Others in fear of modernization destroyed textile machinery. The government did, however, manage to buy guns and warships for the beginnings of a navy.

But the Self-Strengthening Movement failed to communicate any sense of urgency to the common people. Even the reclusive Empress Dowager Ci Xi (Tzu-Hsi) seemed unaware that China was disintegrating—losing territory to the "foreign devils."

DOWNFALL OF THE MANCHU

Figure 9-17 shows China's territorial losses. Vietnam was lost to France in 1885 and Burma fell into British hands the following year. Russia, desperate for an ice-free port in the Pacific, gained access to Manchuria.

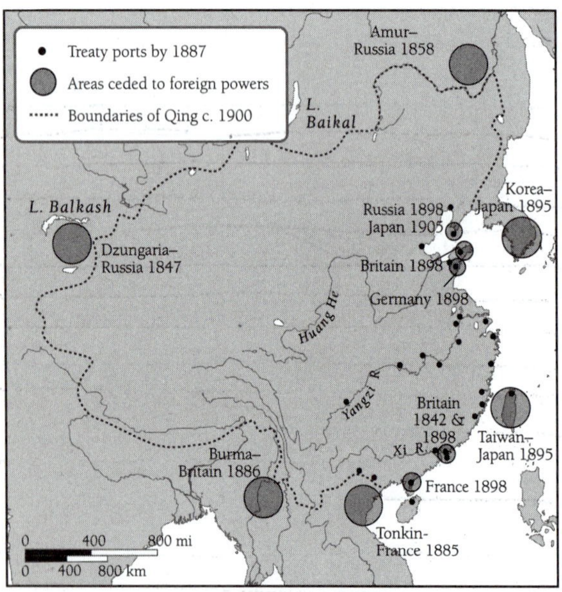

Figure 9-17

The Qing (Manchu) Empire in the nineteenth century. This map shows the extent of the Qing Empire including areas that paid tribute to China. It also shows treaty ports and other areas forcibly ceded to foreign powers in the era of Imperialism.

It also occupied Xinjiang border regions, where chaos had ruled for years. In fact the Russians stayed in the Ili Valley for 10 years. To get rid of them, the Chinese had to give them navigation rights on Manchurian river systems. The Russians then began road and railway construction in Manchuria.

Between 1875 and 1880, Japan forced its way into two more tributary states: Korea and the Ryukyu Islands. This and related events precipitated the Sino–Japanese war of 1894–1895. Chinese troops were decimated at land and sea while the Empress Dowager spent money repairing the Summer Palace.

The war ended with the Treaty of Shimonoseki (1895), whereby China recognized Korea as an independent kingdom. China also ceded the island of Taiwan, the Pescadores, and the Liaodong (Liao-tung) peninsula to Japan. Later, the Russians pressed the Japanese to leave the peninsula because of their interest in the ice-free port of Dalian (Dairen or Port Arthur). The Japanese obliged, and with that move, the Chinese saw the Russians as their friends. A secret treaty was signed giving the Russians rights to build railroads across Manchuria as far as Vladivostok.

After the Treaty of Shimonoseki concessions, there was a torrent of requests for land, mining rights, and other trade deals. The Germans took territories in Shandong in 1897, Russia took Dalian the following year, and France claimed Guangxi as a French sphere of influence. Britain wanted control of the Yangzi and a region north of Hong Kong known as the New Territories.

Most Chinese wanted the foreigners to disappear and were prone to support a secret anti-foreigner society called The League of Harmonious Fists—the Boxers. The Boxers were anti-Christian and believed themselves to be invincible to bullets. Efforts at reform by the Manchu government were dismissed by the Boxers.

The country was in a state of disruption and distress worsened by Huang He floods. In 1900, the Boxers entered the capital and stormed the foreign legations. After 55 days of seige an international relief force arrived. The Empress Dowager fled in a cart to Xian, leaving the Imperial Palace open to looters.

Again the foreign powers humiliated China with war reparations, and scores of new treaty ports fell firmly into the hands of foreign capitalists: bankers, shipping and railway companies, merchants, and missionaries. The interior was open—Western gunboats patrolled China's rivers and Western soldiers guarded settlements. Then the Empress Dowager ensconced herself back in power in Beijing. Clearly the Manchu Dynasty had to be overthrown.

A frantic Empress Dowager entertained Western diplomats, promised to outlaw foot binding, permitted Manchus to marry Chinese, and abolished the state examinations that had supplied the Imperial Civil Service for 2,000 years. She sent military officers to train with the Japanese who had just won the Russo–Japanese War of 1904–1905. But it was too late for the "Old Buddha," as she was known in the north. Ci Xi died in 1908 as did her nephew, the new emperor. His successor was the two-year-old Pu-yi, whose father became Regent. The Manchu Dynasty was near its end. A modern revolution was at hand.

Asia's First Nationalist: Dr. Sun Yat-sen

Sun Yat-sen was born of a peasant family in 1866. He attended a mission school in Hawaii and became a Christian. He then studied at Hong Kong Medical School and graduated as a doctor in 1892. In 1895, he ventured into the business of revolution, but this ended with the arrest and execution of most of his compatriots. Sun spent the next 16 years in exile trying to drum up support for the liberation of China. He promoted democracy, nationalism, and equalization of land rights.

Between 1908 and 1911, the Manchus encouraged European firms to build new railways. This policy infuriated the people of Sichuan, and a peasant revolt ensued on October 10, 1910, later referred to as the *Double Tenth.* Revolution spread to central and southern China. Sun Yat-sen returned from abroad and was quickly elected as provisional president and proclaimed a Chinese Republic on January 1, 1912. But Sun Yat-sen stepped down after six weeks in favor of a powerful warlord whom the Manchus had named as their successor.

In 1910, Japan had annexed Korea and in 1914 seized Shandong and extended its influence in Manchuria. China subsequently was immersed in rebellions, with multiple warlords pitted against one another for territorial gain. Interminable military campaigns ruined the harvests, and starvation and depredation were widespread.

Europe was locked in the battles of World War I (1914–1918), and China hoped that by allying itself with Britain and France against the Germans, it would gain their support against the Japanese. Therefore, the 1919 Treaty of Versailles came as a shocker. Japan was allowed to keep its territorial gains in China, and the Europeans kept all their treaty ports and privileges both along the coast and along the Yangzi.

Students led a nationalistic demonstration that sparked a nationwide strike. One of the strike leaders was a young teacher named **Mao Zedong** (Mao Tse-tung). Communism attracted many young students with its promises of equality, and the logic of Marx and Engels appealed to many Chinese Confucian intellectuals. A Chinese Communist Party (CCP) was created and held its First National Congress in Shanghai on July 1, 1921. Warlordism continued as Chiang Kai-shek's Kuomintang (KMT) tried to revitalize itself with Russian assistance. Unfortunately, Sun Yat-sen, the first great Asian nationalist, was unable to complete his mission. Suffering from cancer, he died on March 12, 1925. KMT leadership was assumed by **Chiang Kai-shek.**

The Road to Communism

At least outwardly allies, Mao Zedong's CCP and Chiang Kai-shek's KMT proceeded to fight the warlords. Strikes, riots, and rebellions sacked the countryside, yet foreign powers continued to pour money into urban industrialization. Disruption in the rural areas drove millions of refugees to these cities which became breeding grounds for secret societies, strike units, and rebel organizations. Meanwhile the CCP was working toward a Communist revolution and had formed peasant organizations in the countryside. These would eventually rise up and annihilate the KMT.

Following a northward sweep by the combined KMT and CCP, which conquered most of southern China by 1927, Chiang Kai-shek decided to rid himself of the CCP. Launching a reign of terror in newly captured cities such as Shanghai, he routed out and executed as many Communists as possible. The Communists fled the cities and vowed to garner the loyalty of the peasants and regroup. Chiang set up his government in Nanjing and proclaimed it capital of the Nationalist Republic of China in 1928.

Chiang Kai-shek and his troops tore after the Communist forces, determined to exterminate them. But in 1931, the Japanese invaded Manchuria, with profound consequences for all factions. The following year they made a savage attack on Shanghai. Because Chiang was unwilling to give up his campaign against the Communists to fight the Japanese, the Japanese were able to set up the puppet state of **Manchukuo** in Manchuria—China's most important industrial region.

The Nationalists continued their extermination campaign against Mao and his forces. In eastern China, the Communists, who were not faring well in battle, decided to move elsewhere to reorganize and reconsider their options. In October 1934, 85,000 made the now famous **Long March** (Figure 9-18). Pursued relentlessly by the Nationalists, they trudged 6,000 miles (9,660 km) through jungles, swamps, hostile tribal areas, torrential rivers, and 16,000-foot (4,800 m) mountains. About 7,000 men completed the march and set up their capital at Yenan in the loess hills of Shensi Province in October 1935.

The Long March became a legend, an epic story of the heroism of soldiers and loyal peasants who placed national interests before individual concerns. The Long March exemplifies struggle for an ideal.

In 1937 the Nationalists and Communists decided to form a temporary united front against Japan. As the Japanese advanced, Chiang moved his capital to Chongqing (Chungking), which he proceeded to industrialize with dismantled factories from the east.

The Japanese pressed on with "Kill all, burn all" tactics. In the now-infamous *Rape of Nanjing* they butchered an estimatd 200,000 people. Nevertheless, Chiang Kai-shek continued to expend much of his energies rooting out Communists. After a failed attack against the Japanese, Mao and his troops fell back to Yenan. The war in China was a stalemate in 1940. The British and French were occupied fighting Germany, and the Japanese pressed southward to French Indo-China.

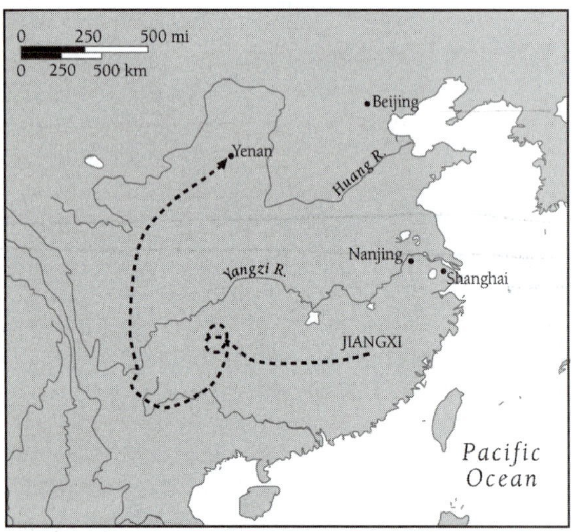

Figure 9-18

The Long March, from October 1934 to October 1935, was one of the most arduous journeys ever made by people. The Long Marchers traveled 6,000 miles (9,660 km) across snow-covered mountain ranges with passes at 16,000 feet (4,800 m), into lands inhabited by wild tribal people, and through deserts and swamps. Only 7,000 of Mao's forces completed the route to the relative safety of the loess caves of Yenan.

Guerilla Behavior and Tactics

Mao Zedong advocated strict rules of behavior for his Red Army: 1. Speak politely to the people and help them when you can; 2. Return doors and straw matting to the owners; 3. Pay for any damage you cause; 4. Pay a fair price for any goods you buy; 5. Be sanitary—establish latrines well away from houses; 6. Do not take liberties with the women folk; 7. Do not ill-treat prisoners; and 8. Do not damage the crops. Unlike the relatively undisciplined Nationalists, Communist soldiers were much better able to endear themselves to the peasants.

Mao's guerilla tactics can be summarized as follows: "The enemy attacks, we retreat; The enemy camps, we harass; The enemy tires, we attack; and, The enemy retreats, we pursue." The Nationalists were unable to cope with this type of warfare.

The Japanese attack on Pearl Harbor on December 7, 1941, changed the face of the war because

Figure 9-19

This picture represents both old and new China. The tile art shows a Silk Route caravan while the bicycles represent the major form of transportation in China today. I shot this photo in Dunhuang, a major gateway to the Silk Route.

China became allied with Britain and the United States. With U.S. aid, Chiang prosecuted the war with the aid of a motley group of warlords from Chongqing. Much U.S. support was wasted in battle losses and mass defections to the Communists. However, Chiang's propaganda machine spouted nothing but success to Western supporters. Meanwhile, Mao was consolidating his position in Yenan. While masterminding guerilla warfare against the Japanese, he spent most of his energy wooing the peasants of North China with Communist propaganda. With the Japanese war's end at hand in 1945, American efforts to bring Communists and Nationalists together failed. Their war continued.

Industrial Manchuria was regarded as an essential prize of war by both Mao and Chiang. In 1946, Chiang moved north. It was a fatal decision because at that point, fed up with bogus stories of military successes and not wanting to take sides in what was now a civil war, the Americans decided to cut off aid to the Nationalists. Meanwhile, inflation and constant threat of conscription ruined morale among civilians. Their despair spread to the troops, who defected to Mao's Red Army by the millions. Chiang Kai-shek, conceding defeat on the mainland, nevertheless transferred his "government" to Taiwan, where he remained until his death in 1975.

By September 1949, Mao was ready to declare a new Republic of China (Figure 9-19). Speaking to the Chinese people from the Gate of Heavenly Peace in Beijing, he said, "Our work shall be written down in the history of mankind, and it will clearly demonstrate the fact that the Chinese, who comprise one quarter of mankind, have from now on stood up . . . we announce the establishment of the People's Republic of China."

Main Point Summary

1. East Asia has been dominated by China for around 7,000 years. China has impacted Japan, Korea, and Taiwan, culturally, politically, and economically.
2. Each country of East Asia is at least two-thirds mountainous. Mountains have acted as barriers to diffusion but have also formed cultural sanctuaries. Mountain passes, river valleys, and ocean channels have encouraged the diffusion of people and ideas throughout the region.
3. China is commonly divided into two parts: the sparsely populated, rugged regions of China Frontier and the densely populated lowlands of China

Proper. Water availability is critical to settlement in both regions.

4. The North China Plain and the Chang Jiang (Yangzi) lowlands are the most important regions of settlement, agriculture, and urbanization in China.
5. East Asia's monsoon is modified by the presence of both continental land mass and ocean expanse. As precipitation decreases northward, seasonal extremes increase. In China, double-cropping is only possible south of the 20-inch isohyet approximated by the Qinling Shan and the Huai He.
6. Populations in East Asia are diverse; many groups are of non-Mongoloid origin. This fact is reflected in linguistic and other cultural differences.
7. Historically, China perceived itself as the *Middle Kingdom,* at the center of the world. The Chinese "People of the Han" were regarded as superior to non-Han peoples. These perceptions fostered the idea of the civilized Han versus the noncivilized or barbaric "other."
8. Ancient philosophies, such as Daoism, instilled the ideas of harmony between people and nature and the proper balance among all forces. These ideas are expressed in the concepts of feng shui, the art of placement. Confucian philosophy promoted government by men of virtue. It also stressed the maintenance of proper relationship between different levels of a hierarchical society, ranked from ruler to female children.
9. Relationships with "barbarians" along the northern frontier and elsewhere beyond the core of China have pervaded the history of the Chinese and other groups in East Asia. The Chinese came to regard tribute-paying regions as legitimate parts of their empire. Landscape evidence of the ever-changing nature of Chinese–barbarian relationships is provided by the Great Wall.
10. In China, time frames of rule were called *dynasties.* Rulers were said to rule by the *Mandate of Heaven.* When conditions were positive, the ruler had Heaven's mandate; when conditions were negative, the mandate was lost. Thus, each dynasty witnessed a cycle of rise and fall. At the same time, borders expanded and shrunk, depending on the strength of the dynastic rule.
11. Barbarians ruled all of China on two occasions: the Mongols (Yuan Dynasty) 1279–1368 and the Manchus (Qing Dynasty) 1644–1911.
12. Buddhism entered China from India in the first century A.D. and diffused along the Silk Route from western Xinjiang to Xian.
13. During the Sui Dynasty (589–618) the Yangzi region was linked with the North China Plain by the Grand Canal. During this dynasty the Koreans accepted their vassal status with China.
14. Under the Tang (618–907), China became the richest and most powerful empire in Asia and Ch'ang-an became the world's largest city. But the Tang forces were defeated by the Turks and Arabs in the far west, thereby closing the Silk Route and redirecting China's attentions southward.
15. Southern regions and cities such as Canton and Hangzhou flourished under the Song rulers (960–1279). Large urban populations were sustained by a shift to double-cropping rice instead of single-cropping wheat and millet as in the north. Technological progress was stayed by the onslaught of the Mongols under Genghis Khan.
16. The Mongols created the largest empire in the world. Merchants and traders traversed the Silk Route to exchange goods and ideas between Europe and China. Kublai Khan attempted, unsuccessfully, to invade Japan. The glories of Kublai Khan's *Pax Mongolica* did not last. Under succeeding incompetent rulers and natural disasters, the Yuan Dynasty collapsed to the Ming in 1368.
17. The Ming (1369–1644) led the world in science and technology. They repaired the Great Wall and redeveloped their capital at Beijing. Population growth and technological improvements transformed the natural landscape into an increasingly cultural one.
18. The Ming were known for sea exploration. Malacca (Melaka) became the main regional entrepot, and Chinese merchants flocked to Southeast Asia. Between 1400 and 1600, many new crops were introduced and the country experienced an agricultural revolution.
19. In 1514, the Portuguese arrived at Canton. Contacts with the West increased from that point on. In

the latter part of the sixteenth century, the Chinese helped the Koreans fight Japanese invaders. As Ming troops dealt blows to the Japanese on land, they were faced with both internal rebellion and Manchu invasions. At the same time, foreigners were carving out spheres of influence and Christian missionaries were pursuing converts.

20. One rebel leader, Coxinga, challenged the Manchu from his escape haven in Taiwan. Eventually, in 1683, Coxinga was defeated and Taiwan became part of the rapidly expanding Qing Empire.
21. By the eighteenth century, the Europeans had established several factories at Canton. Although the Chinese elite were not interested in trade, millions of Chinese were anxious to become part of this growing industry. The British forced China to import opium from India, and this ultimately led to war and Britain's acquisition of Hong Kong.
22. One of several attempts to oust the Europeans was the Taiping Rebellion, which resulted in the loss of 20 million Chinese lives. This rebellion demonstrated the potential power of the peasantry.
23. Particularly galling to the Chinese was the Doctrine of Extraterritoriality, which allowed that foreigners were subject to their own laws in Chinese territory.
24. Many Chinese left to study or emigrated abroad for work. Chinatowns emerged in Australia and North America, where Chinese often faced discriminatory behavior and policies.
25. In the latter part of the nineteenth century, the by-then corrupt Manchus were forced to yield increasing territorial privileges to both the Europeans and the Japanese, who were on their own imperialistic drive. The Boxer Rebellion in 1900 failed, and China was further humiliated as more treaty ports were opened to foreign capitalists, missionaries, and the like.
26. In the fray of a failing Manchu Dynasty, three individuals emerged who would shape the future of China and East Asia. An early leader of the Kuomintang, Sun Yat-sen, became the first Asian nationalist. Another nationalist, General Chiang Kai-shek, inherited the Kuomintang Nationalist leadership after the death of Dr. Sun. A young revolutionary, schooled in the ways of communism, became the leader of the Chinese Communist Party. His name was Mao Zedong.
27. Internal war broke out among various war lords, the Nationalists, and the Communists. Meanwhile, Japan was forging its own empire in East Asia. Communists and Nationalists presented a united front against the Japanese, but upon Japan's defeat in 1945, returned to battling each other.
28. With strong peasant support and expertise in guerilla warfare, Mao and his Communists were able to thwart the inept and corrupt Kuomintang forces. Chiang Kai-shek and his supporters sought refuge in Taiwan. Mao Zedong proclaimed the People's Republic of China in September 1949.

Terms to Know

China Frontier
China Proper
artesian basin
aquifer
karez
loess
Sinicize
Pan-gu
Yi jing
oracle bones
Pinyin
well-field system
Confucius
The Analects of Confucius
filial piety
Mandate of Heaven
dynasty
dynastic cycle
Laozi (Lao Tze)
Dao de Jing
yin
yang
Feng shui
ch'i
geomancers
foot binding
Genghis Khan
Kublai Khan
kamikaze
Pax Mongolica
extraterritoriality
Taiping Rebellion
Sun Yat-sen
Mao Zedong
Chiang Kai-shek
Manchukuo
Long March

Bibliography

Ballas, Donald. 1968. An Introduction to the Historical Geography of Han China. *The Professional Geographer,* 22: 155–62.

Blunden, Caroline, and Mark Elvin. 1983. *Cultural Atlas of China.* New York: Equinox.

Cotterell, Arthur. 1988. *China: A Cultural History.* New York: Penguin.

Gernet, Jacques. 1983. *A History of Chinese Civilization,* 2nd edition. Trans. J. R. Foster and C. Hartman. New York: Cambridge.

Levathes, Louise. 1994. *When China Ruled the Seas.* New York: Oxford.

Levy, Howard. 1966. *Chinese Footbinding: A History of a Curious Erotic Custom.* New York: W. Rawls.

Lip, Evelyn. 1989. *Feng Shui: A Layman's Guide to Chinese Geomancy.* Union City, CA: Heian.

Murphey, Rhoads, 1996. *A History of Asia,* 2nd edition. New York: HarperCollins.

Nemeth, David. 1995. Feng Shui. *Asian American Encyclopedia,* 2: 414–16.

Ronan, Colin, and Joseph Needham. 1981. *The Shorter Science and Civilization in China:* 2. (Abridged). New York: Cambridge.

Spence, Johnathan D. 1993. Confucius. *Wilson Quarterly,* 17: 30–38.

———. 1999. Paradise Lost. *Far Eastern Economic Review,* 15 (April): 40–45.

Temple, Robert. 1986. *The Genius of China: 3,000 years of Science, Discovery, and Invention.* New York: Simon and Schuster.

Waldron, Arthur. 1991. *The Great Wall of China: From History to Myth.* Cambridge: University Press.

Wills, John, E. Jr. 1994. *Mountain of Fame: Portraits in Chinese History.* Princeton: University Press.

Chapter 10

China: Great Dragon Rising

"The comrades must be helped to preserve the style of plain living and hard struggle."

(Mao Zedong 1949)

"To get rich is to be glorious."

(Deng Xiaoping 1980)

These contradictory quotes illustrate the political and economic transition that China has made in the latter half of the twentieth century. A grim situation faced the new government in 1949. The economy was severely disrupted; 40 percent of the arable land was flooded; and industrial and food production were well below prewar levels. In this chapter I will tell you about China's rollercoaster ride from the strains of rural-based communism to the relative prosperity experienced under the current policy of market socialism.

Agrarian and Industrial Revolution

In 1950, the government abolished the "land ownership of feudal exploitation" and redistributed land to landless peasants. Two years later, 300 million peasants owned a sixth of an acre apiece and thousands of landlords had been eliminated. The following year witnessed **collectivization**, which put an end to land ownership.

Collectivization involved several stages. Farmers first were organized into *mutual-aid teams*, pooling their implements and helping one another during planting and harvest seasons. The second stage involved *producers' cooperatives*, which pooled implements, labor, and land, although theoretically retaining individual land ownership. The third stage was the *fully socialized cooperative*, modeled after the Soviet collective farm in which all members collectively owned the land. By 1957, there were close to 800,000 cooperative farms, each averaging 160 families, or 600 to 700 persons.

Large industry was seen as the foundation of socialist society. Plans were made for the building of hundreds of industrial projects, many with Russian assistance. Spectacular advances were made in both industrial and agricultural production, and by 1962, the national income had increased by half.

The Foolish Old Man and Dazhai (Tachai)

Landscape modification was a major theme in creating a socialist landscape. Diking, damming, and terracing have always been part of Chinese land management, but Mao wanted the land to have social utility—to become a cohesive force unifying untold numbers of laborers of the soil. To exhort the masses to struggle against the uncertainties of nature, Mao turned to an ancient fable, "The Foolish Old Man Who Removed the Mountains."

The Foolish Old Man of North Mountain had a south-facing house that was obstructed by two mountains. He and his sons began to dig away these mountains but were observed by the Wise Old Man who criticized this silly attempt to complete such an impossible task. The Foolish Old Man responded that when he and his sons were done, their sons and grandsons would continue the work, and subsequent generations would finish it. Moved by this thinking, God sent two angels, who immediately removed the mountains, which became the weights of imperialism and feudalism on the backs of the Chinese people. Like the Foolish Old Man, the CCP was determined to dig up these burdensome mountains.

Dazhai is located in Shanxi Province in the loess lands of northern China. Situated in a region of high relief, subject to both floods and droughts, and devastating erosion, Dazhai had a particularly precarious history. Like other villages, Dazhai went through reorganization through production brigades, mutual aid teams, and communization. Through cooperation, more efficient use of tools and labor, and increased use of organic and chemical fertilizers, the workers transformed Dazhai into a productive unit with sturdy irrigated terraces and dramatically increased yields. The extraordinary independence, perseverance, and self-reliance, even through natural disasters, led Mao to elevate Dazhai as China's model village. The most popular slogan in the 1960s was, "In agriculture, learn from Tachai." As with the Foolish Old Man, never-ending cooperative labor can challenge even the most difficult landscape transformation projects. With the success of Dazhai, the legend of the Foolish Old Man and the story of the amazing accomplishments of Dazhai were combined in a new exhortation: "Dare to think; dare to act!"

THE GREAT LEAP FORWARD

The **Great Leap Forward** was announced in 1958. Cooperatives were reorganized into **People's communes.** There were 26,000 of these, embracing 98 percent of the farm population. Each comprised about 30 cooperatives, totaling approximately 5,000 households or 25,000 people. Communes took on the responsibilities of villages, collecting taxes, operating schools, banks, and so forth. The commune was seen as "the morning sun above the broad horizon of East Asia." However, with unrealistic goals, lack of expertise, ill-founded programs, and general mismanagement, agricultural production actually declined.

Communization was expanded to include urban and industrial regions (Figure 10-1). Collective living, public eating halls, child care centers, and the like served workers who were organized into brigades, battalions, and platoons to raise every conceivable kind of output.

Everyone was to participate in industrialization. Thousands of *backyard steel furnaces* sprung up across the country. Production quotas were raised sky-high, and outputs burgeoned. However, quality was sacrificed for quantity—in 1959, the government admitted that almost a third of the steel made was unfit for industrial use.

THE SINO–SOVIET SPLIT

The problematic circumstances of the Great Leap were compounded by the Sino–Soviet split. A formal Moscow–Beijing axis had been established in 1950, and

Figure 10-1

This is Hohhot, the "Green City." Hohhot has been the capital of Inner Mongolia since 1952. Part of Mao's decentralization and industrialization plans, the city now has more than a million residents. These are workers' apartments. Chimneys attest to the presence of industry, mostly iron and steel and related products. Hohhot is one of China's most polluted cities. Hohhot is dominated by ethnic Chinese. Why do you think this is the case?

thousands of scientists, technicians, and advisors had swarmed into the PRC. In 1957, the Soviets agreed to help China in its nuclear development. It sent them a heavy-water-type reactor the following year. But by the end of the decade, strains were evident in the relationship.

The Sino–Soviet pact, once described as "lasting, unbreakable, and invincible," degenerated into a bitter ideological and territorial dispute. While Mao counted on the rural masses as the basis for communist transformation, Soviet Marxists scorned their potential, stressing the importance of the urban proletariat in organizing the revolution. Disagreement over the path of international communism meant that Mao was challenging Soviet hegemony over the movement.

Compounding the problems of ideological disputes was contention over the 4,150 mile (6,682 km) Sino–Soviet border. Nineteenth-century treaties had divested China of significant chunks of territory, including lands in southern Siberia and Central Asia. In 1963, China demanded that the Soviets recognize that their frontier was a product of unequal treaties. Military clashes ensued along the border of the Amur and Ussuri Rivers.

These policy and territorial conflicts led to the withdrawal of Soviet technicians—along with tools, supplies, and blueprints—in 1959. China, determined to go it alone, successfully completed the major construction projects. In 1964, China exploded its first atomic device in Xinjiang. However, all was not well. Bad weather, poor harvests, and the departure of the Soviets created severe economic dislocations.

Overall, the Great Leap Forward was disastrous. Normal market mechanisms were disrupted, distribution systems became dysfunctional, and products were shoddy and useless. Most serious was the fact that famine and disease swept the countryside, resulting in anywhere from 14 million to 26 million deaths between 1959 and 1961.

The Great Proletarian Cultural Revolution

The Great Leap, communes, and the Sino–Soviet rift fomented dissention in the CCP. Some questioned the wisdom of campaigns stressing "redness" over technological expertise. Others espoused that Soviet aid was essential for continued economic, military, and scientific development. In 1966, ideologies collided in the **Great Proletarian Cultural Revolution**—a giant social, political, and cultural upheaval.

Pure communization in agriculture was altered, with a policy of *Three Privates and One Guarantee*. This allowed peasants to cultivate their own plots (about 5 percent of the arable land), operate private handicraft enterprises, and sell their products at rural free markets. The *One Guarantee* called for the fulfillment of agricultural quotas set by the government.

Worried about the lack of revolutionary zeal within the party apparatus, among young people, and within the educational system, Mao had ordered a Socialist Education Movement in 1962. Suspected anti-Maoists were "sent downward" to the countryside to experience peasant living and working conditions. This was known as the ***hsia-fang* (downward transfer) movement.** Thousands of cadres and intellectuals were "rehabilitated" in this manner. In addition, millions of urban-based middle-school students were dispersed to "strike roots firmly" in the countryside. By May 1975, nearly 10 million youths had been resettled in rural areas.

Cultural revolutionary groups were established at all levels to promote Mao's mass line and class struggle. The youthful Red Guards saw themselves as true revolutionaries, dedicated to the obliteration of old thought, customs, and habits. Knowledge was deemed the source of reactionary and bourgeois thought and action. Eliminating their perceived enemies, they rampaged cities, ransacked private property, and attacked anything remotely modern. Ultimately, Mao employed the army to restore order.

Although the Maoists proclaimed the Cultural Revolution a great victory, it actually ushered in a decade of turmoil and civil strife that drove the country to the edge of bankruptcy. The CCP was decimated through purges. Members were killed or imprisoned. Industrial and agricultural production suffered major setbacks, and disruptions in education reduced the availability of trained manpower. It is estimated that nearly 100 million people were targeted during the Cultural Revolution. Tens of thousands were killed. Added to this was a series of natural disasters. The Huang He flooded seven times, and a major earthquake wiped out Tangshan and killed some 470,000 people.

Mao Zedong's death in 1976 set off a scramble for succession. Purges, including that of Mao's wife Jiang Qing and her associates, culminated in 1977 with the installment of **Deng Xiaoping** as party chairman. Deng, a colleague of Mao and a survivor of the Long March, had been purged twice during the Cultural Revolution. His "rehabilitation" ushered in a new era for the PRC. But the China inherited by Deng exhibited widespread changes in light of agricultural and industrial policies pursued since 1949.

Altered Spatial Patterns

Spatial change was evident in both rural and urban landscapes. Communes, although reduced in size, had transformed the face of agriculture. Significantly, private plots were producing the bulk of vegetables and pork. Rice acreage had been dramatically increased, acknowledging rice's significance as China's most important crop of production and consumption (Figure 10-2). But, as peasants devoted more energy to private plots, rice had begun to decline in favor of more profitable cash crops (Figure 10-3).

Industrial expansion resulted in changing spatial patterns in the distribution of both industry and infrastructure. From the beginning the government sought to decentralize industry to the interior. At that time, egalitarianism in regional development was deemed more important than efficiency. In 1949, most industrial centers were concentrated in the northeast from southern Manchuria through Shanghai. Subsequently, other centers were developed in line with new resource finds in such key items as coal and oil and the share of industrial output increased in the interior provinces. China's single oil well at Yumen became only one of hundreds across the nation. Xinjiang became a cornerstone of natural resource exploitation (Figure 10-4).

Railroads were extended into the interior. No longer was the network largely concentrated in northern and northeastern China. New links were established between north and south and extensions were built to the northwest and southwest. Railroads, although built for strategic reasons, were critical facilitators of population redistribution in concert with the *hsia-fang* program and industrialization strategies. They were also important channels of rural–urban migration streams deriving from the dislocations of the Great Leap Forward. Lanzhou, a city of 190,000 people in 1949, mushroomed to nearly 2 million by 1960. Urumchi, at 80,000 in 1949, burgeoned to 200,000. Towns became cities, and even new cities appeared on the landscape.

Although Chinese definitions of *city* have been inconsistent, the number of designated cities increased

Figure 10-2

Threshing rice on communal land. Rice and other grain quotas are set by the government. Farmers work in groups to fulfill these quotas before spending time on their private plots. This threshing machine is operated by a foot pedal. The harvesting operation involves cutting the rice stalks; threshing to remove the grain; bundling the stalks for animal feed and other uses; sacking the rice kernels; and carrying the sacked rice to the road for truck pickup to a rice mill.

from 132 in 1949 to 188 in 1976. Within this framework of flexible definitions, the percentage of urban population changed from 5 percent in 1949 to 8 percent in 1976. Between 1953 and 1961, 39 new cities were established in context of industrialization and decentralization. This trend toward urbanization has continued—in 1990 there were 300 more Chinese cities than in 1953 (Figure 10-5). Fifteen percent of these were rootless, in that they had no historic foundation and were built at or near sites of rich natural resources.

The 1990 census indicates that China is one-quarter urbanized. It also reveals the fact that although enormous urban growth has occurred in the past half century, the broad spatial pattern has not changed dramatically. More than 90 percent of the population continues to reside in the eastern half of the country.

Under Deng Xiaoping's platform of reform and economic liberalization, China has become an economic and political force to be reckoned with on a global scale. China is now one of the world's top ten economies. Whereas Mao Zedong decried foreign interference and

Figure 10-3

In this privately owned market garden outside the city of Guilin, women are pouring "night soil" on their vegetables. Night soil is a mixture of human waste and water. Although the use of chemicals has increased dramatically, night soil use remains common in many places. Never eat unwashed vegetables in China.

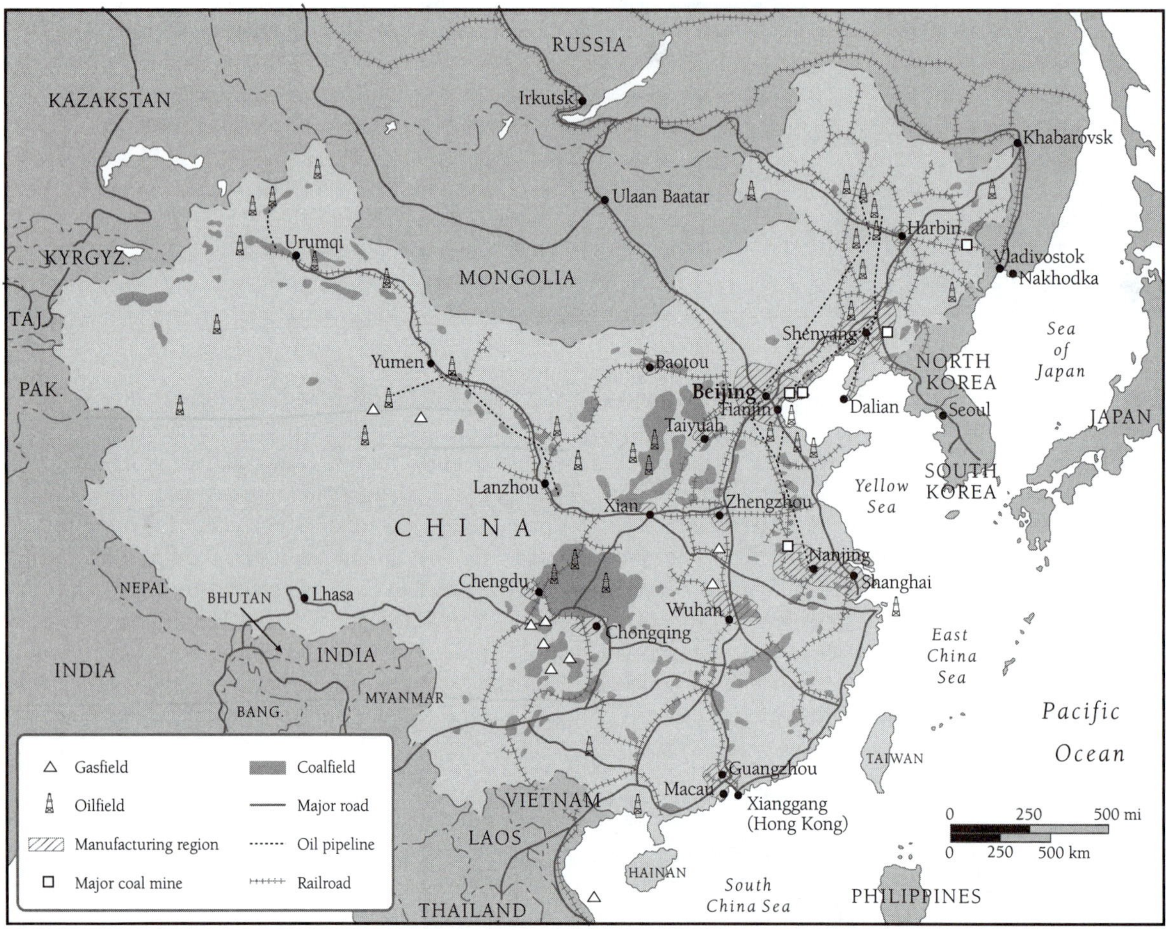

Figure 10-4

Note how surface communications are concentrated in China Proper. Why is China Frontier so devoid of road and rail transportation?

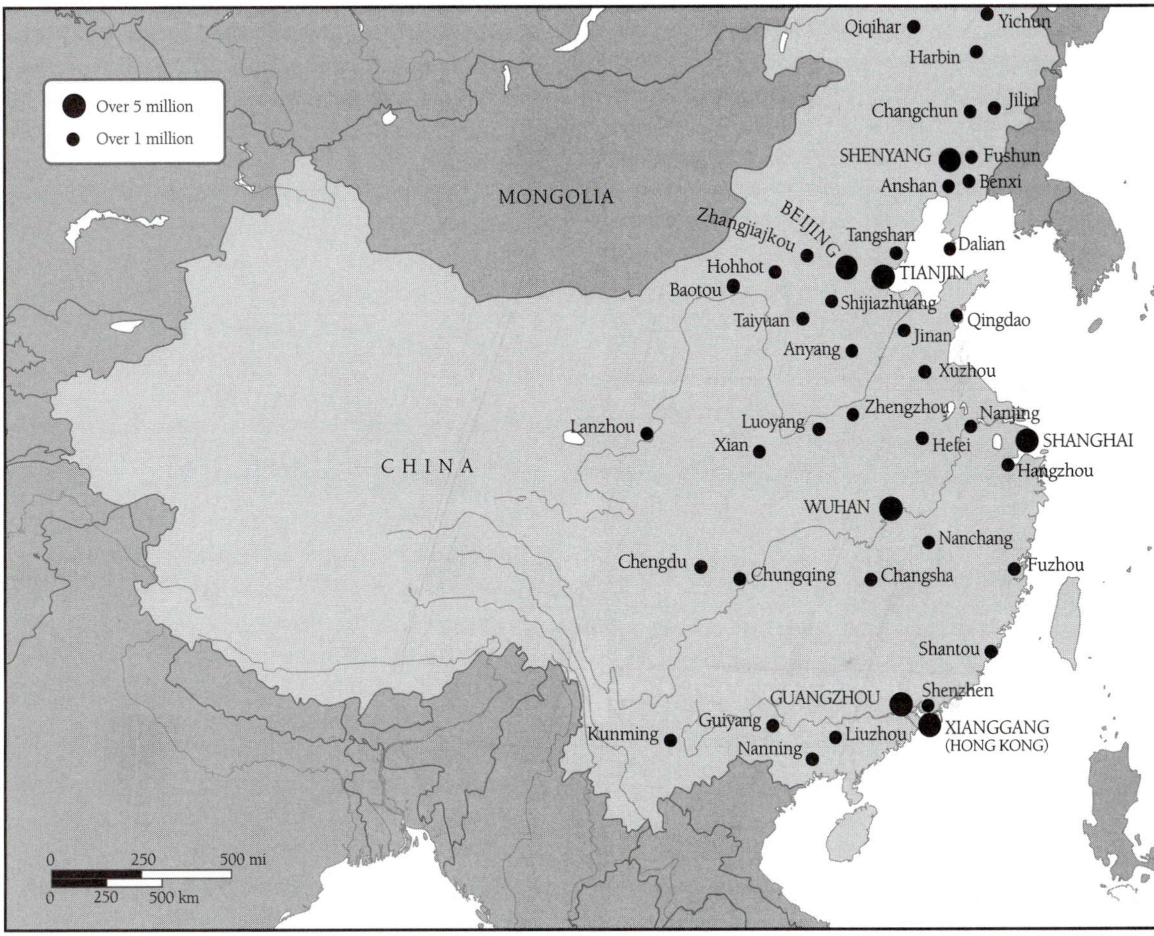

Figure 10-5

Compare this map with Figure 10-4. How many over a million cities does China have?

stressed "self-reliance," Deng Xiaoping pursued more pragmatic, flexible policies and opened the gates to the outside world. Deng said, "I don't care whether the cat is black or white as long as it catches mice." Foreign investment and even capitalism would propel China into the twenty-first century.

Emerging Modern China

ADMINISTRATIVE FRAMEWORK

Figure 10-6 shows that the PRC has four categories of administrative structure. There are 22 provinces; 5 autonomous regions (ARs); 4 central government controlled municipalities known as *Shi's;* and 1 special administrative region (SAR)—Xianggang (Hong Kong).

Autonomous regions were created to recognize the existence of non-Han peoples. There are 55 officially recognized minority groups in China (Figure 10-7). Although minorities constitute only a small percentage of China's overall population, they occupy large regions such as Xinjiang and Nei Mongol ARs. China's population is 94 percent Han Chinese. Minorities make up a mere 6 percent, yet they are spread over 60 percent of China's territory, mostly concentrated in the international border regions (Figure 10-8).

Autonomous regions have a legacy of population transfers in context of Sinicization policies. Han Chinese actually outnumber minorities in some places

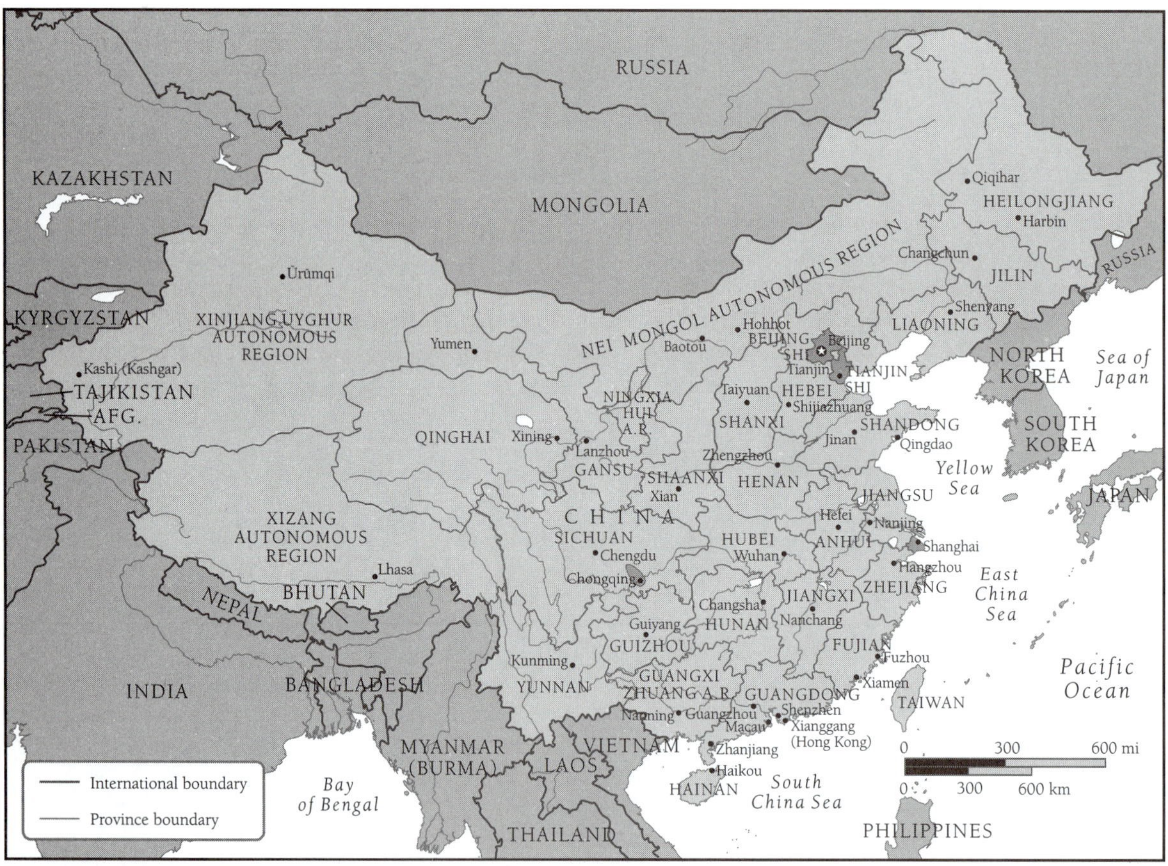

Figure 10-6

Note how China's Autonomous Regions are on the periphery. Why do you think this is the case? Which cities are designated as shi?

Figure 10-7

This Kazakh woman and daughter were spending the summer in the Tien Shan where they were tending animal herds. Their way of life is being altered as the Chinese government is promoting tourism in this region. I asked the daughter what she was going to do with her life. Her eyes filled with tears as she sadly proclaimed, "I don't know what will happen to me."

such as Nei Mongol AR (Inner Mongolia). Only the larger minority groups such as the Uyger and Chuang have their own ARs. Numerous other non-Mandarin speakers such as the Hakka, the Min, and the Yue (Cantonese) live in the provinces, especially in southern China.

The four *Shi's* are the capital, Beijing; its port, Tianjin; the largest city, Shanghai; and, the Chang/Yangzi river port of Chongqing. When Chongqing was designated a *Shi* in 1996, its municipal boundaries were extended to incorporate a huge hinterland totaling 30 million people! In reality, the central urban area has around 4 million inhabitants. However, because Chongqing *Shi* is separate from the provincial administrative area, its creation actually reduced the population of Sichuan Province.

Figure 10-8

This man is clearly not Chinese. In fact, he is a Uygur and speaks a Turkic language. Uygurs are one of China's largest minority groups and dominate the Xinjiang Uygur Autonomous Region. I took this photo at the Sunday market in Kashgar.

Xinjiang: New Frontier

Occupying one-sixth of China, the Xinjiang Uyghur Autonomous Region is the country's largest administrative area. Divided into a series of basins and ranges (Chapter 9), it is home to more than 40 different ethnic groups. The Uyghurs are a Muslim, Turkic-speaking people who constitute almost half of Xinjiang's population. Other groups include the Mongols, Hui (Chinese Muslims), Kazakhs, Tajiks, Tatars, Russians, and Xibo.

Due to Sinicization efforts, Han Chinese now make up one-third of its population. In 1949, the Han numbered 200,000. In 1993, there were 6 million Han in a population of 16 million. About 250,000 to 300,000 Han enter the region yearly. Sinicization is important to keep any notions of separatism at bay. Such movements on the part of the Uygurs in the 1990s have been ruthlessly crushed.

In 1992, the capital of Urumchi was decreed a port and given tax incentives like other port cities such as Shanghai. This is interesting because Urumchi is one of the most landlocked cities in the world. However, joint-venture economic development makes sense in light of the fact that Xinjiang gained three new nearby trading partners with the collapse of the Soviet Union: Kazakhstan, Kyrgyzstan, and Tajikistan. Joint markets and factories have emerged along these borders.

Because of its dry climate, Xinjiang relies almost entirely on irrigation for cultivation. Both spring and summer wheat are grown, along with corn, rice, cotton, millet, and *kaoliang* (sorghum). Xinjiang is also noted for its Hami melons, seedless Turfan grapes, and Ili apples. Sheep and horses are the principle livestock. While Han Chinese tend to dominate business, minority peoples dominate agriculture.

Unfortunately, the Chinese government has deemed nomadic ways of life politically impractical and economically inefficient. Consequently, it has made every effort to sedentarize groups such as the nomadic Kazakhs, many of whom have resisted (Figure 10-9). Provision of electricity is one of the tactics employed, as people are more likely to stay settled in an electrified, permanent dwelling. Only by settling down can the nomads "shake off poverty." However, since 1996, the government is focusing more on tourism to engage the Kazakhs in the regular economy. Increasing levels of cultivated land, encroachment by Han farmers, tourism, and other entrepreneurial activities all ensure the decrease in opportunity to pursue a nomadic lifestyle.

Similar to schemes of the eighteenth century, in the 1950s, soldiers organized the Production and Construction Corps. Continuing to the present day, this structure of state farms runs an empire of 2.2 million mostly Han Chinese with direct loyalties to Beijing. State farms are complete entities with their own factories, oil refineries, processing plants, and marketing entities. Many of the workers are under contract.

Xinjiang is rich in natural resources, especially in the Junggar and Tarim Basins with their wealth of oil. The Junggar Basin is of particular significance because of its rail links to Kazakhstan and Central Asia's wealth of oil and gas. Pipelines between the two regions are already under construction.

Xinjiang is a potentially strategic region especially in light of its new connections with the Central Asian Republics. Pan-Islamic movements threaten stability, and historic Russian interests in this region always weigh in the background. This is also the heartland of China's space program and nuclear weapons development. This is one of many regions of China worth keeping an eye on.

THE FOUR MODERNIZATIONS

The year 1978 was a pivotal point in China's experience, for at that time the leadership called upon the people to achieve the **Four Modernizations.** Agricul-

Figure 10-9

The Kazakhs practice transhumance, dividing their time between the alpine slopes of the Tien Shan in summer and the lowlands near Urumchi in winter. Their summer homes are yurts made of wood, felt, and canvas. Yurts are divided into male and female sections with a common area in the middle.

ture, industry, education, science, and defense were to be developed through readjustment, reform, and improvement. The proposed alterations implied dramatic change. In agriculture, collectivization was to be practically eliminated and top-down planning was to be decentralized. State control of capital and industry was to be reduced. Self-reliance would sway in the wake of an infusion of foreign investment and expertise under an *open-door policy*. Spatial change was imminent.

Agricultural China

Overall crop patterns have remained essentially the same (Figure 10-10) but within this larger framework, landscapes are being notably transformed. Now, food needs are not only aligned with an increasing population but also with heightened consumption, including more meat, eggs, and poultry.

Large-scale commodity production has become the essence of Chinese agriculture. Double-cropping (two crops a year) is practiced almost everywhere south of Beijing, where rice is the dominant crop. While two rice crops are garnered in the south, in the Yangzi valley rice is alternated with wheat. Wheat dominates the north. On the North China Plain, where soils do not retain water well, wheat is grown through the winter. It is replaced by cotton, maize, millet, tobacco, or sweet potatoes in the summer. In the north and northeast, it is too cold for winter wheat so maize and millet are single-cropped. Vegetables are largely concentrated in market gardens around settlements.

Crop yields are higher in the south than in the north but in both areas diminish westward. Nevertheless, some eastern provinces such as Shandong, Hebei, and Henan, have made spectacular gains of as much as 20 percent under the new policies.

Chinese Pigs

China has the world's largest agricultural economy and one of the most varied. One of the most significant elements is the pig. In 1955, China had an inventory of 88 million. This number had increased to 331 million by 1985. Nearly 71 million pigs live in Sichuan Province alone, almost three times as many as in the United States hog belt of Illinois, Indiana, Iowa, and Missouri. Today, China has 126 of the 300 breeds of pigs in the world and 51 percent of the world's pig population. Of the world's total

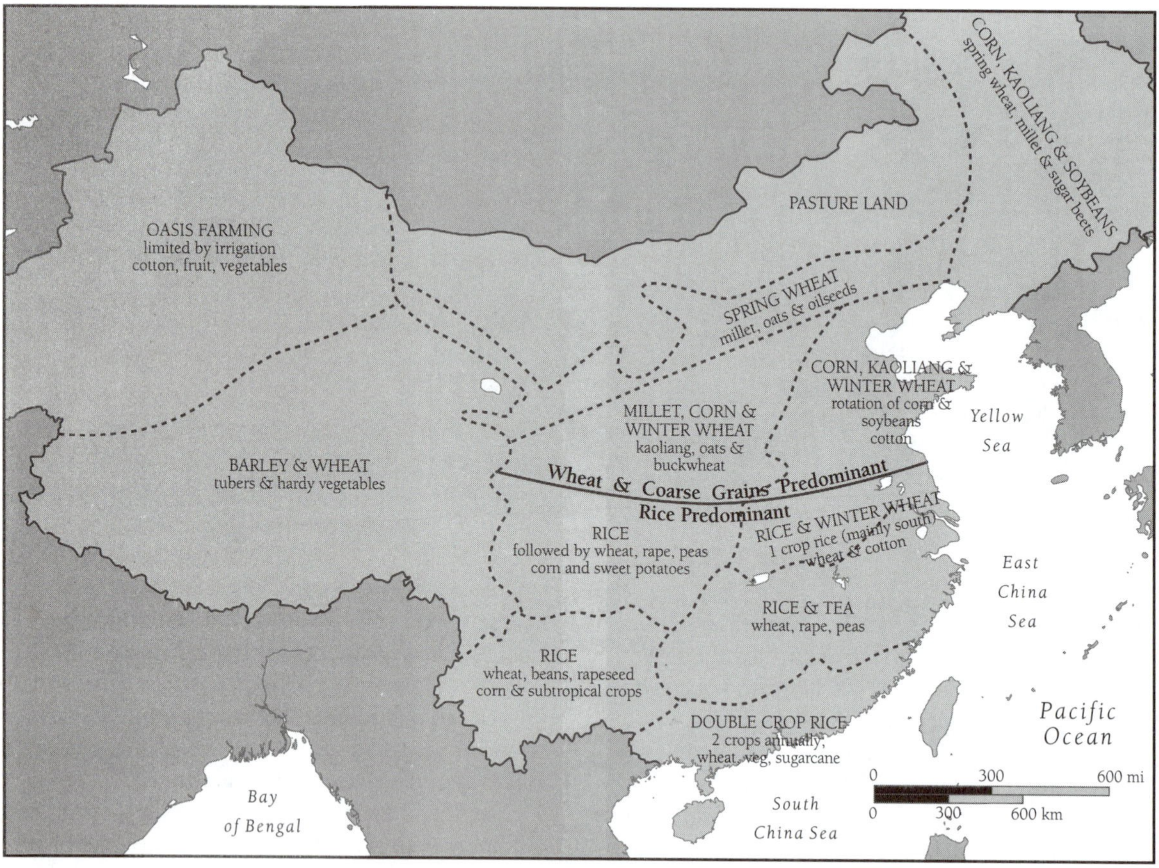

Figure 10-10

This agricultural map of China shows the division between north and south in terms of crops grown and the ability to double-crop. The north grows more wheat while the south grows more rice. The division coincides with the 20 inch isohyet and the Qinling Shan and the Huai He.

meat increase in recent years, China's pork accounts for two-thirds.

China has had a long-standing relationship with the pig's ancestor, the wild boar. Farmers hated the boar for destroying their vegetable crops but admired it for its strength, speed, and the perceived therapeutic value of its body parts. Once the boar was domesticated as the pig, it provided meat, bristles, fertilizer, and countless other by-products. It required no large space and ate anything. Its fast growth and reproductive ability made it a symbol of wealth and prosperity. Even today, you can purchase gilded golden pigs to bring prosperity to your home or office.

Before 200 B.C., social stratification defined what one could eat. Pork could be eaten by the emperor and lower ministers. Others, such as feudal lords, state ministers, generals, and commoners were required to eat other meats. Pigs also played a role in life and death rituals, and pig effigies and bones were often included in burials. Pigs were also thought to be efficient guardians of children.

Archaeological studies at Neolithic burial sites in Shandong show that intensive pig production was critical not only to the human diet, but also for ritual, and the display of wealth and inequality in the rise of political elites. Pigs functioned as means of subsistence as well as objects of social production—ceremonies, displays of status, feasts, exchanges, and so forth. In pre-history, pigs were important as beings as well as pork.

Although the most salient product in China's livestock sector, the family farm pig could decline in importance because of constraints in available feed grain. Furthermore, increasing numbers of commercial feed-lot operations are appearing in urban hinterlands.

Agriculture is rooted in the **household responsibility system,** representing a return to family farming. Under the reforms, families are able to lease land for up to 30 years. This land may also be inherited. In return, farmers must fulfill state quotas in grain or industrial crops. Surpluses can be used or sold to the state or free markets for even higher prices. Such incentives have increased crop yields and rendered labor more efficient and productive. Many farmers have become relatively rich and are constructing new and larger homes (Figure 10-11). Countless farmers have gone into rural business enterprises such as pig farming, fisheries, or small factories.

Figure 10-11

In the era of market socialism, numerous people have accumulated new wealth and have put it on display. This is the home of a farmer-businessman on the fringes of Guangzhou. It shows a lot of traditional symbolism, things that would have been forbidden a decade ago. The dragon stands for power and change; the phoenix for virtue and grace; the clouds for fortune and happiness; and the pearl for purity.

THE *HUKOU* SYSTEM

Socioeconomic position and opportunities for mobility are controlled by a household registration system known as ***hukou.*** Every Chinese citizen is registered as nonagricultural (mostly urban) or agricultural (mostly rural). *Nonagricultural* persons are heavily subsidized by the state in grain rations, housing, medical care, and so forth. Registration includes one's particular geographic location, and it is very difficult to alter one's classification. *Hukou* denotes one's identity and place in society.

Until the 1980s, the system rooted people to their birthplace and allowed little movement elsewhere. A migrant was excluded from most desirable jobs and subsidized housing and benefits elsewhere. It is extremely difficult to change one's *hukou*, especially from agricultural to nonagricultural and to thereby shift location from rural to urban. Nonagricultural *hukou* are considered superior to agricultural *hukou*.

Since market reforms have produced a sizable labor surplus, the state has allowed temporary migration to towns and cities, although permanent change of status remains strictly controlled.

Geographers Cindy Fan and Youqin Huang (1998) have discovered a trend of **marriage migration.** Many women in poorer areas employ kinship and friend networks as well as marriage brokers to marry someone located in a more desirable area. Women gain access to better economic opportunities, and socially or otherwise disadvantaged men are able to get brides. Marriage migration streams have been demonstrated from poorer southwestern provinces such as Yunnan, Gweichou, Sichuan, and Guangxi to the more affluent coastal provinces of Guangdong, Hebei, Jiangsu, Zhejiang, Fujien, and Anhui. Brides then take on the *hukou* of their spouse and are able to participate in the local economy.

TOWNSHIP AND VILLAGE ENTERPRISES

Rural industry is springing up around cities and towns, and it is estimated that 60 to 100 million have left the land for better pay in small industries around towns and along transportation lines. Others work in industry on a part-time basis, as in Korea and Japan. Farm types, based on income sources, can be categorized as traditional agricultural, mixed activity, and industrially focused.

These trends illustrate the formal policy of "Leave the land but not the township or village." This results in a pattern of partially urbanized rural people who engage in nonfarm sector employment without migrating to cities. In the 1990s, rural industry contributed more than 40 percent of rural total output value. Agriculture contributed 45 percent. Moreover, the growth in nonfarm activities was significantly greater than farm ones.

Systemic changes and population shifts have also resulted in millions being displaced, and many of these individuals have joined the ranks of the urban unemployed. A sizable **floating population** moves from city to city seeking work on new construction projects. This population is now estimated at 120 million.

Free-flying industrial development without the building or maintenance of rural infrastructure has induced serious problems. Accessible lowland is becoming scarce in the east because of demands of state and city occupance, peasant house-construction, soil erosion, and consumption by local industry. Farmland in the east has been lost at the rate of 6,700 hectares annually since 1979. Another problem is the return to fragmentation of land with family farming. Pollution, from increased use of chemical fertilizers and pesticides, as well as the accumulation of refuse, especially plastics, has mushroomed in small towns and villages in association with new economic enterprises. Increasing infringement on farm space is a critical issue in the question of feeding China's burgeoning population.

Feeding China

Up to the mid-1990s, China managed to feed almost a quarter of the world's population on about 7 percent of the world's agricultural land. Since then, grain imports have become increasingly significant, leading agriculture experts to question China's ability to feed itself as well as the world's role in feeding China.

Lester R. Brown (1994) of World Watch points out that China's population is growing by 14 million a year, that incomes are dramatically rising, and that the consumption of meat and eggs is surging. In 1978, only 7 percent of grain was being used for livestock feed. By 1990 that figure had risen to 20 percent, mostly for pork. Vast amounts of grain are now being directed to poultry and fish farms and increasingly to beer factories.

The country's capacity to produce food is shrinking. Cropland has been sacrificed to modernization and urbanization. Application of fertilizers has reached the point of diminishing returns. Both rice yields and wheat yields are starting to level off. Diversion of water to nonfarm uses has generated water shortages, and expansion of irrigated areas cannot keep pace with ballooning demand. Aquifers are being depleted. Now more than 300 Chinese cities are short of water. For example, the aquifer under Beijing has fallen from 16 feet (5 m) below the surface in 1950 to 164 feet (50 m) below in 1993. Waterlogging and salting are also reducing productivity in irrigated areas. Brown calculates that China's grain production will fall by one-fifth a year until 2030, with a resulting huge grain deficit. China will then require imports greater than the world can produce.

Other studies are not as pessimistic. Chinese experts criticize the fact that Brown's report assumes Chinese consumption patterns equal to those of Europe. They also point out that the production level of nearly two-thirds of China's arable land is medium or low, leaving room for significant yield increases. Chinese authorities expect, with technological adjustments in such areas as dryland farming, to improve yields by more than half over the next three decades. They also note that importing some 3 percent of its total grain consumption is quite normal for a country the size of China.

The International Food Policy Research Institute concludes that while China's grain imports will continue to increase, domestic production will also rise. Grain imports should level off by 2020 and even in a high population growth scenario, demands will not create a global food shortage.

Urban and Regional Development

Since 1978, China's development policy has stressed efficiency over equity and an open door instead of self-reliance. However, in order to defray the impact of radical change on the political system, Deng Xiaoping focused his new economic policies on the country's Pacific Rim. The plan incorporated a system of **special economic zones (SEZs)** designed to attract technologies and foreign investment from abroad.

SPECIAL ECONOMIC ZONES

SEZs are founded on a variety of incentives such as relaxed import and export regulations and low taxes. Land can be leased and labor can be contracted. Goods produced can be sold abroad and even within China. Investors can reap profits at home.

The SEZs include a series of *open coastal areas* in addition to a number of *open cities.* Initially, there were four SEZs along the coast (Figure 10-12). Each of these possesses particular locational characteristics. Shenzhen is adjacent to Hong Kong (Xianggang) on the Pearl River estuary in southern Guangdong Province. In the same region is Zhuhai, adjacent to the former Portuguese colony of Macau. Shantou and Xiamen are former treaty ports across from Taiwan. Hainan Island, approaching Southeast Asia, was added in 1988 and Pudong, linked to Shanghai, was added in 1990.

Fourteen additional cities were opened to foreign enterprise in 1984. Most of these were once treaty ports with a history of foreign relations. They also had established industry and transport infrastructures as well as local labor pools. Moreover, all the open cities had connections to overseas Chinese communities.

In 1988, three *open economic regions* were designated as special zones for foreign investment: the Yangzi (Chang Jiang) Delta Economic Region around Shanghai; the Pearl River (Zhujiang) Delta Economic Region around Guangzhou; and the Minnan Delta Economic Region around Xiamen. These represent efforts to diffuse the benefits of an open policy from SEZs to other parts of the country.

More freedom and opportunity for development and profit generated unrest. University students and other dissidents decried the greed and corruption they saw among CCP officials and affiliates taking advantage of the new economic situation. In 1989, a huge protest demonstration took place in Beijing's Tienanmen Square. Deng Xiaoping called in the military and the demonstrations were ruthlessly crushed with a significant loss of life. Whether this incident is interpreted as a "cry for democracy" or a "counterrevolutionary rebellion," it is a black spot on China's modern history.

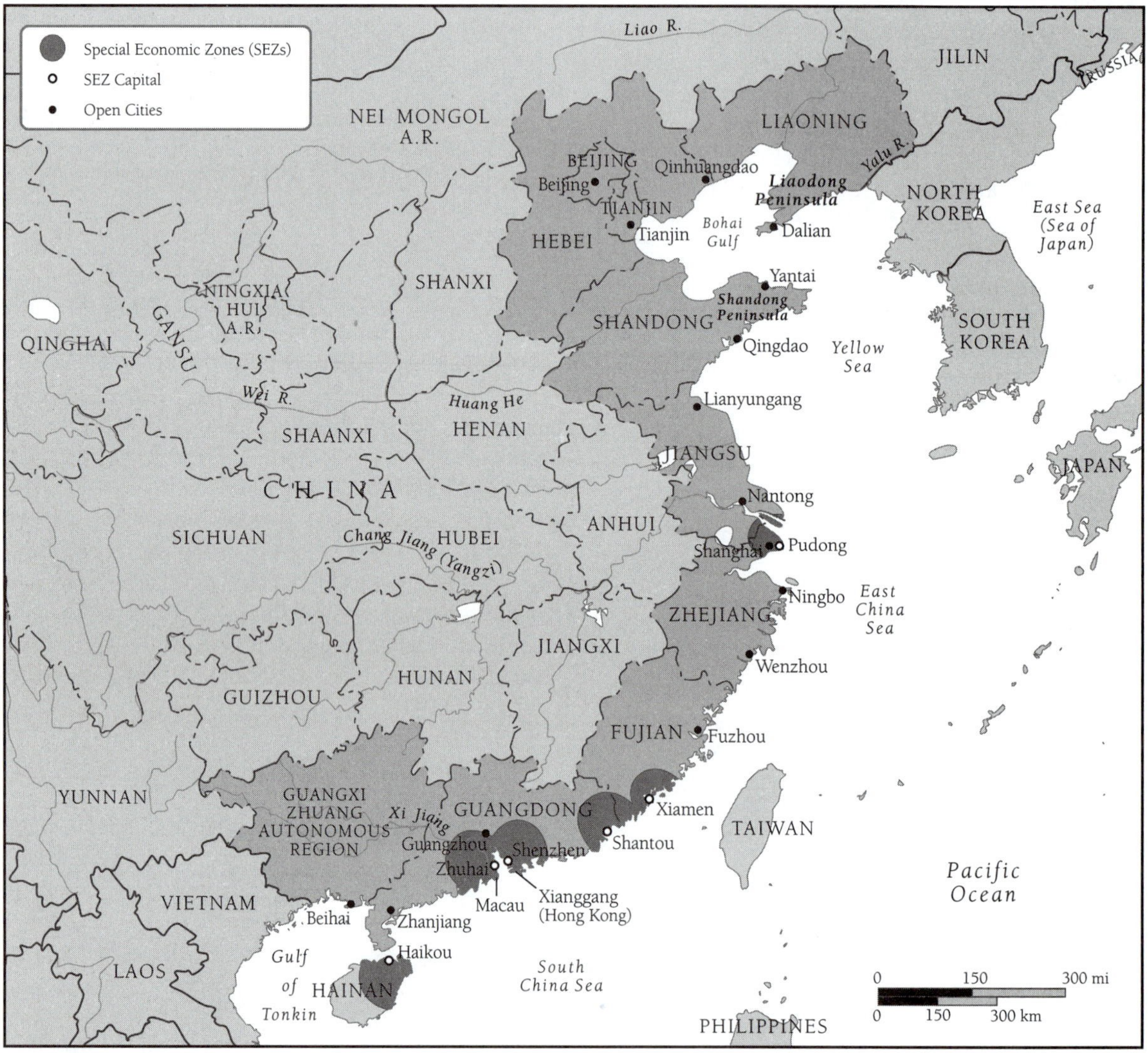

Figure 10-12

Economic zones of China.

FOREIGN INVESTMENT

These policies opened China to the global economy and spurred sharp increases in tourism, imports and exports, and foreign investment. Non-Chinese tourists increased from 0.5 million in 1980 to 1.74 million in 1990—a 248 percent rise. The number of Chinese tourists from Hong Kong, Macau, and Taiwan increased 397 percent to more than 27 million in 1990. Imports and exports have risen dramatically since 1978. Since 1987, there has been a trade surplus because of China's growing exports, especially to Hong Kong and Japan.

Exports make up 80 percent of China's foreign currency earnings, and there has been a shift in the structure of these as light industrial products overtake agricultural products. In 1990, light industrial products made up over half of China's exports and heavy industrial products increased to more than a third.

Initially, foreign investment was concentrated in the south but ultimately spread northward as the 14 cities were opened. A second spatial pattern emerged with the infusion of investment in interior cities, especially along the Chang Jiang and Huang He. Developments in Wuhan and Chongqing have contributed to the Yangzi valley being called *China's Soaring Dragon.* Most of the foreign capital has come from Asian sources, directed into places with Chinese family ties. Most Western investment pours into large cities such as Shanghai, Beijing, and Tianjin.

Shanghai: Head of the Dragon

The "Paris of the East." Asia's most sophisticated city. The world's most crime-ridden city. This was Shanghai before 1949. Today, Shanghai is China's largest city, its foremost economic center, and a rival of Beijing as cultural capital. In fact, Shanghai rivals Hong Kong in importance.

Shanghai thrived long before the arrival of Europeans. It had a population in excess of a million in the Ming Dynasty (1368–1644) and 3 million by 1816. From 1895 to 1945, the city's population increased tenfold. Migrants accounted for a significant portion of this growth because the city was safely distant from most battlefronts and had economic opportunities associated with the multiple foreign concessions. While migrants came from the surrounding provinces, they also came from Guangdong to pursue business enterprise. This connection remains significant in China's new economic landscape.

In 1949, Shanghai accounted for one-quarter of China's industrial production. The city housed a highly educated population and continued to be a migration source in Mao's era. However, when large numbers of people were forcibly moved out of Shanghai to populate China Frontier and as part of the *hsia-fang* movement, from 1958–1976, the city experienced a net out-migration of more than 1.2 million. With its educated population, Shanghai had a tremendous impact on development in the northeast and western regions. With the end of the Cultural Revolution the city once more experienced net in-migration.

Since 1992, when liberal policies were "given" to Shanghai, the city has undergone substantial reform and is in the midst of an economic boom. One advantage Shanghai has is that it is of sufficient size to become the "Dragon Head" of the Huangpu and Yangzi River commercial ribbon. It is already a key commercial center with proximity to both the north and south coasts. It has the largest pool of experts in science, technology, and management, as well as workers.

These advantages led to the 1990 decisions to promote both Shanghai on the west bank of the Huangpu River and Pudong on the east side of the river. Pudong was to be developed and opened to foreign banks, finance, and insurance companies, department stores, and supermarkets. Since 1992, Shanghai's growth rate has exceeded that of all China!

While the Shanghai economy is mainly industrial, the service sector is becoming increasingly important. The tertiary sector, accounting for 30 percent of the industrial product by 1990, is designed to serve the entire nation. Within the industrial sector, there are six pillars of industry: automobiles, iron and steel, petrochemicals, power equipment, telecommunications, and electrical. The Jinqiao Export Processing Zone in Pudong has played an essential role in this development with income multiplying fiftyfold from 1993 to 1994.

The high cost of land and labor in Shanghai has forced it into high-tech and high-value industry. Other kinds of industry are gravitating to the Nanjing and Hangzhou areas, where a new freeway is expected to enhance their competitiveness. A new Ring Road, six-lane elevated highways, and subway lines now facilitate spatial interaction in the web of industry and commerce.

Housing and services are of great concern. Shanghai has five times the density of Beijing with 12,000 persons per sq km (Figure 10-13). The plan is to have 10 square meters per person by 2000. The new housing plan is based on the Singapore Model, whereby the residents own their own housing. An owner must occupy the dwelling for five years before renting or selling it. New houses and apartments are rising everywhere, especially in the former market gardening suburbs (Figure 10-14).

Shanghai is one of the greatest urban renewal stories in history. Nearly 17 billion dollars in foreign investment has transformed the once drab skyline to one of glittering skyscrapers and neon. But all is not perfect. There is a vast influx of "floating" and "temporary" populations. There is a growing income disparity between rich and poor. State-run industrial enterprises are money-losers generating massive layoffs. Government regulation still hinders essential quick decision making. Therefore, market strategy shifts are slow and competitiveness sluggish. Some are looking to Hong Kong as a model for new policy.

Observers use Pudong as an analogy. *Hardware* in Pudong means infrastructure; *software* means services. Both must exist in tandem for Pudong to work. Change is in the offing and as Shanghai becomes more affordable for foreign investors, even greater progress is expected.

Figure 10-13

This is a traditional housing area in Shanghai known as a lilong. *These apartments do not have modern facilities. Bathrooms are in a building down the street. Laundry is hung outside for lack of space inside.*

Figure 10-14

These new apartments in Shanghai illustrate the impact of market socialism on the Chinese landscape. These blocks are gate-guarded and have a small kiosk for the purchase of basic essentials, the delivery of mail, and use of a public telephone. This building is located on former farmland.

NEW CITIES AND RURAL URBANIZATION

One of the most spectacular achievements in the 1980s and 1990s is the increase in the number of cities in China. Policy was relaxed to permit the formation of more cities, and city status became the mechanism to attract foreign investment. The total number of cities rose from 194 in 1978 to 467 in 1990, with a concomitant increase in city population. A relatively large number of small- and medium-size cities were added to the system.

The *household responsibility system* effectively raised agricultural productivity, stimulating growth in rural commercial and manufacturing activities and the revival of market towns. Rural savings are reinvested in rural enterprises, fostering town growth. In some regions, such as the Pearl River Delta, small towns are growing faster than cities. Hovever, it must be noted that rural urbanization is highly concentrated in the eastern coastal region.

NON-PERMANENT POPULATIONS

I have already mentioned the *floating population* displaced by economic, systemic changes. But, there is another group of **temporary residents,** residual farm labor, who obtain permission to stay in a city for a fixed amount of time to work in factories or on construction sites. These individuals make up a quarter of some cities' populations. In Guangzhou they comprise 38 percent, and they outnumber permanent residents in Shenzhen. Most of the temporary people work in the building industry, followed by retailing, especially individual street enterprises (Figure 10-15, p. 268). Temporary populations place great stress on social facilities, transport, and housing.

REGIONAL DISPARITIES

Geographer Cindy Fan (1995) notes that, "The paradox of a post-Mao regional policy is that a policy designed to favor China's most developed eastern region has reduced inequality among all provinces in the nation." This happened because policies favored a number of regions not already highly developed and these, from Shandong to Guangdong, constitute China's *growth corridor.* Excluded from this corridor is the older, relatively stagnant region of the northeast.

Figure 10-15

Selling soap and shampoo as a private enterprise in Shanghai—another example of market socialism on the landscape.

China's Rust Belt

Home of the Manchus, founders of China's last dynasty; a region of historic contention among China, Japan, and Russia; a northeastern extension of the agriculturally productive North China Plain. This is former Manchuria, resource rich and responsible for 25 percent of China's industrial output in the 1970s. Today the northeastern provinces of Heilungjiang, Jilin, and Liaoning are trying to overcome their rust belt status.

In the early years of communism, planners concentrated state-run industries around the coal and iron fields of Shenyang, Fushun, and Anshan. Shenyang became the nation's leading steel making complex. Discovery of the Daqing oilfields west of Harbin rendered China self-sufficient in oil. Directed on giant state farms, agriculture boomed and further illustrated what communist planning could accomplish.

With the new market reforms of the post-Mao era, state-run industries stood as monoliths of obsolescence, unable to cope with privatization and restructuring. Now the northeast contributes a mere 10 percent of China's industrial output.

Nevertheless, there is light on the horizon for former Manchuria. Its resource wealth and its proximity to Russian, Korean, and Japanese ports make it a potential core of development in East Asia. The launching of an economic development zone at Dalian, as part of the Liaoning Economic Development Area, has drawn the attention of foreign investors, especially the Japanese.

In 1994, one-third of Japan's investment in China was pumped into Dalian to manufacture everything from motors to chopsticks. Spatial association, economic incentives, a huge harbor, and a cheap labor pool have drawn more than 800 firms to Dalian.

Although much of the northeast still struggles with its rust belt heritage, some economic geographers see economic potential in light of future possibilities. A reunified Korea or a prosperous Russian Far East could alter northeastern fortunes.

Professor Fan (1995) further observes that neoclassical development theories, while influential in shaping policy, do not satisfactorily explain recent spatial trends in regional development because they downplay the influence of the state. State decisions have shaped development policies and resultant spatial patterns. Improvement of interprovincial equality has occurred and regional imbalances between coastal and interior regions are being redressed in context of new policies encouraging interior investment. Furthermore, studies in Guangdong Province exemplify this significant spatial restructuring at the intra-provincial level.

Guangdong

Guangdong fits the definition of a **regional state.** It is a natural economic zone shaped by the global economy in which it participates. Guangdong Province—and especially the Pearl River Delta's agglomeration of Guangdong, Shenzen, and Zhuhai—has become a place apart from the rest of China. Guangdong's southern location has particular political–economic and sociocultural implications. Proximity to Hong Kong and the incentive of low production costs have transformed the Zhujiang Delta into an outward processing zone. Since many Hong Kong residents once resided in Guangdong, business ties are facilitated. This explains the flows of labor-intensive industries and investment into Guangdong from Hong Kong. Economic structural change has seen a decline in agricultural participation in favor of industrial participation.

Shenzhen has become Guangdong's wealthiest area and the most important channel for foreign investment, accounting for 24 percent of the province's total investment in 1990. Shenzhen was destined to align with Hong Kong in its transition from a British colony to a socialist state. It would be a showcase for exhibiting the concept of "one country, two systems."

The state has continued to expand open zones in Guangdong. More rural and peripheral counties are being drawn into the picture as investors continue to seek lower wages and subcontracting opportunities in context of established social relations and linkages.

These development dynamics have generated new migration streams in Guangdong and elsewhere. Although migration was once strictly controlled by the government, recent economic reforms have created new push and pull forces fostering population movement. The *household responsibility system* with its boom in agricultural productivity has released a surplus labor supply. Foreign investment has instituted a plethora of labor-intensive industries looking to recruit workers. In 1984, a State Council directive allowed rural residents to move to market towns and even larger urban areas, provided they did not require the state-rationed grains to which the nonagricultural population was entitled. The eastern region is now the recipient of migration streams seeking economic opportunity in the industrial production system.

If this growth continues in Guangdong, the Pearl River region will have three cities with more than 5 million inhabitants, five with more than a million, and six with more than half a million. Per capita incomes in the province will be 10 times the national average.

Hong Kong (Xianggang)

Here on 400 square miles (1000 sq km) of fragmented, rugged land live 6 million people, many of them refugees from the mainland. During British colonial rule, Hong Kong was divided into three segments: Hong Kong Island with the capital of Victoria; Kowloon Peninsula and an array of other islands including the largest, Lantau; and the mainland section known as the New Territories. While the Hong Kong island and Kowloon were ceded to Britain in 1841 and 1860, the New Territories were leased by Britain in 1898 for 100 years.

The expiration of that lease in 1997 impelled Britain to negotiate the return of the entire colony to China. Then, Hong Kong, British Crown Colony, became the Xianggang Special Administrative Region, although the name Hong Kong remains in general use. Hong Kong is governed by a mini-constitution known as the Basic Law. Basic Law gives Hong Kong a high degree of autonomy except in matters of foreign policy and defense, which are under the control of Beijing. Beijing is committed to preserving Hong Kong's capitalist system and lifestyle for 50 years. Capital is free flowing in and out of the region; Hong Kong is a member of the

World Trade Organization, the World Bank, and a variety of other economic and trade structures.

As a free port, unencumbered by tariffs or other restrictions, with a magnificent, deepwater harbor, Hong Kong soon became a hub of economic activity in East Asia. It has become one of the world's largest container ports, a giant in banking and finance with its own stock exchange, a major processing and re-exporting center, and the world's tenth largest trading economy (Figure 10-16). However, China's economic policies in adjacent Guangdong Province have changed the structure of Hong Kong's economy.

Figure 10-16

This scene is in the CBD of Hong Kong. It shows the density of land use, orderly traffic, and a giant symbol for the city's spotless metro. Hong Kong is a very crowded city, but several forms of public transportation make getting around quite easy.

When Shenzhen SEZ was established, thousands of factories moved there from Hong Kong to take advantage of cheap land and labor. Workers poured in from the mainland, especially from Guangdong and Guangxi-Zhuang Autonomous Region. The former fishing and duck-farming community grew to nearly 4 million, and Shenzhen took on the appearance of high-rise Hong Kong with corporate offices, hotels, department stores, and gambling casinos.

Geographer Chi Kin Leung (1993) found that the subcontracting activities of Hong Kong were widely distributed in the Zhujiang Delta with Shenzhen and Guangzhou two of the major centers. Leung discovered that existing kinship and established business ties were even more important than proximity in forging these links. Another factor was the efforts of Chinese authorities to attract foreign investment into the Delta. An interdependency has arisen, with Hong Kong relying on low-cost labor to maintain its export competitiveness, and the Delta depending on Hong Kong's technological and marketing expertise to procure overseas markets.

In the 1990s, a quarter of China's foreign trade was funneled through Hong Kong. Today, more than 75 percent of the Hong Kong work force is in the service sector with only 13 percent remaining in manufacturing. More than 5 million Chinese laborers work in Hong Kong–owned factories on the mainland. However, costs have risen in Shenzhen, so many Hong Kong factories are moving elsewhere in Guangdong stimulating the growth of other cities and triggering more intra-regional migration.

An economic downturn in East and Southeast Asia in the late 1990s quelled this rapid regional development. Hong Kong's inflated real estate market collapsed, and service industries such as tourism endured significant losses. For now, Hong Kong's fortunes are unstable, and its status as an "economic tiger" questionable.

Population Issues

In the beginning, Mao Zedong believed that people were his greatest resource, that to reduce population growth was a capitalistic, anti-communist plot. Therefore, initial efforts were to reduce mortality. A program of sending minimally trained "barefoot doctors" into the countryside, coupled with antisterilization and antiabortion policies, forged this goal. Only in the 1970s, after disastrous famines associated with the Great Leap Forward, did consensus shift from the idealistic view of people as producers to the realistic understanding of people as consumers.

In the early 1970s, China's yearly natural increase rate was about 3 percent. Couples were encouraged (often by coercive means) to marry later, have fewer births, and space children farther apart. A more vigorous population control program was pursued after Mao's death in 1976. Families were ordered to have only one child with severe economic and social consequences for failure to follow this rule. By the mid-1980s, China's natural increase rate was reduced to 1.2 percent. That rate has been further reduced to an officially claimed 1.0 percent. Nevertheless, China's population approaches 1.3 billion—the largest population in the world (Figure 10-17)! Figure 10-18 shows population concentrations in relatively humid lowlands.

The one-child policy has been most successful in the cities, where it is rigorously enforced (Figure 10-19). Policies vary in rural areas, depending on local needs. Minorities are permitted to have several children. Consequently, both rural populations and minority populations are growing faster than urban and Han populations.

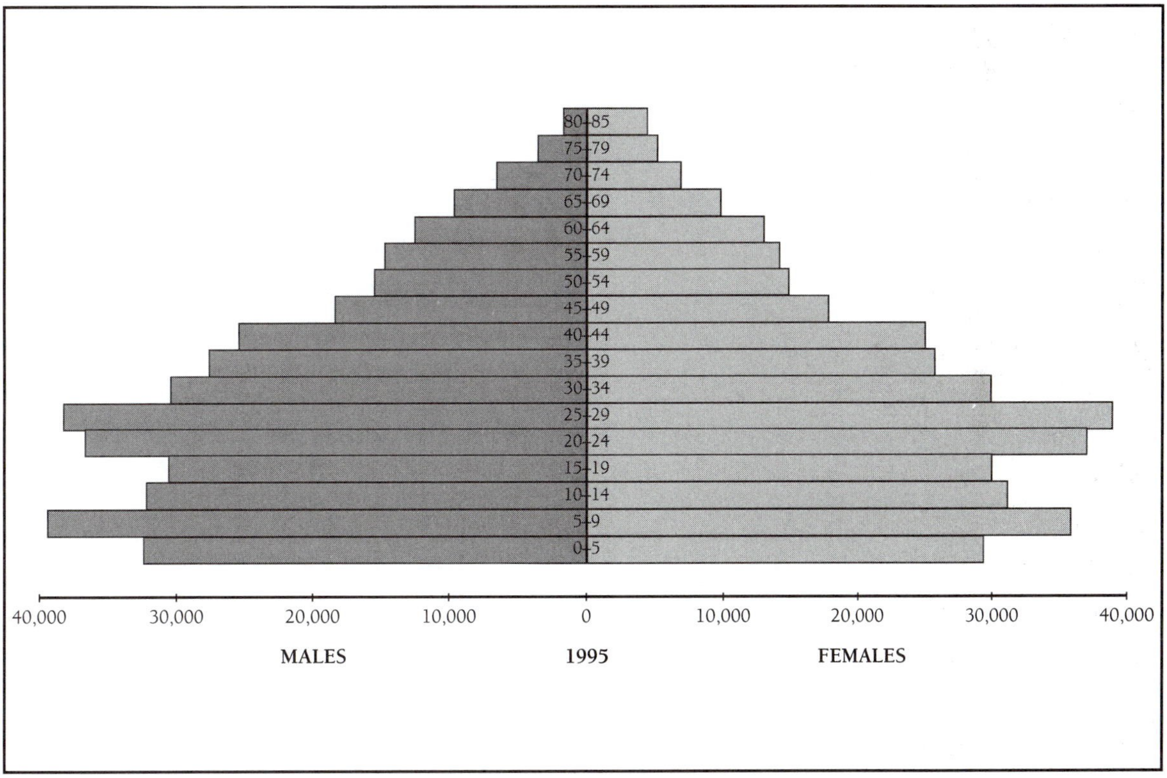

Figure 10-17

This population pyramid reflects the impacts of major events. See if you can select the age cohorts for events such as the Great Leap Forward, The Cultural Revolution, and the one-child policy. How have these events affected population numbers and sex ratios?

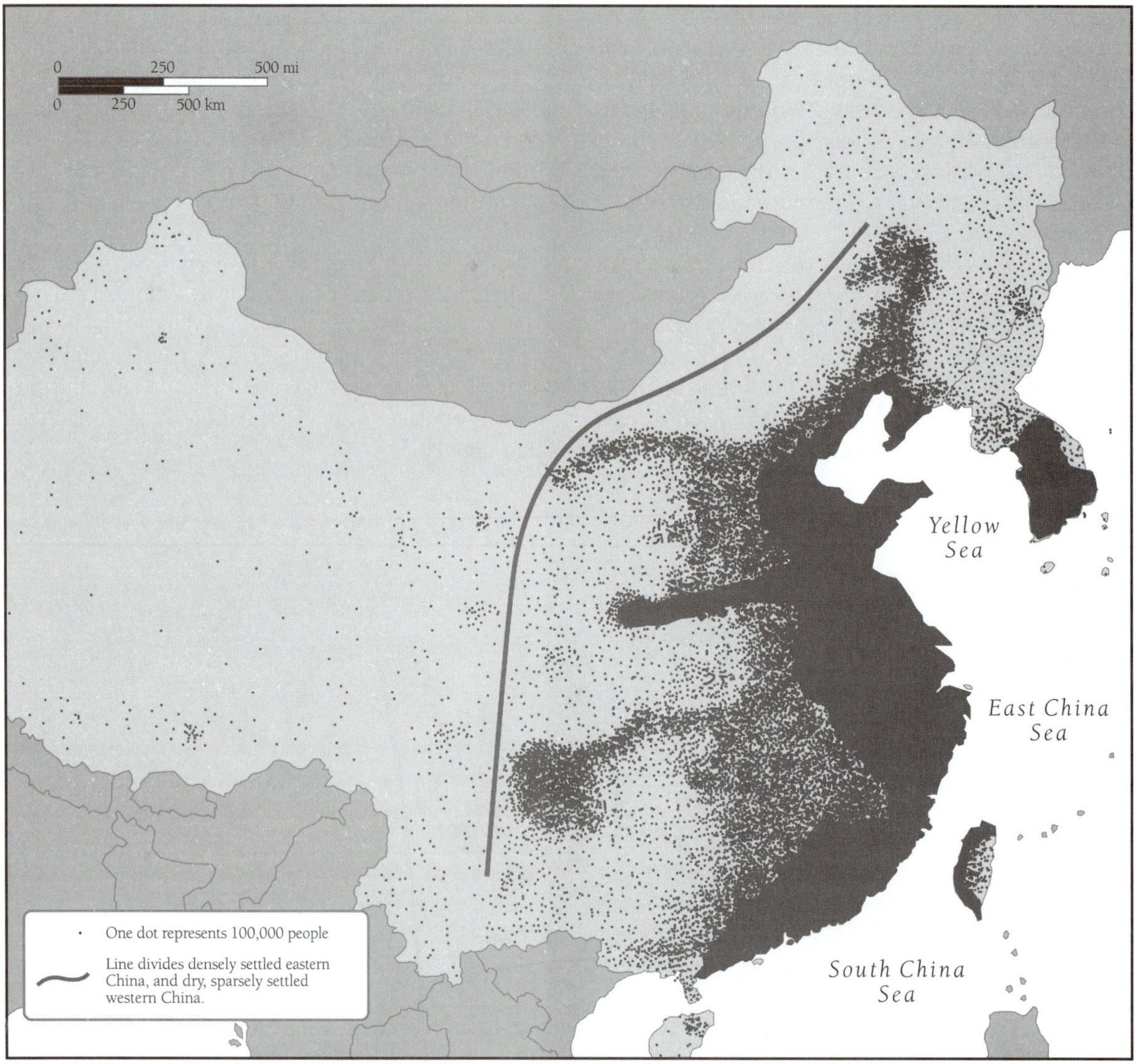

Figure 10-18

China's population distribution. Notice the difference between China Proper and China Frontier. What roles have history and current policies played in shaping this distribution?

China's population policies have engendered a variety of controversies. The preference for boys has fostered sex-selective abortion and female infanticide, especially in the countryside. Moreover, parents tend to hide or undercount their daughters. The discrepancy in the sex-ratio is real, as illustrated by a 1995 sample census that counted 118 boys under the age of five for every 100 girls. Many single children, boys in particular, are indulged to the point of their becoming overweight, spoiled brats called *little emperors*. There is now concern that not enough young people will be available to support the large number of elderly in

Figure 10-19

I met this lady and her daughter in Guangzhou. When I asked to take the photograph, the lady proudly pointed to her child and announced in English: "One child—girl."

future years. Whereas this is traditionally a boy's responsibility, girls are now encouraged to take on the duty of looking after aging parents.

China's population is graying in that the percentage of older people is outpacing that of younger people. By 2025, the number of people in their late twenties is expected to drop by 20 million. Persons over 65 are expected to increase at about 3.5 percent a year accounting for over three-fifths of China's population between now and 2025. This situation begs the questions of who will take care of these people, and how will the country handle a potential manpower shortage?

Women's Work

"Women hold up half the sky." So said Mao Zedong in the 1950s. Women achieved many gains under communism, especially during the Cultural Revolution (1966–1976) when people tried to be gender neutral and participation of women in political and economic activities reached nearly 100 percent. Now, as communism gives way to capitalism, women are falling back into old situations of inequality in line with traditional Confucian and Daoist thinking where women were defined as culturally inferior to men.

In China, dyadic thinking is common and is reflected in the fact that experience is classified in paired concepts that are frequently polar opposites. Furthermore, contrasting concepts often have one that is more highly valued. Gender differences reflect this system.

One fundamental dichotomy is that women are associated with "inside" domains of work and men are associated with "outside" domains of work. Women do "light" work while men do "heavy" work. "Skilled" work is done by men, and "unskilled" work is done by women. The introduction of new technology has exacerbated these divisions. Men dominate labor-saving machinery while women are relegated to more menial, labor-intensive tasks. Today, women work in the fields while many men leave the village to work in a new factory enterprise. Agriculture has been reinvented as lesser "inside" work. Factory work is more important "outside" work. Moreover, women are responsible for family members and interactions within the household but external representations of family business to the outside world falls to males.

China expert Tamara Jacka (1997) found that although male–female work patterns have changed with recent reforms, "expectations embodied in the traditional saying 'Men rule outside, women rule inside' . . . continued to exert an important influence on work patterns." It is true that modernization has given women opportunities to work outside the home. Still, she is entirely responsible for household affairs and so carries a double burden of responsibility. Women are still expected to be mothers and keepers of social stability in the home. Although the content of male and female work is changing, the conceptual divisions remain.

Extra assembly or piece-work done in the household courtyard is also incorporated in the inferior "inside" domain. Commodity production in the household realm has become important, yet women are poorly paid for this work. Even if the "courtyard economy" expands into a larger enterprise, management is typically seized by men.

The government mandates equal opportunity, but things are different in practice. At present, large numbers of women work in township and village enterprises yet they are poorly paid, do the most menial tasks, and are the first fired. Very few are involved in management over male workers. As many as 70 percent of laid-off workers are women, and women are forced to retire at 50, five years earlier than men.

In the SEZs, female workers are desirable because they will work longer for less. Many women are working in deplorable conditions, locked in their unsafe factories at night. Hundreds have died in fires, unable to escape the barricaded structures. Managers find that women, who make up nearly half of the floating population streaming into the coastal regions, especially Guangdong, are subservient and docile, and easy to replace.

Newfound affluence has also revived the notion of women as property. This has led to an upsurge in prostitution, concubinage, and the abduction of women for forced marriages.

Professor Jacka notes that gendered dichotomies have been bolstered over the years by state-sponsored shifts in social and political values. Mao's notion of "What men can do, women can do too," has been rejected in the wake of promoting gender divisions of labor as essential to the success of development and modernization. The critical role of women as mothers and promoters of spiritual civilization is also a central element in modern thinking.

Environmental Dragons

China has a spate of environmental problems, of which it is possible to touch on only a few here. In industrialized China Proper, the air is black with soot and smoke. Only about five of China's hundreds of cities are deemed to have clean air. People in the iron and steel city of Baotou have brittle bones and teeth due to decades of hydrogen fluoride being pumped into the air. Up to 30 percent of acid rain in Japan is linked to sulphur dioxide from coal burning in China, where such emissions will exceed those of the entire world by 2035.

Nearly half of China's major river systems are seriously polluted—in part because 80 percent of industrial waste is dumped untreated into them. Shanghai's drinking water is contaminated with oil, ammonia, and nitrogen among other chemicals. Huge quantities of silt from intensive farming pour into the Yellow Sea. This silt is laced with heavy metals from industry. In 1989, the Huang He dumped 751 tons of such minerals as cadmium, zinc, arsenic, and chromium, along with 21,000 tons of oil into the Bohai.

In China's coastal waters, metals in fish rose fourfold from 1981 to 1984. Oysters have ten times acceptable chromium content. In many areas, fish have disappeared altogether. The more coastal development and migration, the more marine habitat is lost or destroyed.

Deforestation and desertification are also serious problems. Throughout history, China's forests have been decimated through human occupance and farming, the requirements for home construction, and the need for domestic fuel. Forests were also ravaged to make industrial charcoal and to manufacture ships. During the last three decades alone, a quarter of China's forests have vanished. With deforestation, soil erosion and desertification are direct consequences.

One of the most severely eroded regions is the Loess Plateau of north central China, which is the most extensive loess region on the face of the earth. Wind and water erosion have taken their toll over the centuries but human activities aggravate the situation. Grazing of sheep and goats on semiarid steppe lands removes the already sparse vegetative cover. Wind and water then dessicate and disect the region, and thousands of tons of loess blow or flow away.

Unwise use of water for poorly planned or inadequate irrigation schemes also contributes to land degradation. This will probably get worse as most global warming models predict increasing aridity.

Efforts have been made to reforest areas, erect wind breaks, and lock the sand in place with layers of straw or clay basins. If successful in soil retention and irrigation, these areas can be reclaimed for such crops as apples and walnuts. Aerial seeding of grasses such as alfalfa (lucerne) has been employed to vegetate bare hills. But salinity is an ongoing hindrance as high potential evaporation rates cause salts to rise to the surface through capillary action. Then, flooding and special drainage are required, making the whole process very costly.

Two weak links in China's economy are transportation and energy. Around 30 percent of the country's industrial capacity is idle due to lack of energy. Many factories operate on a three- or four-day workweek. When I visited the last steam-locomotive factory in the world in Datong, I was told how very lucky I was to find it open. Interestingly, luxury hotels in tourist and business centers are rarely short of electricity. Furthermore, China's energy supplies are not necessarily where they are needed, and transportation is woefully inadequate to link source and destination. Most of China's industry is concentrated in the coastal provinces but they possess only 10 percent of the energy stock.

China has the world's largest hydroelectric power potential yet it is developed to less than 10 percent. One massive project has come to world attention for its controversial impacts. This is the **Three Gorges Dam** on the Yangzi.

The Chang Jiang (Long River) is the longest river in Asia and third longest in the world (Figure 10-20). Known as the Yangzi to foreigners, the Chinese use that name only for the last 300 or 400 miles (480 or 645 km) of its 3,937-mile (6,300 km) course. Yichang, 1,000 miles (1,600 km) from the sea, is the head of navigation for river steamers. Ocean vessels may navigate the river to Hankou, about 600 miles (1,000 km) from the sea. Between Yichang and Chongqing, at an altitude of 130 feet (40 m), are the spectacular Three Gorges, noted for their geological magnificence and cultural significance.

The river and its tributaries drain 650,000 square miles (1,683,500 sq km) of territory. The entire region is subject to floods. From the Han to the Qing Dynasties (206 B.C. to 1911), devastating floods occurred once every 10 years. During the last 300 years, dams have been breached more than 60 times and flooding continues to be a problem all along the river. About one-third of China's population lives along its banks.

The Three Gorges Dam project is the world's largest water conservation project including the world's most powerful hydroelectric dam. The normal water level of 574 feet (175 m) will be reached in 2009, with a reservoir covering 408 square miles (1,045 sq km). Its generating capacity will be 18.2 kilowatts with a yearly output capacity of 84.7 billion kilowatt hours. This is equivalent to burning 50 million tons of coal or 25 million tons of crude oil. Should water power actually replace mineral power, massive amounts of carbon dioxide, sulphur dioxide, and dust would be eliminated from the air. The hydroelectric power is intended for use in the energy-deficient eastern and central regions as well as in Sichuan Province.

Shipping is expected to improve markedly and affect Chongqing's status as a *Shi*. In effect, the city would be tied to the Pacific Rim economy. However, its greater fortunes would appear to lie westward with the rapidly developing Sichuan Basin, a densely populated, resource rich, agricultural, and industrial region virtually isolated by topography from the rest of China. Chongqing might evolve as the most important growth pole in China's interior.

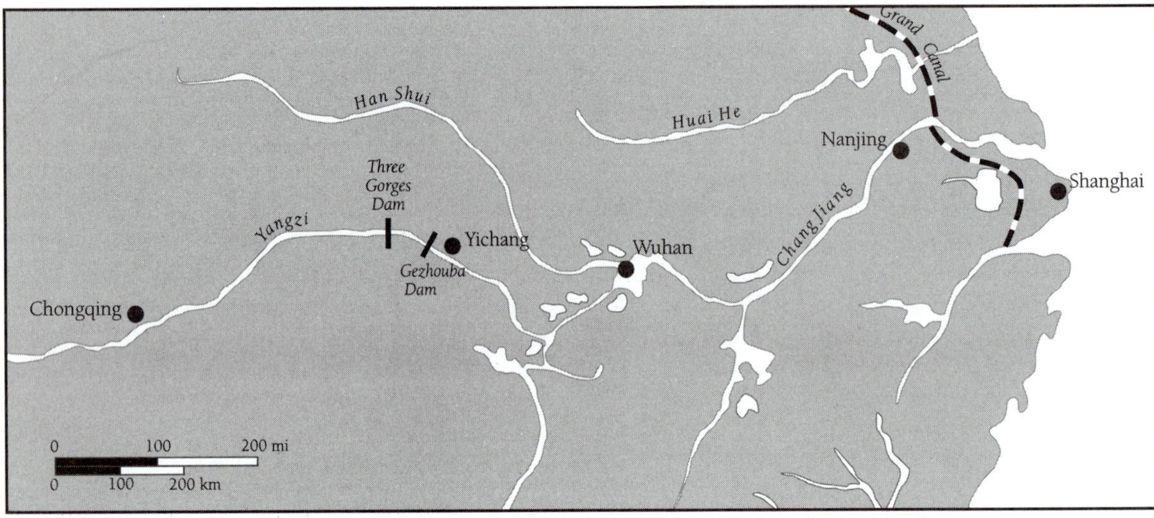

Figure 10-20

The Yangzi River and the Three Gorges Dam. Currently ocean-going vessels can reach Wuhan. The new dam will permit larger ships to reach Chongqing. What impact do you think this will have on Chongqing? How will the Three Gorges Dam impact this whole region?

Between 1.2 and 1.9 million people will be forced to leave their homes, and some 620,000 acres (248,000 ha) of farmland will be lost. Approximately 8,000 cultural sites and some 1,000 industrial enterprises will be inundated as well.

The Three Gorges Dam is extremely controversial. Forced removal of people is seen by many as a human rights issue. Although China claims that people are treated fairly with appropriate compensation and special privileges, many Chinese and others claim that resettlement is racked with corruption and unfairness. Compensation is inadequate, promised farmland is unproductive, and guaranteed jobs do not exist.

Scientists are also concerned. Some fear that the dam will ultimately collapse. In 1975, 62 dams in Henan Province gave way, killing anywhere from 86,000 to 230,000 individuals. Disease and famine affected over 10 million in the aftermath. A Three Gorges flood would be 40 times as great! Environmentalists claim that the Yangzi River ecosystem will be irreversibly upset. The Chinese alligator, the finless porpoise, the river dolphin, and the Chinese sturgeon will be severely endangered or even wiped out. Hydrologists postulate that sedimentation will cause sewage backups and floods upstream around Chongqing. Downstream, people worry about a lack of silt for agriculture.

Another aspect worth consideration is the fact that control of the dam by the northern power structure means control over water supply to the south. Water could be used as a weapon against any separatist movements in the south. What are your thoughts on this possibility?

The debate over the potential benefits and detriments of the Three Gorges Dam project continues around the world. Both the United States and the World Bank have elected to withdraw their support of the project.

Xizang/Tibet

Tibet, according to geographer Antonia Hussey (1992), "is indeed a special place, at the top of the world, a landscape of light and shadow dancing over towering mountains, gravel plains and lush green valleys." A mysterious place, isolated by massive topographic barriers, richly landscaped with Tibetan Buddhism, the region is centered on a forbidden city—Lhasa, the soul of the nation. But, Professor Hussey declares that Tibet is no longer mysterious. "The distinctive landscape of this ancient theocratic state with its Buddhist traditions and institutions has been shattered and replaced with the utilitarian socialist landscape of China and renamed Xizang."

Back in the seventh century, Tibet was a great military power expanded far beyond its current borders. In fact, the Chinese paid a tribute of 100,000 rolls of silk to Tibet. By the tenth century Tibet had foresworn the arts of war and had withdrawn from its Chinese and Indian territories.

In 1207, Tibet fell under Mongol domination but maintained an amenable priest–patron relationship with the Mongol court, many of whom had adopted Tibetan Buddhism. During this period, the Dalai Lama system emerged. A specially chosen child would become the Dalai Lama, spiritual head of the Tibetan people. Great monasteries such as Drepung and Potala were built, eventually housing thousands of monks.

By the mid-seventeenth century, Mongol power weakened. In 1720, the Qing Dynasty sent a garrison of Chinese soldiers to Lhasa. The next year they proclaimed that Tibet had always been a vassal of China.

By the late ninteenth century, Tibet was caught up in an imperialist power struggle between the British and the Russians. British troops were placed in Tibet, and Tibet agreed to keep other powers out and trade with British India. When Chinese forces entered Tibet in 1910, the Tibetans asked the British for help and Britain refused. Meanwhile in 1911, corrupt China had fallen to the Nationalists. In 1912, the Dalai Lama proclaimed Tibet's independence.

The Tibet of the early twentieth century was dominated by powerful monasteries and feudal landlords. There was no secular education, and the masses lived in dire poverty. The Dalai Lama's attempts at reform were crushed by the traditionalists. The victory of the communists in China in 1949 had little meaning to most ignorant and isolated Tibetans.

The story in China was different—in 1950, Radio Beijing announced, "The task of the People's Liberation Army for 1950 is to liberate Tibet." On October 7, 84,000 Chinese soldiers entered Tibet. India and Nepal expressed sympathy but did nothing, and Britain again refused to get involved. The 15-year-old Dalai Lama was forced to agree to return Tibet to the "Motherland." The resistance movement headed to the mountains to pursue guerilla warfare.

In 1959, Tibetan resistance exploded and the Dalai Lama went into exile in India. Chinese retaliation reaped anywhere from 87,000 to 430,000 deaths. Torture and humiliation were ubiquitous. Peasants and nomads, forced into communes, were made to grow wheat instead of their traditional barley. Tens of thousands starved to death in the famines following the Great Leap Forward. Tibet was declared a Xizang Autonomous Region of the People's Republic of China in 1965.

The Cultural Revolution, led by the Red Guards, decimated Tibet. Thousands were killed or sent to labor camps. Monasteries and religious artifacts were destroyed. By 1978, the number of monasteries had plummeted from 2,700 to 8. Holy texts were turned into toilet paper. Only Mao's death in 1976 ended this hideous onslaught.

In the 1980s, the Chinese apologized and proceeded to rebuild monasteries and restore the cultural landscape. However, about 130,000 Tibetans lived in exile and the Dalai Lama, headquartered in Dharamsala, India, began to seek understanding and aid from abroad. Support for Tibet's independence has grown worldwide. But, in the absence of the Dalai Lama, socialist transformation had begun in earnest.

China now sees Tibet as a storehouse of natural resources and is determined to Sinicize it. Coal, lithium, uranium, gold, chromium, and tin are only some of the minerals sought by the Chinese. The Qaidam Basin (once part of Tibet) is of special interest for oil.

Chinese settlers have streamed into Tibet, a country of barely more than 2 million inhabitants. Lhasa and other towns are becoming dominated by Han Chinese. Out of some 13,000 shopkeepers in Lhasa, for instance, only 300 are Tibetans. Tourism is being promoted, and huge hotels and related tourist attractions are being built. Much of traditional Lhasa has disappeared. The caretaker monks of the Potala Palace have been replaced by Chinese-trained "guides." The Drepang monastery on the outskirts of Lhasa, once housing 10,000 monks, now has around 700. Most of the monastery lies in ruins.

Crackdowns on religion continue. As late as 1997, the Chinese authorities announced that religion would have to bow to communism. "Development" has meant that half of Tibet's forests have been felled and more than a quarter of its mineral resources have been extracted since 1959. A fiber-optic cable has been laid between Lhasa and Shigatse, and plans are underway for a rail link with Sichuan.

It is clear that the Chinese are attempting to marginalize the Tibetans and make Tibet more livable for the Han. Tibet's new name is a metaphor for its relationship with China: Xizang—Our Western Storehouse.

China's Conundrums and Contradictions

China is a land of conundrums and contradictions. It is the world's most populous nation. Yet, it has to feed more than 10 million new mouths every year. It also has to create more than this number of jobs every year. The foundation of the revolution was the peasantry, yet the government predicts 137 million farmers will become redundant in the next 5 years. China is a rural country but it has more cities of over a million than the rest of the world combined. Thirty more are likely to appear in the next decade. China has many provinces but just one of these—Sichuan—would rank as the world's seventh largest country. Nine additional provinces rank among the globe's top 20 populous nations. Clearly, scale is a significant issue, making change far more difficult and complex than it might be otherwise.

China has a highly limited resource base. Constraints derive from the fact that the country must support 20 percent of humanity with less than one-fifteenth of the world's farmland. Now, China is a food importer. In 1949, China had a mere dribble of oil. Thirty years later it was self-sufficient in oil. Today it is a net importer of oil. Self-sufficiency is dead; China is inexorably tied to the world economy. Moreover, the shortage of arable land, rising population, and overbearing pollution push the notion of sustainable development into the frontline.

Even if China could slow its economic growth considerably, annual pollution increases would still equal that of the entire United States. China has seven of the ten most polluted cities in the world. Half the rain in Guangdong is acidic. China's water table is dropping drastically, and its waters are fouled with industrial effluents. The fact that more and more Chinese are acquiring refrigerators and cars guarantees rampant increases in carbon dioxide emissions.

China is changing rapidly within the constraints of a culture notable for its longevity. Autocratic leadership, saving face, distrust of merchants, the subservience of women, nationalistic and xenophobic leanings, and a sense of endless time are all as much a part of modern China as they were in the past. Change is sweeping the landscapes of China Proper while emerging as mere oases in China Frontier's sea of tradition. Politics are liberalizing, yet change is taking place within the existing system.

China's character is centered in continuity. So much has changed, yet so much remains the same. While capitalistic ventures are making some people very rich, this happens under the umbrella of socialism. It is the vision of socialism that is changing. As one Chinese economic planner stated: "Socialism used to mean government planning. Now it means common prosperity." While the past carries into the future, policies will continue to evolve. As the Vice Mayor of Beijing said: "It's like a fried egg. Once you've tried it, it tastes good, and if you wanted to make it raw again, you can't."

Mongolia: In Range of the Dragon's Breath

Mongolia, a landlocked country of 2.4 million and one of the world's most sparsely populated countries, is three times the size of France or four times the size of California. Here in this remote and arid land, livestock outnumber people 12 to 1. Mongolia, lacking in infrastructure, modern industrial skills, and managerial knowledge, is struggling to develop its economy in light of its 1990 freedom from Soviet control. It is the first country in Asia to experience a bloodless transition from communism to democracy.

THE PHYSICAL LANDSCAPE

Mongolia's physical environment is characterized by extremes as well as fragility. Well-watered mountains in the west and northwest descend to steppe on the eastern plateau. Steppe lands graduate into *gobi*—semidesert and desert. Rivers are concentrated in northern Mongolia but these flow to the Arctic, the Pacific, or into landlocked basins. The *gobi* has no perennial rivers, therefore underground wells are important. But, most of these freeze in winter. Soviet-installed pumps have fallen apart, and spare parts are not available. Getting water is problematic in the arid regions. At other times, dry river beds are subject to flash floods.

During the Mezozoic Period, much of the *gobi* was swamp and favored by dinosaurs. In the early 1920s, the world's first known dinosaur egg was discovered here. In 1993, a 75-million-year-old fossilized embryo of a meat-eating dinosaur was found in the western *gobi,* a region rich in vertebrate fossils.

Soil erosion is common throughout Mongolia. In many areas, soils are only a few inches thick and overlain on granite. Dessicating winds easily whip up the earth, thereby filling the air with dust and leaving the ground barren. Overgrazing contributes to the problem.

Mongolia has an extreme continental climate with sharp fluctuations in temperatures both seasonally and diurnally. Temperatures average 15°F (−9°C) in January and 64°F (18°C) in July. Winter is long, lasting from October to May. Ulaanbaator (Ulan Bator) is the coldest capital in the world, with winter temperatures sinking to as low as −50°F (−45°C). Precipitation is limited, varying between 10 and 20 inches (25–51 cm). The Chinese used to refer to Mongolia as *han-hai:* "dry sea."

HISTORICAL BACKGROUND

Cultural Mongolia is much larger than indicated by current political boundaries. There are many Mongol tribes speaking various dialects, and several of these live outside Mongolia and China's Inner Mongolia. These various warring groups were unified under Chinggis Khan in 1206.

Inner and Outer Mongolia are political designations dating from the Manchu Dynasty (1644–1911). In 1622, the tribes of South and East Mongolia allied with China. This is why China claims legitimacy over Nei Monggo. In 1689, the Treaty of Nerchinsk laid down the frontiers between Russia and China. At that time Mongolia was part of China. Two years later, the Mongols agreed to pay a yearly tribute known as the "Nine Whites"—eight white horses and one white camel. Mongolia then became known as Outer Mongolia. These historic connections provide rationale for China's notions of hegemony over this region.

Outer Mongolia persisted until 1911 when it became an autonomous monarchy ruled by a religious leader, a Lama-King. But, it soon fell under the influence of Soviet communism and in 1921, was taken over by the Mongolian People's Party. With the help of the Russians, the People's Republic of Mongolia was proclaimed in 1924, and Mongolia became the world's second communist nation. Mongolia remained a Soviet satellite until the collapse of the U.S.S.R. in 1990.

SOVIET IMPACTS

The Soviet's chief aim was to transform pastoral Mongolia into a source of raw materials for Siberian industrial concerns. Concerted efforts were also made to settle both rural dwellers and urban workers out of their traditional *gers* and into multistoried buildings associated with industrialization and regional development. By 1981, more than half of Mongolia's population was classified as urban as compared to less than 22 percent in 1956.

A major contribution to this rapid urbanization was the expansion of Ulaanbaatar with multistoried suburbs encroaching on *ger* communities. Second, administrative centers of Mongolia's 18 provinces acted as magnets for regional growth. Third, the establishment of new towns mediated the dispersal of industry.

Darkhan is an example of a new town that burgeoned from a railway station of 2,000 people to an industrial center of 25,000 by 1970. Being close to the Soviet border, thermal power based on local coal, brick works, meat-packing plants, and other industries easily served Soviet needs. Other centers were developed to exploit and process new resources such as molybdenum and copper. Some administrative centers added manufacturing functions.

Mongolia got its first motor road in 1937 from Ulaanbaatar to Altanbulag on the Soviet Frontier. The Ulan Bator Industrial Combine was also developed in the 1930s, producing carpets, wool garments, felt, and leather goods.

To further effect control, the Communists established state farms and required the permanent settlement of the nomadic herders. Needs were subsidized by the state. When the subsidies ended after 1990, the herders returned to their nomadic lifestyle, moving frequently across the steppe in summer and clustering in more sheltered valleys in winter.

Another Soviet goal was to transform Buddhist Mongolia into a secular state. Most of the country's 700 temples were destroyed in the 1930s, and more than 100,000 lamas dispersed or killed. Five hundred truckloads of religious art treasures disappeared into the Soviet Union. Since 1990, monasteries are being rebuilt and the country's Tibetan-style Buddhism is experiencing a revival.

MODERN MONGOLIA

Mongolia is undergoing its own version of Perestroika—*shinechiel,* meaning "renewal." Free-market economy has replaced state-controlled enterprises. People are studying English as well as their traditional Uygur-Turkish script, which had been replaced by the Russian Cyrillic alphabet. Trade that was as high as 90 percent with the Soviet Union has dropped to 50 percent and has increased to China, South Korea, and Japan. Foreign aid is pouring into the country, and Japan is the largest aid donor.

Not all aid projects are successful. One crossbreeding program resulted in Mongol-European cattle that died in the winter cold. A prefabricated mini-mill for steel production was brought from Japan. It was to manufacture steel from scrap metal. However, no one realized that Mongolia, with few cars or machines, possessed virtually no metal. Japan has also introduced windmill, battery, and TV assemblies. People once isolated in their round, felt *gers* on the steppe now can choose from 52 channels, including the Cartoon Network and MTV. What impact do you think this will have on Mongolia's development?

Even with exploitation of mineral resources such as coal, oil, molybdenum, copper, and uranium, animals and animal products account for half of industrial output and 90 percent of exports. Material infrastructure is either lacking or obsolete, and few Mongols are educated in modern business and industrial management.

Three-quarters of Mongols are livestock herders practicing transhumance and living either in their large, permanent winter *ger* or in their smaller, portable summer *ger*. Because animal pastures are sparsely covered with low-yield grasses, families must move around 30 times a year, covering distances of 125–185 miles (200–300 km).

One of the most important animal products is cashmere from cashmere goats. China was buying over half of Mongolia's raw wool, to the detriment of Mongolia's processing industry. Consequently, the Mongolian government has banned the sale of raw wool, but now it is smuggled into China.

Development, in context of transition from rigid control under communism to freedoms presented by democracy, has its downside. Mongolia is short of food, and there are thousands of destitute people, mainly women and children. In 1996, the World Bank estimated that only 15 percent entitled to government benefits were receiving any. In addition, literacy has fallen dramatically and school dropouts have increased. Maternal mortality has doubled. Although Mongolia's medium- to long-term economic prospects are good, social reforms are sorely needed.

Relationships with China are difficult, as Mongols do not wish to be dominated by China, as is the case in Nei Mongol where millions of Chinese have settled. China is anxious to intensify relations however, as it needs lumber, minerals, and animal products as well as markets for its farm produce and low-end manufactures. Now that Mongolia is out from under Russian control, China is exerting its authority based on historic relationships. For example, China has loudly protested visits of the Dalai Lama to Mongolia and has discouraged connections with the Mongols of Nei Mongol. For decades, Mongolia has served as a buffer state between China and the Soviet Union with its strongest links to the U.S.S.R. Today, Mongolia is a buffer state between China and Russia with increasing ties—wanted and unwanted—to China.

Main Point Summary

1. In 1949, in order to prevent famine and chaos, the Communist government took drastic action to rectify a severely damaged environment and disrupted economy.
2. An agrarian and industrial revolution that included collectivization reached its zenith with the Great Leap Forward in 1958. The failures of the Great Leap were compounded by the Sino–Soviet split, and millions perished in subsequent famines.
3. Policy conflicts led to the Great Proletarian Cultural Revolution in 1966, a giant social, economic, and political upheaval. Millions were forced to move to frontier regions to encourage development, and millions were "sent down to the countryside" to learn the ways of the peasants and become rehabilitated. Revolutionary groups such as the Red Guards ransacked the land and killed thousands.
4. After the death of Mao Zedong in 1976, a "rehabilitated" Deng Xiaoping took the helm and promoted the Four Modernizations stressing agriculture, industry, education, science, and defense in a context of market socialism. Deng moved away from Mao's "self-reliance" edict and promoted an "open-door" policy along with economic liberalization.
5. By the late 1970s, agriculture had improved and private plots were the most productive parts of the farmland. Industry had been diversified and decentralized to some degree. Railroads diffused westward and southward, further decentralizing population and economic activity into China Frontier.
6. Rural to urban migration streams and decentralization led to the growth of towns to cities and the formation of completely new urban places. New cities tended to be associated with natural resources. Nevertheless, most of China's population continued to live in the eastern provinces.
7. There are four categories of administrative structure: provinces such as Sichuan, autonomous regions such as Xinjiang, municipalities known as *Shi* such as Tianjin, and special administrative regions such as Hong Kong.
8. There are 55 officially recognized minority groups in China. Most live in border regions such as Guangxi Chuang A. R. or Xinjiang Uyghur A. R. Xinjiang is strategically significant because of its wealth of natural resources, its proximity to the new Central Asian states, and its historic relationship with Russia.
9. China's overall cropping patterns have not changed drastically but modernization is improving crop yields. There is now greater emphasis on the production of grain, meat, and poultry. Land is used intensively, with double-cropping the rule wherever climate or water supply allows. Efforts are being made to open new crop lands in China Frontier.
10. Agriculture functions by the household responsibility system whereby farmland can be leased and even inherited. Families must fulfill quotas for

grain or industrial crop production, otherwise using the land as they wish. Many farmers are becoming quite wealthy under this system. However, greater efficiency has released millions from the land who must seek employment elsewhere.

11. All Chinese citizens are registered in a particular location and classified as non-agricultural or agricultural. Movement of location and change of classification are very difficult. Since the 1980s, the government has relaxed this *hukou* system to some degree and there is a large rural floating population seeking work in cities. There are also marriage migrations of brides from poor locations seeking economic opportunities in richer areas through arranged marriages.
12. Many seek work in township and village enterprises. Others become part of a floating population, moving around to find work. Some become temporary residents of towns and cities while holding a permanent job.
13. The loss of farmland begs the question as to how China will match food production with its ever-growing population. Pessimists say that there is insufficient suitable land and that crop yields have essentially peaked. China is already importing grain, and this demand will increase exponentially with population growth. Optimists say that China is already bringing new land under cultivation, that yields can be raised significantly higher, and that there is nothing terrible about having to import some food.
14. Special economic zones (SEZs) were created in the 1980s to open up regions to foreign investment and technology. Open zones and cities have diffused into the interior, especially along river arteries. Many of these regions have strong connections to overseas Chinese communities. Most of the foreign capital has come from Asian sources.
15. Shanghai is China's largest and most sophisticated city, the economic head of the Yangzi Dragon. Shanghai has been both a source and destination for migrants, and has played an active role in facilitating development of the interior. Pudong development is intended to draw significant foreign investment and technology in tertiary activities. Shanghai is becoming strongly integrated with Nanjing, Hangzhou, and other centers linked by new transportation corridors. Agricultural hinterlands are giving way to housing and industrial developments.
16. The number of cities in China have more than doubled since 1978. Small- and medium-sized cities are experiencing rapid growth. Urbanization remains concentrated in the eastern coastal region.
17. Temporary residents make up significant proportions of city work forces. Most work in the building industry. Others operate as street entrepreneurs. All place strains on urban infrastructures.
18. Regional disparities exist in China. For example, the northeast is suffering from industrial obsolescence and interior provinces have not taken the same economic strides as the coastal provinces. Classical economic theories do not satisfactorily explain these spatial trends as they fail to account for the intrusive role of the state in China's economic development.
19. Guandong fits the definition of a regional state—pursuing its own lines of development with global enterprises. Proximity to Hong Kong and long-standing business and family connections have facilitated economic enterprise. Shanzhen has become the province's wealthiest area.
20. Hong Kong, no longer a British colony, has strengthened its ties with the mainland. Hundreds of industries now locate in Guangdong and Guangxi to take advantage of tax incentives and cheap land and labor. Shenzhen, Guangzhou, and Zhujiang are becoming increasingly integrated in the Hong Kong S.A.R. and Guangdong regional state phenomena.
21. Major efforts are being made to slow China's population growth. Vigorous population control programs have been pursued in recent years. One child per family is held up as the ideal. However, this has been successful mostly among the better-educated urban populations. Rural people want

more children, and their preference for male offspring has led to female infanticide and sex-selective abortion.

22. China's population is aging. Given the population structure, there could be a shortage of labor and not enough young people to care for a burgeoning elderly population.
23. Women are not equal in China. Dyadic thinking polarizes duties and responsibilities in "inside" and "outside" categories, with outside being superior. Even as more women work outside in agriculture and courtyard enterprises, these roles are reclassified as inferior inside work. Even though the nature of women's work is changing under economic restructuring, inequitable conceptual divisions remain.
24. China has a spate of environmental problems, exacerbated by continued development. Air, water, vegetation, and soil are all under severe stress with little relief in sight. Many factories operate only a few days a week for lack of power.
25. The Three Gorges Dam on the Chang Jiang is intended to ameliorate the power shortages in China's southeast and improve navigation on the river to Chongqing. However, the environmental and human consequences of this world's largest water conservation project and most powerful dam will be devastating. Certain organizations such as the World Bank have withdrawn their support for the project.
26. Xizang A. R. (Tibet) is experiencing drastic change as traditional landscapes are being replaced with Han Chinese, socialist ones. Even though Tibet was once a vast empire to which the Chinese paid tribute, China claims that Tibet is inexorably a part of China. China is determined to Sinicize the region and harvest its wealth of natural resources by force if necessary. Chinese invasions have subdued the Tibetans, and their culture is constrained and practiced only under the watchful eye of the Chinese military. The Dalai Lama and other Tibetans living in exile have brought world attention to the plight of Tibet.
27. China is a land of conundrums and contradictions. While so much has changed in recent years, much has remained the same particularly in the realm of cultural values. While China's landscape is being irreversibly altered, China's character is centered in continuity.
28. Mongolia is struggling to develop its economy in light of its recent freedom from Soviet control in 1990. A sparsely populated land, it was exploited by the Soviets for natural resources. Nomadic culture and Buddhist religion, then squashed, are being revived in a situation of newfound freedom. Three-quarters of Mongolians are livestock herders practicing transhumance. Mongolia remains a very poor country with social reforms sorely needed.
29. Proximity to Nei Mongol with its influx of Han migrants puts Mongolia in a potentially precarious position in relation to China. China wants access to timber, minerals, and other resources but does not want to facilitate any pan-Mongol movement. Mongolia remains a buffer state between Russia and China.

Terms to Know

collectivization
Great Leap Forward
People's communes
Great Proletariat Cultural Revolution
Hsia-fang (downward transfer) movement
Deng Xiaoping
Four Modernizations
household responsibility system
hukou
marriage migration
floating population
special economic zones (SEZs)
temporary residents
regional state
Three Gorges Dam

Bibliography

Allen, Thomas B. 1996. Xinjiang. *National Geographic* 189: 2–43.

Baird, Vanessa, ed. 1995. Tibet A Cause For Courage. *New Internationalist,* No. 274.

Benson, Linda, and Ingvar Svanberg. 1998. *China's Last Nomads: The History and Culture of China's Kazaks.* Armonk: M. E. Sharpe.

Brown, Lester R. 1994. Who Will Feed China? *World Watch,* 7: 10–19.

Brunn, Ole, and Ole Odgaard, eds. 1996. *Mongolia in Transition: Old Patterns, New Challenges.* Richmond, Surrey: Curzon.

Burstein, Daniel, and Arne deKeijzer. 1998. *Big Dragon.* New York: Touchstone.

Cannon, Terry, and Alan Jenkins, eds. 1990. *The Geography of Contemporary China: The Impact of Deng Xiaoping's Decade.* New York: Routledge.

Chang, Kuei-sheng. 1961. The Changing Railroad Pattern in Mainland China. *The Geographical Review,* 51: 534–48.

Fan, C. Cindy. 1996. Economic Opportunities and Internal Migration: A Case Study of Guangdong Province, China. *Professional Geographer,* 48: 28–45.

Fan, C. Cindy. 1995. Of Belts and Ladders: State Policy and Uneven Regional Development in Post-Mao China. *Annals of the Association of American Geographers,* 85: 421–49.

Fan, C. Cindy, and Youqin Huang. 1998. Waves of Rural Brides: Female Marriage Migration in China. *Annals of the Association of American Geographers,* 88: 227–51.

Farley, Maggie. 1995. China's Women Pay Cost of Change. *Los Angeles Times* (August) 31: A4.

Fullen, Michael, and David Mitchell. 1991. Taming the Shamo Dragon. *Geographical Magazine,* 63: 26–29.

Gately, David. 1997. Growth in China's Grain Imports an Opportunity for Exporting Countries, Not a Threat to World Food Supply. *International Food Policy Research Institute News Release.* Washington, D.C.: IFPRI.

Gong, Hongmian. 1995. Spatial Patterns of Foreign Investment in China's Cities, 1980–1989. *Urban Geography,* 16: 198–209.

Goodman, David, S. G. ed. 1989. *China's Regional Development.* New York: Routledge.

Hall, Derek. 1987. Economic and Urban Development in Mongolia. *Geography,* 72: 73–76.

He, Bochuan. 1991. *China on the Edge: The Crisis of Ecology and Development.* San Francisco: China Books.

Hsu, Immanuel, C. 1995. *The Rise of Modern China,* 5th edition. New York: Oxford.

Hsu, Mei-ling. 1994. The Expansion of the Chinese Urban System, 1953–1990. *Urban Geography,* 15: 514–36.

Hussey, Antonia. 1992. The Landscape of Lhasa, Tibet. *Focus,* 42: 8–12.

Jacka, Tamara. 1997. *Women's Work in Rural China: Change and Continuity in an Era of Reform.* New York: Cambridge.

Jacobs, J. Bruce. 1997. Shanghai: An Alternative Centre? In *China's Provinces in Reform,* ed. David S. G. Goodman, pp. 163–93. New York: Routledge.

Judd, Ellen R. "Men Are More Able": Rural Chinese Women's Conceptions of Gender and Agency. *Pacific Affairs,* 63: 40–61.

Kaye, Lincoln. 1992. Faltering Steppes: China Looms Large in Ethnic Upsurge Among Mongols. *Far Eastern Economic Review,* 9 (April): 16–18.

Lai, David Chuenyan. 1995. *Land of Genghis Khan: The Rise and Fall of Nation States in China's Northern Frontiers,* 30, Western Geographical Series. Victoria, British Columbia: University of Victoria.

Lattimore, Owen. 1955. *Nationalism and Revolution in Mongolia.* New York: Oxford.

Leeming, Frank. 1993. *The Changing Geography of China.* Oxford: Blackwell.

Leung, Chi Kin. 1993. Personal Contacts, Subcontracting Linkages, and Development in the Hong Kong-Zhujiang Delta Region. *Annals of the Association of American Geographers,* 83: 272–302.

Li, Dun J. 1978. *Modern China: From Mandarin to Commissar.* New York: Scribner's.

Lu, Jonathan J., and Ma Shengtang. 1990. Changes in China's Rice Area: 1966–1985. *Asian Profile,* 18: 391–414.

Ma, Laurence J. C. 1976. Anti-Urbanism in China. *Proceedings of the Association of American Geographers,* 8: 114–18.

McColl, R. W. In the Shadow of Chinggis Khan. *Focus,* 44: 1–9.

Nemeth, David J. 1998. Privy-Pigs in Prehistory? A Korean Analog for Neolithic Chinese Subsistence Practices. *MASCA Research Papers in Science and Archaeology,* 15: 11–25.

Pannell, Clifton, W. 1988. Regional Shifts in China's Industrial Output. *The Professional Geographer,* 40: 19–32.

Pannell, Clifton W., and Jeffrey S. Torguson. 1991. Interpreting Spatial Patterns From the 1990 Census. *The Geographical Review,* 81: 304–17.

Pearce, Fred. 1995. The Biggest Dam in the World. *New Scientist,* 28 (February): 25–9.

Platt, Anne E. 1995. Dying Seas. *World Watch,* 8: 10–19.

Salter, Christopher L. The Litany of Tachai and the Foolish Old Man: Agricultural Landscape Modification in Mainland China. *The Professional Geographer,* 24: 113–17.

Sattaur, Omar. 1990. Thoroughly Modern Mongolia. *New Scientist,* 1: 45–50.

Smil, Vaclav. 1993. *China's Environmental Crisis: An Inquiry into the Limits of National Development.* Armonk: M. E. Sharpe.

Thayer, Nate. 1997. Forward Steppes. *Far Eastern Economic Review,* 27 (March): 18–23.

Tien, H. Yuan, et al. 1992. China's Demographic Dilemmas. Population Bulletin 47: 1–44.

Topping, Audrey, R. 1995. Ecological Roulette: Damming the Yangtze. *Foreign Affairs,* 74: 132–46.

Veeck, Gregory, and Clifton W. Pannell. 1989. Rural Economic Restructuring and Farm Household Income in Jiangsu, People's Republic of China. *Annals of the Association of American Geographers,* 79: 275–92.

Vriens, Hans. 1997. Lhasa Lost. *Far Eastern Economic Review,* 29 (May): 44–5.

N. A. 1996. *White Paper—The Grain Issue in China.* Information Office of the State Council of the People's Republic of China. October: 1–12.

Wong, Shue Tuck, and Sun Sheng Han. 1998. Whither China's Market Economy? The Case of Lijin Zhen. *The Geographical Review,* 88: 29–46.

Yabuki, Susuma. 1995. Trans. S. M. Harner. *China's New Political Economy: The Giant Awakes.* Boulder: Westview.

Yatsko, Pamela. 1997. Work In Progress (Shanghai). *Far Eastern Economic Review,* 7 (August): 66–69.

Yeh, Anthony Gar-on, and Xueqiang Xu. 1996. Globalization and the Urban System in. China. In *Emerging World Cities in Pacific Asia,* eds. Fu-chen Lo and Yue-man Yeung, pp. 219–65. New York: United Nations.

Zich, Arthur. 1997. China's Three Gorges. *National Geographic,* 192: 2–33.

Zimmerman, Tim et al. 1996. China Takes a Deep Breath. *U.S. News and World Report,* 9 (September): 36–39.

Chapter 11

East Asia's Tigers: Japan, Korea, and Taiwan

"Our greatest glory is not in never falling but in rising every time we fall."

(Confucius 551–479 B.C.)

In this chapter I will tell you about three tigers of East Asia. I include Japan as a tiger and will talk about it first, as it is the most developed country in all of Asia. Then, I will discuss South Korea, and Taiwan. I will talk about North Korea, but it definitely is not an economic tiger. As you read on, the fact that the historical evolution of these four regions are not only related to each other, but also related to developments in China, will become increasingly evident.

Japan: A Tiger Risen

Rich, highly urbanized, densely populated, and one of the world's largest users of natural resources, Japan is Asia's leading tiger. In the 1990s, Japan was the world's largest creditor nation. The Tokyo–Yokohama–Kawasaki urban area is the largest metropolitan region on earth. Japan gobbles up more tropical forests than any other nation, a reminder that Japan is virtually devoid of the natural resources it is willing to exploit.

Having achieved full status as an economic tiger, Japan is reckoning with problems looming large for all developed economies in the twenty-first century. How did Japan reach this pinnacle of economic development? What has gone awry, and how is this affecting economic structure and spatial organization? And, what social changes are occurring in context of economic transformation?

DYNAMIC ARCHIPELAGO

Four main islands dominate the Japanese archipelago: Hokkaido; Honshu; Kyushu; and Shikoku. These and hundreds of lesser islands lie at the junction of three tectonic plates: the Pacific, Philippine, and Eurasian (Figure 11-1). The Pacific plate dips between the Eurasian and Philippine plates in the Japan trench north of Tokyo. The Philippine plate is being consumed by the Eurasian plate in the Nankai trench south of Kyushu and Shikoku. The Pacific plate drops at a rate of 2.9 inches (7.5 cm) a year.

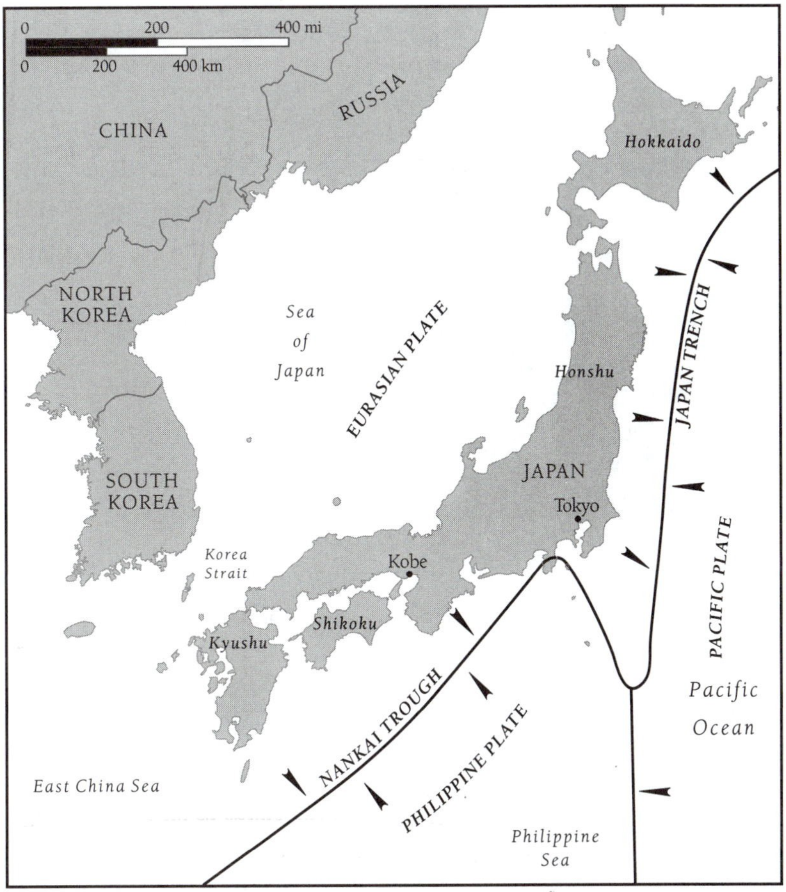

Figure 11-1

The earth in action. Note how the Pacific and Philippine Plates push against the Eurasian Plate. This geological movement makes Japan one of the most unstable places on earth.

Japan is the most seismically active country in the world. Such instability results in earthquakes. In Tokyo, there are an average of three noticeable quakes every month. In 1923 an earthquake registering 8.3 on the Richter scale devastated Tokyo and the Kanto Plain around it. Cooking stoves fired the mostly wooden and paper structures. The quake, the fires, and a 36-foot (10.8 m) tsunami claimed 140,000 lives. This was known as the *Great Kanto Quake.*

Another severe tremor shook the Kanto region in 1983, resulting in much property damage. All Japanese know that major quakes strike roughly every 70 years. With acres of landfill to liquify, thousands of skyscrapers to crumble, miles of freeways to collapse, and miles of underground gas lines to rupture, Tokyo is a scary place.

In 1995, another ruinous earthquake struck the city of Kobe (Figure 11-2). This quake was caused by a "strike-slip" (sideways) motion along a minor fault. Older, concrete buildings collapsed, while those of flexible construction or on shock absorbers sustained less damage, although some still had severe damage. Fires raged out of control for days; 5,000 people were killed, and 250,000 were made homeless

Volcanic activity is another hazard in Japan, which boasts more than 200 volcanoes. Sixty of these have been active in historic time. Mount Fuji, a dormant volcano towering to 12,385 feet (3,776 m), is regarded

Figure 11-2

As part of the disastrous Kobe Quake, the Hanshin Expressway fell over. Fortunately, this was early morning and there was little traffic. Even in the midst of chaos, the orderly Japanese waited for lights to change and used crosswalks.

as sacred by the Japanese. Japan is also the world's foremost hot springs country, with close to 20,000 springs and some 2,000 resort spas, many in Kyushu.

Japan is a very mountainous country, with only 15 percent lowland (Figure 11-3). Landforms show evidence of tectonic uplift with coastal terraces, and glaciation with ice-carved features in the Japanese Alps. In many places, the mountains descend directly into the sea. Plains are restricted to narrow river floodplains and deltas. The most extensive plains on the main island of Honshu are the Kanto Plain around Tokyo and Yokohama, the plain around Nagoya, and the Yamato Plain around Osaka, Kyoto, and Kobe. Needless to say, these areas are very densely populated.

ORIGINS

There is a popular myth that the Japanese and the first emperor are direct descendants of the Sun Goddess Amaterasu. The first emperor took the throne in 660 B.C. But recent archeological evidence allows a different story.

About 400 B.C., people from the Korean peninsula migrated to the island of Kyushu. Bringing with them rice culture from China, their numbers rose about 2 percent a year. In a few hundred years, their culture overwhelmed that of the indigenous Jomon hunters and gatherers who had inhabited the region for at least 10,000 years. Using metal tools and the techniques of large-scale rice cultivation, they established a regional core on the Yamato Plain. Japanese culture diffused from the Kansai Plain northward throughout the islands. Even though this story is based on scientific evidence, it is difficult for many Japanese to accept the fact that they are related to Koreans, whom they long have considered inferior.

The language that originally diffused to Japan was not Korean as we know it today. It is thought that the Japanese language diverged from the Korean language at least 4,000 years ago, as neither has words in common. However, both Japanese and Korean employ common Chinese words. All evidence points to the fact that culture advanced from China to Korea and from Korea to Japan.

HISTORICAL DEVELOPMENT

Opening Closed Doors

Fearing foreign intrusion that was changing other parts of East Asia, Japan essentially closed its doors from 1603 to 1854. Some trading was permitted through the port of Nagasaki. Then, an event occurred that

Figure 11-3

Physical features of Japan. What is the significance of 37° north latitude?

would change Japan forever: the Americans arrived. In 1854, Commodore Perry landed at Yokohama. With intimidation, Perry engineered the first unequal treaty and opened Japan's doors to outsiders.

It was readily clear to the Japanese that they had missed the Industrial Revolution that had rocked the rest of the world; they were technologically inferior. Britain, an island nation in the Atlantic that had acquired the largest colonial empire in the world, became a model for the Japanese. Confucian values were called upon to educate and advance Japan into the twentieth century.

In the rush to modernize, militarize, and industrialize, military-industrial conglomerates emerged on the scene. These were based on hierarchically structured loyalty and were called ***zaibatsu***. The most powerful of these were Mitsubishi, Mitsui, and Sumitomo, which dominated heavy industry.

At the same time, Japan's traditional religion, ***Shinto***, became embroiled in promoting worship of the emperor. In this Cult of the Emperor, the ruler became god-like, and all Japanese were forbidden to lay eyes on him. This situation continued until the end of World War II in 1945. Only when the people heard Emperor Hirohito's voice on the radio announcing surrender were they convinced he was not a god.

Expansionism and Empire

Flexing its new muscles, Japan launched wars with Russia and China. They got Taiwan in 1895 and annexed Korea in 1910. In 1931, they invaded Manchuria. This was the beginning of their creation of the Greater East Asia Co-Prosperity Sphere (Chapter 1). Infuriated by the fact that the United States cut off their oil supply lines, the Japanese bombed Pearl Harbor in 1941. This event brought the Americans into the ongoing Pacific War. Although the Japanese were able to sweep through China and Southeast Asia, the war was brought to an abrupt halt with the Americans dropping atomic bombs on the cities of Hiroshima and Nagasaki in 1945. This marked the end of GEACPS but a new beginning for Japan.

Phoenix from the Ashes

Japan's postwar recovery was the most rapid of any nation in history. Relying on the Confucian values of hard work, loyalty, and cooperative efforts and with American assistance, infrastructures were rebuilt. A democratic form of government was put into place, the *zaibatsu* were broken up, divorced from a reduced military, and restructured. Land reform gave land to the war-ravaged peasants.

Grounded in a fresh technological base, government, financial institutions, and industry banded together to build the new Japan. Another form of industrial conglomerate emerged, known as ***keiretsu.***

Did you know that Toyota Motor Group is a keiretsu *with ties to Mitsui and that Honda has very close relations with Mitsubishi?*

Keiretsu

Keiretsu means something close to "link" or "affiliated with." A *keiretsu* is a hierarchical organization of companies. There are "horizontal *keiretsu*" and "vertical *keiretsu*."

Horizontal *keiretsu* revolve around a financial core including a major bank, a giant trading company, and a huge manufacturer. For instance, there is Mitsubishi Bank, Mitsubishi Corporation, and Mitsubishi Heavy Industries. There are also about 30 companies and more than 200 affiliates (10 percent equity ownership) that are part of this horizontal *keiretsu.*

Many vertical *keiretsu* lie within the borders of the major horizontal *keiretsu* such as Mitsubishi, Sumitomo, or Mitsui. Vertical organizations are either in production or distribution. A collection of firms assembles parts for a manufacturer who directs its end product through a distribution network of wholesalers and retailers. Most production *kereitsu* concern automobiles and electronics. Most distribution *keiretsu* are involved in consumer electronics.

A member, but very independent, of the Mitsui Group is the Toyota Motor Group. Toyota was founded in 1894 by the Toyoda clan that had its start in the manufacture of wooden spinning machines. Toyota, with its automobiles, real estate firms, insurance companies, and other interests is becoming a combination vertical-horizontal *keiretsu.*

A key reason why Japan was able to rise like a phoenix from the ashes of war is its ability to borrow and adapt. Chinese characters, chopsticks, city grid layout—these and many other things were borrowed from China and changed to suit Japanese tastes. China's Chan Buddhism became Zen in Japan. Japan borrowed ideas of empire from Britain and technology from Europe and America. Japanese ability to absorb different ideas is evident in lifestyle: marry Shinto; die Buddhist; and celebrate Christian holidays. Christmas, for example, calls for decorations, and is a peak shopping period in Japan. In addition, Japan's favorite drinks are *saki* (rice wine) and scotch, and the favorite sports are Sumo wrestling and golf.

Japan's spectacular growth made it the richest country in Asia and one of the richest in the world by the 1970s. By 1990, it had the world's six largest banks and over 100 companies listed in *Fortune* magazine's top 500. It is the number one investor in East and Southeast Asia and one of the top investors in the United States, Canada, Australia, and the rest of the world.

AGRICULTURAL JAPAN

In 1920, 51 percent of Japan's labor force was in agriculture. Now less than 10 percent is farming. About 90 percent of these individuals are part-time farmers who earn most of their income from non-farm activities. In fact, much of the farm work is done by women and older people; men work other jobs, and young people move to the cities.

Japanese farms are very small, averaging 2.5 acres (1 ha). Since agriculture competes with other activities for level land, farmland is very expensive. Many farmers are independently rich after selling their land to industry. However, resistance to parting with land imbued with centuries of family tradition has made acquisition of farmland for urban and other development difficult and expensive.

Climatic factors and limited space call for intensive land use. The growing season varies from 260 days a year in Kyushu to 150 days in Hokkaido. The northern limit for double-cropping is 37° latitude.

The primary crop is rice, planted in summer. Rice is rotated with wheat or barley in the double-cropping zone. Tea and fruit such as apples grow on hill slopes. Much of Japan's agriculture is **market gardening:** producing vegetables and poultry products for the vast urban market. For this purpose, greenhouse agriculture is very important and organic farming is on the increase. Nevertheless, agriculture is highly mechanized and chemical fertilizers and pesticides remain widely used. Consequently, crop yields are very high with rice yields three times the average Asian yield.

Animal herds are rare other than in Hokkaido, where extensive plains make dairying feasible. An ordinance of 1896 established nonmilitary settlements on Hokkaido, and these were modeled on the township system in the United States. Here dairy farms look like American farmsteads. Some beef cattle are kept near Kobe. These animals are beer fed and hand massaged daily. A Kobe beef steak can cost hundreds of dollars.

Japan's major source of protein is fish—the country has the world's largest fishing fleet. In addition, **aquaculture** is common, especially around the Inland Sea where shrimp, oysters, and seaweed are cultivated. Fish are raised in ponds and in ocean-based fish farms.

Agriculture is facing an array of problems in Japan. Government subsidies pay for surplus crops. Rice is six times more expensive than elsewhere to grow, yet it is a tradition supported by the government, which also prevents imports of cheaper *Japonica* rice from places such as California. Yet, food preferences are also changing. Many Japanese are turning to imported meat, vegetables, and dairy products. Traditional meals are often replaced by packaged instant noodles, McDonald's, and other fast foods.

While farmland can be sold for housing or industry, no corporations are allowed to purchase it. By preserving the family farm, Japan has sought to save the ***mura***—rural village as a self-sustaining community. But demographics indicate that the mura are declining rapidly. There is a growing shortage of farmers as young people are recruited by business concerns. Over half of Japan's farmers are now over 65 years old. As few children intend to carry on with farming, the *mura* are dying out.

AGING SOCIETY

Japan's population is declining. At 126 million in 1998, this number is expected to peak in 2007 and plunge by half at the end of the twenty-first century. The average life expectancy for women is 83.6 years, for men 77 years. For the first time in history, Japan has more people over 65 than it does children under age 15. By 2035, the government estimates that more than 30 percent of Japanese will be 65 or older. Japan is preparing to launch a public nursing insurance program, and several companies, such as Sanyo and Mitsubishi, are jumping on the bandwagon to produce senior-oriented products and services.

People are putting off marriage and having fewer children than in the past. The average age of marriage has risen to 26.3 for women and 28.5 for men. Furthermore, many women are seeking careers and are less likely to have or even want children. In 1998, there were 9.5 births per 1,000 people, compared with 28.1 births per 1,000 people in 1950. The average Japanese mother has about 1.4 children. The working age group began to decline in 1997. Now, numerous local governments are actually giving financial incentives for a first child, and day-care centers are becoming commonplace.

Changing population dynamics have serious ramifications throughout both economy and society. The Japanese government is concerned about its tax base and available work force. It must also face the fact that family structure is changing from extended to nuclear. In 1960, 87 percent of Japanese over 65 lived with adult children. By 1990 this number had dropped to

60 percent, and it has been falling ever since. It must cater to older people because they are the most likely to vote. Aside from this, Japan is still a country where the elderly are revered and taken care of.

Did you know that minority people are not well taken care of in Japan?

The Dark Side of Japan

A myth of Japan is that it is an ethnically and culturally homogeneous society. In fact, Japan has three classes of minorities: native ethnic groups such as the **Ainu**; foreign ethnic groups such as the Koreans; and social outcasts known as ***Burakumin.***

Native to Hokkaido, the Ainu are racially unrelated to the Japanese. Currently, about 30,000 identify themselves as Ainu. Their culture has been undermined by assimilative pressure, they suffer discrimination in all walks of life, and they live in relative poverty. Recent public awareness of the Ainu plight has not reduced discrimination, an issue most Japanese do not face. Media portray the adorable and folksy Ainu living close to nature. According to historian David Howell (1996), recognizing some aspects of Ainu identity may generate pride but this "also shields Japanese society at large from the need to re-examine its basic premises."

There are about 750,000 ethnic Koreans living in Japan, descended from the 2.5 million drafted as soldiers or forced laborers during Japan's 35 years of colonial rule in Korea. Most are classed as resident aliens; until the 1990s, even those Koreans born in Japan were routinely fingerprinted. About 20,000 Koreans died in the bombing of Hiroshima, but the needs of bomb-diseased survivors are virtually ignored. Discrimination in housing, employment, and education is rampant. Consequently, the majority of Japan's Koreans are trapped in poorly paid, unskilled jobs.

The largest minority group is the *Burakumin,* who are Japanese descended from underclasses defined in medieval times. The *Hinin* (nonpeople) were beggars, prostitutes, entertainers, and the like. The *Eta* (filth) were a subclass of untouchables regarded as spiritually unclean because of their occupations such as burying the dead or slaughtering animals. Since myths of spiritual and racial purity still abound in Japan, *Burakumin* are shunned and remain cloistered in urban ghettos.

Many companies, including major ones, will not hire minorities. Students from the best schools will not work for companies that do. To marry a minority is unthinkable.

Several minority action groups deal with government agencies for improved conditions, but not all minorities are equally represented. Moreover, some minorities (e.g., the Koreans) eschew full participation in Japanese society. This results in a paradox: sustaining group identity in the face of discrimination and relying on organizational intermediaries to deal with major issues shields Japanese society and individuals from confronting their prejudicial behaviors and practices.

TIGER'S TRANSFORMATION

The world cities of Tokyo and Osaka both have their origins as medieval **castle towns**—fortified towns bracketing a transportation corridor known as the Tokaido Road (Figure 11-4). The road became a ribbon of commerce after the capital was moved from Kyoto to **Edo** (Tokyo) in 1603. By the middle of the eighteenth century, Tokyo had 1 million inhabitants. With rapid postwar development, this urbanized and industrialized region, including Nagoya, became known as the **Tokaido megalopolis.**

Techno-economic expansion from 1955 to 1970 witnessed the movement of more than 30 million peo-

Figure 11-4

This upscale residential area in Yokohama illustrates the density of Japanese living space.

ple, 9 million into Tokyo, Nagoya, and Osaka. This led to the convergence of multiple connurbations into a continuous megalopolitan region extending from Tokyo along the inland sea to Northern Kyushu (Figure 11-5).

Depopulation of rural areas and congestion in urban areas stimulated decentralization: construction of dormitory towns; development of growth poles; and increased public spending and industrialization in smaller, micropolitan areas. Outward expansion was explosive.

Industry also changed. Newly emerging high-tech activities began to replace leading manufacturing industries. The new business of information technology began to chart the restructuring process. Growth in the Capital Tokyo Metropolitan Area (CTMA) began to exceed that of Osaka and Nagoya. In fact, these two cities became out-migration areas for the CTMA.

Tsukuba: Science City

Tsukuba Science City, 40 miles northeast of Tokyo, was conceived in the 1960s, originally as a government-sponsored scheme. It is an extensive development covering 112 square miles (28,559 ha) and has at least 48 research and educational institutions. Until very recently, the city was regarded as spatially and psychologically isolated from Tokyo. Another major complaint was that too many institutions were working independently; there was insufficient communication among disciplines. With the expansion of the CTMA and greater attention paid to organizational linkages, Tsukuba is now enmeshed in Japan's larger urban system.

CTMA's phenomenal growth paralleled Japan's integration in the global economy. Its geographical location between New York and London put it conveniently on line in the global, 24-hour, international telecommunication and financial transaction system. As a result, Tokyo was expected to provide an array of businesses and services not directly related to the domestic economy. As demand rose for office space in the city center, land values skyrocketed. Nevertheless, concentration of economic activity in the CTMA accelerated.

Other sectors of the economy increasingly demanded specialized services in order to maintain their competitive edge. The crucial role of research and development was soon recognized. **Kansai Science City** began to take shape in the 1980s. Kansai Techno-

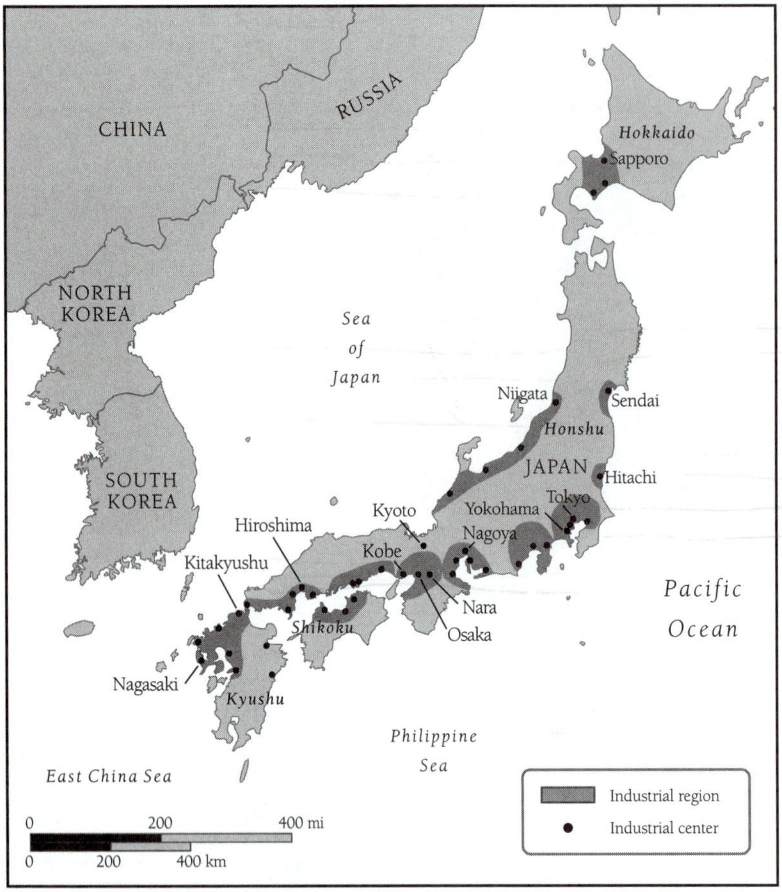

Figure 11-5

Industrial Japan. Why do you suppose that most of Japan's industrial activity is along the southern coast?

Research Complex includes five cities and three towns in the Kyoto–Osaka–Nara prefectures. It is an integrated, multinodal development with a projected population of 380,000 by the year 2000. This structural format models plans for urban Japan of the twenty-first century.

The new Japan will possess a multicentered regional and urban system founded on advanced information technology. The system is reorganizing from an assemblage of hierarchical structure to a unicentered, functional network with CTMA as the central node.

The **technopolis** is integral to the master plan that conceives a vast network of science cities across the nation. Fourteen industries such as biotechnology, robotics, software, and space have been targeted for development in tandem with research and development and other service functions. These are to fuel additional industry in the region. Each technopolis centers on a "mother city" and is to be linked by rapid transit to Tokyo, Nagoya, or Osaka by no longer than a one-day return trip. A new undersea tunnel to Hokkaido and a bridge to Shikoku along with more expressways and high-speed bullet trains are intended to facilitate this goal. This evolving **Techno-Archipelago** is Japan's landscape of the twenty-first century.

In 1990, 30 million people lived in CTMA (Figure 11-6). It has become the cog of national and regional

Figure 11-6

Shibuya is a crowded commercial center at the intersection of several transport lines in western Tokyo. It is a multifunctional area with offices, department stores, traditional shopping streets, and a nighttime entertainment district. © Barry Cronin/Gamma Liaison.

functions, as well as for international techo-economic transactions. Other cities are restructuring in light of this new reality. For example, Nagoya, home of Toyota automobile industry, is redefining its metropolitan function as high-tech design and innovation. Osaka has a new airport and is being developed as a "technoport" based on information technology.

Did you know that at rush hour, men are employed to push people in and pull them out of commuter trains?

The Tokyo Crush

Unprecedented agglomeration has caused the CTMA administration to examine its infrastructure. Transportation problems are enormous—a "commuter's hell." As work places are increasingly concentrated in the central business district and residences increasingly dispersed, 4- and 5-hour one-way journeys are commonplace. The automobile is becoming a nonviable mode of transport in midday, working Tokyo.

Land issues are critical because demand exceeds supply. Space for "intelligent office buildings" equipped with the state-of-the-art telecommunications is being accessed through land reclamation in Tokyo Bay. Moreover, the government is encouraging high-rise building in Shinjuku, Shibuyu, and other subcenters of the CTMA.

Housing is another key issue. Most people in Tokyo feel that their housing conditions are inadequate: too distant, too small, and too expensive. Fewer people are homeowners in the

CTMA than anywhere else in Japan. Old, traditional, urban communities tend to vanish in the wake of placeless offices, and many citizens complain that their sense of place has been wiped out. As the landscape is geared for the modern worker, aging populations become alienated. Further, as techno-jobs call for highly educated workers, traditional workers are cut out of the loop, and the gap between the haves and have-nots is widening.

One of the most serious problems in the CTMA is industrial and household waste disposal. NIMBY (Not-In-My-Backyard) attitudes are prevalent. The average per-person daily waste has doubled since 1970. In Tokyo, two-thirds of the waste is burnt and one-third used for landfill in Tokyo Bay. The waste problem is already getting out of control.

A NEW GEACPS?

As domestic manufacturing enterprises have lost competitiveness with rising labor costs, many traditional industries, such as steel production and shipbuilding, were forced to shut down, downsize, or seek cheaper labor elsewhere. As this type of employment shifted overseas, many Japanese secondary industries were said to be **hollowed-out.**

Over the years, Japan's foreign direct investment has increased around the world. Initially, investment was in primary and secondary activities but by the 1980s, three-quarters of this investment was in non-manufacturing activities, especially in real estate and services. Too-rapid expansion, based on excessive loans and speculation, came to a crashing end in 1990 with appreciation of the Japanese yen. Unable to meet payments, multiple enterprises were forced to restructure. Investment was pulled back from North America, Singapore, and Australia and redirected to China, Taiwan, and Southeast Asian countries. Between 1985 and 1994, the number of Japanese firms in Asia increased from 58,000 to 120,000. As its economic involvement advances, Japan can be expected to play a stronger role in the region.

Korea: Two Countries, One Nation

PHYSICAL SETTING

Korea is one of the world's oldest known land areas dating back 1,600 to 2,700 million years! Made of granite and limestone, the peninsula is tilted westward to the Yellow Sea where hundreds of islands dot the coast. Only 20 percent of Korea is level land. The rest is low-rising, sharp, and often bare mountain ridges. These ridges never reach above 9,200 feet (2,800 meters). The mountains are more concentrated in the north and east. Along the east coast, the Taebaek Range runs like a spine with southwestern spurs that cut the peninsula into narrow valleys and alluvial plains (Figure 11-7).

Korea is divided from China by the navigable Yalu and Tumen Rivers, both of which flow from Paektusan, the highest point in Korea at 9,000 feet (2,744 m). A range of extinct volcanoes leads from Paektusan to the Sea of Japan.

Like the rest of East Asia, Korea has a monsoon climate with warm, humid summers and cold, dry winters. While North Korea is influenced by continental extremes, the south is warmed by the Japan Current. Southwestern Korea and Cheju Island are, in fact, subtropical and have considerably more rainfall than the rest of the peninsula. About half the rainfall comes in June, July, and August, about 30 to 40 inches in the south and less in the north, which is much colder and dryer.

North–South differences are also apparent in resource availability. Although the peninsula is only moderately endowed with mineral resources, the north is favored with substantial deposits of coal, iron, copper, zinc, and other ingredients for heavy industry. The north also has extensive hydroelectric power resources, particularly along the Yalu, Tumen, and Taedong rivers. In the case of the Yalu and Tumen, the power is shared with China. In South Korea, minerals are similar to those in the north but, with the exception

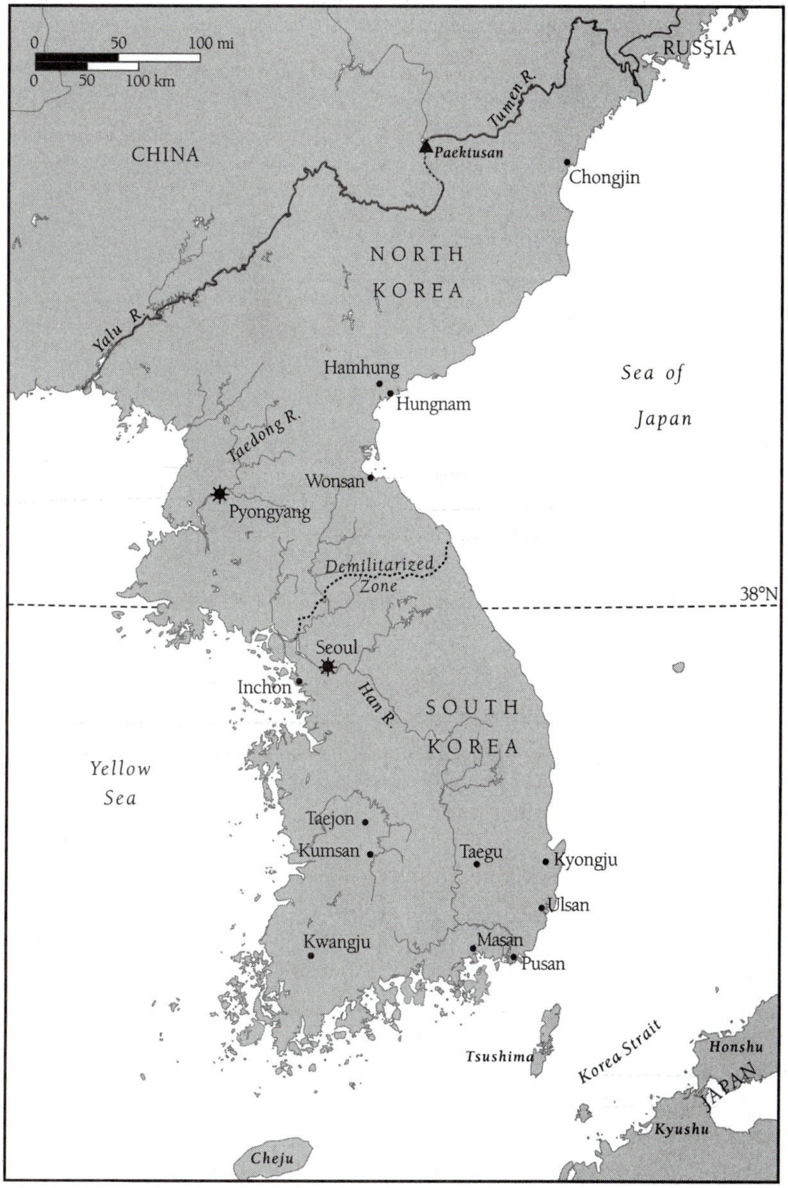

Figure 11-7

Geography of North and South Korea.

of graphite and tungsten, exist in smaller quantities and are of less quality. Consequently, the South relies on imported raw materials for its heavy industrial base.

Only about a sixth of mountainous North Korea is suitable for cropping. Coastal lowlands produce rice, corn, wheat, millet, and soybeans but food sufficiency is a problem. The hills have rich timber reserves, extensive orchards, and livestock grazing.

South Korea is the peninsula's traditional rice bowl, with about 20 percent level land. However, greater productivity is offset by a considerably larger population, almost double that of North Korea's 24 million in 1996. South Korea's agriculture is intensive, focused on rice, vegetables, fruit, and other types of market gardening. Kumsan is notable for its ginseng production (Figure 11-8). Cattle are raised on Cheju Island, where it is just warm enough to grow oranges and pineapples. Greenhouse agriculture is ubiquitous in South Korea.

South Korea is more urbanized than North Korea, which has only one city over a million people. This is the capital of Pyongyang. The twin cities of Hamhung-Hungnam contain fewer than a million people. In the South, six cities have more than a million people. The connurbation of Seoul and its port Inchon is the largest, with a combined population of 13 million people. Seoul, which means "capital," has been Korea's (now South Korea's) capital for 500 years.

The quintessential difference between North and South is political. North Korea's strict communist system has produced a harsh landscape of rigidity and poverty, isolated from all except China and Russia. Variation in South Korea's government from authoritarian to democracy has produced a landscape of free enterprise, modernization, and economic development to the status of Tiger in East Asia.

HISTORIC EVOLUTION

Archaeological evidence suggests that the Korean Peninsula was first settled 5,000 years ago when Tungusic tribes migrated to the area from Manchuria and northern China. Koguryo, the first state, developed around 100 A.D. The kingdoms of Paekche in the southwest and Silla in the southeast emerged around 250 and 350 A.D., respectively. These three kingdoms vied for control of the peninsula until the seventh century.

Aided by the Chinese, Silla was able to conquer Paekche and Koguryo by 668 A.D. Trading centers developed in Kyongju, the capital, as well as in provincial towns. Korean ships engaged in the region's sea trade. Buddhism (from China) was established as the state religion, and lavish temples and tombs were constructed (Figure 11-9). Woodblock printing was developed for the production of religious texts. From this time forward, Korea had an important cultural influence on Japan, including the introduction of Buddhism.

By 918, the kingdom of **Koryo** emerged in the central region of the peninsula and overthrew Silla in 935. Korea is the westernized version of Koryo. Buddhism remained the state religion but Confucian-

Figure 11-8

These are fields of ginseng outside Kumsan, the main marketing center for ginseng in Korea. Here the root is known as insam. *The soil is a rich mulch of chestnut or oak leaves, and the plants are shaded with thatch or fiber mats. Once used, the soil is not cultivated again for another 10 to 15 years. Ginseng, especially the red variety, has been part of East Asian pharmacopoeia for centuries.*

Figure 11-9

This corner eave of the great Buddhist temple at Pulguksa illustrates the detailed craftsmanship of Korean temple builders. The central dragon is meant to ward off evil. Built in the sixth century, this is one of the oldest surviving Buddhist monasteries in Korea.

ism emerged as a major force in government ethics and political structure. Even today, Korea is considered the most Confucian of all Asian nations. Check out its flag. Where do these symbols originate?

In 1231, Mongol armies moved into the peninsula. After 30 years of war, Koryo became a vassal state of the Mongol Yuan Dynasty in China. Following the fall of the Yuan, Korea's Yi Dynasty continued the tributary relationship with China's Ming Dynasty.

The Yi moved their capital to Seoul on the Han River and renamed their kingdom Chosun in 1392. Under Yi rule, Confucianism replaced Buddhism as the state belief system. The world's first movable, metal type was developed. More importantly perhaps, a Korean lettering system was developed in the fifteenth century. Known as **Hangul**, it reduced dependency on Chinese characters and made literacy achievable for the general public.

Hangul

In the fifteenth century, King Sejong commissioned scholars from the "Hall of Worthies" to develop a Korean alphabet system for those who had not mastered Chinese characters. After analyzing the phonetic structure of the language, they distinguished three separate elements of words: initial, medial, and final sounds. Of these, the medial element is a vowel. The new alphabet consisted of 29 letters—22 consonants and 7 vowels. Consonants and vowels are grouped into a syllabic block form to create a word. This new system, called *Hangul*, differed radically from the Chinese ideographic system where a concept is conveyed by each character. King Sejang presented his "correct sounds for the instruction of the people" in 1443 and literacy became widespread.

Tangling with a Japanese Tiger

Japanese forces invaded Korea in 1592 and 1597. With the help of China the Koreans fended off the Japanese but were forced to send artisans and scholars to teach Chosun's advanced technology in Japan. The Japanese ceramic and laquerware industries owe their origins to the Koreans. Later, the Japanese were given residence rights in a small area of Pusan.

The seventeenth century brought Manchu control and subsequent decline. After much struggle, Korea fell to Manchu power 30 years before the fall of China's

Ming Dynasty in 1644. Significant change was on the horizon. Factional strife within Korea weakened the political structure. Christian ideas, introduced via official tribute missions to China, clashed with Confucianism. New agricultural techniques fostered a commercial economy, despite aristocratic disdain for such activity. Corruption and intrigue hampered orderly administration. By the mid-1800s, Korean politics and the economy were in shambles.

In 1864, the Korean government closed the country to all except the Chinese and the Japanese at Pusan. All other foreign interests were spurned until Japan forced Korea to enter into its first foreign commercial treaty in 1876. It was during this period that Korea became known as the *Hermit Kingdom*. This period also marked a great divide in the nation's history. Prior history was China-centered. The future would be determined by Japan and the West.

Korea fell prey to Japanese imperialism in the late nineteenth century. The Japanese engineered the murder of the anti-Japanese Korean Queen Min in 1895. The Russo–Japanese War of 1904–1905 saw Japanese troops operating on the Korean peninsula as they moved the war front into Manchuria. Korea became a Japanese protectorate in 1905, and despite anti-Japanese efforts, Japan was able to annex the country in 1910. Japanese rule was highly resented by the Koreans, who viewed themselves as culturally superior. Japan relied on the use of force to transform its colony.

While the traditional, Confucian agrarian order was retained, a modern administration was established, with Japanese functioning in key positions. Both agriculture and industry were structured to meet the needs of Japan and its plans for aggrandizement. Transportation, communication, mining, and other infrastructure were modernized and expanded. Thousands of Koreans were sent to Japan to work in factories. In Japan, the Korean language was banned and Koreans had to take Japanese names. In Korea, resistance movements were brutally crushed, deepening Korean hatred of their Japanese overlords. Japanese rule of Korea did not end until Tokyo surrendered to the Allied Forces in 1945.

North– South Division

In 1945, the Russians and the Americans divided the peninsula into two spheres of influence and military occupation at the **38th parallel**. Free elections were supposed to follow, but disagreement on this issue ensued in the next two years. In the face of adamant refusal of the Soviets to subscribe to United Nations supervised elections, the Republic of Korea was established south of the 38th parallel on August 15, 1948. The Soviet response came in September, with the formation of the Democratic People's Republic in the north. The United States withdrew its military forces the following year.

On June 25, 1950, North Korean armed forces attacked the South across the 38th parallel. The United Nations responded with the support of sixteen nations, led by the United States, and the Korean War ensued. The conflict was prolonged by the entry of Chinese troops in support of North Korea in 1950. After three years of fighting, an armistice was signed, with an agreement to find a political solution to the problem of division. The 1953 agreement led to the creation of a Demilitarized Zone (DMZ), which has become heavily militarized on both sides.

The DMZ is an interesting site. You can visit the southern flank and look across a "no-man's-land" to a phony "ideal" North Korean village and propaganda signboards touting the superior lifestyle of North Koreans. The route to the DMZ, just 25 miles (40 km) north of Seoul, is marked with military bunkers and bomb-filled overpasses ready to crush any invasion from the North. The North Koreans have made several tunneling attempts into the South, but these have been detected and destroyed.

The Korean War devastated the country, especially the north, where systematic bombing eradicated the industrial plants. Some 4 million perished, and millions of refugees remained trapped on the wrong side of the divide, most never to see their families again. Around 1 million refugees came to the South with the retreating U.N. forces. Subsequently, the Russians and Chinese aided the North and the Americans supported the South in recovery efforts. Korea was now two countries, each to pursue its own path of development.

NORTH KOREA: A CLOSED SOCIETY

The Democratic People's Republic of Korea (DPRK) is one of the world's most closed and rigid societies. Under the absolute power of "Great Leader" Kim Il Song, the north was brought from the brink of disaster into an industrial economy based on central planning and state ownership of the means of production. Individual needs are subjected to the will of the state, and ownership is limited to individual and household possessions.

Great Leader espoused the ***juche* philosophy**, meaning self-reliance or self-sufficiency. Juche is based on the proposition that humans are masters of all things and that they, among all creatures, are endowed with the capability of self-reliance and independence. However, the masses cannot succeed in their struggle for these goals without a supreme leader to give form to their causes and direction to their actions. This leader was Kim Il Song, followed by his son "Dear Leader" Kim Jong Il in 1994.

North Korean society is stratified according to Communist Party loyalty; level of commitment determines job prospects, and access to housing, education, and health care. Citizens can be detained arbitrarily; there are an estimated 150,000 political prisoners. Listening to foreign radio broadcasts is punishable by death. Subjugation to a top-down, tightly structured administration is a familiar mode of rule in Korea, and this reinforces the absolute authoritarianism that North Koreans now experience.

This is a highly socialized command economy, where farm land is collectivized, and agriculture accounts for 36 percent of the work force. State-owned industry produces 95 percent of manufactured goods, which focus on heavy industrial products and armaments. Light industry lags far behind. In 1992, output dropped sharply because of disruption in relations with the Soviet Union and China and reduction in their economic and technical support. There is a shortage of arable land and the country has not achieved self-sufficiency in food production.

North Korea needs at least 6 million tons of feed grain a year to get by. Its last full crop in 1994 produced 4.1 million tons. Torrential rains in a badly deforested environment wrecked the harvests of 1995 and 1996. Reports from people who manage to sneak across the Chinese border to the Korean community there and from aid workers report that starvation is widespread. Hillsides are stripped of vegetation. There are no cats, dogs, or chickens. It is said that the military and the Party elite remain well fed, given food aid from China, Russia, Japan, and the United States.

In 1995, peasants were allowed to cultivate private plots, but this resulted in reduction of yields at the collective level. The DPRK estimates that famine claimed at least 220,000 lives between 1995 and 1998. International organizations estimate that 1.5 million North Koreans have perished due to famine-related causes. Although climatic conditions improved in 1999, the nation is still significantly short of food and lacks the minimal calories required each day to stave off malnutrition. The government has broadcast recipes for boiling grasses.

Data on urbanization in North Korea is notably unreliable. However, according to DPRK information, urbanization has increased from less than 20 percent in 1953 to more than 60 percent in 1987. The definition of *urban* is unclear. Most urbanites live in 23 cities of more than 89,000. Pyongyang has over a million. Hamhung, the second largest city, is about a third the size of Pyongyang. The port of Chongjin is another large, industrial city.

Beijing has cast its relationship with North Korea as being "close as lips and teeth." It is in China's interest to prop up the North Korean regime while at the same time effecting its redirection along a more liberal path according to the China model. China wants regional stability. It does not want a floundering country on its northeastern border next to its own Korean population, nor does it want a unified Korea, friendly to the West and housing American troops.

North Korea with its population of a mere 24 million, has the fifth-largest standing army in the world after China, the United States, Russia, and India. More than 1 million troops hover over the half-million South Korean forces.

In 1994, the DPRK suggested that it could build a nuclear bomb by extracting plutonium from one of its test reactors. In response, the United States, South Korea, and other countries agreed to provide the North with oil until the South Koreans completed two nuclear power reactors in the North. This has stopped further nuclear bomb research, for now.

The DPRK also has an arsenal of missiles capable of striking Japan and beyond. It sells these to the Middle East for badly needed foreign exchange. In 1998, while engaged in talks with the United States regarding normalization of relations and food aid, it proposed that Washington pay U.S.$500 million to stop missile sales. North Korea's actions have peaked concerns in the region and have revived support for Star-Wars-type missile defense systems in Japan and Taiwan. China is not pleased with this turn of events. Clearly North Korea's actions are detrimental to any consideration of reunification of North and South.

SOUTH KOREA: KOREA, INC.

South Korea has transformed itself from a poor, agrarian society to one of the world's most highly industrialized nations. Land reform, education, and economic expansion have eliminated the *yangban,* landlord elite. The growth of a middle class has been accompanied by dramatic population growth and urbanization. Because of the influential role played by the government, South Korea is said to be a product of **state capitalism.**

State Capitalism

South Korea's transition from an agricultural to an urban and industrial society followed three phases. Government and business targeted specific industries for development. Initially, emphasis was placed on light manufactures and textiles. In the 1970s, heavy industries such as iron and steel, and chemicals were emphasized. Then, the focus was shifted to making automobiles, ships and electronics.

The lynchpin of industrialization is the ***chaebol.*** A chaebol is a giant corporate conglomerate. Hyundai, Daewoo, Samsung, and LG are leading *chaebols.* In the early years of development, preferential investment in exports discouraged the production of consumer goods. This policy was reversed in the 1980s. Then labor unions demanded wage increases and domestic consumption rose. Higher wages saw a decline in South Korea's international competitiveness in labor-intensive manufacturing, and overseas investment for cheap labor sources became essential.

Chaebols

Chaebols are the equivalent of the Japanese *Keiretsu.* Both terms are written with the same Chinese characters meaning "fortune cluster." Owned and managed by family groups, they are enormous conglomerates of diversified businesses, each having hundreds of ancillary firms. Of about 30 such concerns, Samsung is the largest. Samsung manufactures products ranging from airplanes, to semiconductors, to flour and paper. It also operates an array of service firms such as insurance, hotels, and department stores.

The *chaebols'* rise to economic dominance can be attributed, in part, to Confucian ideals and loyalties. The *chaebol* is like a family with its structured loyalties intensifying toward the family head: the boss. Unlike Japan, where consensus rules, in Korea consensus is what the boss says it is.

In the Hyundai *chaebol,* Confucian spirit became the "Hyundai Spirit." A manual given to new recruits observes: "The indomitable driving force, a religious belief in attaining a goal, and a moral diligence tempered with the frugality of the Hyundai group have materialized as a major part in the heavy industry of Korea."

Another important *chaebol* anchor is the support of state-owned banks and guaranteed loans. Government-infused capital has fostered expansion and diversification. According to *The Economist* (1996), the four top *chaebols* account for 80 percent of GDP and 60 percent of exports.

More recently, with rapid wage increases under democratic reforms, *chaebols* have become less competitive. Disbanding or selling off numerous unprofitable "family members" (ancillary firms), venturing more into research and development, and procuring cheaper labor overseas are part of recent restructuring efforts.

South Korean enterprises are found worldwide. For example, Daewoo, which means "Great Universe," has poured billions of dollars in planet-circling investments including electronics and components, hotels, autos, construction, finance, textiles, and heavy industries. Daewoo's globalization plans include hiring 250,000 foreign workers at 1,000 overseas subsidiaries. Daewoo Motors has plants in such countries as Poland, India, Iran, Uzbekistan, and the Philippines. Daewoo Electronics hopes to capture 10 percent of the world market in TVs, VCRs, PC monitors, refrigerators, washing machines, and microwave ovens.

Urbanization

South Korea's population doubled from 20 million in 1949 to 44.5 million by 1994. A major contributor was the influx of 3.3 million Koreans who had been living in Japan, China, and the Soviet Union. An additional 2 million refugees migrated from North Korea. A population control program launched in 1965 curbed the birth rate, but improved living standards lowered the death rate and increased life expectancy. Rural–urban migration depleted the countryside and provoked a major urban housing crisis. Millions of apartment units (flats) have been constructed to house this 75 percent urban society.

Urbanization and apartment living have altered Korean traditional family structure. Most Koreans now live in nuclear family settings rather than in extended family systems. However, Confucian values prevail. Women now have more control over their lives but are far from equal to men in decision making or economic opportunity. Women are found mostly in service or "nimble-finger" jobs. Corporations are overwhelmingly male enterprises at management levels.

Two-thirds of South Koreans live in cities of 50,000 or greater. Over half the population lives in the six largest cities. Greater Seoul has sprawled eastward and southward with an agglomeration of new satellite cities (Figure 11-10). Formerly focused on a single CBD dating from Japanese times, Seoul has evolved from a monocentric to a multicentric city. Despite rapid construction of apartments, the average occupation rate in 1990 was 193 percent. This means that most flats house two families—one owner and one renter. In older, more densely populated areas, occupation rates are as high as 600 percent!

Much of the world of downtown Seoul is underground. Miles of shopping arcades lay under the streets. All are connected to a network of subway lines and linked to major department stores and hotels above ground.

Another modern development is the rejuvenation of the Han River. Formerly a trash-filled, dry ditch in summer, its water level is now kept constant by a

Figure 11-10

This is a southern expansion of Seoul looking past the Han River and the Olympic Expressway to the suburb of Tongjakgu. The proportion of Seoul's population living south of the Han River increased from 17 percent in 1960 to 48 percent in 1990.

series of dams. The river is criss-crossed by bridges, sailing and passenger boats, and lined with parks and high-rise housing developments. Whereas 80 percent of raw sewage used to flow into the Han channel, new sewage systems and treatment plants now treat all sewage.

Inchon and Taegu are cities of more than 2 million, while Taejon and Kwangju have more than a million residents apiece. The port of Pusan has a population of close to 4 million. With their concentrations of industry, these largest cities continue to grow, while medium size and smaller places tend to stagnate. However, their rate of growth is decreasing as traditional industries have begun to decline.

With the bulk of industry still concentrated in the capital region (Seoul), the Korean government has devised a **technobelt** system whereby research and industry are to be linked via communication and telecommunication. The key nodes of this system are Seoul and Taedok, a new research and development center near Taejon. Each belt has designated specializations. For example, the Seoul to Kwangju technobelt will focus on energy and food production. The Seoul–Taedok–Pusan technobelt will concentrate on electronics and textiles. Other areas designated for growth (e.g., Masan and Changwon on the south coast) have become engineering centers linked with the Pusan–Ulsan heavy industrial region. Ulsan is dominated by Hyundai car manufacturing and ship building.

Overseas Koreans

Nearly 5 million Koreans live overseas. The largest group outside the peninsula resides closest to it in China. As many as 2 million live in China's Manchurian region. The strongest concentration is in Yanbian Korean Autonomous Prefecture just north of the Tumen River. Here, Koreans are able to exercise a degree of autonomy and speak Korean in accordance with the rights given to all minority nationalities in China.

Koreans were pioneers in opening up the cold Manchurian region to paddy rice farming. Rice was first grown in Yanbian in 1877, and Koreans are still responsible for producing much of the rice crop in the most northerly province of Heilungjiang.

Large contingents of Koreans also live in the Central Asian republics of Kazakhstan and Uzbekistan and on Sakhalin off the east coast of Siberia. Unofficial estimates range up to 750,000. These populations are important links for South Korea's recent Central Asian investments.

As discussed earlier, the Korean community in Japan makes up 85 percent of its foreign population. Despite their status as citizens of the Japanese Empire in colonial times, the majority of Koreans continue to be classified as aliens and suffer various forms of discrimination.

Emigration to the United States began in the early twentieth century, from 1903 to 1905, when a group of 7,226 Koreans were allowed to live in Hawaii and work on sugar plantations. Under Japanese pressure, this migration was stopped and many Korean settlers abandoned the plantations and became shopkeepers. Some 2,000 moved to California. This was the seed of a subsequent migration stream to the United States. The majority of these are Christian.

About a third of America's Koreans live in California, mainly Los Angeles. Other significant concentrations are in New York and Chicago.

REUNIFICATION'S PROSPECTS

North and South Korea are entering the twenty-first century as two diametrically opposed systems with divergent paths of development. Both Koreas remain pledged to reunification and occasionally present proposals. A strong emotional commitment to reunification exists among many Koreans, especially those with cross-border family ties.

Extant differences make reunification problematic. North Koreans are not even an eighth as rich as South Koreans. South Koreans are twice as numerous. North Koreans have virtually no idea of how to function in a modern, technologically advanced society.

There are only 24 telephone lines between Seoul and Pyongyang, and these are rarely used. In the 1980s, a visiting DPRK delegation accused the South of rounding up every car in Korea to create traffic jams in Seoul. Even the language has evolved differently, with the South's technological vocabulary unknown in the North.

Raising the North's productivity level to stem a southward flow of migrants would cost billions. Although the South has traded with the North since 1990—food for raw materials—investment has been severely limited by the DPRK government.

South Korea wants to reunify in stages: diplomatic contacts to build trust; investment in special economic zones in the North; and, down the line when the economic gap has been narrowed, full unification.

In 2000, an historic event took place. President Kim Dae Jung of South Korea met with President Kim Jong Il of North Korea to discuss reunification prospects. As a result of this meeting, both sides have stopped propaganda broadcasts across the border and allowed a limited number of cross-border visits for family reunification. Agreement was made for the construction of a railway from Seoul to Pyongyang. Symbolically, North and South Korean athletes marched together under one flag at the Olympic Games in Sydney, Australia. More talks are scheduled in the future. Kim Dae Jung has been awarded the Nobel Peace Prize for his efforts towards peace in the region.

Figure 11-11

Taiwan.

Taiwan: Little Tiger and Big Dragon

Small but rich, economically powerful but politically isolated: this is Taiwan, an island just 100 miles (165 km) from the big dragon, Mainland China (Figure 11-11). The history of Taiwan is inextricably tied to that of the People's Republic, and its current development remains couched in that relationship. It is one of East Asia's NICs and has achieved tiger status in terms of economic development.

THE PHYSICAL LANDSCAPE

Taiwan incorporates the main island and 64 smaller islands called the Penghu Islands. The Penghu were colonized by the Portuguese, who named them the Pescadores (Fishermen). Here, fringing reefs provide good shelter for ships and there is an important naval base at Makung. Because of their position in the Taiwan Strait, the Penghu Islands are of great strategic importance to Taiwan.

The island of Taiwan can be divided both vertically and horizontally. The vertical division is created by the Chungyang Shan (Central Mountains) ranging from 10 to 35 miles (16–56 km) in width. The Chungyang are extremely rugged with more than 30 peaks reaching over 10,000 feet (3,000 m). The highest is Yu Shan at 13,113 feet (3,997 m). River valleys and canyons cut the mountainsides, which are coated with coniferous and mixed forests. With the exception of the Taitung rift valley, cool temperatures, thin soils, and severe erosion hinder cultivation. Still, the mountains are important for forest products and hydroelectric power.

Most people dwell on the western coastal plain, which can be divided horizontally by the Choshui River. To the north are hills and basins, including those of Taipei and Taichung. To the south is a low, flat, alluvial plain. The river also marks the climatic boundary between north and south. In the north, the rainy season is from October to March; in the south it is from May to September. Typhoons are common in July and August. Agricultural patterns differ as well with such crops as rice and tea in the north and soybeans, pineapples, sugar cane, and other tropical crops in the south.

Agricultural output has steadily declined over the years as farmland and employment shifts to urbanization and industry. Taiwan is 75 percent urban with major population clusters in the capital of Taipei, and in the ports of Chilung, Tainan, and Kaohsiung.

FROM *BAO DAO* TO FORMOSA TO TAIWAN

Ancient Chinese maps portray the island of Taiwan as the island of **Bao Dao.** Written accounts referred to it as "a mudball across the sea, not worthy of China." Bao Dao was known to Chinese fishermen and pirates but was inhabited by non-Chinese people. These aboriginal people are Malay-Polynesian and are related to the indigenous people of the mountains of northern Luzon, the closest island of the Philippines. There are nine distinct ethnic groups among the aborigines, each with its own language and culture. However, immigration from China pushed these people into the mountains and subsequent discriminatory policies of the Taiwanese government have rendered their cultures almost obsolete.

Immigrants from China began arriving in the fourteenth century, a migration stream that continued until World War II. These were Han people from the Fujian coast who spoke the dialect known variously as Amoy, Minnan, or Hoklo. This dialect is called *Taiwanese,* and these people are known as **native Taiwanese.** Native Taiwanese are 70 percent of Taiwan's population. In recent years the term **Taiwanese** is used to mean everyone who lives on the island.

In the sixteenth century, the Portuguese were struck by the beauty of the forested mountains and pristine beaches of the island and called it *Ilha Formosa,* beautiful island. Taiwan was known to the world as **Formosa** for much of the twentieth century.

Japanese traders also visited the island and even established a settlement near present-day Tainan. They called the island "*Tai Wan,*" meaning Big Bay.

In 1624, the Dutch established Forts Zeelandia and Providentia near the by-then defunct Japanese settlement at Tainan. Spain established colonies in 1626 but was later ousted by the Dutch. The Dutch established sugar plantations and encouraged increased immigration from China as fieldworkers. The Dutch were thwarted by a Ming general in 1662.

In the seventeenth century, Hakka speakers from Guangdong arrived and settled in the foothills where they developed Taiwan's forest products industry including camphor wood and oil. Hakka account for 10 to 15 percent of the population and make up a disproportionate share of Taiwan's industrial workers.

Emigration from China had been illegal since the Ming Dynasty. Even so, people left to avoid taxes and seize new opportunity elsewhere. The Chinese regime regarded them as people who renounced civilization. Taiwan was, to China, uncivilized.

In 1684, the Qing Dynasty achieved control of Taiwan for the first time. But the island's population rebelled against them repeatedly over the next 200 years. One official Qing document complained: "Every three years an uprising! Every five years a rebellion!" Chinese immigration continued, and by the late nineteenth century, the island had a population of 2.5 million.

Western traders reappeared after the Opium Wars of the 1840s. Britain and Canada expanded agricultural exports, sent Presbyterian missionaries, and established hospitals and schools.

Safety of traders was a major issue. China denied responsibility for incidents of violence because these were taking place "beyond the boundaries of China." Nevertheless, China made Taiwan a province of China in 1887. These unclear relationships between China and Taiwan were preludes to decades of uncertainty and difficulties continuing to this day.

JAPANESE RULE

At the end of the Sino-Japanese War (1894–1895), the Treaty of Shimonoseki gave Formosa and the Pescadores (along with Korea) to Japan. On June 1, 1895, a resistance movement proclaimed Asia's first republic: the Republic of Formosa. This republican group was quickly crushed, and the Japanese began their era of harsh rule.

Japan accelerated Taiwan's economic development. This was meant to provide Japan with agricultural and later, industrial products. The colonial authorities expanded electric power, and improved port facilities, internal transportation, and communication. A health system was established, and an educational system promoted Japanese language and culture. Many Taiwanese attended Japanese universities. The groundwork for future progress was laid at this time.

When the Qing Dynasty was overthrown in 1911, both the Chinese Nationalists and the Communists wanted Taiwan liberated from the Japanese. Both groups voiced support for an "independent and democratic" Taiwan.

Chiang Kai-shek became increasingly interested in Taiwan as territory as the Second World War ensued. Fearing that Chiang's corrupt and incompetent troops would surrender to Japan, the allied powers issued the Cairo Declaration in 1943. This was one of several communiqués promising to restore Taiwan and the Pescadores to China. Ever since, despite their earlier support of Taiwan's independence, the Nationalists and Communists have cited this declaration to justify their view that "Taiwan is a sacred and inseparable province of China." But the Cairo Declaration was, in fact, a statement and not a legally binding treaty. The island's uncertain status was furthered by the Allied–Japanese peace treaties that left the fate of the island to its inhabitants.

Another element of the island's population are immigrants who arrived from the mainland between 1945 and 1949. These are referred to as **mainlanders**, although their "mainlander" descendants are now born in Taiwan. In fact, the 1990 census shows that more than 90 percent of Taiwanese were born on the island.

The "One China" Question

In 1949, Chiang Kai-shek and his 2 million loyalists set up what they claimed to be the true government of the Republic of China (ROC) (in exile). In that same year, Mao Zedong renamed China the People's Republic of China (PRC). Both Mao and Chiang claimed to head the government of "one China," of which Taiwan is a province.

In Taiwan, Mandarin was declared the official language by the KMT government. However, Taiwanese is widely spoken in informal settings, and current demand has resulted in the production of Taiwanese language media.

The United States was entrenched in the Cold War against communism and consequently supported Chiang and the KMT. This support met with world realities in 1971 when the ROC lost its seat at the United Nations in favor of the PRC. The seating of the PRC in the world body quickly eroded Taiwan's status as representative of China. In 1972, U.S. President Nixon signed the Shanghai Communiqué that "acknowledged" the PRC view as well as the KMT view that "All Chinese on either side of the Taiwan Strait maintain there is but one China and that Taiwan is a part of China." It gave no credence to the Taiwanese desire for self-determination.

In 1978, President Carter established formal diplomatic relations with the PRC and reiterated the Shanghai Communiqué. The United States also severed formal relations with Taiwan, although economic and cultural offices maintain informal ties. The United States wants the issue resolved peacefully. The PRC says that the issue of reunification is an internal affair.

Regional considerations are also critical. If the PRC controlled both flanks of the Taiwan Strait, it would have a chokehold on main sea lanes to South Korea and Japan.

As PRC-international relations increased, Taiwan became represented only as "Taipei" in international organizations. Taiwan has no official embassies abroad but, rather, trade offices that handle similar functions. In the 1990s, the island began promoting itself as "China Taipei."

Both Taiwan and the mainland cling to their notions of one China. The PRC hopes that the examples of Hong Kong and Macau and their "one country, two systems" policy will stimulate the ROC to formally rejoin the PRC. But many Taiwanese, aware of conditions on the mainland, have little interest in this goal. There is also talk of changing the name PRC to simply "China." However, while everyone could agree that they belong to "China," political, economic and other practical matters would remain at issue. The China–Taiwan problem remains unresolved.

The tenuous political situation has not precluded ongoing and intensifying economic relationships between China and Taiwan. In fact, big dragon China is of significant economic consequence to little tiger Taiwan.

RISE OF THE TIGER

Taiwan's achievements as an economic tiger are underpinned by its excellent educational system, in-place infrastructure, and American connections. Initially, KMT land reform gave land to tenant farmers and labor-intensive industries employed thousands. The state confined itself to key sectors such as finance and energy, leaving most industrial opportunities to private entrepreneurs. Farmers now focus on specialty crops such as mushrooms and aquaculture products such as eels. Industry has moved steadily up the value-added ladder from garments, to footwear, to electronics.

Taiwan has long met the criteria of a *kotadesasi* (discussed in Chapter 4) through its economic transformation and evolving spatial structure. Growth of population in the four metropolitan regions of Taipei, Taichung, Tainan, and Kaohsiung and the construction of a north–south freeway from Taipei to Kaohsiung have created a metropolitan corridor similar to that between Tokyo and Osaka. With its multiplex of service, industrial, and agricultural activities, linked by transportation and communication networks, this part of Taiwan exhibits all the features of a *kotadesasi* zone.

Hsinchu Science Park, established in 1980, has become the world's third largest high-tech industrial center with strong links to California's Silicon valley, where hundreds of U.S.-schooled Taiwanese are now employed. A second science park is being built near Tainan. Taiwan now leads the world in the production of notebook computers and is among the top three makers of desktop computers and peripherals. But a small domestic market, increasing labor costs, shortage of land for expansion, and lack of raw materials have compelled structural and spatial changes.

Taiwan has taken advantage of **commoditization** of the personal computer (PC) industry. This is the division of the PC market into two sectors: big name-brand marketing and sales, and hundreds of no-name component suppliers. Taiwan has garnered the latter—the Original Equipment Manufacture market. Three-quarters of the country's electronic products is sold under another brand name.

A penchant for expansion, combined with cost efficiency, has led Taiwan to extend its operations overseas, where both space and cheap labor are available. More than 80,000 firms have moved offshore—half to Southeast Asia and half to the PRC. For example, Paochen footwear could handle 10,000 workers in Taiwan but employs 120,000 in the PRC, where it is the world's largest producer of Nike, Reebok, and other name-brand shoes.

The China Connection

Taiwan is the second largest investor in China after Hong Kong. It had well over 30,000 investment projects by 1994. Along with Hong Kong, it shares half the investment projects in Shanghai's Pudong development project.

China is Taiwan's third largest trading partner after the United States and Japan. Until recently, because of the political situation, most of this trade was indirect via Hong Kong. But Hong Kong, even with its SAR status, is now part of the PRC. In 1997, transshipment services commenced between Kaohsiung and Xiamen and Fuzhou in Fujian Province—the first authorized cross-strait shipping links.

To further ease restrictions, vessels from China or Taiwan are no longer required to transfer cargo at a third port. Instead, they only have to pass through the waters of a third port on their way across the Strait. Taiwan is also creating an SEZ to handle the China trade. Foreign ships are not allowed to participate in this trade because the PRC considers cross-strait traffic to be domestic. Foreign ships can ply the Hong Kong–Shanghai route but not the Shanghai–Taiwan route.

As part of its efforts to lure Taiwan into the PRC fold, China is deepening and widening harbors to accommodate Taiwanese ships. It also gives preferential treatment to Taiwanese investors on the mainland who have built factories from Suzhou to Urumchi. Some fear that Taiwan's increased foreign investment and plant location in the PRC could ultimately hold the country hostage to the whims of the Communist government in Beijing.

Southward Policy

Taiwan's **southward policy** involves increasing investment in Southeast Asian countries to offset investments in China. By 1994 the ASEAN nations including Vietnam comprised Tawan's third most important export destination after Japan and the United States. Subic Bay in the Philippines, Batam, Indonesia (just south of Singapore), and Vietnam are key targets for investment, construction, and development loans. Southeast Asia supplies Taiwan with raw materials including minerals and forest products.

This investment strategy is intended to upgrade Taiwan's industrial capacity, internationalize its business operations, and maintain its competitive advantage and economic growth in context of regional and global dynamics. Moreover, strong economic impacts can only strengthen Taiwan's *de facto* political influence in the region.

THE DOWNSIDE OF DEVELOPMENT

Development has brought a lot of benefits, but many Taiwanese are not happy with its negative side effects. For instance, there is a growing rich–poor gap as money becomes increasingly concentrated in the upper echelons of society. By 1994, the richest 20 percent controlled more than half the wealth in Taiwan. Business is notorious for corruption and nepotism. Gambling, prostitution, and other criminal activities are serious issues. Individualism, materialism, and hedonism are striking blows at traditional Confucian values, to the dismay of the older generation. Americanization is often blamed for these trends.

Environmental quality has sadly deteriorated with unchecked air, ground, and water pollution. Traffic congestion is out of control, and the infrastructure cannot keep up with demands for electricity and power. One good sign is that people are forming issue-oriented public interest groups and that environmental organizations have multiplied. However, environmental and social campaigns have not kept pace with rapid change and the uncontrolled growth of Taiwan's material culture.

Regional Integration

Significant changes have taken place in the economies of East Asia. The primary product economy is no longer directly related to the industrial economy, thereby diminishing the role of natural resources. Production has become detached from employment as industries shift from being labor intensive to knowledge intensive. Capital movement has become the engine of the world economy, with financial flows and transactions overarching the "real economy" of production and trade in goods and services. FDI has fostered the entanglement of countries and cities in Pacific Asia in dynamic webs of economic interdependence. Japan is a critical cog in this system.

What has emerged in Pacific Asia is a **functional city system** in which the activities or functions of a city determine its importance in a hierarchy of expanded networks. Simple factors of location, or population size, or a raw material are no longer paramount in regional supremacy. For example, the populations of Osaka, Seoul, and Taipei are smaller than those of Calcutta, yet they are considered "world cities" and Calcutta is not. Increasing numbers of transnational corporations have made East Asian cities their home, furthering urban ties to the world city network.

Global restructuring can be observed at several levels. First, there is a tendency for capital to be concentrated in fewer cities in relation to their key functions, as in the case of Tokyo. Manufacturing concerns are increasingly located in suburban areas including industrial parks, science cities, and technopolises. Several urban corridors with *kotadesasi* features have developed. Imagine this: An inverted S-shaped 930-mile (1,500 km) urban corridor stretches from Beijing to Tokyo via Pyongyang and Seoul, connecting 77 cities of over 200,000 inhabitants apiece. More than 97 million urbanites reside in this corridor.

Various economic hubs have emerged to take advantage of complementarity, especially in labor supply, across national boundaries. Certainly, Taiwan is now enmeshed with the economy of China, especially the south and Hong Kong. North Korea and South Korea are tied in a **growth triangle** with Northeast Asia. The growth triangle that has evolved between Japan, Korea, and Taiwan is called the *Jakota Triangle.* If trends continue, the growth triangles of East Asia will merge with those of Southeast Asia. China, with its tremendous growth potential, will play an increasingly significant role in the larger trans-Pacific and global scheme. The following chapters on Southeast Asia will reveal even more about regional and global dynamics.

Main Point Summary

1. Rich and highly urbanized, Japan is one of the world's largest users of natural resources. The Tokyo urban area is the largest metropolitan region on earth.
2. The islands of Honshu, Hokkaido, Kyushu, and Shikoku dominate the Japanese archipelago.
3. Japan is the most seismically active country in the world. Its most recent devastating quake was in Kobe in 1995. Japan also has more than 200 volcanoes.
4. Japan is very mountainous. Settlement clusters on the 15 percent that is lowland.
5. Evidence points to the fact that Japanese migrated from Korea. Japanese culture diffused from the Kansai Plain northward.
6. Fearing foreign intrusion, Japan closed the country to outside contact from 1603 to 1854. The Americans were the first to open Japan to outsiders.
7. Once Japan realized its backwardness relative to Europe, it raced to modernize, industrialize, and militarize. Industrial-military conglomerates called *zaibatsu* emerged. A cult of the emperor developed as well.
8. Japan launched a series of wars with Russia and China. They got Taiwan as a prize of victory, annexed Korea in 1910, and annexed Manchuria in 1931. Their plan was to create an empire known as the Greater East Asia Co-Prosperity Sphere, and they managed to sweep through China and Southeast Asia to Indonesia. The bombing of Pearl Harbor in Hawaii in 1941 brought the Americans into the war.
9. After its defeat in 1945, Japan made a rapid recovery grounded in a fresh industrial base, government, and financial institutions. Another form of industrial conglomerate emerged, known as the *keiretsu.*
10. Japan has a tradition of borrowing and adapting goods and ideas from other countries. Chopsticks from China and technology from the United States are examples. By 1970, Japan was the richest country in the world.
11. Less than 10 percent of Japanese are in farming. Many of these are part-time farmers working additional jobs. Japanese farms are very small and agriculture very intensive. The primary crop is rice, which is alternated with other grains and vegetables depending on the season. Since Japan is highly urbanized, market gardening is very important.
12. Agricultural patterns and food habits are changing rapidly. Former farmland is used for industry and housing. Small farming mura are dying out. The Japanese are turning increasingly to fast food.
13. Japan has an aging population with increasing numbers of people over 65. Changing population dynamics have implications for the country's tax base, care of older and perhaps infirm people, and upcoming labor pool.
14. The notion of Japan having a homogeneous society is a myth. There are numerous minority peoples such as the Koreans, the Ainu, and the less easily defined *Eta.* These and other minority groups are discriminated against.
15. Japan is a land of mega-urban landscapes. The best known is the Tokaido megalopolis. Rural-to-urban and urban-to-urban migration continues the process of the development of urban agglomerations focusing on already established major cities.

16. Tsukuba Science City was an early effort to decentralize economic activity from major core regions. Tsukuba is now enmeshed in the greater Tokyo region. Other "science cities" are being developed contributing to Japan's multinodal urban pattern. The concept of the *technopolis* is integral to a master plan conceiving a network of science cities across the nation.
17. The Tokyo region remains central to urban development, which is linking Hokkaido with the other islands in a Techno-archipelago.
18. Housing is a major problem in cities, as traditional quarters disappear in the wake of placeless high-rise offices. People complain of loss of their sense of place.
19. Many secondary industries are "hollowed-out," as industries seek cheaper employment outside the country. Investment is increasingly in the tertiary sector, although the financial crisis of the 1990s forced a pullback in this area. Many industries are in the process of restructuring.
20. Korea is one of the world's oldest land areas. Only 20 percent of the land is level.
21. North–south differences are apparent in climate, resource and farmland availability. The north has a shorter growing season, less land suitable for agriculture, but has greater natural resource wealth. South Korea is the peninsula's rice bowl.
22. South Korea is highly urbanized with several cities of over one million people. Only Pyongyang, the capital of the North, has more than a million inhabitants.
23. The regions also vary politically, with the North being communist and the South being democratic.
24. The Korean peninsula was settled more than 5,000 years ago. Subsequent kingdoms were influenced particularly by the Chinese. Today, Korea is considered the most Confucian country in Asia.
25. Hangul, an alphabet for spoken Korean, was devised in the fifteenth century. This enabled the Koreans to become highly literate.
26. In the seventeenth century Korea fell under Chinese control. Internal dissention led to the closing of the country to all but the Chinese and the Japanese. Korea became known as the "Hermit Kingdom."
27. Korea was occupied and annexed by Japan in 1910, and an elaborate infrastructure was put into place. Thousands of Koreans moved to Japan to work where they were discriminated against. Japanese rule of Korea did not end until 1945.
28. After the Second World War, Korea was divided at the 38th parallel. The intended reunification never happened, and different cultures emerged in the northern and southern parts of the peninsula. In 1953, a demilitarized zone was created, further separating the two Koreas. As a result of the Korean War and subsequent division, millions of families were broken and divided between the new countries of North Korea and South Korea.
29. North Korea is essentially a closed society. It adheres to the *juche* philosophy, which means self-reliance. Society is stratified according to Communist Party loyalty. The population lives under absolute authoritarianism. There are food shortages, and the 1990s have witnessed severe famine. Even so, North Korea has managed to devise missiles capable of striking Japan and beyond.
30. China props up North Korea because it does not want instability along its border.
31. South Korea has transformed itself from an agrarian society to one of the world's most highly industrialized nations. It has been designated as an Asian Tiger.
32. South Korea is a product of state capitalism. The economy is founded on giant corporate conglomerates known as *chaebols*. These are similar to the Japanese *keiretsu*. Samsung is the largest. *Chaebols* are run as families with strong Confucian regulations and loyalties. Industries and banks are closely related within *chaebol* complexes.
33. Two-thirds of South Koreans live in cities of 50,000 or greater. South Korea is meeting its high demand for housing by constructing thousands of new apartment units. Many flats house two families. Apartment living has altered the traditional extended family structure.

34. South Korea has established a technobelt system linking research and industry via telecommunication. The key nodes of this system are Seoul and Taedok.
35. Nearly 5 million Koreans live overseas, the largest group in mainland China. Other large contingents are in Central Asia, Japan, and North America.
36. There is potential for the reunification of the two Koreas but the probability is small. Most North Koreans would be unprepared to live in a modern, technologically advanced society.
37. Rich and economically powerful Taiwan is another designated Tiger of Asia. Its history is closely tied to that of the PRC.
38. The island is very mountainous, with a western coastal plain. Here is where most development has taken place. Agricultural output is steadily declining as farmland gives way to urbanization and industrialization.
39. Taiwan has a long history of relationships with the PRC. Known as Bao Dao on ancient Chinese maps, it was a migration destination for opportunists and refugees from southeastern China. These native Taiwanese make up 70 percent of Taiwan's population. The island was known as Formosa for many years until after the Japanese occupation.
40. The Japanese accelerated Taiwan's development by building an infrastructure geared to the needs of Japan.
41. Various proclamations were made by both the Chinese and the Japanese regarding the independence of Taiwan. However, the end of WWII saw the occupation of the island by Chiang Kai-shek and his Nationalist troops from the mainland. Many "mainlander" refugees migrated there as well.
42. The distinction between Taiwanese and mainlanders is becoming blurred. More than 90 percent of inhabitants were born on the island.
43. Both Taiwan and the PRC claim to be the legitimate government of all of China. Realities reveal sharp differences between the two entities—politically, economically, and socially. Nevertheless, the PRC insists that Taiwan is a province of China and Taiwan is hinting at its being an independent entity.
44. The United States does not want to see Taiwan as a part of the PRC because that would give the PRC complete control over the shipping lanes through the Taiwan Strait. The issue of unification or separation remains unresolved.
45. Taiwan's achievements as an economic tiger are underpinned by its excellent education system, infrastructure, and American connections. The state is heavily involved in finance and energy but most businesses are privately run. Electronics are of increasing significance. Farmers now focus on specialty crops.
46. Taiwan meets the criteria of a *kotadesasi* through its economic transformation and evolving spatial structure. There is an evolving metropolitan corridor stretching from Taipei to Kaohsiung. This includes science parks such as those associated with Hsinchu—Taiwan's silicon valley.
47. Taiwan has "commoditized" its computer industry—making components for big-name brands.
48. Taiwan has extended its operations overseas to take advantage of space and cheap labor. Many of these new companies are in the PRC.
49. Taiwan is increasingly doing business with and investing in the PRC. Taiwan also has a Southward Policy whereby it is focussing investment in Southeast Asia.
50. Rapid development has its downside with even more noticeable pollution, congestion, hedonism, and a rich–poor gap.
51. Capital movement has become the engine of development in the ever-changing economies of East Asia. Production has become detached from employment as industries become more knowledge intensive. A functional city system has emerged incorporating production networks unrelated to population size or access to raw materials. Urban restructuring has generated webs of world-class cities, each with their own hierarchies of various functions.
52. Growth triangles such as the Jakota Triangle have emerged. These transcend national boundaries and are part of larger global schemes.

Terms to Know

zaibatsu	castle towns	Koryo	Native Taiwanese
Shinto	Edo	Hangul	Taiwanese
keiretsu	Tokaido megalopolis	38th Parallel	Formosa
market gardening	Tsukuba Science City	*juche* philosophy	mainlanders
aquaculture	Kansai Science City	state capitalism	commoditization
mura	technopolis	*Chaebol*	southward policy
Ainu	Techno-archipelago	Technobelt	functional city systems
Burakumin	hollowed-out industry	Bao Dao	Growth Triangle

Bibliography

Baum, Julian. 1998. A Long Wait Ahead: Commercial Concerns Hold Back Cross-Strait Shipping Links. *Far Eastern Economic Review,* 30 (April): 68.

Castells, Manuel, and Peter Hall. 1994. *Technopoles of the World.* New York: Routledge.

Chan, Gerald. 1997. Taiwan's Economic Growth and its Southward Policy in Asia. In *Asia Pacific: New Geographies of the Pacific Rim,* eds. R. F. Waters, and T. F. McGee, pp. 206–222. Vancouver: UBC.

Chanda, Nayan et al. 1997. On Borrowed Time (North Korea). *Far Eastern Economic Review,* 26 (June): 22–28.

Cheng, Allen, T. 1997. And Now For Taiwan. *Asia, Inc.* (June): 38–45.

Cybriwsky, Roman. 1988. Shibuya Center, Tokyo. *The Geographical Review,* 78: 48–61.

Diamond, Jared. 1998. Japanese Roots. *Discover,* 19: 86–94.

Eberstadt, Nicholas. 1995. *Korea Approaches Reunification.* Armonk, N.Y.: M. E. Sharpe.

1996. In the Shadow of the Chaebol. *The Economist,* 6 (July): 57–8.

1994. Kim Jong Il's Inheritance. *The Economist,* 16 (July): 19–21.

1990. The Sun Sets on Rice. *The Economist,* 21 (April): 44.

1998. Taiwan Survey. *The Economist,* 7 (November).

Edgington, David, W. 1997. The Rise of the Yen, "Hollowing Out" and Japan's Troubled Industries. In *Asia Pacific: New Geographies of the Pacific Rim,* eds. R. F. Watters and T. G. McGee, pp. 170–189. Vancouver: UBC.

Evans, Richard. 1990. Japan's Untouchables. *Geographical Magazine* (November): 46–47.

Fujita, Kuniko, and Richard Hill, eds. 1993. *Japanese Cities in the World Economy.* Philadelphia: Temple.

Garrett, Banning, and Bonnie Glaser. 1995. Looking Across the Yalu: Chinese Assessments of North Korea. *Asian Survey,* 35: 528–45.

Harris, Chauncy, D. 1982. The Urban and Industrial Transformation of Japan. *The Geographical Review* 72: 50–89.

Hoon, Shim Jae et al. 1993. Rural Exodus: Seeds of Despair. *Far Eastern Economic Review,* 4 (March): 20–21.

Howard, Keith, ed. 1996. *Korea: People, Country and Culture.* London: School of Oriental and African Studies.

Howell, David, L. 1996. Ethnicity and Culture in Contemporary Japan. *Journal of Contemporary History,* 31: 171–90.

Hsieh, Chiao-min. 1964. *Taiwan—Ilha Formosa.* Washington D.C.: Butterworth.

Kirk, Donald. 1994. *Korean Dynasty: Hyundai and Chung Ju Yung.* Armonk, N.Y.: M. E. Sharpe.

Lamont-Brown, Raymond. 1995. Earthquake—The Ultimate Japanese Nightmare. *Contemporary Review,* 266: 117–19.

Landers, Peter. 1998. Arthritic Nation (Japan). *Far Eastern Economic Review* 16 (July): 10–12.

Landers, Peter et al. 1998. Hard Target (North Korea). *Far Eastern Economic Review.* September 24: 20–21.

Lee, Charles. 1997. Tomorrow the World (South Korea). *Far Eastern Economic Review.* May 1: 43–45.

Macdonald, Donald S. 1990. *The Koreans: Contemporary Politics and Society,* 2nd edition. Boulder: Westview.

McCarthy, Shawn. 1997. The Aging of Japan (*The Toronto Star*). *World Press Review* (August): 34–35.

Miyashita, Kenichi, and David Russell. 1996. *Keiretsu: Inside the Hidden Japanese Conglomerates.* New York: McGraw-Hill.

Rimmer, Peter, J. 1997. Japan's Foreign Direct Investment in the Pacific Rim. In *Asia Pacific: New Geographies of the Pacific Rim,* eds. R. F. Watters and T. G. McGee, pp. 113–32. Vancouver: UBC.

Rothery, David A. 1995. Kobe—Anatomy of an Earthquake. *Geographical* 67:54–55.

Shambaug, David. 1998. *Contemporary Taiwan*. Oxford: Clarendon.

Smith, Anne, and Tony Smith. 1986. Tokyo On a Plate. *Geographical Magazine,* 58: 106–107.

Takahashi, Junjiro, and Noriyuki Sugiura. 1996. The Japanese Urban System and the Growing Centrality of Tokyo in the Global Economy. In *Emerging World Cities in Pacific Asia,* eds. Fu-chen Lo and Yue-man Yeung, pp. 101–43. New York: United Nations.

Tsai, H. H. 1996. Globalization and the Urban System in Taiwan. In *Emerging World Cities in Pacific Asia,* eds. Fu-chen Lo and Yue-ma Yeung, pp. 179–218. New York: United Nations.

Weiner, Michael, ed. 1997. *Japan's Minorities: The Illusion of Homogeneity*. New York: Routledge.

Yeung, Yue-man, and Fu-chen Lo. 1996. Global Restructuring and Emerging Urban Corridors in Pacific Asia. In *Emerging World Cities in Pacific Asia,* eds. Fu-chen Lo and Yue-ma Yeung, pp. 17–47. New York: United Nations.

Chapter 12

Southeast Asia: Transition Among the Nagas

"To understand the present and anticipate the future, one must know enough of the past, enough to have a sense of the history of a people."

(LEE KUAN YEW 1980)

"A more highly fragmented part of the globe could scarcely be imagined," said geographer Norton Ginsburg (1972). Southeast Asia, land of the ***nagas***, is a realm of contrasts in both physical and cultural geography. Fragmented landscapes have formed the backdrop for the establishment of glorious kingdoms and trading states, while hills and vales have complicated the imposition of boundaries and the demarcation of people and territory. Separation has enhanced cultural diversity but thwarted regional unity. Nevertheless, several Southeast Asian states have become prominent on the global economic map. Malaysia, Thailand, and Singapore are designated as economic tigers. Indonesia has tiger potential even though weighted with political troubles. Vietnam boasts of its new label: *baby tiger*. In contrast, countries such as Myanmar, Laos, and Cambodia lag far behind in both the human condition and regional stability.

The Naga

Naga, in Sanskrit, means serpent. The *naga* is an aquatic symbol permeating the daily life of people everywhere in the watery realm of Southeast Asia. They appear in art form as dragons and snake-like creatures, and frequently as mutations of different beings. For instance, Thailand has a *sang*, a cross between a *naga* and a *singh* (lion). Water symbolism in *naga* form is ubiquitous in Thai and other cosmological representations. The Khmer and numerous other civilizations are portrayed as emerging from water. Sometimes whole populations have been identified as the *naga* people or form. In almost everything of significance, the *naga* symbol asserts itself (Figure 12-1).

Figure 12-1

Most temples in Laos exhibit nagas *as guardian figures.*

Many geographers stress the region's diversity over unity. Geographer Jonathan Rigg (1991) notes, "Southeast Asia represents a residual region: it owes its regional identity not to internal coherence, but to external incoherence." In this chapter, you will learn about differences, but also similarities, among the landscapes and cultures of this complex region.

Physical Landscapes

The difference between land and water is clearly evident. In fact, there is a mainland Southeast Asia and an island or archipelagic Southeast Asia (Figure 12-2). The mainland includes Burma (Myanmar), Thailand, Laos, Cambodia, and Vietnam. I will include Malaysia with the insular region as it not only shares the island of Borneo with Indonesia but also is connected to the mainland only by the narrow Kra Peninsula. Consequently, Malaysia, Indonesia, Brunei, and the Philippines form the insular or archipelagic region.

All told, there is four times more water area than there is land area. While civilizations arose in the mainland interior, marine landscapes and coastal locations dominated the fortunes of the island realm. Control of ocean passageways marked the rise and fall of many a power group and fixed Singapore as the dominant port of the entire region.

Mountain formation remains active—crustal instability is a feature affecting the lives of millions (refer to Chapter 2). Throughout this unstable region, mountains preside over plains. The hills of Myanmar are cut by the Irrawaddy and its major tributary the Chindwin, which juncture southwest of the city of Mandalay. The Sittang valley lies to the east. Both the Sittang and Irrawaddy deltas are used extensively for rice cultivation. The highly dissected Shan Plateau of eastern Myanmar rises over 4,000 feet (1,200 m) in many places.

To the south in Thailand is the delta of the Chao Praya. Since this river is confined in a geological trough, its delta is elongated rather than fan-shaped. Alluvial lowlands extend nearly 310 miles (500 km) upstream from the Gulf of Thailand.

Southeast of the Shan Plateau is the Khorat Plateau of eastern Thailand, lying at an elevation of just under 1,000 feet (300 m) but edged by higher escarpments, maintaining its separateness from the rest of the country. South of the Khorat Plateau lies the lowland of the Cambodian Basin with Lake Tonle Sap in the lowest, middle portion. In the wet season, Mekong River drainage rolls inward to fill the lake up to six times its average depth. Tonle Sap extends from 1,600 square miles (2,580 sq km) in the dry season to 12,400 square miles (20,000 sq km) in the wet.

The Mekong River flows to the east of the Cambodian Basin. The longest river of Southeast Asia, the Mekong has constructed an expansive delta starting at Phnom Penh in Cambodia and extending 175 miles (282 km) into the sea. This vast, alluvial delta has developed as the heartland of southern Vietnam.

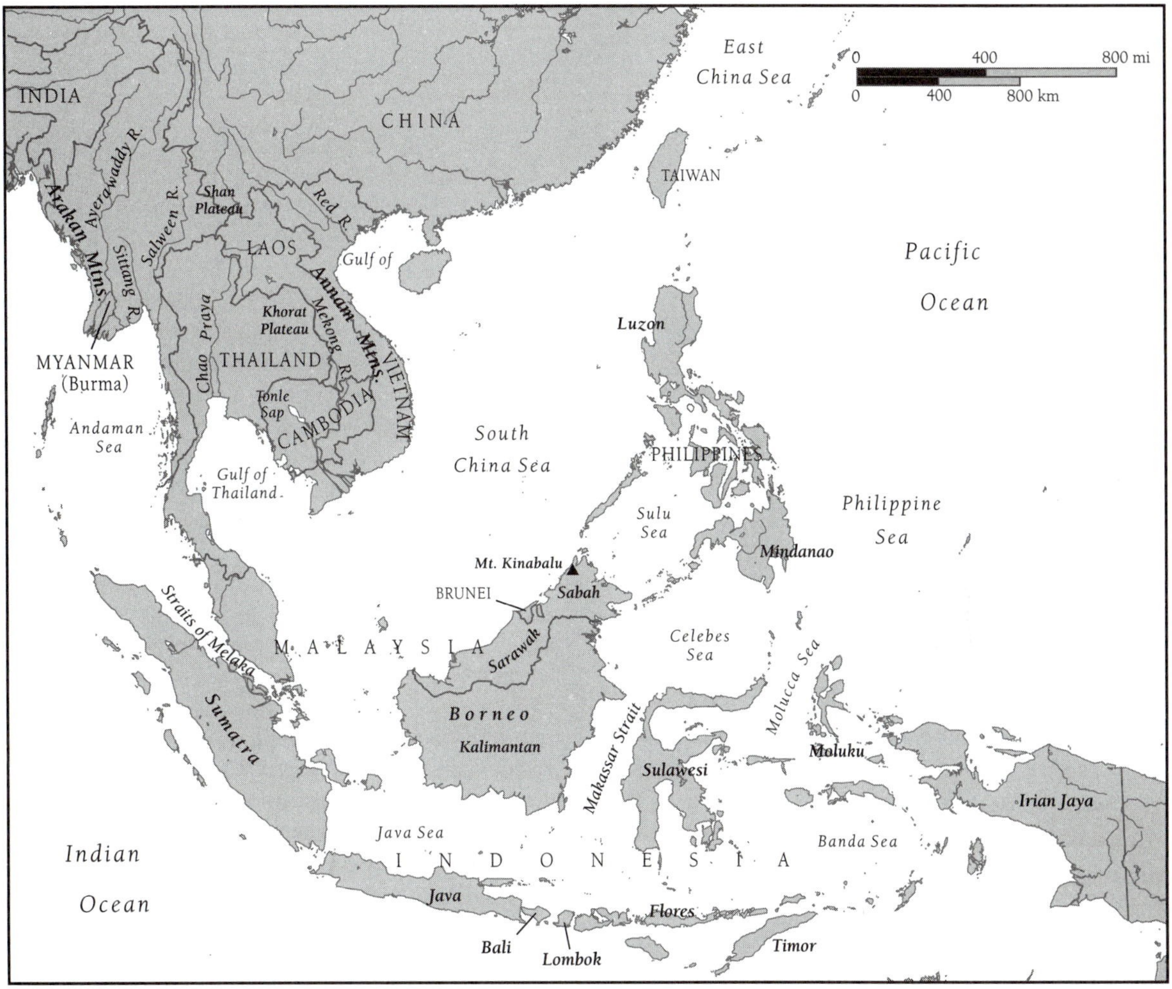

Figure 12-2

Physical features of Southeast Asia. Note how rugged the region is.

The Mekong

The Mekong is the twelfth longest river in the world. Originating in the Tibetan Highlands of China's Qinghai Province, the river flows 2,600 miles through southwestern China, along the borders of Myanmar, Laos, and Thailand, across Cambodia, and down through Vietnam where it sprawls across the 26,000-square-mile (67,340 sq km) Mekong Delta to the South China Sea. One-third of the population of the latter four countries lives in the lower Mekong Basin.

The Mekong cultural realm is overwhelmingly rural. Most of the region is poor, with people engaging primarily in rice cultivation and fishing. However, as foreign aid and investment increase and modern development ensues, both landscapes and lifestyles are being transformed rapidly. Increased population pressure and resource demands, coupled with uncoordinated policies and regulations, are threatening both human and physical environments.

Although the river is an artery of human activity, upstream rapids and waterfalls have prevented any large-scale development for international shipping. Smaller seagoing vessels can travel upstream to Phnom Penh, but upstream sailing is limited to local traffic. A bridge connecting Laos to Thailand was constructed in 1994.

Cooperative efforts to develop the river basin have been discussed since 1957. Several hydroelectric power, navigation, flood control, and irrigation projects were planned, but war and political upheaval have allowed only some to be implemented. For example, dam construction has made Laos a key exporter of power to Thailand. More recently, many plans have been scaled down because of both cost factors and environmental concerns.

The Annamite Ranges run through Vietnam between the Mekong and the South China Sea. In northern Vietnam, the Song Koi (Red) River flows in a trough-like valley to the Gulf of Tonkin, where it forms a 5,460-square-mile (14,000-sq-km) delta. The southern Annamites give way to the Mekong delta.

The shallow sea enclosed by Cambodia, Malaysia, and the island of Borneo is less than 656 feet (200 m) deep, as it is part of the Sunda Shelf (Chapter 2). Borneo is the chief emergent part of the shelf, having several mountains above 6,560 feet (2,000 m). Mount Kinabalu in northern Sabah is the highest peak at 13,451 feet (4101 m).

Southeast Asia's relief offers little opportunity for human settlement and populations are crushed along rivers. Most of the Sunda Shelf, which could have offered livable land, is overlain by the sea. Extensive coastal areas are mangrove swamp. The most densely populated areas in the island realm are Java in Indonesia and Luzon in the Philippines, both of which possess fertile volcanic soils.

Even within this tropical realm, straddling the equator, there are apparent temperature variations. Pinang, Malaysia, at 5° 21′ N averages 80°F (27°C) for twelve months, Hanoi, Vietnam at 21° 04′ N might drop to 60°F (16°C) in December or January. Altitude also plays a role. With temperatures falling 3 degrees per 1,000 feet, Southeast Asia had its hill stations comparable to those of the Raj. Java's Bandung and the Malay peninsula's Cameron Highlands are just two examples of where temperatures might be 10 to 20 degrees cooler than in the sweltering lands below.

Precipitation varies both in amount received and seasonality (Figure 12-3). Although 60 inches (152 cm) will support tropical rainforests or monsoon forests, areas with less than this amount can be considered moisture deficient. Most moisture-deficient areas are on the mainland and on the islands east of Java. Distinct wet and dry seasons increase away from the equator. In equatorial regions, the wet and dry monsoons are affected by convectional rainfall. Even so, topographic barriers exhibit seasonally humid and dry sides, depending on the monsoon and other wind patterns.

Length of dry season varies as well, lengthening both north and south of the equator. And, some "dry seasons" are merely less wet ones. For instance, there is no real dry season in Borneo, and in peninsular Malaysia it lasts only a month, but on the mainland the dry season lasts four to five months.

THE ROLE OF WATER

According to Rigg (1991) "Of the environmental elements, water has possibly been the most instrumental in forging Southeast Asia's distinctive personality." Fairly predictable winds, wood supplies close to shore for ship building, and endless waterways made this region most favorable to maritime activity. Extensive trade existed well before the fourteenth century.

In prehistory, waves of migrants filtered southward from the mainland to the Indonesian Islands and on to the Pacific Islands and Australia. At the beginning of the first century A.D., there was an infusion of Hindu and Buddhist culture when Indian sea traders ventured for gold and spices. This Indianization of Southeast Asia peaked with the establishment of fantastic empires centered on great complexes such as Srivijaya on Sumatra and Angkor in Cambodia.

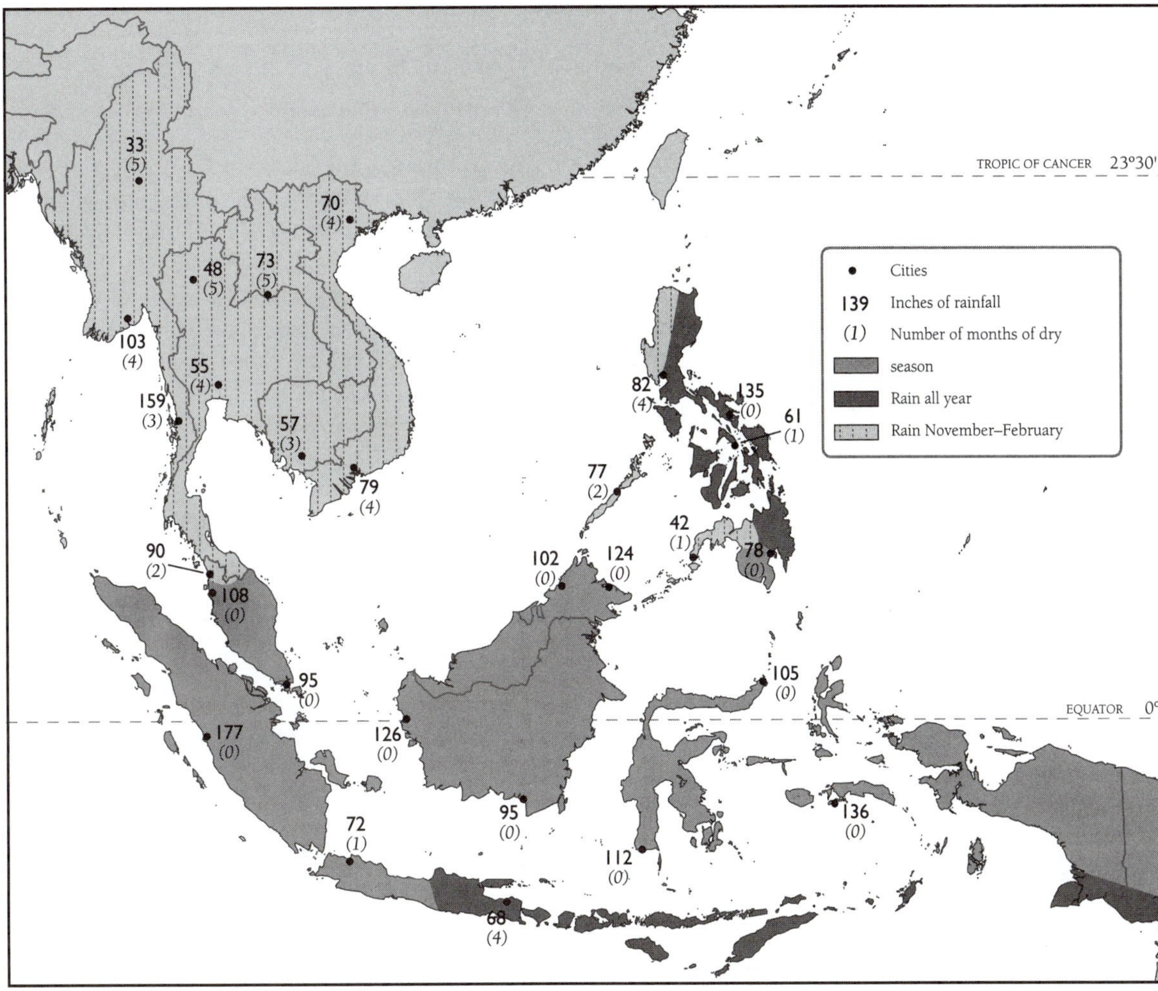

Figure 12-3

Length of dry seasons and precipitation. Note that dry seasons increase in length toward the interior of the mainland.

In the second century, both Arab and Indian ocean traders brought Islam into the island realm. Meanwhile, monks from Sri Lanka disseminated Theravada Buddhism to the mainland.

In the modern era, colonials introduced new technology and power into the region and created conditions conducive to the immigration of large numbers of Indians and Chinese. Commercial interaction ensured diffusion of products and ideas and linked the peoples of the region with each other. Interestingly, with the possible exception of Vietnam, no Southeast Asian country established strong ties with India or China.

Amount and availability of water are crucial to life in Southeast Asia. Virtually all outdoor economic activities are regulated by precipitation. Water is also central to rice cultivation (Figure 12-4). In fact, there is no other staple crop so dependent on its presence. Many regions have rituals surrounding the "gift of water." In Thailand, female cats (the personification of dryness) are carried through the streets and soaked

Figure 12-4

Irrigated rice terraces (sawah) in Indonesia. Water is moved by gravity flow.

with water while people chant "Hail *Nang Maew* (Lady Cat) give us rain." In parts of northern Thailand, Bali, Java, Sumatra, and the Philippines, elaborate irrigation systems have been constructed to ensure a stable water supply.

Bali's Subak Irrigation

In Bali, rice is a gift of the gods and the traditional rice cycle commences on a day set according to a religious calendar. Water regulation is an essential aspect of this highly regimented pattern of activities.

Rivers cut through Bali's volcanic mountains to the sea, and the Balinese have settled the mountain ridges and covered the slopes with ***sawah:*** terraced paddies. Water runs by gravity flow but must be fairly distributed and regulated according to the needs of the system. For example, irrigated *sawah* are drained up to three times to encourage root development and allow weeding. Fields are drained for harvesting and then reflooded to allow other plants to grow. Water hyacinth is popular because it can be worked back into the soil or used as pig feed.

Balinese farmers belong to cooperatives called ***subak.*** Every farmer who owns or rents more than a quarter of an acre (one-tenth of a hectare) must join. Bali has more than 1,000 *subak*. Each *subak* elects a leader who runs the operation and liasons with agricultural and other agencies. *Subak* law (*awig-awig*) meetings must be held on one auspicious day each month of the religious calendar. Monies are derived from fees, fines, and land sales.

Cooperation is the key to success. *Subak* members repair and maintain the irrigation system after the harvest in accordance with the amount of water they have received. Those not working pay a fee. The average farmer gives the communal system about 15 days a year. An important aspect of the *subak* is its religious connection. Each keeps a temple, which must be cooperatively maintained by the members.

The *subak* system diffused from Java a thousand years ago. It has survived in Bali because of its isolation as an island and because the *adat* system of local regulations, stressing personal responsibility and communal consensus, remains strongly entrenched.

It now seems clear that rice was first domesticated in Southeast Asia and diffused from there to China and India, replacing millet as the staple crop. Around 4000 to 3500 B.C., a new red-on-tan pottery form appeared in northern Thailand. This is known as Ban Chiang pottery. Shortly thereafter, bronze-making emerged, carried out at the village level.

UPLANDS AND LOWLANDS

From about 1000 B.C., major changes began to take place. With the expansion of the Han Chinese in person and influence, small towns began to emerge. In addition, entrepots, to facilitate trade between India and China, were established along the coasts of the Malay peninsula and what is today southern Vietnam. These entrepots became centers of diffusion of Indian culture throughout Southeast Asia.

As these events ensued, primitive and tribal people began to lose their dominant positions. The new civilizations centered around the major river valleys and Cambodia's Tonle Sap basin (see Chapter 13). These lowlands are separated by sparsely populated mountainous areas—the uplands. Hill people became increasingly differentiated from lowland people, yet they remained part of their cultural realm as "holders of the wild" or "people of the upland fields." Ceremonial exchange relationships validated hill people–lowlander bonds, and tangible trade was important as well. Lowlanders also viewed hill people as potential laborers. Slavery was not unusual. Hill tribes were often pitted against each other in slave raids to sell to lowlanders. In some regions such as Vietnam, efforts were made to reduce contact with the hill people to simple tribute payments and restricted trade.

The mountains have acted as barriers between larger settlements. Over time, successive immigrant groups have settled in the lowlands, but more recent and more populous groups have driven earlier, less-sophisticated groups into the upland regions.

Lowland areas are inhabited by the dominant ethnic group of each country—Thais in Thailand, Khmers in Cambodia, Javanese in Java, and so forth. Upland, forested areas are inhabited by minority groups such as the Shans in Burma and the Yao and Miao in Thailand.

In these contrasting settings, different ways of life have evolved. Lowlanders primarily practice wet rice cultivation while uplanders focus on some form of dry field agriculture or shifting cultivation. These general patterns continue today and affect both patterns of development and the resolution of environmental issues.

The Impact of Indian Culture

Mountain barriers diverted the diffusion of Indian culture into Southeast Asia almost entirely by sea in sailing ships furled by the monsoon across the Bay of Bengal. As early as the sixth century B.C. Indian traders were searching the coasts of Myanmar, the Malay peninsula, and western Indonesia. The Indians impressed local leaders with lavish gifts, their knowledge of medicine, and their frequent portrayal of themselves as royalty. They also married into Southeast Asian society. Temple forms, styles of art, music, dance, and tales of the *Ramayana* are woven into the cultural tapestries of most Southeast Asian countries to this day. The impact of Indian culture was greatest during the Gupta Empire—India's Golden Age (320–535 A.D.).

Hinduism and Buddhism were the most salient of multiple media of Indianization. Missions were sent back and forth to Ceylon (Sri Lanka), the core area of Theravada Buddhism. Mahayana Buddhism was accepted in central Java during the eighth century and in Cambodia's Angkor in the twelfth century. Sanskrit was the first written language of the region. Its related Pali provided the basis for developing local writing systems.

The Hindu concept of kingship, law codes, economic treatises, and other canon were absorbed by Southeast Asian rulers. However, the caste system and inferior standing of women were not adopted. Although there is a patriarchal veneer to Southeast Asian society, couples tend to live with the wife's parents, or at least relate most closely to them. Women have much more social and economic power than in other Asian societies.

The earliest Indian states emerged along the coast, with their nuclei at river mouths. The first of these, called Funan by the Chinese, became an imperial power in the third century A.D. This was followed by the Khmer kingdom in Cambodia. The kingdom of Champa (1192–1471) was established along the east coast of Vietnam. Another early sea power to accept Indian influence was Srivijaya with its capital at Palembang on Sumatra. Apparently this was a center of Mahayana Buddhism, as was central Java at the time. Borobudur (eighth century) stands as a testament to the strength of Buddhism (Figure 12-5). All of Southeast Asia, with the exception of the Tonkin region (northern Vietnam), was impacted by Indian culture.

The Chinese Influence

The earliest contact with China can be traced to the Chin Dynasty (221–207 B.C.). Chinese influence was charted overland; the region mainly affected was Vietnam. In fact, northern Vietnam remained a Chinese territory until it broke away in 907 A.D. China referred to the region of the Southern Seas as ***Nanyang.*** Its cultural impact was the direct result of colonization and conquest. Chinese administration style, writing system, and Confucian classics were all superimposed on indigenous culture. Many Chinese cultural influences have persisted in Vietnam.

By the sixteenth century, Chinese trading communities were permanently established throughout the region. However, large-scale Chinese migration did not take place until the turmoil of China's nineteenth century, when thousands of Chinese migrated from southern China to work in mines, on plantations, and in other European operated enterprises. Others came as farmers, fishers, traders, retailers, or skilled craftsmen. Increasing numbers of Chinese came to play a critical role in the region's commerce.

With the extant plethora of ethnic groups and the influx of Europeans, Indians, and Chinese, each functioning in their own realm of expertise, Southeast Asia became what is referred to as a **plural society.**

Overseas Chinese in Southeast Asia

Overseas Chinese are prominent in Southeast Asia beyond their numbers. For example, in Malaysia ethnic Chinese make up 29 percent of the population but control 69 percent of share capital by market capitalization. In Thailand, ethnic Chinese make up 10 percent of the population, yet they control 81percent of listed firms by market capitalization. In the Philippines 2 percent control half of the share market, and in Indonesia 3.5 percent ethnic Chinese control 73 percent of the listed firms. Clearly the Chinese wield a potent clout in Southeast Asian economies.

Figure 12-5

Built by Buddhists on Java in the eighth century at the height of the Sailendra Dynasty, Borobudur's construction took about 75 years. It comprises 1,600,000 volcanic, andesite stones and is ornamented with at least 1,500 bas reliefs and 500 statues of the Buddha. No mortar was used in its construction. The large base platform was added at a later date and hides a panel of reliefs known as the "hidden foot." Borobudur stands as a stupa, a replica of the cosmic Mount Meru, center of the universe, and as a mandala to assist meditation.

Southeast Asia's Chinese have a reputation of being rich and influential but loyal to their own group. They are said to do business with each other and hire their kin rather than the local people. Suspicion and tension are rife in several countries, the least being in Thailand and Singapore.

Chinese family businesses, taking advantage of their kin networks and regional connections, often prosper in the worst of times leading to more jealousy and even hatred. Chinese have frequently been the scapegoats for any sort of trouble. Thousands have been killed, raped, and tortured and their homes and their businesses were destroyed. The last blast against the Chinese was in Indonesia during the economic crisis of 1997.

Linguistic Pluralism

Hundreds of distinct languages and dialects are spoken in Southeast Asia. These can be classified into four main language families: Malayo-Polynesian (e.g., Malay and Javanese); Sino-Tibetan (e.g., Thai, Burmese, and Cantonese); Austro-Asiatic (e.g., Khmer and Annamese, or Vietnamese); and Papuan (e.g., New Guinea, or Irian Jaya). Although the insular region's languages harken mainly to the Malayan division of the Malayo-Polynesian group, the mainland's languages form a complicated mosaic. There are multiple Sino-Tibetan and Austro-Asiatic languages and dialects spoken throughout the region—at least six versions of Chinese for example. Hill regions, with their isolated pockets, have developed a myriad of dialects. Since the eighteenth century, English has been employed widely in the fields of education and administration in many areas such as the Philippines, Malaysia, Thailand, and Myanmar. In Indo-China (Cambodia, Laos, and Vietnam), French was the lingua franca. Dutch served Indonesia until after World War II, when English fell into favor. The prevalence of English as a means of general communication is readily apparent on signage in many places.

Religious Pluralism

The mainland is largely Buddhist with the exception of the Vietnamese lowlands, where "Chinese religions" and Roman Catholicism are apparent. Chinese religions are a mixture of Mahayana Buddhism, Daoism, and ancestor worship, a social practice affiliated with Confucianism. The island world is largely Islamic with the exception of Bali, which retained Hinduism after the last Hindu dynasty in eastern Java had been overthrown in the seventeenth century. Christians predominate in the Philippines except on the island of Mindanao, where Islam prevails.

Early Kingdoms

Prior to the coming of the Europeans, Southeast Asia boasted a number of kingdoms whose achievements are notable to this day (Figure 12-6). Kingdoms were either land or maritime based. Funan, one of the oldest regional states entered the historical record primarily through Chinese sources as early as the second century B.C.

FUNAN

Funan, in Indochina, developed from a core area in the Mekong delta. Its capital was Vyadhapura (Banam near Phnom Penh), and it thrived on the growing maritime trade from China to the Middle East and Europe. Most ships stuck to coastal routes, and Funan's safe harbor, situated between China and Malaya (the Malay Peninsula), made it an excellent port of call. Indians came to trade for spices such as cloves and cardamom, and precious items such as rhinoceros horn and ivory. These items were shipped to Europe and the Middle East. The Indians sailed the monsoon winds. In between seasons, they established large colonies in the Mekong delta. Thus began the Indianization of Funan.

Funan had a productive agricultural hinterland and was able to increase production by draining mangrove swamps and introducing multiple canal irrigation. The canals kept sea salt from the delta plain, allowing these alluvial tracts to be cultivated. Ships

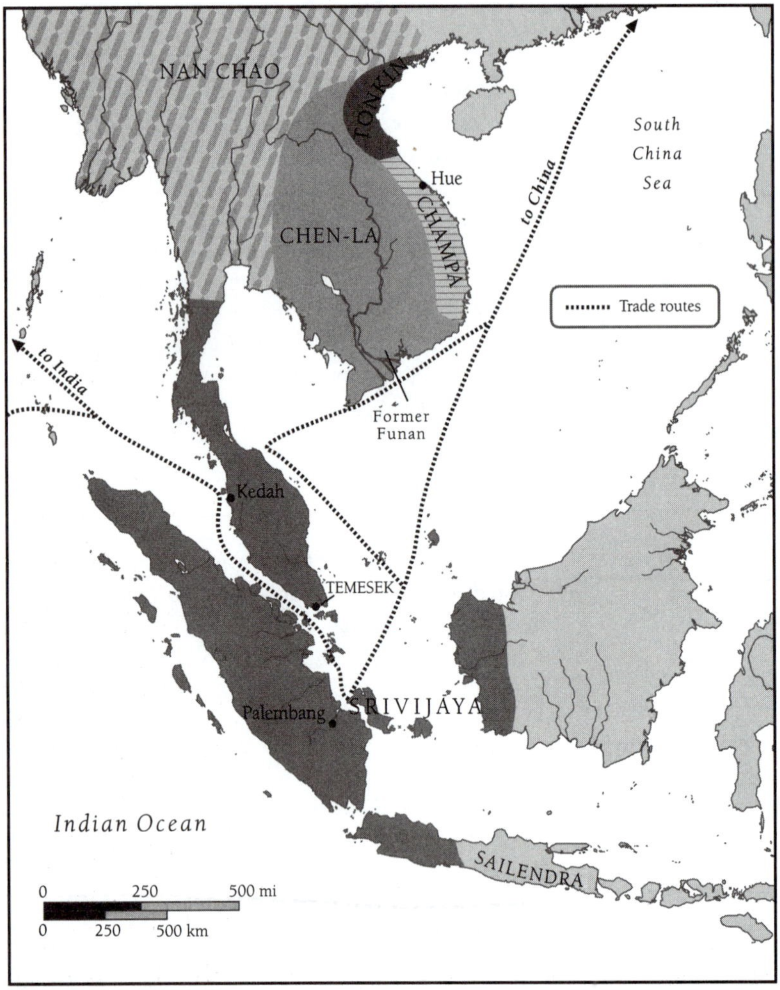

Figure 12-6

Note how the physiography of the region fosters both land- and sea-based kingdoms.

could even sail to inland towns that were built on artificial or natural levees, surrounded by earthen embankments or moats. Water was the national and international lifeblood of the empire.

Funan surpassed every other state in the area. It ultimately extended from southern Vietnam, through Cambodia, central Thailand, northern Malaya, and southern Burma. It was the most important maritime intermediary in the Sino-Indian trade. Funan continued to expand through the fifth century A.D., when international trade began to shift toward to the Strait of Malacca.

Better ships and navigational techniques allowed coastal routes to be replaced by direct routes across the South China Sea, and the fortunes of Funan fell. Chenla, a former vassal of Funan, grew in power. With its land base and army, it was able to annex Funan before the mid-sixth century.

While Funan was a tributary state of China, after the fifth century, Chinese records began to include references to Javanese and Sumatran emissaries, pointing to the shift in trading patterns and the rise of Srivijaya on Sumatra.

CHAMPA

Champa, or Lin-yi in Chinese records, was contemporaneous with Funan but centered near the present city of Hue, Vietnam. The Chams were Malayo-Polynesian people who settled on the east coast of Vietnam before the Christian era. Champa's first capital at Hue was probably founded around 192 A.D. It sent its first emissary to the Chinese court in 284 A.D., with whom it had a relationship of alternating subservience and hostility. The Champa kingdom was split by mountains; nevertheless, it was able to challenge the Funanese, Chinese, Khmer, and Vietnamese. The Chams readily adopted Indian culture and were noted for their Pallavan-style architecture. Cham society was and is matriarchal; daughters have the right of inheritance.

Champa's fortunes rode on its sea connections with India and China because possibilities for land expansion were limited by the Annamite Cordillera and the powerful Khmer empire. The mountains became the Cham source of precious woods and minerals, which they exported. Champa fleets took advantage of the monsoon prevailing winds for northern journeys between May and August and southern travels from October to February.

The Chams developed a strong naval force, which supported their existence as a sea power for 1,300 years. They engaged in a sort of piracy whereby they would exact tolls from coastal plying ships. They also dealt in the slave trade. In the fifth century, they attacked Tonkin, southern Chinese provinces, and eventually sacked Angkor in 1177.

For more than a thousand years, the Chams withstood the political and cultural pressures of China and Vietnam. Without an agricultural land base, Champa was eventually squeezed out by the northern Viets, who had a very productive agricultural base in Tonkin. After years of conflict, the Vietnamese finally occupied Champa in 1471. About 60,000 Chams were slain and a similar number were abducted into Vietnamese slavery. Today, about 40,000 Vietnamese and 85,000 Cambodians claim Cham ancestry.

THE KHMER

The mainland successor to Funan was the agriculturally based Khmer empire. As Funan disintegrated, the Khmer inherited much of its territory, including its complex irrigation system. This controlled the monsoon rain by using grass dams to hold water in natural basins. The Khmer also used Hindu rituals to legitimize the position of the king as bearer and transmitter of life power. They eventually exerted their authority over Lower Burma (Myanmar), the upper Malay Peninsula, central Thailand, Cambodia, and southern Vietnam.

Agricultural and water management techniques, exploiting Tonle Sap, became increasingly complex with the construction of waterways, basins, channels, and fountains, many of which were lined with brick. Immense storage tanks called *barays* were devised, one of them with a capacity of 30 million cubic meters. More than 12.5 million acres (5 million ha) of land were brought into production, yielding three or four rice harvests a year.

The world-renowned temple complex at Angkor was founded in 802 A.D. As territorial conquest enlarged the empire, more temples and rice paddies were established until the complex became one of the largest in the world (Figure 12-7).

Recovering Angkor

Angkor was the capital of the Khmer empire, a civilization that flourished from about 800 until about 1370. A mixture of sandstone, brick, and laterite was used to build an immense complex of temples, each one a symbolic representation of Hindu or Buddhist cosmology. The whole of the Angkor complex covers an area of more than 15.5 miles (25 km) east to west and almost 6.2 miles (10 km) north to south.

Angkor Wat, dedicated to the Hindu god Vishnu, is just one temple. Erected as the funerary memorial of King Suryavaram II in the twelfth century, it covers 1 square mile (2.5 square km) and it is one of the largest religious edifices in the world. Angkor Thom is slightly larger. It is entered by four axially aligned gateways that define a square almost 0.62 mile (1 km) across.

Angkor has escaped the ravages of time but it is not unscathed. Certainly the jungle has taken its toll, trees splitting walls and lichens etching away illustrious carvings. French and Indian archaeologists have worked to clear the forest from much of the monument. But worse is the toll of robbers who have absconded with carvings, statue heads and even whole statues. And, more recently, Cambodia's war has resulted in the complex being shot full of bullet holes, artifacts being smashed, and political slogans being carved into bas reliefs.

Fortunately, international assistance is supporting archaeologists and the local people in their restoration efforts. But, treasure hunting raids still occur and certain sections can only be visited in the company of armed guards.

Figure 12-7

The main wat at Angkor is reflected in a moat surrounding the complex. This temple, built of laterite, covers one square mile (2.5 sq km) and is one of the largest edifices in the world.

Building projects laid a heavy burden on the Khmer economy. For example, under one king alone, 102 hospitals, 101 pilgrim rest houses, and 20,000 shrines were built. Countless temples were erected, and 300,000 priests and monks were supported by the state. Wars against Champa and other vassal kingdoms further drained the treasury.

Sukothai, in the upper Chao Praya, declared its independence of Khmer rule in 1219. Thais were pressured by the Mongols in China to move further into Southeast Asia. Both the Thais and the Chams took large chunks of the Khmer empire. In 1431, Angkor was captured by the Thais. Khmer authority declined, and the civilization moved toward Phnom Penh.

SRIVIJAYA

Meanwhile, shifting trade routes through the Strait of Malacca saw the rise of new and larger maritime kingdoms in the adjacent archipelago. The first of these appears to be Srivijaya around Palembang on Sumatra. It is thought to have controlled the Strait of Malacca from the seventh through the twelfth centuries A.D. Chinese writers referred to Srivijaya as *Jinzhou,* meaning "Gold Coast." They described a hill that was covered with gold and silver images of the Buddha. Palembang provided an excellent harbor sheltered from the fury of the northeast and southwest monsoons by the mountains of both Malaya and Sumatra.

Unlike Funan, Srivijaya had no large agricultural hinterland. It was purely a maritime kingdom and controlled even the coasts of the mainland, where the Mons and the Khmers held sway. It survived on revenues from passing trade. It also benefited from China's increased trade during the Tang Dynasty from the seventh to tenth centuries selling resins and gums to supplement China's demand for frankincense and myrrh from Arabia and eastern Africa. Porcelains, jades, and silks from China; camphor, sandalwood, spices, and resins from the Moluccas (Maluku); and textiles from India found eager customers in Palembang. The fortified city, apparently of bamboo, was surrounded by a complex of canals and water tanks. It was also a center of Buddhist learning. Furthermore, it controlled piracy in the region. Along with other competing states, Srivijaya paid tribute to the Chinese court. Srivijaya's tribute included golden lotus bowls filled with pearls and other precious objects. It eventually fell under the power of India's Chola empire and gave way to new powerful trading and agricultural kingdoms on Java, such as the Sailendras and Mataram.

Numerous additional states and kingdoms rose and fell throughout the early history of Southeast Asia. Both unity and disunity prevailed under various authorities over space and time. This was the situation when the Europeans first set foot in the region in the sixteenth century.

Coming of the Europeans

European political contacts with Southeast Asia commenced with the Portuguese conquest of Malacca in 1511. Still, there was no major European political or cultural impress on the region until well into the nineteenth century. The Philippines is an exception in that European penetration here was greater than anywhere else in Southeast Asia. Indeed, there were a string of fortified ports in the region (see Chapter 1), but even the Dutch, who controlled the western third of Java centered on Batavia (Jakarta), were unaware of Borobudur, perhaps the greatest Buddhist monument in the world.

The Spice Islands

Long known as the Moluccas and now Maluku, the Spice Islands are a volcanic mini-archipelago between Sulawesi and Irian Jaya north of Timor. Ambon, Banda, and Ceram are only 3 of the 1,000 islands. Some of the islands were known for their spices, especially nutmeg and cloves. Spices, important for preserving and flavoring meat, were traded for Indian cotton, Chinese porcelain, and Javanese rice for at least 2,000 years before the Europeans came on the scene.

The arrival of the Portuguese in 1511 initiated decades of conflict among the local sultans and the competing colonial powers. The Dutch established settlements in 1599 and by 1667, with their superior sea power, they had completed their conquest of the islands. Their first treaty with Ambon gave them monopoly over all the cloves. The Dutch East India Company, founded in 1602, put an end to individual profiteering by ousting everybody else by force and intrigue. At one point, they eliminated the entire population of Banda.

To guarantee control, the Dutch destroyed all the nutmeg and clove plants except for certain areas approved for cultivation. Even the number of trees was restricted. Cloves were removed from all of the islands except Ambon—a cruel move because a clove tree was planted at the birth of a child, and to destroy a tree portended ill-will for the child. The penalty for cultivating nonregulated trees was death.

Even sago palms were eliminated from Ceram, then known as the "mother island," because of its supply of this food staple. The area became uninhabitable and today it is said that Ceram is populated by ghosts.

Restrictive policies were enforced until 1824, and the Mollucans were reduced to a subsistence existence. The Company's monopoly was broken when a Frenchman smuggled nutmeg and clove seedlings to Zanzibar. Moreover, the spice trade declined with the availability of fresh meat and changing dietary habits in Europe.

The nineteenth century ushered in the Industrial Revolution, creating increased demand for raw materials and markets. The Caribbean, focus of eighteenth-century trade, fell into decline with the abolition of slavery and the slave-labor system. The introduction of steamships and the opening of the Suez Canal in 1869 shortened the journey to Asia, and increasing interest in China put Southeast Asia once more at the crossroads of activity. European missionary zeal fostered additional ventures to the region. From about 1825 to 1870, most of Southeast Asia came formally under European control (refer back to Figure 1-11).

EUROPEAN IMPRESS

The various powers had their own agendas, which diversified their influences on people and landscapes. In 1896, Britain and France decided to preserve Siam (Thailand) as a buffer state to protect their valuable colonial dependencies of Burma (Myanmar) and Indochina. Britain treated Burma as an extension of its Indian Empire and pursued a policy of development of resources and building the economy. France's doctrinaire policy of assimilation of Indochina into France left most of that region undeveloped. The Dutch commercial exploitation policies left most of Indonesia undeveloped and unprepared, economically or politically, for self-government. In the Philippines, American anticolonial attitudes assured that country's direction toward self-rule.

The Dutch Culture System

The **Dutch Culture System** (cultuur stelsel) in Indonesia exemplifies the penetration of capitalism and commercialization into the landscape. First established by the Dutch East India Company, the system required that every farm household allocate one-fifth of its land to the production of a cash or export crop. The system worked well in the sugar-growing areas of Java but proved burdensome where land was sadly needed for food crops. The system was rife with exploitation by local officials who received commissions for production they delivered to authorities. When rice was added to the list of export crops in the 1840s, famine struck one area of Java where hundreds died of starvation and thousands left the area.

Even after the Dutch relinquished their monopoly control to private capital after 1877, compulsory labor was still required, substantial taxes were called for, and reasonable-rate credit was not available.

The powerful position of European and Chinese entrepreneurs added to the exploitation and disruption of traditional Indonesian society.

The British, who ruled Burma from Calcutta, brought in Indian money-lender castes to bank and fund the expansion of rice cultivation in the Irrawaddy delta. Indians often foreclosed on the Burmese, who were hit especially hard during the world depression. Anti-Indian sentiments in Burma derive from these tactics fostered under the British.

Economic concerns also stimulated population movements. For instance, the cultivation of new crops and the opening up of new lands attracted masses of migrants. The rapid expansion of rice cultivation into the Mekong, Irrawaddy and Chao Praya deltas are testament to increased global demand for rice in the late nineteenth century. In lower Burma, where the rice area burgeoned from 988,000 acres (400,000 ha) to 9,880,000 acres (4 million ha), populations increased from 1.5 million in 1855 to 8 million in 1930.

Populations were also drawn to other commodity regions such as rubber and tin. Rubber was planted over large areas of Malaya, Sumatra, and Java, literally transforming these landscapes. Malay and Chinese smallholders accounted for 39 percent of the rubber industry by 1939, yet it was the capital- and labor-intensive plantations that dominated the process. As Malays were not keen on wage labor, companies imported large numbers of indentured Indian and Ceylonese workers. More than 700,000 Indian workers entered Malaya between 1907 and 1917. The tin industry in Malaya, Indonesia, and Thailand was dominated by Chinese mining enterprises employing cheap Chinese immigrant labor. Between 1890 and 1899, almost 165,000 Chinese laborers a year were entering the Straits Settlement of Singapore and Pinang alone.

With the influx of Europeans, Indians, and Chinese, plural societies emerged with each group occupying their own residential quarters but amalgamating in the marketplace. The immigrants were concentrated in urban places and zones of commercial agriculture or mining.

Cities were now founded on commerce and, with the exception of Bangkok, were dominated by external forces. Cities such as Jakarta, Manila, and Kuala Lumpur were largely foreign places, alienated from the circumstances of the rural masses who made up 90 percent of the population (Figure 12-8).

Transportation—river, road, and rail—was immensely improved to serve the colonial presence. Routes linked raw materials with coastal ports, which served as entrepots for outgoing and incoming goods. This selective transport development meant that many areas were devoid of communication systems and isolated from the larger economy. Dual economies emerged, only to be exacerbated as development increased. Moreover, the rural–urban divide widened as education and health care were made available to urban elites. Only the Philippines had widespread educational facilities.

POLITICAL PATTERNS

Politically, territorial units had little relevance to historical and ethnic circumstances. Territorial forms were derived from European considerations: the need to access raw materials; the need to avoid conflict with each other; and administrative conveniences. Boundaries did not coincide with patterns of settlement of ethnic groups, which were often divided by boundary constructs.

One of the few instances where boundaries were drawn according to ethnic differences is between parts of Laos and Vietnam. Villages where the majority of houses were built on piles with entrances facing east were assigned to Laos. If the majority of houses were built on the ground, facing south, that village went to Vietnam.

Burma and Thailand each have long tails trailing down the Malay Peninsula. Laos is landlocked, and what eventually became Malaysia is fragmented between the mainland and island realms.

Although the colonials frequently turned internal hostilities to their own advantage, no attempts were made to integrate territorially based ethnic identities into unified wholes. In particular, Chinese and Indian migrants remained largely unassimilated and resented.

As the Portuguese, British, and Dutch introduced all manner of changes that shredded the fabric of indigenous societies, the net result was the creation of social systems that could not survive without the paternalism of colonial governments. France, Spain, and the United States also played roles in the colonial period but their activities did not as directly alter regional systems.

Figure 12-8

This photo shows the traditional Chinese/European styled shophouses that once dominated Kuala Lumpur. These stand in sharp contrast to today's modern architecture.

Rising Nationalism and Independence

Geographer Jonathan Rigg (1991) suggests, "The uneven pattern of economic development, and the subordinate position of the local populations, engendered a great deal of bitterness. In addition, the paternalistic attitude of many colonial administrators and their occasionally harsh treatment of the local inhabitants created a reservoir of resentment that nationalist leaders were only too happy to exploit in order to achieve their aims."

Nationalistic leaders were available because certain elite had been given the opportunity to study abroad. It was these individuals, such as Sukarno of Indonesia, Aung San of Burma, and Ho Chi Minh of Vietnam, who spearheaded independence movements. New ideologies were framed in context of the Boxer Rebellion in China (1899), the Chinese Revolution of 1911, and the rise of Japan and its defeat of Russia in 1905, events that proved that the Western powers were not invincible.

After the Japanese invasion of 1941, Japan promoted "Asia for the Asians," encouraged emerging nationalist movements, and even transferred some powers to local leaders such as the Sultans of the Malay States. The Japanese occupation proved to be no better than that of the former colonial powers, but the period between 1941 and 1945 was important in laying the groundwork for subsequent independence.

After the defeat of Japan in WWII, the colonial powers returned to discover they had to negotiate with well-organized and informed nationalists with popular support. Not all efforts were peaceful, and there was often much acrimony between the negotiating parties. The Philippines gained independence in 1946, Indonesia in 1949, and Malaya along with British North Borneo and Sarawak (today's Malaysia) in 1957. The process in some of the other states was prolonged. For example, Singapore did not gain independence until 1963 and Brunei in 1984 (see Figure 1-11). Whatever the case the colonial period left what Rigg (1991) calls "an institutional and ideological legacy" of Westernization that became superimposed on or integrated with former modes of governance.

A Bitter Battle: French Indochina

French Indochina—Cambodia, Laos, and Vietnam—became nominally independent in 1949, although political and economic power was retained by France. Various nationalistic leaders including Ho Chi Minh resisted the French with aid from communist China and the Soviet Union. A guerilla war ensued in Vietnam, and the French were finally defeated at the Battle of Dien Bien Phu in 1954.

The United States, panicked over the further spread of communism, stepped into what became known in America as the "Vietnam War" and in Vietnam as the "American War." The American-supported noncommunist government of southern Vietnam waged battle against the communist forces of northern Vietnam. After years of terrible conflict, loss of life, and increasing corruption in South Vietnam, public opinion in America forced U.S. withdrawal in 1973. The war finally ended in 1975. Divided Vietnam was then unified by the communists under the leadership of Ho Chi Minh.

The Vietnam/American war was a human and environmental travesty. More than 4.5 million died in Indochina, including more than 58,000 Americans. Bombs, napalm, and chemical defoliants destroyed millions of acres of forests and cropland. Thousands have been killed or maimed by unexploded ordnance throughout Vietnam, Cambodia, and Laos. As you will see in the next chapter, the ramifications of this war were severe, widespread, and ongoing.

Changing Urban and Rural Worlds

Although there were a number of very large cities in 1900, most of mainland and insular Southeast Asia remained sparsely populated. However, the twentieth century has witnessed unprecedented population growth in concert with economic, social, and political transformation (Figure 12-9).

Historically, settlement patterns were geared to coasts and rivers and were based on seaborne interchange. Inland transportation was difficult until the era of road and railroad building in the nineteenth century.

Settlement has been tied to rice cultivation, which has, with improvements, supported ever-larger populations. Centralized policies, in places such as Angkor and the Red River delta, were requisite to command and control the expansion of rice production. Given the fractured terrain, for most of history, small societies and local economies dominated the Southeast Asian world. This situation was changed in the late nineteenth century as the tentacles of European imperialism reached beyond port cities into the interior. Large-scale plantations, mines, and administrative cities were constructed in any potentially profitable niche.

With the growth of colonial enterprises, urban centers including traditional entrepots and new colonial cities exhibited spurts of growth. By 1910, there were eleven Southeast Asian cities of more than 100,000 people: Mandalay, Rangoon (Yangon), Bangkok, Hanoi, Saigon, Georgetown (Pinang), Singapore, Batavia (Jakarta), Surakarta, Surabaja, and Manila. These cities were primarily administrative and commercial centers without large industrial bases. Moreover, the colonial economy did little to stimulate economic development beyond the export sector.

The rural sector exhibited three categories. The largest was the peasantry tied to a subsistence economy. The second was the commercialized peasantry who participated in the growing market economy based on urban needs. The third was the enclave economies of plantations and mines that were opened up to Western and Chinese capital and imported wage labor.

A demographic picture of Southeast Asia at the turn of the twentieth century reveals two primary settlement patterns. At one extreme were the dense populations engaged in irrigated rice cultivation. Java, with its population of 30 million in 1901, stands as an example. At the other end of the spectrum were vast insular and mainland areas sparsely settled by shifting cultivators. Around 1900, irrigated rice was diffusing, and several areas such as the remaining frontiers of Java, lower Burma, central Siam (Thailand), and the Mekong delta were transitioning from low to high population densities. The first three decades of the twentieth century continued this pattern of opening up the region to export industries and tightening the links between the colonies to the imperial powers overseas.

The end of the Japanese interregnum in 1945 ushered in a period of transition into independence. During this time, international trade withered, urban and industrial employment fell, and many were forced to revert to a subsistence existence. Nevertheless, there were factors of continuity as well as those of change.

The transition to national independence increased population mobility, only part of which was rural to urban migration. Many moved from overcrowded rural areas to less crowded ones. Others moved to cities as squatters, and still others populated the first suburban and even exurban areas outside large cities.

Out-migration of Western colonial administrators and business people—along with ethnic Chinese from several areas, and all Indians from Burma, plus Vietnamese from Cambodia—were other significant movements. Plantation and commercial rice regions were the areas mainly affected by these out-migrations. Many foreign managers and overseers were replaced by indigenous people.

METROPOLITAN EXPANSION

In Southeast Asia, it is the largest metropolitan areas that are growing much faster than other cities. For example, Kuala Lumpur has grown at almost double the rate of Singapore in the postwar period, reflecting its position as the capital of the new state of Malaysia. Once French Indochina split into Vietnam, Laos, and Cambodia, Phnom Penh grew rapidly in contrast to the former Indochina capital of Hanoi. Primate cities such as Bangkok, Jakarta, Manila, and Ho Chi Minh City (Saigon) continue to dominate the urban hierarchy.

Urbanization in general is on the increase. For example, the proportion of Indonesians living in urban areas of more than one million has risen from 32.6 percent in 1971 to 41.5 percent in 1990. In the Philippines, the proportion of people living in the Manila

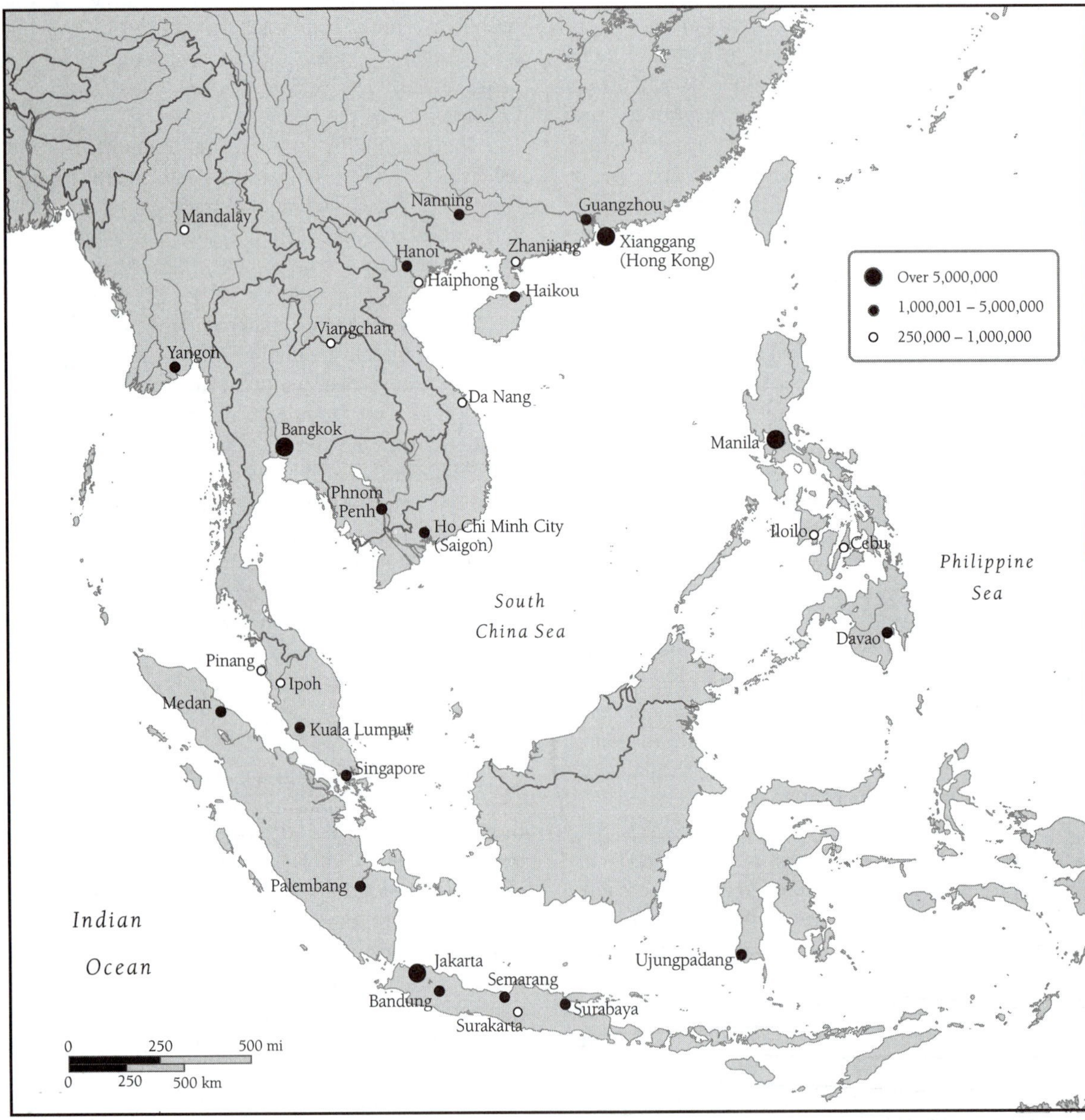

Figure 12-9

Note that the majority of cities in Southeast Asia are either coastal or on major rivers. Why is this the case?

Metropolitan Region has gone up from 29.4 percent in 1970 to 33 percent in 1990.

Much of these increases is related to processes of the transactional revolution, whereby flows of people, commodities, capital, and information are no longer hampered by the limits of space. While cities continue to grow, they do so over larger areas. This is called **region-based urbanization,** and it produces **extended metropolitan regions (EMRs).**

It is in the areas adjacent to currently defined urban areas that urbanization is occurring most rapidly, and this is reflected in the growing proportion of populations engaged in nonagricultural activities (refer to *kotadesasi* settlements in Chapter 4). EMRs have two main components: the central city and the metropolitan region containing adjacent urban settlements. With the exception of Singapore, it is these outer rings that are experiencing the most rapid growth. Meanwhile, city cores have maintained substantial shares of metropolitan populations. In fact, city populations continue to grow with increasing density and congestion (Figure 12-10). EMRs also continue to attract rural–urban migrants. At the social level these areas are extraordinarily heterogeneous with multiple types of land use existing side by side. You will learn more about these trends and individual cities in the next two chapters.

NEW RURAL WORLDS

There have always been market dependencies and interactions between rural and urban worlds but the rural world of today is different in many ways. While *village* may describe life for the majority of Southeast Asians, millions now spend much of their time in urban areas both living and working and millions spend time working in urban-type activities that have come to the countryside.

There is an assumption that people who live in villages are farmers (i.e., that "rural" is synonymous with "agriculture"). But this is not necessarily the case. Urban residents from rural areas often still identify themselves as villagers and farmers. In other words, *domicile* does not always mean "home" and *rural* and *urban* are frequently melded.

Recent studies of village economies consistently indicate that between 30 and 50 percent of total household income is derived from nonfarm or off-farm sources. Many households have different family members participating in either rural or urban livelihoods, making it problematic to classify the household as rural or urban.

The transport revolution has been instrumental in expanding networks of contact and communication. The construction of new roads, the easing of restric-

Figure 12-10

More than 10 percent of Thailand's population lives in the capital city, Bangkok. Bangkok is also known as Krung Thep, *meaning "City of Angels." Here the crowded old city is surrounded by new office and apartment towers. Hundreds of people pour into the city every day seeking employment opportunity. Densities are as high as 5,800 people per square mile (3,600 per sq km), and an estimated 400 new vehicles are registered every day. Bangkok is one of the most congested cities in the world.*

tions on travel, and the advent of the bicycle, motor scooter, and other cheap forms of transport encourage rural and urban interchange in terms of both job transfers and industrial location (Figure 12-11). This increases economic opportunities and supplements incomes. There is a definite link between poor infrastructure and high rates of poverty.

The merging of rural and urban worlds has had other impacts. The use of chemical fertilizers, wage labor, and mechanization has increased. More emphasis is placed on raising animal stock and cash crops for the urban market. More people are engaging in nonfarm activities on the farm, such as mat and basket weaving, pottery and furniture making. Off-season, full-time employment in factories or public sector jobs has increased, and there is increasing migration for employment beyond local areas.

There are also age and gender considerations. Agriculture remains the dominant occupation among older age groups. Younger groups are more likely to engage in wage-laboring, which points to the availability of such opportunities in or near rural areas. A second change is the growing importance of female employment. Rising female labor is pronounced in export-oriented activities such as the garment industry, electronics, and footwear. Declining fertility, later marriages, more education, and increased mobility have fostered conditions for women to work beyond the confines of the village.

Rural and urban changes function in tandem. There is more to rural life than agriculture, and at any time in the year different members of different households may be engaged in rural- or urban-type occupations. Although better accessibility allows rural dwellers to live and work in cities, governments are encouraging factories and the like to locate in nonurban settings where a pool of workers are ready to take up nonfarm employment.

Population Dynamics

Prior to the intensification of the European role, Southeast Asia was an apparently healthy region. But, by the nineteenth century, conditions had deteriorated. Western advances in health provision and medical techniques had spurred population growth, meaning that increased concentration and contact and influx of new groups reduced public health effectiveness. Epidemics became a problem. Paradoxically, populations increased with growth rates climbing from 0.2 percent to well over 1 percent in many areas.

Figure 12-11

I met these two students in Viangchan, Laos. Both were very proud of their new motor bikes.

The Spread of Disease

When the rainforest was cleared over vast expanses to plant rubber trees, water bodies were exposed to sunlight. This caused the *Anopheles maculatus* to breed. This mosquito is one of the most dangerous epidemic vectors of malaria in the world. Epidemics were so severe that they had to carry Tamil workers off in droves and replace them with new ones. These deadly mosquitoes continue to plague settlement schemes, logging operations, agricultural clearance, and the like.

In another case, the British accidentally introduced the *Aedes aegypti,* another mosquito originally from Africa. As a super vector of dangerous viruses, it caused dengue fever to spread explosively in urban areas. Today it is responsible for the emergence and spread of the deadly dengue hemorrhagic fever.

Interregional connections and land conversion policies of the Europeans changed the biotic ecology of disease, especially vectored ones. Millions of victims have suffered or died as a result.

This growth is thought to be attributable to colonial activities. For instance, the spread of Christianity in the Philippines increased birth rates considerably. The expansion of settled agriculture and general peace provided an atmosphere of stability and security in which to raise families. Improvements in agriculture meant improvement in nutrition encouraging longer life expectancies and lower mortality rates. By the 1960s, fertility rates had increased dramatically. For instance, Indonesia's total fertility rate (TFR) was 5.67 and Thailand's TFR was 6.4. However, with family planning, these rates had fallen to 2.7 and 2.0 respectively by 1998. The achievements of these two countries in reducing fertility is to be noted for their noncoercive nature (unlike China).

Ratios of 1950 to 1990 populations were impressive. Thailand's population increased from 20 million to 56 million, an increase ratio of 2.81. Ratios ranged from 2.93 in the Philippines to 1.98 in Cambodia. The average ratio of population change was 2.8.

In the twenty-first century, Southeast Asia's population is more than half a billion. In 1900, the population was a mere 80 million. Even with birth control programs on the part of such countries as Thailand, Indonesia, and Malaysia, population dynamics are uneven. On the one hand, richer countries such as Singapore and Thailand have low fertility rates. On the other hand, poorer countries such as Cambodia and Laos have high fertility rates (refer to Table 3-1). Further, different groups within countries have different rates. For example, Malaysian women of Chinese ancestry (38 percent of the population) tend to have about two children each, while Malays and other indigenous groups average about four children each. In Southeast Asia, the use of birth control is at 55 percent with contraceptive use rising with educational level. A major problem is access to contraception, as in Indonesia, Vietnam, and the Philippines.

Environmental Concerns

The most severe environmental problem in Southeast Asia is deforestation. While slash-and-burn cultivation and clearing for migrant settlement are important causes, perhaps the most important is commercial logging. The forests of Southeast Asia have high commercial value in addition to high yields. Most of today's imported wood products originate in these forests. More than half of tropical wood exports come from the island of Borneo alone.

Commercial logging is a lucrative activity. The entire process has involved considerable corruption, especially in Indonesia, Malaysia, the Philippines, and Thailand. Much forest is logged illegally. In other cases, timber exports are underinvoiced and the logs smuggled overseas. Most logs out of the Philippines are exported illegally. Finally, timber concessions are granted for political or money favors. Who has access to and control of forests is one of the most salient aspects of their use today.

Southeast Asian sources of tropical timber products are concentrated in the Philippines, the Malaysian states of Sarawak and Sabah, and in Indonesia's outer islands. The primary destination of logs and related products is Japan, which has been the world's largest importer of tropical timber since the 1960s. Forest devastation has been so great that today only Sarawak continues to export raw logs legally.

Timber extraction in Southeast Asia is related to Japan's investment and development plans in the region. For example, Japan's early role in the timber industry in Kalimantan began with technical assistance in exchange for timber export agreements. After 1967, most of the major Japanese trading companies such as Mitsubishi, Mitsui, and Sumitomo were involved in loaning money for infrastructural development, including access roads into timber areas. In 1971, Indonesia replaced the Philippines as Japan's first source of raw logs, which were largely processed into plywood in Japan.

By 1980, after rampant forest destruction, Indonesia shifted its efforts to develop its own forest products industry. It also began a phased ban on the export of raw logs, which was fully implemented by 1988. Most Japanese companies withdrew from the Indonesian timber industry, but Japan remains the largest importer of Indonesian plywood. This example shows the relationship between one country's political and economic activities and environmental consequences in another.

Although several countries have taken steps to control wholesale forest destruction, illegal logging continues. Deforestation, with its accompanying soil erosion, and habitat elimination (refer to Chapter 2), continues as one of the major issues in the region.

Conflicts at Sea

Territorial conflicts have laced the history of Southeast Asia. Given this ocean-bound world, it is no surprise that some conflicts would be over ocean spaces. Control over ocean space becomes critical when natural resources are involved as in the case of the Paracel and Spratly Islands (refer to Figure 2-5). Furthermore, control over ocean space becomes paramount when piracy is involved.

DISPUTED ISLANDS

The Paracels are disputed islands and reefs approximately 16 degrees north of the equator and about 233 miles (375 km) east of Da Nang in Vietnam. They were a part of French Indochina until the Japanese held them in World War II. Although these islands are closer to Vietnam than they are to China, the Chinese took the islands from Vietnam in 1974, claiming they were always a part of China. The Paracels are also claimed by Vietnam and Taiwan. The interest centers on potential oil reserves beneath the ocean.

The Spratly Islands are another group of a hundred or so islands and reefs in the South China Sea lying about two-thirds of the way from southern Vietnam to the Philippines. They were used as a submarine base by the Japanese in World War II. Their natural resources are fish, guano, and undetermined natural gas and oil potential. They are also situated near several primary shipping lanes. About 45 of the islands are claimed and occupied by China, Malaysia, the Philippines, Taiwan, and Vietnam. In 1984, Brunei established an exclusive fishing zone in the southern Spratlys.

In 1994, the American company Exxon signed a multibillion dollar deal with Indonesia to develop the Natuna gas fields in the southern Spratlys, an area partly claimed by China. There were clashes between China and the Philippines in 1995 because China occupied one island within 124 miles (200 km) of the Philippine island of Palawan. That same year, Indonesia expressed concern over China's claim to the natural gas fields. Some have suggested joint development of this region, but no agreements have been reached.

PIRACY

Traders have been harassed in the Southeast Asian waters for centuries. The nomadic seafaring *Bugis* from Sulawesi were once infamous as pirates and continue to ply goods throughout Indonesia. But piracy is not a thing of the past. Grim harvests continued to be reaped from the Straits of Malacca to the Celebes Sea. Cargo ships, fishing vessels, passenger liners, and refugee boats are all victims.

The situation was particularly tough during the Vietnam War when thousands of Vietnamese and Cambodians were trying to escape persecution. From 1980 to 1986 alone it is estimated that around 1,500 "boat people" (escapees) were killed, and hundreds of women were raped, abducted, and sold to brothels in Thailand (see Chapter 3 for more information on the sex trade in Thailand).

The Law of the Sea has unfortunately put piratical victims beyond the reach of international law. Piracy is only piracy when attacks occur within jurisdictional waters of coastal states, and most states lack the resources to police their vast marine areas effectively. Many countries now claim straight base lines along their coasts within which the marine areas become coastal waters and pirates are subject to arrest.

Commercial piracy occurs in several regions. It is most common in the Strait of Malacca and the Strait of Singapore. Some 6,000 vessels pass through these passageways every month. Such attacks irritate relations among Indonesia, Singapore, and Malaysia. Endemic overfishing in the Gulf of Thailand and poverty levels also foster piracy in that region. The Sulu and Celebes seas, between Malaysia and the Philippines, are frequent scenes of attacks on coastal freighters, passenger ships, and fishermen. Some freighters have taken to blowing the pirates away with fire hoses. The Moro National Liberation Front (insurgents in Mindanao) have hindered suppression efforts and increased tensions between Malaysia and the Philippines.

Because modern pirates in Southeast Asia are land-based, and not rovers like the ancient *Bugis,* coastal states are expected to take on the responsibility of stopping them. However, the implements of piracy are similar to those of smuggling, also a chronic regional problem. Thus far, it seems that maritime frontiers are virtually impossible to police effectively.

Shapes of Countries

You have probably noticed that Southeast Asian countries take on many different shapes, some rather odd. Country shapes are significant in that they may affect a country's ability to consolidate its territory and control transborder interactions. The most desirable shape is circular, because that provides the shortest possible border in relation to territory and allows equal access to all places from the center. Although this implies stability, stability is not necessarily the case. Both Singapore and Cambodia have relatively circular or *compact* shapes, but Cambodia does not have a history of stability. *Elongated* countries such as Vietnam and Laos tend to have unity and control problems.

Prorupt states are almost compact but have at least one extension of territory. This makes control problematic, especially in the prorupted regions. Thailand and Myanmar, with their long territorial tails, are examples of prorupt states.

Some countries are broken up into pieces. Archipelagos are therefore referred to as *fragmented* states. Both the Philippines and Indonesia are examples of fragmented states, and both have trouble with unity and have separatist movements. However, there are exceptions, such as Brunei. Brunei is a fragmented state yet does not suffer from instability. Why do you suppose this is the case?

Country shapes provide grist for discussion but do not determine a country's political or economic circumstances. Other factors such as resource distribution, physical features, and historic events play even more important roles.

ASEAN

The Association of Southeast Asian Nations (ASEAN) was formed in 1967 to promote social, economic, and political cooperation within the region. After 1945, the newly independent countries had little to bind them together in terms of interstate cooperation, and since economic development was the foremost concern, an international organization seemed prudent. The original members were Thailand, Malaysia, Singapore, Indonesia, and the Philippines. Brunei joined after its independence in 1984, and Vietnam, Laos, Myanmar, and Cambodia joined in the 1990s.

ASEAN's main goals are to promote and integrate economic development, encourage social and cultural progress, and guarantee regional peace and stability.

Actually member countries trade more with the outside than with each other, as so many of them produce the same things. Moreover, international trade activity is dominated by Singapore, Malaysia, Indonesia, and Thailand. Intra-ASEAN trade is also dominated by Singapore. Minerals and fuels account for more than half the volume of exchange.

Three obstacles stand in the way of increased intra-ASEAN trade. The economies of the members are competitive rather than complementary, which supports trade outside the organization. Second, nontariff barriers that are employed to protect domestic industries serve as barriers to intraregional exchange. Third, the productivity of countries such as Vietnam or Brunei cannot be expected to match the volume and diversity of output of Singapore or Thailand.

According to geographer Antonia Hussey (1991), economically, "ASEAN has been long on rhetoric and short on concrete results; all projects sponsored by the organization have fallen short of their goals." ASEAN has been most effective in the political arena, curbing hostilities and acting as a forum to solve numerous disputes. Finally, a strong regional identity has been created among its members. Hussey (1991) wrote, "The sense of identity and cooperation espoused by ASEAN is not found elsewhere in the third world."

Main Point Summary

1. Southeast Asia is fragmented into mainland and insular regions. Fragmented landscapes have supported the establishment of states and kingdoms either land-based on agriculture or sea-based on maritime trade.
2. Populations have been separated by mountains and tend to concentrate in river valleys and on river deltas. Mountain and lowland people have led a symbiotic existence yet have evolved in separate ways.
3. The Mekong is the longest river in Southeast Asia. It is an artery of human activity but not suitable for international shipping. One-third of the populations of Laos, Thailand, Cambodia, and Vietnam live along the Mekong.
4. Although Southeast Asia is tropical, there are temperature variations, depending on latitude and altitude. Precipitation also varies, with some regions experiencing dry seasons derived from the monsoon and prevailing wind directions. Dry seasons increase in length with increasing latitude.
5. Water and control of ocean passageways have been most important in the historical development of the region. Arab and Indian ocean traders, Hinduism, Buddhism, Islam, the Chinese, and the European colonials all arrived primarily by sea. Commercial interaction ensured diffusion of products and ideas and linked the people of the region with each other.
6. Water is essential to agriculture in its amount and distribution. Elaborate irrigation systems have been developed in many areas. Bali exemplifies how water control can be woven into the tapestry of life in a place.
7. Mountain barriers diverted the diffusion of Indian culture almost entirely by sea. Southeast Asian societies adopted many aspects of Indian culture, including religion, architectural styles, and concepts of kingship. All of Southeast Asia except the Tonkin region was affected by Indian culture.
8. Buddhism that diffused from Sri Lanka also exerted a powerful influence. Borobudur on Java testifies to the strength of Buddhism in the region.
9. The Chinese referred to Southeast Asia as *Nanyang*. Its impact was the direct result of colonization and conquest. Vietnam remained a Chinese territory until 907 A.D. By the sixteenth century, Chinese colonies were established throughout the region, but large-scale immigration commenced with China's internal turmoil in the nineteenth century.
10. With the coming together of so many different ethnic groups, languages, religions, and lifeways, most of Southeast Asia evolved as plural societies.
11. Prior to the coming of the Europeans, the region boasted a number of kingdoms with notable

achievements. One of these was Funan, which capitalized on the growing maritime trade from China to the Middle East and Europe. Centered in the Mekong delta, it ultimately extended from northern Malaya to southern Burma. It began to fail as ocean routes moved away from coasts to more direct routes across the South China Sea.

12. Champa, centered on Hue, Vietnam, readily adapted Indian culture and withstood the political and cultural pressures of both the Chinese and the Viets for more than a thousand years.
13. The Khmer kingdom was the mainland successor to Funan. Supported by elaborate irrigation systems and agriculture, the Khmer exerted their authority over much of the mainland. Their legacy is the huge temple complex at Angkor in Cambodia. The Khmer empire collapsed to outside pressures, and their relic core moved toward Phnom Penh.
14. Shifting trade routes through the Straits of Malacca saw the rise of a new kingdom known as Srivijaya, centered at Palembang on Sumatra. It thrived on ocean trade and piracy and eventually fell to the pressures of the Indian Cholas and other trading groups on Java.
15. European political contacts began with the Portuguese capture of Malacca in 1511. With the exception of the Philippines, there was no major impress on the region until the nineteenth century.
16. One hot spot of activity was the Spice Islands (Maluku), where various European powers contested for spices such as cloves and nutmeg. Eventually, the Dutch wrested control from others and tightened its grip by killing off plants except for specific regulated areas, devastating the native populations.
17. Southeast Asia became the crossroads of activity with the demands of the Industrial Revolution, opening of the Suez Canal, and decline of Caribbean commerce. From 1825 to 1870, most of the region fell under European control.
18. Although the various European powers had their own agendas, Thailand was preserved as a buffer state between British and French interests on the mainland. The strongest rule was exerted by the Dutch Culture System, which forced Indonesians to grow specific amounts of cash crops.
19. The opening of new lands such as the Chao Praya and the Mekong deltas attracted thousand of new migrants. Indians and Chinese were drawn to tin mines and rubber plantations. Plural societies emerged whereby different ethnic groups had their own residential quarters but came into contact in the marketplace. Cities were largely foreign places, alienated from the circumstances of the rural masses.
20. Transportation and other infrastructure components were designed for foreign interests. Coastal entrepots serving shipping left much of the interiors to its own devices. Dual economies emerged and became exaggerated as uneven development increased.
21. Political units, designed for the needs of Europeans, had little to do with historical or ethnic circumstances. Many groups were divided by arbitrary boundaries. Myanmar and Thailand are protruded and Laos is landlocked. Indonesia and the Philippines are fragmented archipelagos.
22. In the late nineteenth and early twentieth centuries, revolutions proved that colonial powers were not invincible. Moreover, Japan's occupation, while as exploitative as the Europeans, promoted Asia for the Asians. After 1945, nationalistic leaders, educated abroad, orchestrated independence movements. They were not typically peaceful, and many colonies did not achieve independence until the 1960s and 1970s. Vietnam was one of the most prolonged conflicts, drawing in even the United States.
23. Under colonial dominance, infrastructure and settlement patterns changed with more primary activity in the interiors. Large-scale plantations, mines, and administrative cities were constructed in every potentially profitable niche. More of the peasantry participated in a growing market economy based on urban needs.
24. Two population patterns existed at the turn of the twentieth century: a dense one linked with intensive rice cultivation and a sparse one associated with vast insular and mainland areas inhabited by shifting cultivators.

25. An array of migration streams were encouraged by new-found independence and related events.
26. Larger metropolitan areas are growing much faster than other cities. Urbanization, in general, is on the increase. This region-based urbanization is fostering extended metropolitan regions that incorporate growing numbers of rural populations indulging in nonagricultural activities. While *village* may describe life for the majority of Southeast Asians, millions now spend time living and working in urban areas.
27. Rural and urban activities exhibit age and gender differences with older people sticking to agriculture and younger people seeking other forms of employment. Females are especially prominent in export-oriented activities.
28. Even in context of poor sanitation, Western advances in medical techniques and health provision, along with improvements in agriculture, spurred population growth. Population growth ratios from 1950 to 1990 were high, causing giant leaps in national populations. Population dynamics remain uneven—countries such as Cambodia and Laos have much higher growth rates than others where family planning programs have been more effective.
29. Environmental concerns are dominated by deforestation, in which Japan continues to play a major role.
30. Territorial conflicts and piracy continue to disturb Southeast Asia. The Paracel and Spratly islands, with their undersea oil potential, are the source of territorial claims of several nations including China. Piracy plagues the South China Sea, the straits of Malacca, and other critical ocean passageways.
31. The Association of Southeast Asian Nations is an intraregional organization intended to promote social and economic development and maintain regional stability. With Cambodia's recent acceptance, ASEAN now includes all Southeast Asia's countries.

Terms to Know

naga	*subak*	plural society	region-based urbanization
sawah	*Nanyang*	Dutch Culture System	extended metropolitan region (EMR)

Bibliography

Brierley, Joanna Hall. 1994. *Spices: The Story of Indonesia's Spice Trade.* New York: Oxford.

Dauvergne, Peter. 1997. *Shadows in the Forest: Japan and the Politics of Timber in Southeast Asia.* Cambridge: MIT.

Duchin, Dian. 1986. The Subak Agriculture of Bali. *The New York Botanical Garden* 10: 6–10.

Dutt, Ashok K. 1985. Indochina: Evolution of the Core Areas of Champa, the Mekong Delta, Angkor, and the Upper Mekong Valley. In *Southeast Asia: Realm of Contrasts,* ed. Ashok K. Dutt, pp. 179–201. Boulder: Westview.

Dutt, Ashok K, ed. 1985. *Southeast Asia: Realm of Contrasts.* Boulder: Westview.

Dzurek, Daniel J. 1989/90. Piracy in Southeast Asia. *Oceanus* 32: 65–70.

Ginsburg, Norton. 1972. The Political Dimension. In *Focus on Southeast Asia,* ed. Alice Taylor, pp. 3–12. New York: Praeger.

Hirschman, Charles. 1994. Population and Society in Twentieth-Century Southeast Asia. *Journal of Southeast Asian Studies* 25: 381–416.

Hussey, Antonia. 1991. Regional Development and Cooperation Through ASEAN. *The Geographical Review* 81: 87–98.

Keyes, Charles F. 1995. *The Golden Peninsula: Culture and Adaptation in Mainland Southeast Asia.* Honolulu: University of Hawaii.

Kummer, David M. The Physical Environment. In *Southeast Asia: Diversity and Development,* ed. Thomas R. Leinbach and Richard Ulack, pp. 6–34. Upper Saddle River, NJ: Prentice Hall.

Ma, Laurence J. 1985. Cultural Diversity. In *Southeast Asia: Realm of Contrasts,* ed. Ashok K. Dutt, pp. 53–64. Boulder: Westview.

McCloud, Donald G. 1995. *Southeast Asia: Tradition and Modernity in the Contemporary World.* Boulder: Westview.

McGee, T. G. 1995. Metrofitting the Emerging Mega-Urban Regions of ASEAN: An Overview. In *The Mega-Urban Regions of Southeast Asia,* ed. T. G. McGee and Ira M. Robinson, pp. 3–26. Vancouver: UBC.

McGee, T. G. 1967. *The Southeast Asian City.* London: GF.Bell.

Michell, George, and John Fritz. 1993. Angkor Regained. *Geographical* 65: 33–38.

Minority Rights Group. 1992. *The Chinese of Southeast Asia.* Manchester: Free Press.

Noble, Allen G. 1985. The Physical Environment. In *Southeast Asia: Realm of Contrasts,* eds. Ashok K. Dutt, pp. 36–51. Boulder: Westview.

Rigg, Jonathan. 1991. *Southeast Asia: A Region in Transition.* London: Cambridge.

Rigg, Jonathan. 1997. *Southeast Asia: The Human Landscape of Modernization and Development.* New York: Routledge.

Sanderson, Warren C., and Jee-ping Tan. 1995. *Population in Asia.* Washington, D.C.: World Bank.

Sar Desai, D. R. 1989. *Southeast Asia: Past and Present.* Boulder: Westview.

Shaffer, Lynda Norene. 1996. *Maritime Southeast Asia to 1500.* London: M. E. Sharpe.

Vatikiotis, Michael, and Prangtip Daorueng. 1998. Survival Tactics (Overseas Chinese). *Far Eastern Economic Review* 26 (February): 42–45.

Chapter 13

Mainland Southeast Asia: Turmoil and Peace

"The waves of human passion rise and fall; worldly things are as transitory as passing clouds."

(CAO BA-QUAT 1809–1854)

In this chapter I will tell you about the five countries of mainland Southeast Asia: Burma/Myanmar; Thailand; Vietnam; Laos; and Cambodia (Figure 13-1). As you will discover, the history of each unfolds in context of the history of the others. Landscapes, with their impress of rice cultivation, Buddhism, and battlegrounds are remarkably similar. Yet, each country exhibits its own character in terms of ethnicity, language, political system, level of development, and various other cultural aspects (Figure 13-2). The complexity of mainland Southeast Asia makes it a geographic conundrum leaving us with more questions than answers.

Burma/Myanmar: Paradise Lost

Imagine living in a country where your universities are closed most of the time. Where you cannot stay at a relative's or friend's house without government permission. Where you voted for democracy but your party is outlawed and your leader is under house arrest. Where barbed wire barricades are stacked on streets ready to block any march or protest. Where people meet in secret to tell you about persecution of minorities and slave labor. This is Myanmar today—oppressive for most within and despised by most without. I will use Myanmar and Burma interchangeably, as the use of Burma is becoming popular once again.

From its prewar position as a productive agricultural colony, a major rice exporter and rich source of minerals, Burma has slumped into being the poorest country in Southeast Asia. How did this happen?

POLITICAL BACKGROUND

Burma is the largest and most ethnically diverse state in mainland Southeast Asia (Figure 13-3). Burmans make up 68 percent of the population but there are 11 major and many more ethnic groups in the country. The Burmese dominate the alluvial plains and cities.

Figure 13-1

Defining the Realm: Mainland and Insular Southeast Asia. Note the location of the core regions.

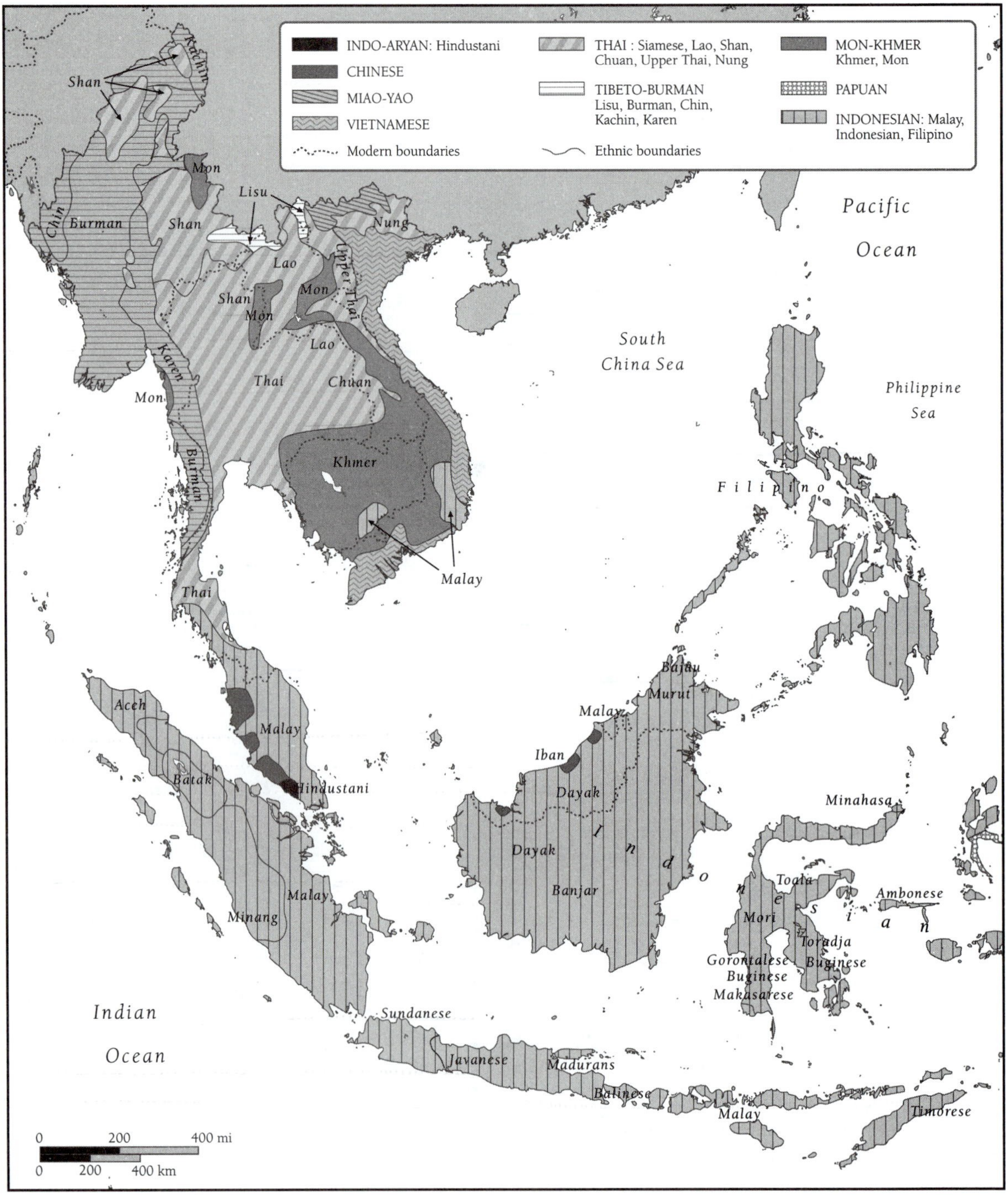

Figure 13-2

This region is a dramatic ethnic-cultural complex. Note the presence of Chinese and Indian communities.

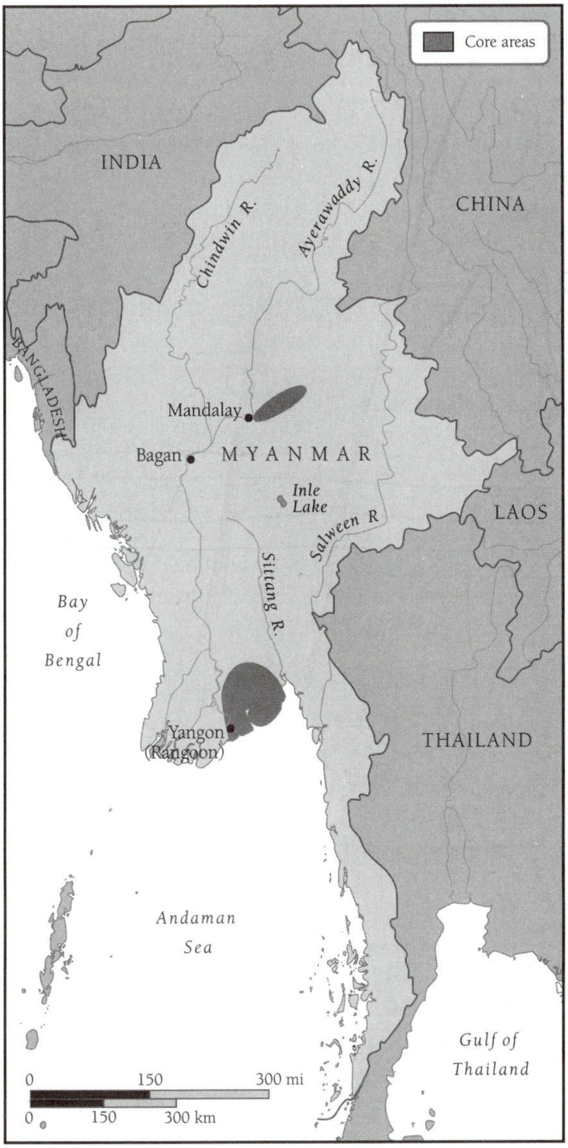

Figure 13-3

Myanmar (Burma) and its core areas. Note the importance of rivers in the location of urban places and core areas.

Tribal groups, flanking the borders of India, Bangladesh, China, Laos, and Thailand, have long resisted Burman dominance. The Burmans are mainly Buddhist. The Shan and Karen are predominantly Christian, and the Arakanese are split between Christian and Muslim. These religious differences have deepened the rifts among Myanmar's ethnic groups.

Burma has always been concerned about the ambitions of its neighbors. War, competition, and animosity between Burma and Thailand are centuries old. The Mongols once invaded Burma, and the Chinese are still seeking access to the Indian Ocean. Myanmar's economy was once dominated by Indian migrants. Thus, historic acrimony shapes the country's attitudes to outside powers.

The territorial boundaries of Myanmar are a British creation, for no single group held a similar territory. The first major kingdom was founded in 1044 around Pagan (Bagan) north of Rangoon (Yangon). It was a Buddhist kingdom with hundreds of magnificent temples stretching 15 square miles (40 sq km) along the banks of the Ayerawaddy. At the end of the thirteenth century, Pagan was sacked by the Mongols from China.

Mandalay became the center of a new Burmese kingdom in the mid-eighteenth century. It exerted its power over much of the hill states and the Shan people and competed with Thailand for territory and people. At the end of the eighteenth century, Burma was the strongest state in mainland Southeast Asia.

Meanwhile, the British East India Company, headquartered in Calcutta, envisioned Burma as a buffer zone. Clashes in Arakan adjacent to Bengal mushroomed into the the first Anglo-Burmese war from 1824 to 1826. Superior arms gave victory to the British, and Burma was forced to cede territory along the Bay of Bengal. A second conflict erupted in 1850, whereby the British acquired more territory in lower Burma. In 1885, the Burman king was exiled to Calcutta (even the throne was removed), and Britain annexed Burma in 1886.

Ruling from Calcutta, the British imposed alien Indian forms of government. Lower Myanmar—the Burman region—fell under direct rule, and the power of local elites was squashed. In Upper Myanmar, indirect rule left the hill tribes pretty much to their own devices. Consequently, the minorities developed a stronger sense of identity in opposition to the British-run region.

In the absence of the king, ritual and symbolism of the court were ended and the authority of the Buddhist hierarchy diminished. These binding forces of society were removed and had no indigenous replacements.

British colonial rule was founded on an Anglo-Burmese and Indian elite. Indians were free to migrate to Burma. By 1931, about 7 percent of the country was Indian, and Rangoon was 53 percent Indian. Chinese were recruited from British Malaya and Singapore and soon added to the population diversity working in Shan state mines and operating small businesses and rice mills.

British efforts to develop the Burma delta regions in the 1850s encouraged large-scale migrations of Burmans from the arid northern regions. The Sittang and Ayerawaddy Deltas became prominent rice-producing regions. The area under rice cultivation expanded from 800,000 acres (320,000 ha) to around 6 million acres (24 million ha) from 1852 to 1901. The associated population increase went from 1 million to 4 million. By the 1920s, population pressure had become a major problem, many farmers were indebted, and tensions among Burmans, Indians, and Chinese were exhibiting violence.

Britain was content to have Burma as a rice, teak, and mineral exporter. There was no attempt to industrialize. This plural society rested singularly on primary economic activities. Moreover, economic position was coincident with ethnicity.

Economic development under colonial rule was accompanied by the spread of Western education and the rise of nationalist leaders. These were not only dominated by ethnic Burmans but also were strongly anti-Chinese and anti-Indian. Foreign capital stimulated the development of a socialist ideology. Placing Buddhism at the core of Burman life further alienated the non-Burman minorities, especially those who were Christian.

Three of the prominent nationalists were Aung San, U Nu, and Ne Win. Aung San was assassinated, U Nu would become the first prime minister, and Ne Win would lead a coup in 1962 placing the military in power, where they remain.

The anti-Western Japanese occupation in the 1940s was welcomed by many Burmans, seduced by an Asian "Co-Prosperity Sphere." But reality did not meet the vision, and the disillusioned Burman nationalists worked toward independence. Burma was terribly damaged in World War II, with most of its infrastructure destroyed.

After the war, things fell into disarray. Minorities found themselves in a new state relentless in its efforts to Burmanize them. A Karen nationalist movement launched several insurrections, which continued into the 1990s. Burma was a democratic state between 1948 and 1962, but its failings were critical. A failing economy, regional insurrections, social urban and rural unrest, and rampant corruption led to Ne Win's overthrow of U Nu. Political and ethnic minority leaders were arrested and opposition from university students and Buddhist monks was ruthlessly suppressed. The Westernized, often Anglo-Burman elite that had run the country under colonial rule and through the 1950s fled, along with large contingents of Indians and Chinese. Muslims on the western border escaped into Bangladesh to avoid persecution in the 1990s.

The Demise of Teak

To finance the civil war, Ne Win's junta sold off much of Burma's teak. An estimated 80 percent of the world's teak is found within Burma's 900-mile (1,500-km) boundary with Thailand. Scores of logging and fishing concessions were sold to Thai firms, along with the rights to gems, minerals, and oil to foreign concerns. Leaders of the Karen and other minorities are being displaced by these concessions and associated road building. Even so, they, too, are selling off teak to finance their insurgencies against the Burmans. However, tribals use elephants for logging, and elephants do not require roads.

Consequently, their operations are less damaging than the mechanized ones that call for forest clearance for road development.

Thailand is hungry for logs as it placed a moratorium on domestic logging in 1988. Burma is losing at least 2 million acres (800,000 ha) of forest cover a year. Forests have been virtually eliminated in coastal areas and in the Ayerawaddy and Sittang valleys.

In 1974 the country was partitioned into administrative divisions: seven minority states and seven Burman divisions based on Western notions of nation and state. But, the regions were not designed for any level of autonomy but rather to facilitate further centralization. Delineations are not consistent with ethnic or residential integrity; groups and villages are splintered and mixed up. This aggravates the existing tensions among different linguistic and religious groups. Moreover, as geographer Curtis Thomson (1995) explains, traditional mapping with bounded concepts "contrasts sharply with the ephemerally bordered, hierarchically centered territories of mainland Southeast Asia." Boundary definitions remain an important focus of rebel movements.

In 1987, the United Nations gave Myanmar "Least Developed Nation" status. However, there are two economies: one legal and the other illegal, focusing on opium, of which Myanmar is the world's major producer.

In 1988, a series of protests and strikes called for an end to military rule. In response, a new organization, the State Law and Order Restoration Council (SLORC), took control and killed thousands of protestors. SLORC decided to hold elections in 1990. Surprisingly, **Aung San Suu Kyi,** daughter of the nationalist hero Aung San, returned from London to lead the National League for Democracy (NLD). She was placed under house arrest in 1989. Nevertheless, the NLD won the election convincingly. But SLORC declared the elections null and void and arrested NLD members. Aung San Suu Kyi remains under house arrest and was awarded the Nobel Peace Prize in 1991. SLORC rules but under a new label: The State Peace and Development Council (SPDC).

Rules of SLORC/SPDC

The rules of SLORC are known as the "People's Desire." They are printed in daily newspapers and on billboards and include such orders as: Oppose those relying on external elements acting as stooges; Oppose those trying to jeopardize stability of the state and programs of the state; Oppose foreign nations interfering in internal affairs of the state; and, Crush all internal and external destructive elements.

SCRAPING THE RICE BOWL

Burma was once known as the "Rice Bowl" of Southeast Asia. Rice production dwindled during decades of socialist rule but is picking up under SPDC's liberalized economic policies.

Rice trade is one of the regime's most important sources of foreign revenue. The government monopolizes rice exports. It buys rice from farmers at low prices, sells it on the world market, and absconds with the profits. Quotas are set for each cultivated area—15 to 20 percent of each farmer's crop is forcibly purchased for less than half the market price. Many farmers cannot make ends meet and wind up losing their land.

There are 7,000 agricultural supervisors posted around the country to encourage harder work. Signs such as "Lazy People Lose Their Land" and "Be Confident in Your Ability" are intended to stimulate productivity.

Paddy production increased around 30 percent from 1992 to 1996, mainly because of expansion of irrigated area and increased supplies of subsidized fertilizer, pesticide, and diesel fuel. Low-cost tractors have also been imported from China.

Pests are a serious problem. Insects are devouring rice because storage techniques are poor. Moreover, more double-cropping with fast-growing seeds means that much rice is harvested in the rainy season. Without pesticide, the wet grain is even more susceptible to pests.

In border regions, quotas and inadequate technology have compelled many farmers to leave their land altogether and migrate to Thailand. Others have been forced off their land by the military. About a million Burmans now work in factories and farms in Thailand.

Inle Lake's Floating Agriculture

Inle Lake in one of the Shan states is Burma's largest lake (Figure 13-4). The lake is home to vast areas of floating gardens anchored in place with long bamboo poles driven into the lake bottom. Adjacent houses are built on stilts and stand in 6 or 10 feet (2 or 3 m) of water.

The gardens are strips roughly 82 feet (25 m) long and 3 feet (1 m) wide cut from floating vegetation at the lake's edge and towed to the garden site. Lake bottom sediment is scooped up and placed on top of these mats to provide topsoil for the planting of vegetables. Plants, with their roots close to the water, absorb nutrients from the lake, which is a repository for all the residential waste in addition to runoff from the surrounding hills. The gardens are tended by farmers who move from strip to strip in dugout canoes. Much of the produce is sold at markets around the lake. Fish farming is also practiced, with the floating gardens acting as nurseries.

Figure 13-4

Inle Lake. This man is tending his gardens from his dugout boat. Note that the paddle is operated by hand and foot.

AUTHORITARIAN DEVELOPMENT

To make way for development projects such as roads and tourist facilities, SPDC has forcibly moved people off their land. Tens of thousands have been moved to satellite towns around Yangon and Mandalay. The entire population of Bagan was displaced to eliminate dwellings from among the temple complexes as part of tourism promotion. In Mandalay, when the government wanted to widen a road, it simply sliced the fronts off all the shop houses lining the route.

One of numerous persecuted groups is the Rohingyas. Burmese Muslims, the Rohingyas began to flee across the border into Muslim Bangladesh in 1978. This exodus has strained relations between Burma and Bangladesh.

Most displaced people still work in the cities but now find transport costly in both time and money. Many have moved to squatter settlements closer to their place of employment. In any case, community and family relationships and village economic systems are broken or strained.

The country has a serious shortage of electricity. In Mandalay, power is rationed. Many homes have electricity on alternate days, and factories can only operate part time. However, government-supported tourist hotels always have power.

The government relies on forced labor for public works and eco-tourism development. In 1996, at least 300 villages were destroyed and many people were killed. There are numerous accounts of people "worked to death." Many eco-tourism projects are in minority regions, as are other developments such as new roads and oil pipeline construction. These projects are controlled by the SPDC military, whose forces have doubled since 1988. Recruitment of 13- to 15-year-olds has boosted numbers to 400,000.

One project of interest is an oil pipeline built from the Yadana oil field in the Andaman Sea to Ratchaburi, southwest of Bangkok. Much of this project used slave labor. Every village along the way had to donate labor, including women and children. On such projects, skilled workers are paid about 50 cents a day; the rest get nothing. Foreign oil companies are heavily involved: Total (partly owned by the French government), the American company UNOCAL, and companies from Myanmar and Thailand. The deal is supposed to give the Yangon generals about U.S.$400 million a year for 30 years. A new railway into the pipeline region is also being constructed with slave labor. UNOCAL denies this, but 5,000 troops have been sent into the region to protect Total personnel.

THE CHINA CONNECTION

The Mekong River and the Burma Road from Kunming, Yunnan, China, to Mandalay now function as major arteries for the influx of Chinese into Burma and elsewhere in Southeast Asia. Thousands of Yunnanese have poured into Mandalay, where they now dominate commercial life.

China refers to the region as the "Great Golden Peninsula." Trade between Yunnan and Burma increased from U.S.$15 million to well over U.S.$800 million annually between 1984 and 1994. Manufactured goods that once came from Thailand now come from China. In addition, the Burma Road has become the main route for timber into China and arms shipments from China. It is also a major conduit for drugs.

The Chinese are anxious for connections to the Indian Ocean. It is surmised that China already has access to Andaman Sea islands for spy operations. Fishing boats with sophisticated monitoring equipment have been detected in the area as well. India has protested to Yangon and Indonesia has expressed concerns over potential Chinese threats to commerce in the Malacca Straits.

LANDSCAPE OF REPRESSION

For decades a tyrannical military regime has kept Burma in political and economic turmoil. Infrastructure is pathetic at best, and thousands of Burmans and minorities are unemployed, engaged in forced labor, or working under oppressive conditions.

I have a vivid memory of Mandalay. I went to a place where they make gold paper to be used for Buddhist temple offerings. Perched at the edge of a muddy lane by a stream was a thatched, wooden structure with a cellar carved out of the earth. The cellar was filled with the stream's murky water. Floating on the water was a platform, and on the platform squatted four women. The only light was that which crept past a tarpaulin shielding the pit's doorway from the sun's intensity. It was 87° F (30° C). Both heat and humidity were tortuous. Each woman had a substantial piece of wood in each hand with which she rhythmically beat the small stack of gold sheets at her feet. The pounding resonated in the enclosure; the noise was deafening. These women work in this manner for eight hours a day, seven days a week. Think—this could be your life!

Unfortunately, Myanmar has been shoved so far into life's cellar that even a change in government will take years to drag the country out of its current state of oppression, corruption, and impoverishment.

Drugs and the Golden Triangle

The production of illicit narcotics has more than doubled over the past few years in the Burmese section of the Golden Triangle where the borders of Burma, Laos, and Thailand intersect. The area under opium poppy cultivation increased from 228,000 acres (92,300 ha) in 1987 to 380,380 acres (154,000 ha) in 1995 (Figure 13-5). Heroin is derived from opium. The potential heroin output rose from 54 tons in 1987 to 166 tons in 1995.

Figure 13-5

Opium poppies in the Golden Triangle—Northern Laos.

For years the drug trade was controlled by the infamous Khun Sa, but in the 1980s, he lost ground to better equipped heroin barons near the Yunnan frontier. Khun Sa surrendered to the Burmese government in 1996 and moved with his heroin-generated fortunes to luxury living in Yangon. Much of his money was invested in perfectly legal businesses such as hotels, real estate, supermarkets, and construction, generating a mini-boom in Myanmar's economy.

Khun Sa's turf has been taken over by the United Wa State Army, a splinter group of the Burmese Communist Party in northern Burma's Shan state. This group maintains a cease fire with the regime in Yangon.

Since the 1980s, a new string of refineries has been established in the Shan State close to the Chinese drug market and with links to new markets abroad. One new market is Taiwan, which in 1989 was virtually drug free. Now drug abuse is a major concern. Drug use is increasing in Vietnam, Bangladesh, and India. There are more than 700,000 Burmese heroin addicts, 70 percent of whom have HIV. There are at least half a million addicts in Thailand, and 14 percent of Malaysian teenagers are now using hard drugs. China counts at least 250,000 addicts, with Yunnan having the highest rate of addiction.

Some 70 percent of all heroin in the United States comes from the Golden Triangle, and 98 percent of that originates in Burma. In Australia, with 30,000 addicts, 75 percent of all heroin comes from Burma, with 5 percent from Laos and Thailand.

For years, Thailand was the main conduit for drugs, but stepped up law enforcement and poppy production alternative programs have altered this pattern. Cambodia and Laos have emerged as major transit routes. Cambodia is also a key center for money laundering. Drug laws are vague in Cambodia, and marijuana is freely available in markets (I've seen it). Cambodia is also a major shipper of marijuana. In addition, after years of war, Cambodia is a major source of arms for the Golden Triangle's drug lords. Cambodia's deep-water ports at Koh Kong and Kompong Som with their hundreds of fishing trawlers are easily able to smuggle the drugs to larger ships in the international distribution web. Cambodia handles about 15 percent of Burma's heroin production. In Laos, drug smuggling is protected by the Mountainous Area Development Corporation, a company controlled by the Lao army.

Another new trail is from Thailand via northern Laos into Vietnam via Hanoi and the port of Haiphong. Danang is another distribution point. Vietnam, like its neighbors, has a long history of opium smoking among ethnic minorities in mountainous areas. Vietnam provides an alternative, and less-policed shipping route. Addiction is becoming problematic in Hanoi among wealthy teens and the disadvantaged.

Profits are huge. A poppy farmer earns U.S.$330 for 7,000 grams of raw opium, which is then converted into a 700-gram heroin brick, worth U.S.$4,250, at a Burmese refinery. In Sydney, the brick will be worth U.S.$53,050. In New York, the brick will be valued at U.S.$80,000 and will be converted into 28,000 packets of adulterated heroin with a total retail value of U.S.$280,000.

Clearly drug production in the Golden Triangle has widespread international implications. Chinese organized crime families from Hong Kong, China, Taiwan, and Singapore show great interest in Cambodia. Various African groups ferry drugs to Nigeria and then on to Europe. The Tamils deal in arms and heroin to finance their war in Sri Lanka.

Democracy alone will not alleviate the drug crisis in northern Burma. Even with a new government, new and old drug lords are unlikely to listen to calls for reconciliation and stop their lucrative activities. In the absence of a comprehensive policy, the fracturing of northern Burma into small, well-armed, and violent drug-trafficking states will continue to create instability in the region and serious drug problems abroad.

Thailand: Tiger in the Metropolis

Like an elephant with its trunk drooping down the Malay Peninsula, Thailand sits in the heartland of mainland Southeast Asia. Because the north–south distance stretches 16 latitudinal degrees, Thailand's climate exhibits great diversity. Mountains in the north give way to limestone-encrusted tropical islands to the south. Rivers and tributaries drain into the Gulf of Thailand via the Chao Praya delta near Bangkok (Figure 13-6).

The country is divided into four main zones:

- The fertile central region dominated by the Chao Praya.
- The drought- and flood-prone, poor, northeast plateau.
- The rugged northern region dominated by mountains and fertile valleys.
- The southern peninsular region characterized by rainforest.

Thailand's population is relatively homogeneous. Uniquely, it avoided the disruptions of Western colonial rule and the upheavals of decolonization. After World War II, which produced minimal effects on the landscape, Thailand pursued capitalist development. By the 1990s it was deemed an NIC and now is one of the tigers of Asia. A brief look at Thailand's history will set the scene for the discussion of this Bangkok centered economy.

THE TAIS

The **Tais** were the principle ancestors of the Thai, Lao, and Shan peoples. *Tai* is a cultural and linguistic term used to denote the various Tai people who became differentiated into a larger number of separate identities. Over many centuries, the Tai migrated southward from western China.

The Tai were wet-rice farmers organized in *muang*—one or more villages under a chieftan. By the thirteenth century, further expansion and alliances transformed *muang* into kingdoms that adapted many of the beliefs of peoples they came into contact with. For example, they probably adopted Theravada Buddhism from the Mon people of central Thailand and the Burmans at Pagan. Angkor served as a blueprint for Tai state builders. From Angkor came the concept of society as a divinely ordained hierarchy under a ruler incarnate of a Hindu deity or Buddhist *boddhisattva*. Angkor provided a model for administering large and scattered populations as well as a range of arts and technologies. These were further accumulated by Tai attacks on Angkor in the fourteenth and fifteenth centuries.

Sukothai

Meanwhile, in the thirteenth century the Tai kingdom of Sukothai had arisen. Modern Thais regard Sukothai as the birthplace of the Thai nation. The people of Sukothai adopted the name "Thai" to distinguish themselves from other Tai-speaking people. The Khmer system of writing was adopted and modified for a Tai

Figure 13-6

Thailand has many cities and towns, but the core area clearly focuses on the primate city of Bangkok. Why do you think this is the case?

ferrous metals, and marble. It has also gone to resource-based industries such as processed foods, palm oil, sugar products, wood-furniture parts, rubber gloves, and paper products.

Thailand remains an attractive destination for foreign involvement. Japan, the major investor, and other countries view Thailand as the hub of a golden pentagon comprising Thailand, Myanmar, and Indochina. Thailand is seen as an ideal site to establish satellite companies into a surrounding hinterland of cheap labor and a potential market of more than 120 million consumers. Thai companies, offshoots of joint ventures with the Japanese, are already trading in Vietnam and have a strong presence in Laos, where they dominate extractive industries.

Thailand is unique among developing countries because industrialization is concentrated in and around the primate city of Bangkok (Figure 13-9).

BANGKOK METROPOLITAN REGION

Why is Thailand a tiger in a metropolis? Consider these facts. Over half of Thais are employed in agriculture, yet agriculture contributes only 16 percent of the GDP. Although only 16 percent of Thais live in Bangkok, it produces more than half of the GDP. Bangkok is Southeast Asia's largest city and is the economic pulse of the region.

A city of more than 10 million, the historic and current primacy of Bangkok relates to its excellent location. It has been the capital since its founding in 1792 and lies in the rich Chao Praya rice region. The government is highly centralized, the city is the spiritual center of the nation, and it is the residence for the highly esteemed royal family.

Bangkok is a newly industrializing city. Economies of agglomeration are supported by a significant informal sector. Urbanization has spread from the Bangkok Metropolitan Area (BMA) into the Bangkok Metropolitan Region (BMR), which comprises five provinces circumscribing the city. Transportation, water, and utility costs are lower in Bangkok than elsewhere in the country. Labor is more productive because of the higher educational levels and greater experience of workers. The city attracts labor-intensive, export activities.

Smaller firms are concentrated in the BMA where land is more expensive. Larger firms prefer the cheaper land of the BMR. In 1989, the BMR accounted for 76 percent of Thailand's gross industrial input and housed 50 percent of all factories in the country.

Ribbon development has extruded onto agricultural land. Rice-producing and other agricultural areas are being displaced by industry, housing estates, and recreational facilities such as golf courses. Thailand has more golf courses than any other Southeast Asian

Figure 13-9

Recent years have seen a dramatic increase in construction in Bangkok. Although money is made in construction, vacancy rates are very high.

country. Although development is clustered in the BMR, numerous benefits flow from this wealth, including upgraded health and education systems and improved social and physical infrastructure throughout the country.

Rapid industrialization has taken its toll on the environment. Air and water pollution are severe and traffic congestion is among the worst in Asia. These seemingly unmanageable problems are constraints on development. Yet construction continues as profits are made in the development process rather than from the final product. Many construction projects are joint ventures with Japanese firms.

Gridlock

In Bangkok, an estimated 3 million vehicles are in the city every day. This figure is rising by 1,000 a month. Nearly half the traffic police are undergoing treatment for respiratory disease. A commuter spends an average of 44 days a year in traffic. Gas stations sell Comfort 100, a portable toilet for when a motorist is stuck in traffic jams.

Road surfaces account for only 8.5 percent of urban space; they should make up 20 percent. Motorcycles, buses, taxis, and trucks comprise the bulk of the traffic. Buses, which account for a mere 1 percent of the traffic contribute more than half of the pollution.

Several mass transit systems are either in the planning stages or under construction. However, lack of coordination, interdepartmental quarrels, and problems with contractors have rendered many of these stagnant.

Bangkok is called the Venice of Asia because of its *khlong* (canal) system linking various parts of the city with the major artery—the Chao Praya. Now many of the *khlongs* have been paved over and quicker boat traffic has given way to slower vehicle traffic.

With development so concentrated in the BMR, the question is this: Is Thailand a newly industrializing country or is Bangkok a newly industrializing city? It does appear that Thailand is a metropolitan tiger rather than a national one.

Vietnam: Baby Tiger

"Two baskets of rice separated by a bamboo pole." This is Vietnam—the densely populated Red River delta strung by 1,000 miles (1,610 km) of mountains and narrow coastal strip to the watery world of the Mekong delta. On these two rich, alluvial plains live two-thirds of the Vietnamese population now numbering 78.5 million (Figure 13-10).

Some minorities are identifiable by their religious affiliation. Roman Catholics are the largest group. At independence in 1954, there were about 2 million. When Vietnam was divided later into two zones, about two-thirds of the 900,000 who left the north were Catholic. Another group is the **Cao Dai,** with its headquarters in Tay Ninh province near Ho Chi Minh City. *Cao Dai* means "high tower" and is a mix of Buddhist, Christian, Confucian, and Taoist beliefs.

Although the population is dominated by the ethnic Vietnamese, the presence of these other groups has made it difficult for any government to create a centralized and highly integrated society. Moreover, regional differences are strong among the northerners, southerners, and the central Vietnamese. Currently, the Communist regime is attempting to reduce these dichotomies by integrating southerners into the more regimented system of the north. Moreover, modernization is moderating differences throughout the country.

ORIGINS AND CHINESE INFLUENCE

It is possible that the ancestors of the Vietnamese originated in southern China, but they first appear in history as *homo erectus* in the Red River delta (*Tonkin* region) at the same time as Peking and Java Man about 500,000 years ago. Available evidence supports the fact that they had mastered primitive agricultural techniques as early as 9,000 years ago. If this is so, the Vietnamese are among the first people to practice settled agriculture. By the second millenium B.C., the Red

Marble Moutain
China Beach

Figure 13-10

Vietnam is part of a region known as Indo-China, including Laos and Cambodia, all of which were French colonies. Note how the Annamese Cordillera dominates the physical geography of Vietnam. With rice-producing core areas at either end of the country, Vietnam is indeed "two baskets of rice separated by a bamboo pole."

River delta inhabitants had created an advanced civilization based on foreign trade and wet rice cultivation.

Then in the second century B.C., the region was absorbed into the Han empire, the Chinese calling it *Nan-Yueh* or *Nan-Viet.* Thus began over 1,000 years of Chinese rule. Chinese writing was introduced along with the Confucian classics, Chinese art and architecture, and Mahayana Buddhism. This differentiated Vietnam from the rest of mainland Southeast Asia with its Indian and Theravada Buddhist ideals.

The peak of Chinese influence was during the Tang dynasty, whose rulers termed Vietnam *Annam,* or "the pacified south." In 939 the Vietnamese asserted their independence and remained free from Chinese dominance except for a 20-year stint under the Ming dynasty from 1407 to 1428.

EXPANSION AND EUROPEAN INFLUENCE

Vietnam was constantly under pressure to find more farm land for its growing population. Between the eleventh and the seventeenth centuries, it gradually extinguished the kingdom of Champa in what is today central Vietnam. Subsequently, it acquired the Mekong delta from the Khmers. During the nineteenth century Vietnam would probably have gotten all of Cambodia had the Thais not challenged their advance and the French not established Cambodia as a "protectorate" in 1863.This was the origin of the anti-Vietnamese sentiments in Cambodia today.

Expansion of the Vietnamese state southward coincided with the appearance of the Portuguese and other European traders in the region. Vietnam's first exposure was in 1535, when a Portuguese ship entered the port of Da Nang on the central coast. Within a few years they had set up a trading post at Faifo (Hoi An). By the seventeenth century several European powers had posts along Vietnam's coast.

The first Catholic mission arrived in 1615, when Jesuits established a mission at Hoi An. Another mission was set up at Thang Long (Hanoi). The Jesuit Alexander of Rhodes organized a society to train missionaries to propagate the faith. However, he was known more for his transliteration of the Vietnamese spoken language into the Roman alphabet. Known as ***quoc ngu*** or "national language," it did not come into general use immediately but was used to translate the Bible into Vietnamese. Consequently, thousands of new converts were gained.

Missionary activities eventually antagonized the authorities, who feared a subversion of Confucianism. Christianity was ultimately barred from both the north and south, and European missionaries were either expelled or executed. In 1697, the French closed down their small factory at Hoi An.

Up until the eighteenth century, the country had been split under two powerful clans: the Trinh in the north and the Nguyen in the south. Peasant rebellions were common, expressing discontent with mandarin corruption. In 1771 the Tay Son Rebellion broke out in the north. It eventually succeeded in taking Saigon. In 1802, a new Nguyen dynasty was founded, with its capital at Hue. Christian missionaries were allowed back for a short time, only to suffer persecution under a new emperor.

FRANCE AND THE *MISSION CIVILISATRICE*

French interests changed the history of the region forever. In France, religious organizations demanded protection for their colleagues overseas. Commercial interests feared the expansion of Britain in Burma and the potential loss of the "China market." The French attacked Vietnam and in 1862 acquired three provinces in the south, the opening of port cities, and the freedom to propagate religion. In 1867, French units seized the remainder of the south and transformed the region into the colony of Cochin China. Further, they assumed Vietnamese rights in Cambodia and turned it into a protectorate. The French finished their conquest of all of Vietnam in the mid-1880s. Shortly thereafter, France completed its "balcony on the Pacific" by making Laos a protectorate as well.

France's civilizing mission was applied to an Indochinese Union comprising five territories or states: north Vietnam, or **Tonkin** (meaning "eastern capital"); central Vietnam, known as **Annam** (from the Chinese "the pacified south"); the colony of **Cochin China** in south Vietnam; Laos; and Cambodia. A hierarchy of French "residents" were installed to run the country.

The primary goals were economic—to exploit natural resources and open up markets for goods produced at home. The French also emphasized the moral aspect of their deeds, and the French citizenry believed that this ***mission civilisatrice*** (civilizing mission) would bring the benefits of modern civilization to the primitive peoples of the world. In fact, the civilizing mission was subsumed by the greater goal of commercial profit. A small colonial elite was educated in Western ways, but the masses of people had only rudimentary exposure to Western culture.

With all its faults, there was economic progress under the French regime. A vigorous commercial and manufacturing system emerged, although dominated by foreigners including Chinese and Indians. Coal mining north of Haiphong was boosted, and Vietnam remains a coal exporting country. Bridges were built over major waterways, and railroads were laid from Saigon to Hanoi and on to Kunming in China. It is said that Vietnam had the best system of paved roads in all of Asia.

The French also expanded agricultural land. In the Mekong delta, they drained marshlands and constructed canals, putting thousands of acres under cultivation for the first time (Figure 13-11). The development of the Mekong delta enabled Vietnam to become one of the world's leading rice exporters by the 1920s. It also established coffee, tea, and rubber plantations along the hill slopes toward Cambodia. However, much of the Mekong delta lands was held by absentee landlords who charged exorbitant rents for the land, and living and working conditions on the French plantations were miserable. Many farmers lacked security as sharecroppers on vast estates. Further, the commercialization of agriculture forced thousands into the cities looking for jobs.

Figure 13-11

Women sifting and cleaning rice in the Mekong Delta.

By the 1930s, there was substantial opposition to the French. However, there was also a new class of native "collaborators" working in tandem with the colonial regime. These French-speaking individuals wore French clothing, ate French food, and lived in French-style houses if they could afford it. They were the middlemen between the traditional masses and the foreign regime and served as evidence of the "success" of *mission civilisatrice.* Nevertheless, Vietnam was a cauldron of social unrest.

THE ROAD TO COMMUNISM

The collapse of Confucian government and the triumph of the West put traditional Vietnamese beliefs into question. What path should be taken into the future? Some opted for Western-style democracy; some looked to China (itself in turmoil) for a restoration of Confucian values. Others became interested in the new Indochina Communist Party (ICP) founded in 1927 by young idealists including **Ho Chi Minh.** But in fact, the most imposing popular movements were *Cao Dai,* with over a million adherents by the 1930s and a Buddhist sect known as *Hoa Hao,* also popular in the south. Christianity had also grown, claiming 10 percent of the population.

The Japanese entered French Indochina in 1940 and set up a nominal government under the emperor Bao Dai. Wartime disruptions in the economy and disastrous weather created terrible famine in Tonkin and Annam. More than 1 million people died. When the war ended, there was a power vacuum in Vietnam, and the stage was set for the August revolution of the **Viet Minh,** a front organization of the ICP.

The Viet Minh took control of most of northern and central Vietnam. They were less successful in the south, where they were opposed by politicians, business people, and religious groups. Ho Chi Minh set up his government in Hanoi. Meanwhile, by late 1945, the French had regained control of the south. War ensued, with the French pitting their conventional tactics against the guerilla tactics of the Viet Minh. With assistance from China and the Soviet Union, Ho Chi Minh's forces were able to subdue the by-now-weary French at **Dien Bien Phu** in 1954. This ended the French efforts to hold Vietnam.

Partition and War

As Dien Bien Phu fell, the great powers were meeting in Geneva to plan Vietnam's future. The decision was made to divide the country at the seventeenth parallel, with the northern Democratic Republic of Vietnam to be ruled by Ho Chi Minh and the southern Republic of Vietnam to be run by Bao Dai. Elections were to be held in 1956 to reunite the country. These never took place. Ultimately, the Americans gave support to Ngo Dinh Diem and Bao Dai left the country.

While Diem spent his aid money on military acquisitions, the north pressed ahead with socialism and agricultural collectivization. Even through the "rich" suffered, popular support for Ho Chi Minh's government remained high. By 1959, opposition to Diem, the "American puppet," took the form of the National Liberation Front (NLF), or the **Viet Cong.** By the early 1960s, the Viet Cong were in control of vast areas of the southern countryside, which compelled the Americans to increase their aid to Diem. By 1963, there were 17,500 American military advisors in Vietnam.

By now Diem and his followers had antagonized all groups, and Buddhist monks were burning themselves alive to protest the regime. Ultimately Diem was assassinated. Several years of unstable government followed, but meanwhile the Americans decided to confront the NLF and had begun to bomb targets in both the north and the south, and American troops landed in the south. Thus began the Vietnam or American war (Chapter 12), which did not end until 1975, with a combined DRV/NLF victory.

Environmental damage was awesome. From 1965 to 1971, in the region of Indochina, an area slightly larger than Texas, at least twice as many bombs were dropped than in all the theaters of World War II. In a typical B-52 mission, seven planes delivered 756,500 pounds of bombs in a pattern that saturated nearly 1,000 acres. B-52s alone were creating about 100,000 new craters a month. The area had as many as 26 million bomb craters. Craters filled in weedy grasses and became breeding grounds for mosquitoes. About 10 percent of the agricultural land of southern Vietnam had to be abandoned. In addition, thousands of tons of chemical defoliants were dropped. Vast tracts of forests, rubber plantations, and mangroves were obliterated or seriously damaged. In these areas, there was a significantly higher rate of pregnancies that resulted in still-births and congenitally abnormal babies.

Thousands fled the country to refugee camps in adjacent countries, especially Thailand. The first migrants used a map torn out of a school atlas to navigate the Gulf of Thailand. More than a million refugees have been resettled since 1975. But more than 50,000 remain in camps in various Southeast Asian countries. Screening continues to determine their actual status. Ninety percent have been "screened out" as economic rather than political refugees. For them, there is no hope of resettlement. These people are repatriated to Vietnam voluntarily or involuntarily.

Aftermath

Since the 1950s, the northerners had lived under an austere form of socialism that reemphasized the traditional regard for social hierarchy and community obligation. The southerners had been introduced to a quasi-capitalistic consumer economy and the trappings of American popular culture. In 1975, the north moved swiftly to subordinate the south to direct rule from Hanoi. In 1976, the country was renamed the Socialist Republic of Vietnam and Saigon was renamed Ho Chi Minh City. Socialization of the southern economy proceeded quickly over the next two years.

However, things did not go as well as planned. There was an acute economic crisis, worsened by floods and other natural disasters in 1977 and 1978. Ambitious industrial targets were not achieved, rice production diminished, and food had to be rationed. In 1978, Vietnam attacked Cambodia to oust the genocidal and anti-Vietnamese regime of Pol Pot. In the 1980s, thousands of Chinese and Sino-Vietnamese fled Vietman as *boat people.* These refugees went to Australia, Canada, the United States, France, and many other countries.

A hard-line communist approach was held in Vietnam until 1985 when reforms in the Soviet Union gave impetus to reforms in Vietnam. In 1986, the Vietnam Communist Party (VCP) instituted the policy of ***doi moi,*** meaning "renovation." Reforms were given further impetus with the collapse of the Soviet Union and the European Eastern Bloc's abandonment of socialism. (This resulted in a 30 percent cut in Vietnam's budget.) Support for reform was also bolstered by China's new policy of market socialism. Today, Vietnam is a hybrid of a one-party government, faltering state enterprises, and burgeoning free enterprise.

Vietnam is anchored by two cities: the capital of Hanoi in the north and Ho Chi Minh City in the south. Hanoi, a quiet city of bicycles in 1995, is now a bustling city of motorcycles and automobiles. Hanoi dominates the Tonkin Basin and its productive agricultural hinterland. Agricultural methods remain traditional—dependent on human and animal labor (Figure 13-12). The port of Haiphong, a mere 60 miles (97 km) away, is a three- to four-hour journey by roads under construction and through a developing *kotadesasi* landscape.

Doi Moi

Vietnam's new economic policy of *doi moi* was first made public in 1979. Measures were introduced to encourage private-sector production, make prices more flexible, and give more discretion to managers on the use of their profits. This was followed by the introduction of a peasant agricultural contract system in 1981. This allowed them to keep anything over their set quotas. State enterprises were permitted to sell production exceeding their quotas.

Some hard liners were distressed by the appearance of small service enterprises such as restaurants and coffee shops. Others expressed concern over the reappearance of a Chinese middle class in Ho Chi Minh City. However, the VCP decided to push ahead with the reforms.

Reforms were applied to prices, wages, and currency. Rations for workers were dropped, and a new wage system was introduced. The currency (Dong) was devalued and later allowed to float. In 1988, a liberal foreign investment code was announced, and price subsidies on staples such as rice and fuel were removed. In essence, Vietnam has made a commitment to an open, deregulated economy.

Figure 13-12

I came across this scene in the Red River delta—two women operating a bucket to lift water from a lower to higher level in the rice paddies.

Figure 13-13

This scene in Saigon shows new hotel construction, as well as evidence of foreign investment.

Ho Chi Minh City (Saigon) and Cholon (Chinatown), with its 8 million people, is an urban agglomeration on the Saigon River. It contains about 10 percent of the country's population. The name Saigon is commonly used. Unlike Hanoi, Saigon can be accessed by ocean-going vessels and a Special Economic Zone has been developed downstream from the port. Saigon is a bustling city of every form of transportation imaginable with traditional districts selling a wide array of consumer goods. High rises punctuate the skyline everywhere (Figure 13-13). Modernizing Saigon is distant in form and fact from the more traditional capital of Hanoi.

With the lifting of a United States embargo in 1994, foreign investors from all over the world began investing in Vietnam. Modern high rises and construction cranes mark the landscapes of Saigon and even Hanoi. Now about 90 percent of the labor force works in the private sector. The primary sector includes rice, corn, potatoes, coffee, tea, rubber, and fish and animal products. Mining is important as is the making of cement and fertilizers, along with food processing, and the manufacture of textiles and clothing. Exports to foreign countries grew by 73 percent in the 1990s. However, reliance mainly on two commodities—rice and crude oil—makes this situation tenuous. Tourism is the main emphasis in the service sector and numerous hotels have been constructed. In fact, the scale of development is such that Vietnam refers to itself as "Asia's baby tiger."

Laos: Land of a Million Elephants

The Lao People's Democratic Republic, now often referred to as simply *Lao* or *Laos* is a mountainous, landlocked, sparsely populated country sharing borders with Thailand, Myanmar, China, Vietnam, and Cambodia. Laos, as we know it today, is a creation of the French, but much of its history has involved conflict with its surrounding countries.

LAN XANG

There are no early documentary records of early Lao history but we know that the region had links to Chiang Mai in the eleventh century and that it was a satellite of the Khmer empire in the twelfth century. The first kingdom of Laos emerged in 1353. This was known as **Lan Xang:** "a million elephants."

Lan Xang stretched from China to Cambodia and from the Khorat Plateau in present Thailand to the

Annamite Mountains in the east. Louang Phrabang, on the Mekong, became the capital of Lan Xang, and the king made Theravada Buddhism the official religion. A census in 1376 indicates that there were 300,000 Thais living in the kingdom. *Muang* administrative districts were organized, and these lasted until abolished by the Communists in 1975.

In the sixteenth century, Vientiane (Viangchan), on the Mekong, became prominent as a religious and trading center. Lan Xang weathered wars with the Vietnamese, the Thais, and eventually the Burmese. Nevertheless, Vientiane reached its acme in the seventeenth century. The first European visitors reported on the city's religious structures and imposing buildings. For lack of a royal heir, the kingdom broke up in 1694 and Vientiane and Louang Phrabang became centers of rival states. The south fell under Thai patronage.

FRENCH TAKEOVER

The nineteenth century was a decade of devastation for a Laos caught in regional rivalries. The surviving kingdom at Louang Prabang acknowledged the overlordship of both the Vietnamese and the Thais, but was really within the Thai orbit of control.

The French takeover between the 1860s and 1885 derived from several factors. They perceived the Mekong as a route to China, although this was an incorrect assumption, as the Mekong is broken by impassable rapids. They feared Thai interests in their territories. They were concerned about the intentions of armed bands of renegade Chinese who were attacking northern Laos and Vietnam. And, they were worried about British encroachment from Burma. By 1885, the French controlled the Vietnamese emperor's claims to overlordship of the Lao territories. Still, Laos remained the center of rivalries among the surrounding states in addition to the competing colonial powers.

The French's attitude to Laos was one of "benign neglect." Few French actually went there, and Vietnamese were placed in administrative positions. Trade was left to the Vietnamese and growing numbers of Chinese. Most people in the region continued as subsistence farmers, the lowlanders growing rice and the uplanders pursuing slash and burn cultivation. The colony's most important products were tin, mined by Vietnamese, and opium grown by the Hmong and other mountain dwellers. The French had a monopoly on the opium trade throughout Indochina.

Prior to the Second World War, modernization was extremely limited. A mere 8,050 miles (5,000 km) of mostly unpaved roads eased communication, but most services, including health care, were limited to the towns, and no Western-style education was available beyond the primary level. However, a small Vietnamese or French-educated elite did emerge, and by the 1940s it would become the core of a Lao nationalist movement.

THE WAR YEARS

Nationalist politicization was a feature of the war years in Laos. After 1945, there were attempts at independence, but the French recaptured Laos in 1946. In the late 1940s, Lao guerilla groups developed along the mountainous Lao–Vietnam border, aided by Viet Minh knowhow and supplies. Meanwhile, the French declared the king in Louang Phrabang king of all Laos, held elections for a national assembly, and declared the country independent in 1949. Nevertheless, the guerilla movement expanded, now known as the ***Pathet Lao.*** By 1954, it controlled large areas of northern Laos.

In 1954, the same year as the French were defeated at Dien Bien Phu, a conference was held in Geneva to settle Indochina disputes. The *Pathet Lao* were not invited. By 1959, guerilla warfare was in full swing, with the United States involved in supporting the Royal Lao Forces. At the same time, the CIA was clandestinely forming links with the opium-growing Hmong, whom they assisted in selling drugs to new markets in Southeast Asia.

Unfortunately Laos became deeply involved in the already well-established Vietnam War. About one-third of the bombs dropped in Indochina fell on Laos, especially on the Plain of Jars, part of the route of the Ho Chi Minh trail. Tens of thousands of unexploded bombs still mar the landscape of northern Laos. When I was in Laos in 1998, seven children were killed by a bomb that had not exploded earlier. It will take many years for these bombs to be removed. Meanwhile, the people collect bomb casings, aircraft fuel tanks, and other materials and incorporate them as practical items in their landscape (Figure 13-14).

By 1972, the *Pathet Lao,* backed by the well-armed DRV, was gaining ground. In 1973, the two sides reached an "Agreement on the Restoration of Peace and Reconciliation in Laos." When, in 1975, communist organizations toppled the power structures in Saigon and Phnom Penh, the *Pathet Lao* supported a "popular revolution." The king abdicated. A total of 300,000 Lao—10 percent of the population—and mostly middle class left the country between 1973 and 1975. Many ended up in the United States and France. Then in December 1975, a National Congress of Peoples Representatives voted the establishment of the Lao People's Democratic Republic.

The Plain of Jars

The Plain of Jars is a plateau 3,280 feet (1,000 m) high covering an area of 390 square miles (1,000 sq km). It is also known as *Plaine de Jarres.* Upon this bomb-crater pocked plateau lie more than 300 stone jars ranging from 3 feet (1 m) to 6 feet (2.5 m) high and around 3 feet (1 m) in diameter. Each 2,000-year-old jar equals the weight of three small cars (Figure 13-15).

Generations of archaeologists, geographers, and others have pondered the origin and use of the jars. Were they used to brew *Lau-lao,* a wicked local drink? Perhaps they were used to store grain or water. The most likely theory holds that they were burial urns with the larger jars being for the elite.

Tools, bronze, and ceramic objects have been found in the jars, indicating that they were used by a sophisticated society. But no one can relate them to any previous civilization.

Figure 13-14

In Laos, a bomb casing makes a perfect herb garden.

Figure 13-15

The Plain of Jars in Laos is an amazing place. I'm sitting on one of the jars to give you an idea of the scale of these objects. Their origin remains a mystery.

The stone that they are hewn from is not found locally.

During the war, the *Pathet Lao* set up their headquarters in a cave near the jars and the region was heavily bombed. Miraculously, the majority of jars remain unscathed.

RURAL SOCIETY

Over 85 percent of Lao is rural, and most people live in villages ranging from 10 to 200 households, or up to approximately 1,200 persons. Towns are mainly administrative and market centers. Most district centers have little more than a middle school and a few officials. Some places have minimally supplied clinics. Since 1975, the government has spent a lot of effort at national unification and instilling a sense of belonging to a Lao state.

One priority has been to improve infrastructure. However, both mountainous topography and torrential rains make access, work, and maintenance problematic. Lao's low population density means that the cost per capita for road construction is often prohibitive. War damage such as bomb craters and the presence of mines and cluster bombs inhibit development of any kind.

Lowland Society

Lao Loum (Laotian of the Valley) has been the dominant group since the founding of Lan Xang in the fourteenth century. The Lao make up half of the population. They live in stilt houses constructed of wood, bamboo, and thatch. Progressive village houses sport a corrugated metal roof. Steamed glutinous rice, or "sticky rice," is the staple food. Eaten with the fingers, it is dipped into a soup or stew dish.

Lao Loum live in stable villages along rivers or streams and indulge primarily in paddy rice cultivation. Where terrain is not level, they practice swidden cultivation. Other crops are cotton, tobacco, or sugar ,cane, but these are grown for personal use. Villagers keep chickens, pigs, and draft animals. Hunting, gathering, and fishing supplement subsistence.

Household tasks are usually divided by gender, but divisions are not rigid. For instance, both men and women cut and carry firewood. Women and children carry water, tend gardens, look after small animals, and are the main marketers for produce. Men look after and market draft animals. In field work, men plow and harrow the rice paddies, women harvest seedlings, and both sexes transplant, weed, and harvest the rice.

Lowland Lao are mostly all Buddhist. Most villages have a *wat,* which serves as a religious and social center. The *wat* also serves as the village meeting house. *Wats* are populated by two to six monks and a few novices, depending on the size of the village.

Midland Society

Lao Theung (Laotian of the mountain slopes) comprise about a quarter of the population and consist of 37 different ethnic groups. They speak languages of the Austroasiatic family. Many are mutually incomprehensible. Characterized as swidden farmers, they are semimigratory, moving their villages on occasion as swidden plots are exhausted. Many of these villagers were disrupted in the war, and many have moved closer to the lowlands. But traditionally, lands were managed in a way to maintain sedentary communities.

Lao Theung villages are generally smaller than *Lao Loum* villages. Some are divided into two segments with a men's common house in the middle. Villages of stilt houses are situated near a stream to access water that may be diverted via bamboo aqueducts.

Swidden rice economy is the basis for most communities. Both regular and glutinous rice are cultivated. Corn, cassava, and wild tubers supplement this diet. Hunting and fishing are also important.

Gender role differentiation is much greater among the hill people. Men clear and burn the swidden fields. Men punch holes for seeds and women follow, dropping seeds in the holes and covering them with topsoil. Both sexes weed, but this is regarded as women's work. Harvest is a joint activity. In the house women cook, gather wood and water, and look after children. Men weave baskets, repair farm tools, hunt for small game, and are much more likely than women to manage finances and indulge in trade. Where villages have schools, both boys and girls attend but girls have significantly higher dropout rates.

Upland Society

Lao Sung (Laotian of the mountain top) include six ethnic groups. The Hmong are among the most numerous. They also live in Vietnam and Thailand. The Lao Sung are among the most recent migrants to Laos, having initiated a series of moves from China in the nineteenth century. All *Lao Sung* villages are located in the north. Only the Hmong live as far south as Viangchan.

Lao Sung typically live above 3,280 feet (1,000 meters) on steep slopes and mountain tops. As swidden cultivators, most groups are semimigratory. Yet some villages have remained sedentary for over 100 years. Hmong houses are built directly on the ground, with walls of vertical wood planks with gabled roofs of thatch or split bamboo.

Hmong swidden farming is based on white (nonglutinous) rice, corn, and a variety of tubers, vegetables, and squash. They also raise pigs and chickens in large numbers. Opium poppies, a cold-season crop, are planted after the corn harvest.

Opium, the sap drawn from the poppies, is a low-bulk, high-value item that is easy to transport. The government has outlawed opium production and instigated a program of alternative crops. Nevertheless, the practice continues in the remoter regions.

Hmong gender roles are strongly differentiated. Task divisions are similar to those of the *Lao Theung.*

Most of the *Lao Sung* are animists, but some have converted to Christianity after contact with missionaries. Most believe that illness is caused by spirits, and shamans play a powerful role in the communities. Shamans are "chosen" by the spirits and may be either male or female.

Lao Sung swidden cultivation is endangered. In the late 1980s, the Thai government imposed a ban on logging, so Thailand began looking to surrounding countries for timber. Logging in Laos is causing deforestation and soil erosion. However, the practice is to blame this on the swidden farmers. The government wants to see the disappearance of swidden early in the twenty-first century. *Lao Sung* are to be resettled in the lowlands.

URBAN SOCIETY

With a population of 400,000 in 1995, Viangchan is the only city of any significant size (Figure 13-16). Other cities such as Louang Phrabang and Pakse are only around 20,000. Lao is 19 percent urbanized, including district centers of 2,000 of 3,000 people. Everywhere, the fundamental village nature is evident with even most residents of Viangchan being rice farm-

Figure 13-16

This is known as Victory Gate in Viangchan. It was constructed using American aid money designated for airport runway construction.

ers. There is very little rural-to-urban migration, as there is not much to migrate to.

NEW THINKING

In recent years, the government has shied away from the centrally planned economy and embarked upon a policy of *Chin Thanakan Mai,* or **New Thinking.** The hammer and sickle were removed from the state crest in 1991, and a series of economic reforms were instigated. With the demise of handouts from Moscow, Viangchan has turned to the West for investment. At first the pace of change was *koi koi bai*—"slowly, slowly." But in 1991, everything became privatized. Limited tourism is being promoted. Nongovernmental organizations are rampant. Thai television, with its materialist message, is beaming across the Mekong. Likewise, more consumer goods, such as children's toys and motor scooters, are appearing in the markets of at least the largest cities. Foreign investment, mainly Japanese, is evident in the capital. Many Thai businessmen have appeared on the scene as well. New businesses are opening every day.

Infrastructure is still very poor. Only a few of the larger towns have municipal water systems. Electrification is in progress. Most roads are not paved, and many are impassable in the wet season. However, there is the Friendship Bridge built by the Australians in 1994. This links Lao and Thailand by road across the Mekong. With upgrading of roads in Lao, soon it will be possible to drive all the way from Singapore to Beijing!

Many Lao see their country's future as a transport hub linking northeast Thailand with the port of Danang in Vietnam, as well as southern China with Thailand, and, from north to south, Malaysia and Singapore. Many fear the effect of rapid development and influences of the surrounding states.

Cambodia: Beyond the Killing Fields

Cambodia is famous for two things. It is highly noted for the wonders of Angkor and it is highly notorious for the genocide that took place from 1975 to 1978. What happened in this small and compact country of Southeast Asia?

This land of the Khmer people is the third smallest nation in the region after Brunei and Singapore. About 5 percent of Cambodia is rivers and lakes. The plains of Cambodia are ringed by the last foothills of the Himalayas, the Cardamom range in the west, the Elephant Mountains to the southwest, and the lengthy

Dangrek chain to the north. The eastern frontier with Vietnam runs up into the high and inaccessible Moi Mountains, which are crossed by only one highway. Rising out of the plains are ranges of hills called *phnom*. These were worshipped, and they sheltered numerous settlements. The Mekong crosses the Cambodian Plain and south of Phnom Penh, splits into the Bassac to the southwest and the lower Mekong to the east. Inland, 340 miles (550 km) from the sea, the Mekong is still navigable year round in Cambodia.

Many thousands of years ago, Cambodia lay at the bottom of the Pacific Ocean, but the Mekong filled in the land to form a fertile plain. This plain, watered with rivers and tributaries, formed natural irrigation systems. Where the land was not uniform, large lakes such as Tonle Sap were created—relics of the former ocean gulf. Here developed the civilization of Cambodia.

Cambodia's history is rife with rivalries of outside forces and riddled with internal power struggles. Although the country has experienced periods of peace, the peace has been enforced. This is the situation today.

THE EARLY PERIOD

By the early centuries of the Christian era, the Khmer and the related Mon peoples occupied a broad swath of land stretching from Burma to Vietnam. This was Funan, which flourished from the third to the seventh centuries (Chapter 12). Early records also mention the states of "water Chenla" in the Mekong delta and "land Chenla" further inland. Chenla brought down Funan.

The consolidation of Khmer society is clearer by the nineth century when Jayavarman II ruled over Angkor and established the concept of *Devaraj:* the god-king. Jayavarman II built temples at widely spaced sites across Cambodia. Eventually, Angkor, near Siem Reap, became the core region with an irrigated area of 13,585 acres (5.5 million ha) and a large population. Then, king Suryavarman II initiated the construction of Angkor Wat, the largest religious building in the world. The complex was enlarged and enhanced by subsequent rulers (Figure 12-7).

The focus of Khmer society slowly altered. Varieties of Buddhism had long existed with Hindu cults, but during the thirteenth century, Theravada Buddhism won general allegiance. Now rulers demonstrated their power through the building of temples, schools, and monasteries.

In the 1440s, possibly because of silting of the irrigation system or even malaria, and surely because of the rise of the Thai state of Ayudhya, Angkor was abandoned in favor of capital sites in the region of Phnom Penh. These sites were closer to the sea and the booming sea trade of the fifteenth century.

Until the late sixteenth century, the Khmer kingdom appears to have been as strong as its rivals Ayudhya and Lan Xang. But thereafter, the Khmers became embroiled in regional machinations of the Thais and the Vietnamese. Vietnamese and Chinese began to dominate Cambodian ports. Lively trade ensued even as Cambodia was caught in the pincers of Thai and Vietnamese expansionism. Repeated incursions left the Thais dominant in the eighteenth century.

In the nineteenth century, the Vietnamese rose to the fore and Vietnamese people were encouraged to colonize the region. Vietnamese language, law, and modes of dress were imposed. A countrywide rebellion broke out in 1840; for five years Thais, Vietnamese, and Cambodian factions fought an inconclusive war.

COMING OF THE FRENCH

The French, having established their colony of Cochin China by 1862, were interested in Cambodia. In 1863 France signed a "treaty of protection" with the king that placed a French resident at Phnom Penh. This gave the French control of foreign policy and opened the door to commercial interests. However, the French soon realized that there was little economic return to be had from Cambodia and so concentrated on Cochin China.

By 1904, the French were able to achieve complete authority over the protectorate and change ensued. In 1921, the population was assessed at 2.5 million. A Chinese rice export business had developed with rice purchased from the Khmer farmers. However, Cambodian rice was regarded as inferior to the rice of Cochin China so did not fare as well in the market. There were also small Chinese timber and pepper industries, and the French financed rubber estates using Vietnamese labor. Maize, kapok, and fish were gleaned from the region of Tonle Sap. The Mekong remained Cambodia's main trade artery but the port of Saigon handled the shipping.

Around 95 percent of Cambodians survived as subsistence farmers. They were characterized by the French, Vietnamese, and others as "lazy," "ignorant," "fatalistic," and "child-like." Western observers condemned them as a "decadent race." As the interlopers took in their profits, no industries of consequence were developed. Towns remained small, with Phnom Penh housing only 100,000 by 1930. Towns were dominated by aliens, and the first Khmer language newspaper only appeared in 1938.

WORLD WAR II AND INDEPENDENCE

During World War II, the French reached an agreement with Japan to retain control of Indochina in return for allowing free movement of Japanese troops. In order to deflect the Cambodian fascination with Japanese rhetoric, the French stimulated a nationalist movement. They tried to institute a new form of writing as in Viet Nam, but the Buddhist community rebelled against what they saw as an attack on Cambodian culture and heritage. When the war ended, Cambodia had no clear nationalist group as existed elsewhere in Southeast Asia, and so the country drifted and the French returned.

While keeping control of key elements of government, the French permitted the formation of political parties but by the 1950s, lack of progress and cooperation was producing acute strains. In 1951, the Khmer People's Revolutionary Party (KPRP) was founded, and organized guerilla operations in outlying areas. Meanwhile, Prince Sihanouk toured France and the United States demanding independence, which the French granted in 1953.

SIHANOUK, WAR, AND REVOLUTION

Sihanouk, who came to power under rigged elections, promoted "Buddhist socialism." He was intolerant of opposition, and hundreds of "dissidents" disappeared. He greatly expanded education but expressed little interest in economic affairs. The Cambodian economy went into decline. Meanwhile the Vietnam conflict was omnipresent and Sihanouk permitted the northern Viet Minh and the southern Viet Cong to use Cambodian territory. He rejected the United States and opened relations with China. This eliminated United States aid and angered conservatives. Communist-led revolts over government seizures of rice were brutally crushed. While Sihanouk was overseas, the National Assembly withdrew their support and installed Lon Nol as the head of the new government of the "Khmer Republic."

The anti-Sihanouk coup polarized the population. The United States supported Lon Nol and his totally inept government. U.S. aid fostered gross corruption in both government and military. At the same time, the United States was carpet bombing Cambodia with more than half a million tons of bombs in connection with the Vietnam War. About 500,000 soldiers and civilians were killed. Communist insurgency was slowed by this devastation but nevertheless, the Lon Nol regime ultimately collapsed and the insurgents took Phnom Penh in April 1975.

The Khmer Rouge (Red Khmer) set up the state of Democratic **Kampuchea.** This name derived from *Kambuja,* meaning "the sons of Kambu"—the ascetic who married a celestial nymph and founded the kingdom of Chenla, forerunner to the great Khmer empire. Their communist leadership was made explicit in 1977 when they announced the existence of the Communist Party of Kampuchea (CPK), which had been founded in 1968. A Paris-educated school teacher rose to take the helm. His name was Saloth Sar, later known as **Pol Pot,** and he declared 1975 as the year zero.

THE KILLING FIELDS

In April 1975 the Khmer Rouge marched triumphantly into Phnom Penh, a city that was swollen by refugees from 600,000 to more than 2 million. Pol Pot immediately pushed his radical Maoist-style agrarian society. Within four days, most of the inhabitants of Phnom Penh had been forcibly relocated to the countryside. Later, hundreds of thousands of people were forced from the southeast of the country to the northwest.

The goal was pure socialism. Everything modern was to be eliminated. A typical slogan was, "We will burn the old grass and new will grow." Money, newspapers, education, and technology were outlawed. Intellectuals were killed. This was to show the strength of the people who, "though bare handed . . . can do everything."

Food was scarce under Pol Pot's inefficient system of farming, and administration was built on fear, torture, and execution (Figure 13-17). Little was known of these atrocities until a few refugees trickled over the Thai border. During the Khmer Rouge reign of terror, more than a million people died; some estimates are as high as 2 to 3 million. The Khmer Rouge turned Cambodia into a terrifying work camp where families were abolished and murder was used as a tool of discipline. Pregnant women were disembowled, babies were torn apart limb from limb, people were buried alive or bashed to death with axe handles. Of 64,000 Buddhist monks, only 2,000 survived. Half of the Cambodian Vietnamese were eliminated, along with thousands from other groups. The CPK even turned upon itself, purging those who had received training in Vietnam.

In 1978, Vietnamese troops entered Cambodia. The Khmer Rouge collapsed before them and they entered the virtual ghost town of Phnom Penh in 1979. However, they failed to capture Pol Pot and his close comrades. The Vietnamese quickly established the People's Republic of Kampuchea (PRK) with a Cambodian government. This included Hun Sen, who would become premier in 1985 and win elections in 1998.

Now the country suffered a devastating famine, and it was not until the mid-1980s that subsistence farming reached an equilibrium and shops and markets began to function normally. Meanwhile the PRK became an international pariah supported by only a few countries, including the Soviet Union. In 1989, the Vietnamese withdrew from Cambodia. Kampuchea was renamed the State of Cambodia and committed itself to a private enterprise economy and the restoration of Buddhism.

The United Nations became involved in Cambodia's internal affairs, and elections were held in 1993. Multiple parties contested and a coalition government was formed. King Sihanouk was reinstalled, although he later died of cancer. The coalition government was heavily financed by foreign aid, but internal dissention and corruption meant that little progress was made toward rebuilding the shattered economy and infrastructure. Phnom Penh University opened in 1988. Only 5,000 teachers survived of the original 20,000 in 1975. A sex ratio imbalance was evident in employment. For example, 70 percent of workers in the textile and tobacco industries are women, mostly widows of pre-1975 workers. A former school in Phnom Penh, where 20,000 died, has been turned into a museum to Pol Pot's genocide.

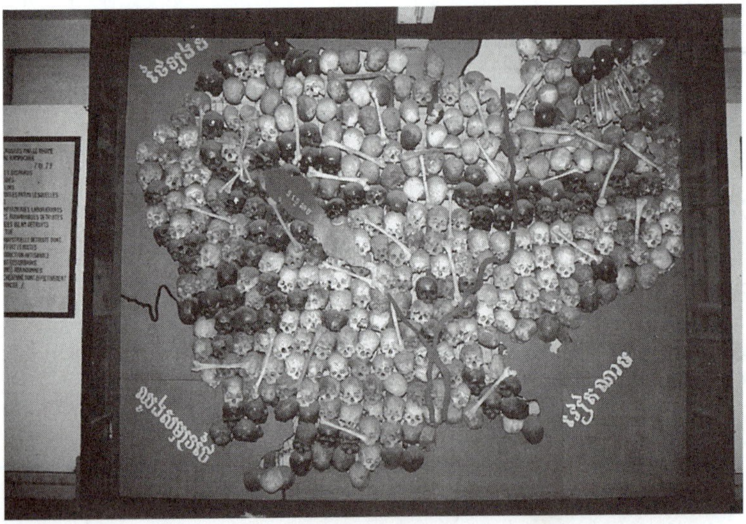

Figure 13-17

This skull map of Cambodia symbolizes the millions of people who died at the hands of the Khmer Rouge.

In disputed elections in 1998, Hun Sen became the head of government. Pol Pot committed suicide in the same year and other Khmer Rouge leaders have either defected or live clandestinely in various parts of the country.

Did you know that Cambodia is the world's biggest minefield?

Death Awaits

Because of the influx of military hardware to various factions during the 1980s, Cambodia has become the most heavily mined place in the world today. There are more than 3 million mines, mostly concentrated in the northwest province of Battembang, a Khmer Rouge stronghold. In addition, there are countless thousands planted along the main roads. When resettlement of refugees began in 1992, there were 7,000 amputees in the camps. Now there are about 300 new amputees every month. One in every 236 Cambodians is an amputee.

Mine clearing is a slow process. A British-based group has been involved since 1990, although fewer than 6,000 mines have been removed. Every week animals or people are maimed or killed by land mines.

CAMBODIA TODAY

Cambodia continues as a very poor country. Forty percent of children under age five are malnourished. Only 13 percent of primary school students complete grade five in five years. Only 19 percent of students in upper secondary school are female. More than 90 percent of births in rural areas are at home attended by traditional midwives. Of the rural population, only 26 percent have access to safe water. A mere 6 percent of rural people have access to improved toilet facilities. Clearly, Cambodia remains one of the least developed countries in the world.

Rice production accounts for nearly 88 percent of the available cultivated area. Farms tend to be very small—around 2.5 acres (1 ha). A farm of this size supports a family of five. Rice is supplemented by fruit and vegetable cultivation. Although rice yields are low, Cambodia became self-sufficient in rice in 1996 with a surplus to export.

Cambodia's rubber plantations are second only to timber as a source of export earnings. In 1995, U.S.$3 million worth of natural rubber was exported to the United States. Expansion of rubber production is planned for the eastern provinces. Cambodia should be able to produce 600,000 tons of rubber annually.

Fishing is important as the main source of protein. Fisheries are both inland and marine, and fish are also raised in ponds, pens, and cages where water is available around the year. Some frozen fish are exported.

Teak and rosewood are among Cambodia's most valuable assets. However, more than 70 percent of Cambodia is deforested and the government has taken measures to stop timber exports. Illegal cutting and exporting continues, nevertheless.

Cambodia is also rich in minerals, including precious stones such as sapphires and rubies. It has oil in the Gulf of Thailand, which is now being exploited by both domestic and foreign concerns.

Foreign aid and investment is critical to Cambodia's development. Light industry is more developed than heavy industry. The more developed industries include household goods, textiles, soft drinks, alcohol, tires, pharmaceuticals, and cigarettes. The extent of foreign involvement can be demonstrated by the textile industry. Of 78 companies, Hong Kong runs 25, Taiwan 15, Cambodia 11, and China 7. Malaysia and Korea each run 6, and the rest are divided up among Singapore, France, Britain, and the United States.

Cambodians are counting on tourism, focusing on Angkor Wat to bring in badly needed foreign exchange. However, the countryside is not entirely safe, and many areas cannot be visited at all. The memory of visiting one of the Angkor temples under the care of armed guards remains vivid. Only political stability and improved infrastructure will bring increasing numbers of tourists to Cambodia.

Main Point Summary

1. Mainland Southeast Asia (MSA) is a complex region with an intertwined history.
2. Burma, or Myanmar, is the largest and most ethnically diverse state in MSA.
3. Although the boundaries of Burma were created by the British, its territory originally centered on Bagan and Mandalay.
4. British colonial rule was founded on an Anglo-Burman and Indian elite. Efforts to develop the delta regions caused a large southward migration.
5. A plural society rested on the fact that the British were only interested in primary products such as teak and not industry.
6. World War II was accompanied by the rise of nationalism, but regional insurrections and rampant corruption led to the downfall of the government.
7. The country was divided into regions not coincident with minority territories.
8. In the 1980s, SLORC took power in Myanmar and imposed a military dictatorship even though elections had been held. The winner of the elections, Aung San Suu Kyi, was put under house arrest.
9. Rice trade is Myanmar's major form of revenue, although quotas and inadequate technology have driven many farmers off the land.
10. The government of Burma has forced thousands of people to move elsewhere to depopulate cities, and to get out of the way of new development projects. Forced labor is widespread.
11. Most heroin (a derivative of opium) in the United States originates in the Golden Triangle shared by Burma, Thailand, and Laos. Burma is the world's largest producer of opium.
12. Thailand has four main zones dominated by the fertile valley of the Chao Praya.
13. The principle ancestors of the Thai were the Tai people who were wet-rice farmers. The Tai founded the kingdom of Sukothai—the birthplace of the Thai nation.
14. Bangkok became prominent in the eighteenth century while Thailand lost territory to colonial powers. Thailand served as a buffer state between the territorial ambitions of Britain and France.
15. Theravada Buddhism and the *sangha* are critical elements of Thai society. A sense of order is central to Thai culture. Shrines for "lords of place" along with *wats* are basic elements of the Thai landscape.
16. More than half of Thai exports are agricultural, and about half the population are farmers. Deforestation has contributed to a shortage of water for irrigation.
17. New Theory farming is community-based farming depending on the cooperation of multiple sectors of agricultural production.
18. Tourism is a leading earner of foreign exchange. Eco-tourism and sex tourism both compromise the cultures of native peoples.
19. Industrialization is focused around the city of Bangkok, where economics of agglomeration are supported by an informal sector. Ribbon development has extruded onto agricultural land. Bangkok has an array of severe environmental problems.
20. Vietnam's population is concentrated in the Tonkin and Mekong delta regions.
21. Although Vietnam's population is dominated by the Vietnamese, the presence of other groups such as Cao Dai and Catholics makes it difficult to engender a fully integrated society.
22. In the second century B.C., the region was absorbed into the Chinese Han empire. This began a period of a thousand years of Chinese domination. Chinese culture had a great influence on Vietnam and differentiated it from other parts of Southeast Asia.
23. The Jesuits created *quoc ngu,* a writing system using the Roman alphabet.
24. The French acquired a colony in the south, which they named Cochin China. Eventually, their *mission civilisatrice* was applied to the Vietnamese regions known as Tonkin and Annam, in addition to Cambodia and Laos.
25. The collapse of the Confucian government and occupation by the Japanese fostered nationalist

and popular movements within Vietnam. The Viet Minh, under Ho Chi Minh, took control of most of northern and central Vietnam.

26. The French were defeated at Dien Bien Phu in 1954, and U.S.-led efforts to prop up a democratic government in the south failed. The Viet Cong emerged to control vast areas of the southern countryside.
27. The Vietnam War took an enormous human and environmental toll. Even after the war, north–south differences prevailed.
28. A hard-line Communist approach eased in the 1980s when the government adopted Doi Moi, economic reforms leading to market socialism.
29. Laos or Lao is a landlocked, sparsely populated country, created by the French as a part of Indochina. It first existed as Lan Xang—a million elephants.
30. The French attitude to Laos was one of benign neglect. Administration was controlled by the Vietnamese. There was very little modernization.
31. During World War II, Communist-inspired guerilla movements emerged, known as the Pathet Lao. They controlled a large area of northern Laos.
32. During the Vietnam War, the American CIA established links with the opium-growing Hmong people to sell drugs in new markets in Southeast Asia to help finance the war.
33. Lao was heavily bombed with enormous damage to the landscape. Many bombs still exist unexploded, and people are killed by them on a weekly basis.
34. Rural Laos can be understood according to lowland, midland, and upland ways of life—all of which have differentiated gender roles.
35. Viangchan is the only city of any significant size. There is very little rural to urban migration.
36. The government is now supporting a policy of "New Thinking," a kind of market socialism. Foreign investment, mainly Japanese, is taking advantage of the more liberal economic policies.
37. Infrastructure is poor, but Lao has possibilities as a transport hub, given its crossroads position in MSA.
38. Cambodia is noted for the temple complex at Angkor, as well as for the genocide that took place under Pol Pot in the 1970s.
39. Cambodia is ringed by mountains and dotted with hills known as phnom. The great lake of Tonle Sap is a remnant of when Cambodia was under the sea millions of years ago.
40. Consolidation of Khmer society occurred in the ninth century with the institution of a god-king and concentration of development of religious sites at Angkor. Hindu influence gave way to Theravada Buddhism.
41. Until the sixteenth century, the Cambodian kingdom rivaled both Ayudhya and Lan Xang. Then it became embroiled in the expansionist machinations of both the Thai and Vietnamese.
42. The French fostered industries such as rice production, timber, pepper, and rubber, and the products were shipped via the Mekong and the port of Saigon.
43. In order to deflect Cambodian interest in the Japanese, the French encouraged nationalist movements. In 1951, the Khmer People's Revolutionary Party was founded. Independence was granted in 1953.
44. The Khmer Rouge set up the state of Democratic Kampuchea in 1975. Pol Pot headed the government.
45. Pol Pot's efforts to create a complete peasant agrarian society resulted in millions of intellectuals and other people being executed. Most Buddhist monks were eliminated.
46. The Vietnamese withdrew in1989 and Kampuchea was renamed Cambodia. Cambodia committed itself to a private enterprise economy and the restoration of Buddhism.
47. In 1998, Hun Sen was elected to head the government and most of the Khmer Rouge defected. Pol Pot committed suicide.
48. Foreign aid is critical to Cambodia's development, and Cambodians are counting on tourism to earn foreign exchange.
49. MSA is a complex region with vast differences in levels of development. Thailand and Vietnam lead the region. Burma, Laos, and Cambodia are far behind.

Terms to Know

Aung San Suu Kyi	Cochin China	Viet Cong	*Lao Theung*
Tais	*Mission Civilisatrice*	*Doi Moi*	*Lao Sung*
Phrapbuum	Ho Chi Minh	Lan Xang	New Thinking
New Theory Farming	Viet Minh	*Pathet Lao*	Kampuchea
Cao Dai	Dien Bien Phu	*Lao Loum*	Pol Pot
quoc ngu			

Bibliography

Birsel, Robert. 1990. Military Regime Minority Insurgence. *Geographical Magazine* 62: 28–32.

Bradshaw, Adrian. 1989. Sequel to the "Killing Fields"? *Geographical Magazine* 61: 22–27.

Church, Peter, ed. 1995. *Focus on Southeast Asia.* ASEAN Focus Group. Sydney: Allen and Unwin.

Cohen, Eric. 1992. The Growing Gap: Hill Tribe Image and Reality. *Pacific Viewpoint* 33: 165–69.

Cummings, Joe. 1997. *Thailand: Lonely Planet Survival Kit.* Singapore: Lonely Planet.

Duiker, William, J. 1995. *Vietnam: Revolution in Transition.* Boulder: Westview.

Eliot, Joshua et al., eds. 1995. *Vietnam, Laos, and Cambodia Handbook.* Chicago: Passport.

Fairclough, Gordon. 1996. Rice to the Occasion: Burma Exhorts Farmers to Grow More Faster. *Far Eastern Economic Review* (August 29): 55–56.

Freeman, Donald B. 1996. Doi Moi Policy and the Small-Enterprise Boom in Ho Chi Minh City, Vietnam. *The Geographical Review* 86: 178–97.

Forbes, Dean. 1996. *Asian Metropolis.* New York: Oxford.

n.a. 1998. Granting A New Life to the Thai People. *Kinaree* 15: 26–27.

Handley, Paul. 1993. Thought For Food: Thai Farm Policy Faces Long-Term Choices. *Far Eastern Economic Review* (April 29): 46–49.

Huke, Robert E. 1998. Myanmar Visited. *Focus* 45: 18–25.

Husssey, Antonia. 1993. Rapid Industrialization in Thailand 1986–1991. *The Geographical Review* 83: 14–28.

Jackson, Karl D., ed. 1989. *Cambodia 1975–1978: Rendezvous with Death.* Princeton: University Press.

Lintner, Bertil. 1997. Drugs, Insurgency and Counterinsurgency in Burma. In *Burma: Myanmar in the Twenty-First Century,* ed. John J. Brandon, pp. 207–45. Bangkok: Open Society Institute.

Lintner, Bill. 1994. Burma: Enter the Dragon. *Far Eastern Economic Review* (December 22): 22–23.

Lintner, Bill. 1998. The Dream Merchants (Narcotics). *Far Eastern Economic Review* (April 16): 26–27.

Mabbett, Ian, and David Chandler. 1995. *The Khmers.* Cambridge: Blackwell.

Mulder, Niels. 1996. *Inside Thai Society: Interpretations of Everyday Life.* Kuala Lumpur: Pepin.

Norman, Colin. 1983. Vietnam's Herbicide Legacy. *Science* 219: 1196–97.

Rigg, Jonathan. 1995. "In the Fields There is Dust." *Geography* 80: 23–32.

Rigg, Jonathan. 1995. Managing Dependency in a Reforming Economy: the Lao PDR. *Contemporary Southeast Asia* 17: 147–72.

Ryan, John C. 1990. War and Teaks in Burma. *World Watch* 3: 8–9.

Savada, Andrea, ed. 1995. *Laos: A Country Study.* Washington, D.C.: U.S. Government.

Shawcross, William. 1994. *Cambodia's New Deal.* Washington, D.C.: Carnegie.

Tesoro, Jose, M. 1998. The Young and the Restless. *Far Eastern Economic Review* (February 13): 23–24.

Thayer, Nate. 1993. Diverted Traffic (Narcotics). *Far Eastern Economic Review* (March 18): 24–25.

Thomson, Curtis N. 1995. Political Stability and Minority Groups in Burma. *The Geographical Review* 85: 269–85.

Wellner, Pamela. 1994. A Pipeline Killing Field: Exploitation of Burma's Natural Gas. *The Ecologist* 24: 189–93.

Westing, Arthur H., and E. W. Pfeiffer. 1972. The Cratering of Indochina. *Scientific American* 226: 21–29.

Chapter 14

Insular Southeast Asia

"There is not one state truly alive if it is not as if a cauldron burns and boils in its representative body, and if there is no clash of convictions in it."

(SUKARNO)

Welcome to the peninsular and island realm of Southeast Asia (refer to Figure 13-1). With five nations including Malaysia, Singapore, Indonesia, the Philippines, and the Sultanate of Brunei, this is one of Asia's most complex regions. Like the rest of Southeast Asia, the region's history is bound up with ocean interactions, foreign intervention, and struggles for identity in the modern world. In this chapter, starting with Malaysia, I will tell you about some of the most intriguing stories of geographic interrelationships in the world today.

Malaysia: 2020

The course of Malaysia's history has been influenced by its strategic position, which made it a natural meeting place for traders from both east and west. The lush tropical forests and abundance of water sustained numerous small, self-supporting communities. But, unlike mainland Southeast Asia, the presence of mountains and the absence of broad plains precluded the development of elaborate systems of water control so fundamental to mainland societies. Sea trade bred contacts from the outside, and this was the foundation of Malaysia's history.

From Whence "Malay"?

Many Malay areas were colonized by Sumatrans long ago. It is conceivable that the word *Melayu* derived from the *Sungei Melayu* (Melayu River) in Sumatra. Then there is the possibility that it came from a Tamil word *malai,* meaning hill.

The geographer Ptolemy called the Malay Peninsula *Aurea Chersonesus,* or "The Golden Chersonese." This was the fabled land of gold.

During the Portuguese and Dutch colonial periods, the whole peninsula was labeled *Ma-*

lacca. There was very little mapping of the peninsula until the early nineteenth century. As you will read shortly, many name changes would occur before reaching the name "Malaysia."

EARLY SETTLEMENT

The earliest of the current inhabitants of Malaysia are the ***Orang Asli,*** meaning "Man of the Land," who inhabited the Malay Peninsula (Figure 14-1). Other groups include the Dayak and Penan of Sarawak and the Rungus of Sabah, people whose nomadic way of life has been almost completely destroyed in the wake of development.

The next arrivals were the Malays, the Proto-Malays establishing themselves around 1000 B.C. These were followed by other immigrants streaming down the peninsula from the mainland region. Many migrated even further to the island realms of Indonesia and elsewhere. The peninsular Malays had their closest links with the Malays of Sumatra, and for centuries the Straits of Malaya served as a corridor for interaction between the two regions.

Together with the *Orang Asli,* all these people share a common culture and make up the ***Bumiputera*** (*Bumiputra*)—"Sons of the Soil." Cultural characteristics are rooted in an agrarian-maritime economy and reflected in village society, where leadership was largely consensual and attitudes were impregnated by a belief in an all-pervasive spiritual world. Although this culture came to be overlaid by Hinduism, Islam, and Western ideologies, elements of the basic culture persist.

Trade with China and India began around the first century B.C., although Hindu and Buddhist elements from India were more influential in terms of impacts on language, literature, and social customs. Numerous temples were constructed throughout the region. For the greater part of this time, Malaysia was subject to either Javanese or Sumatran power structures.

The Peranakan

By 1400, Melaka had become an important outpost for Chinese traders. They arrived from November to March on the northeast monsoon and departed in June with the southwest monsoon. Melaka's sultans even paid tribute at the Ming court. Many Chinese settled in Melaka, especially around *Bukit Cina*—Chinese Hill. In

Figure 14-1

This man is an Orang Asli, *the original inhabitants of the Malay peninsula. He is cutting a sago palm, one of the most useful palms in the forest. The carbohydrate-rich pith is dried as "pearl sugar" and is eaten with fish and fruit. The palm provides fiber for rope, matting, or cloth; the leaves can be used for thatch; and the sap ferments into a liquor.*

fact, today *Bukit Cina* is a historical park housing the world's largest traditional Chinese cemetery.

Subsequent generations of Straits Chinese became known as *Peranakan.* The Peranakans were known for their business acumen and lavish lifestyles. They built classic townhouses and developed all the trappings of a "high society." They also developed their own cuisine, a blend of Malay and Chinese, which is noted for its meticulous preparation and pungent flavors.

Blending easily into British society, they remained divorced from other Chinese. Many became doctors, lawyers, and entered public office. In Penang they were known as "the Queen's Chinese."

The Hindu–Buddhist period of Malaysian history ended with the penetration of Islam. Introduced primarily by Indian and Arab traders, Islam became a major influence after 1400 with the conversion of formerly Hindu rulers of Melaka. Islam diffused along the trade routes into the rest of the insular region. The Sultanate of Melaka and the Kingdom of Brunei became major proponents of the faith.

Both Melaka and Brunei were shattered by the influx of Europeans. The Portuguese, Dutch and Spanish made their presence felt through force of arms. Despite their technological superiority, European influence was limited until the coming of the British at the end of the eighteenth century and their introduction of the resources and organization of the Industrial Revolution.

MAKING MALAYSIA

Under the British, Pulau Pinang (1786), Singapore (1819), and Melaka (1824) became known collectively as the Straits Settlements. Garnering other Malay states under their wing of control, the Straits Settlements were later expanded to include the island of Labuan off the northern coast of Brunei; on the peninsula, a system of Federated Malay States was established with the capital at Kuala Lumpur; several other peninsular states were given British Advisors and became known as the Unfederated Malay States; Sarawak on the island of Borneo became a British protectorate ruled by the Brooke family, and Sabah also became a Protectorate ruled by the Chartered Company of the British North Borneo.

Rajah Brooke

In 1839 British adventurer-explorer James Brooke sailed from Singapore to Sarawak, where a revolt was in progress against the Sultan of Brunei. In 1840, Brooke was recruited by the British to quell the revolt. With the help of the British navy and local Dayak, he suppressed the rebellion and, in return, obtained a feudal title to the area. He made the town of Kuching (meaning cat) his base. Brooke became known as the "White Rajah." From this future capital of Sarawak, he and his nephew Charles expanded the territory eastward. By the end of the century, impoverished and factionalized Brunei ended up as two little pieces of land—enclaves—within Sarawak.

In 1846, the British, true to their tradition of acquiring strategically placed islands among major sea routes, obtained from Brunei the small island of Labuan, conveniently situated only six miles (10 km) offshore to command the entrance to Brunei Bay. Labuan was to function as an anti-piracy base, a port-of-call, and a forward base from which to challenge Spanish incursion into the region. James Brooke was appointed governor of the Crown Colony in 1847. Since 1990, Labuan has served as an international off-shore financial center.

This confusing arrangement satisfied few, and opposition frequently turned violent. In 1948, the British established the Federation of Malaya, including all the nine Malay states of the peninsula. They also committed to the region's independence. Local elections were held in 1951 and federal elections in 1955. The British released their sovereignty in 1957.

The idea of the formation of Malaysia was introduced in 1961. This was to include the peninsular states, the Borneo states, and Singapore. Elections were held shortly thereafter, with only Brunei declining to join. Tension was apparent, however, with regard to Singapore. Its large contingent of Chinese and their potential political power, in concert with Chinese elsewhere in Malaysia, concerned the Malay majority. Singapore was pressured out of the union in 1965 (Figure 14-2).

TIN AND RUBBER

At the beginning of the nineteenth century the population of peninsular Malaysia was no more than 300,000. Most were Malay rice cultivators. A heavily forested terrain deterred the British, who were more interested in their ventures in Pinang and Singapore. However, the success of these entrepôts stimulated migrations from India and China, people who eventually would become intermediaries between the British and the Malays.

Chinese miners revolutionized tin mining, which had previously been on a very small scale. Given a low-cost means of extraction and the location of tin-bearing gravels along the western foot of the Main Range, a mere 25 to 30 miles (40 to 50 km) from the coast, production surged—a "tin rush" ensued in the three western states of Perak, Selangor, and Negeri Sembilan.

Relations between the Chinese and Malays were not good, so the British decided to provide a system

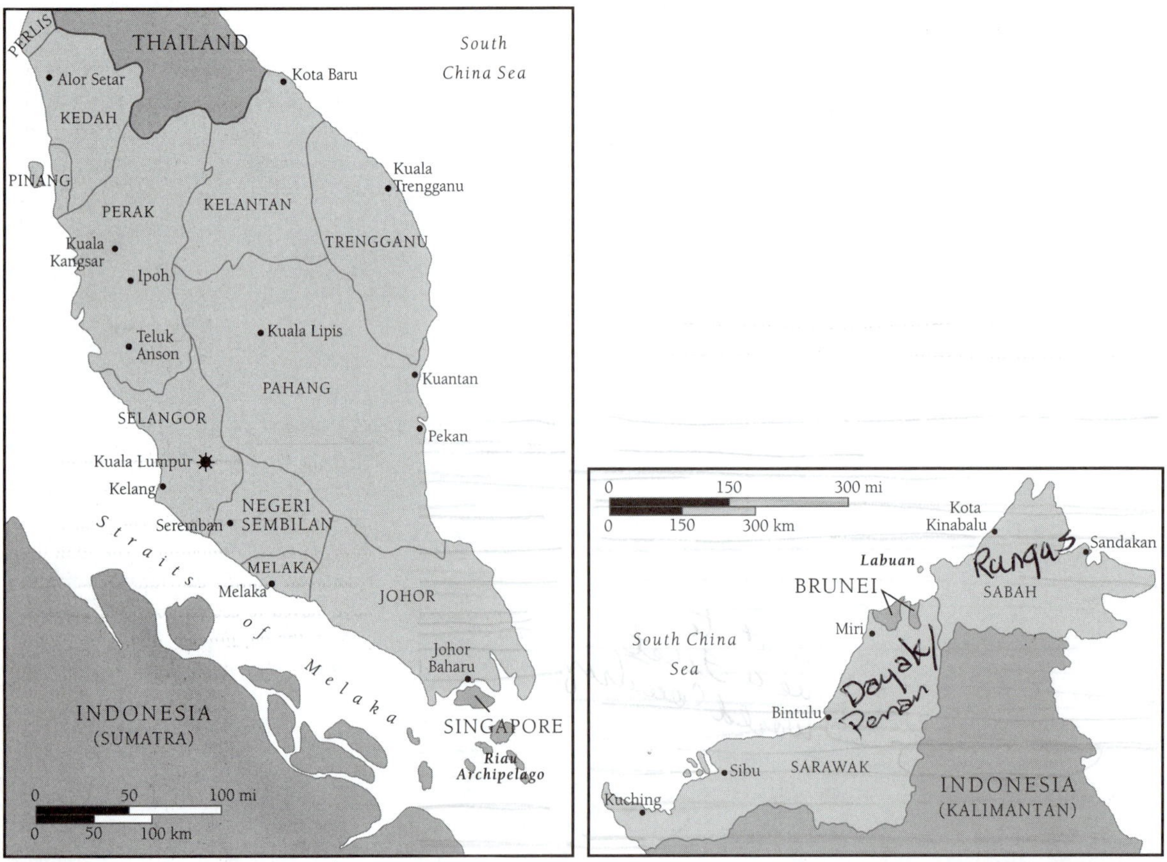

Figure 14-2

Malaysia. Like Indonesia and the Philippines, Malaysia is a fragmented country.

of law and order in the region. On the landscape this took the form of an elaborate infrastructure of road, rail, and port facilities. Smelting and exporting of tin were centered at Pinang and Singapore.

In the early 1900s, an increasingly motorized world was demanding rubber. The climate of Malaysia was perfect for the raising of this commodity. Consequently, new rubber estates were established within the area already possessing an infrastructure. Indian rubber tappers were brought to work the trees, and the rubber industry boomed until the Great Depression of 1930. Even so, Malaysia remains the world's largest rubber producer.

The peninsula's development was severely unbalanced. A rubber and tin belt occupied the southwest. It covered only a quarter of the land area but accounted for 85 percent of its economic activity. Here, too, were the ten largest towns and a large but diverse population of Chinese, Malays, and Indians. The east was largely undeveloped and contained a much lower density of predominantly Malay peasants. When Malaysia became independent, a major goal was to rectify this unbalance.

EAST MALAYSIA

East Malaysia consists of Sarawak and Sabah. Prior to the economic intervention of the Europeans, northern Borneo's trading economy was dominated by indigenous, unprocessed, and luxury products. Jungle products such as bezoar stones, hornbill beaks, camphor, and bird's nests were most important. Other products included rattan, bamboo, other woods, and gutta percha (used to make cable coverings and glue).

This was an economy separate from the world market. Traditional networks of collecting and marketing channeled jungle products to coastal ports where Chinese or Malay traders sold them on the regional market (Figure 14-3).

By the late nineteenth century, the East Malaysian economies had begun to shift toward participation in a world economy, meeting the needs of world markets. Control over production systems was increasingly from abroad as plantation farming, rubber, timber, and oil took precedent over jungle products. Indigenous groups, such as the Penan and the Dayak (engaged directly in accessing these products), fell increasingly under the influence of others, particularly the Chinese. Javanese workers were also recruited to work in the region. However, as jungle produce continued to deliver significant income, indigenous people clung to those activities while outsiders involved themselves in tobacco, pepper, and other new production enterprises. Oil was very important in British North Borneo, where it was the stable backbone of revenue.

Precolonial and colonial trading economies had become intermeshed. Whatever the case, the focus

Figure 14-3

Here the Iban people of Sarawak are selling "jungle products," which they have collected from the forest. The fruit in the foreground is "rambutan." The Iban are traditional shifting cultivators, but many have moved to coastal towns to participate in the oil, lumber, and plantation economy. They make up nearly one-third of the population of Sarawak.

was on environmental exploitation. Native populations became ever-more marginalized as commodity trade increasingly fulfilled the requirements of the growing world economy. This heritage affects the developmental patterns of East Malaysia and Brunei to this day.

ETHNICITY AND DEVELOPMENT

Malaysia has a plural society and development policies have favored indigenous Malays. However, the relationship between ethnicity and status is difficult to define as there are not viable definitions of ethnicity. Ethnic pluralism is a historical construct. For example, in Malaysia the British perceived the Malay as farmers unwilling to do wage labor. Chinese and Indians were inherently suited for mining and plantation work. In other words, Malays were shoved to the margins of development by both perceptions and policy. As geographers Mark Cleary and Brian Shaw (1994) observed,

> Ethnic pluralism in Malaysia might be more accurately represented as a series of overlapping Malay, Chinese or Indian ethnic dimensions (language, business practice, occupation, religion) rather than lines of rigid demarcation. They represent socially constructed rather than "natural" difference.

Just as economic and infrastructural factors were imbalanced, so was population distribution. By the late 1960s, some 15 percent of the Malay population were classified as rural. About 45 percent of Chinese and 35 percent of Indians were classified as urban. Those Malays in the cities were heavily concentrated in government employment giving them little chance to participate in commercial and industrial sectors.

Modern development in Malaysia has been defined by a New Economic Policy (NEP) and subsequent "Malaysia Plans." A fundamental goal of the NEP was to restructure Malay society by raising incomes for all groups and eliminating the association of ethnicity and economic function. This meant an increase in the production of export commodities such as timber and oil as well as export-oriented industrialization. Explicit within the NEP was the intention to create a Malay (*Bumiputera*) business community that would own or manage at least 30 percent of all these activities by 1990.

FELDA

One widely acclaimed land development scheme was the Federal Land Development Authority (**FELDA**) settlement plan. Commencing with a 2,470 to 4,940 acre (1,000 to 2,000 ha) block of land, over virgin forest, contractors were brought in to clear the jungle, plant the main crop—rubber in the early years and later palm oil, cocoa, and coffee—and build basic infrastructure such as roads, houses, and schools. Settlers were selected, moved in, and employed as salaried workers until the crop began to produce. Each family was given title to a plot of land once the mortgage was paid off. By 1987 there were 422 FELDA schemes in all states except Pinang (Figure 14-4).

As oil palms replaced rubber, some of the FELDA schemes such as in Sabah and the Jenka Triangle in Pahang became highly integrated, including their own harvesting, milling, marketing, and shipping facilities. These organizations with their economies of scale and integrated management contributed to the increasing commercialization of all FELDA operations.

Today, FELDA schemes are more likely intensively run "agro-businesses" than traditional Malaysian farming communities. Moreover, the schemes increasingly are becoming part of the Western, mechanized, uniform production system. In addition, there is a scarcity of new land development, environmentalists are pressuring the government, costs of resettlement have doubled, and young people with rising aspirations are reluctant to run their parents' holdings. In the final analysis, it now costs more to run FELDA schemes than similar operations in the private sector. Consequently, the Sixth Malaysia Plan (1991) cut FELDA targets to zero.

Land development is to be reduced and focused on joint ventures and the private sector. More emphasis will be placed on increasing agricultural productivity. The small land holder is now perceived as inefficient. The state will concentrate more on education and training, health, rural roads, housing, communication, and electricity. Land settlement is now seen as the

Figure 14-4

This is the fruit of the oil palm, economic focus of many FELDA schemes. The fruit is crushed for oil which is used for cooking and in such products as margarine and soap. Malaysia is the primary palm oil producer in the world.

concern of the private sector within regional planning frameworks.

A POPULATION ANOMALY

Malaysia's population growth is relatively high, given its level of development. It is also higher than the average rate for Southeast Asia. Malays have significantly higher total fertility rates (TFRs), averaging 4.5 in the 1980s, than the Chinese and the Indian populations, averaging 2.3 and 3.5, respectively, by 1987. The national TFR in the late 1990s is 5.2, of which Malays have a two-thirds contribution.

High Malay TFRs can be related to Islamic fundamentalism and associated decline in the use of contraception. Furthermore, population policy in 1984 promoted five-child families. Malay families receive preferential treatment in university education, government employment, and the like. Malaysia 2020 has pronounced a goal of 70 million people as the optimum population size. This would substantially enlarge the consumer market in light of industrialization ambitions.

The structure of Malaysian population has been altered with Malays now forming a dominant majority. Many Chinese, seeing no future for their children, left for Singapore, Canada, and Australia. From 1970 to 1995, Malays have increased their share from 56 percent to 62 percent. The Chinese have declined from 34 to 27 and the Indians from 9 to 8 percent.

East Malaysia's patterns are similar. However, populations there are concentrated along the coast surrounding urban centers. This contrasts with the sparsely populated interior focusing on river arteries. Sabah's population has been boosted by the in-migration of nearly half a million Filipinos during the 1980s.

URBANIZATION

Migration has taken place in two ways. In the 1960s and 1970s, government efforts were concentrated in redistributing rural populations via land development schemes. Rural-to-rural migration patterns prevailed. The 1980s witnessed a dramatic change, with the largest share of migration being rural to urban. The rural sector was less able to absorb more people due to a fall in commodity prices, and the NEP pointedly tried to reduce rural poverty and dissociate Malays from strictly agricultural pursuits.

The result has been the cultural diversification of urban places once dominated by Chinese. The percentage of Malays residing in urban places rose to 51 percent by 1991. Moreover, the cultural landscapes of the largest urban centers have been noticeably "Islamicized" in recent years. More mosques, the high-rise Islamic Bank, Islamic archways over roads leading into the cities, and the presence of more women wearing head covering are some of the more noticeable landscape elements (Figure 14-5).

Urban population distribution is still unbalanced. The west coast states accounted for 59 percent of the country's urban population in 1991. Historical inertia plays a role in this phenomenon. In East Malaysia, rates of urbanization have been rapid, yet the percentage of urban dwellers is low (37 percent in Sarawak and 33 percent in Sabah). This can be attributed to the absence of an urban-based manufacturing economy.

The paramount urban connurbation is that of Kuala Lumpur and the Kelang Valley (Figure 14-6). Because of the domination of Singapore and Pinang in the colonial period, Kuala Lumpur's population exceeded 100,000 only by 1930. Kuala Lumpur was designated a federal capital in 1974. Rapid growth ensued in the 1970s under NEP policies and today the city is the economic, political, and cultural core of the country.

The city incorporates a number of satellite cities in the state of Selangor, including an east–west corridor between it and the port of Kelang. Petaling Jaya was established in 1952 as a "new town," and by 1991, Petaling Jaya and Kelang were the nation's fourth and fifth most populous urban centers. In the 1970s, a free trade zone was established at Sungei Way. Further west is Shah Alam, the newly planned Selangor state capital that houses Southeast Asia's largest mosque (Figure 14-7). All these are joined by superhighways.

Kuala Lumpur is a modern landscape with a mix of modern and colonial buildings, Chinese shop houses, shopping malls, a fabulous Moorish style-railroad sta-

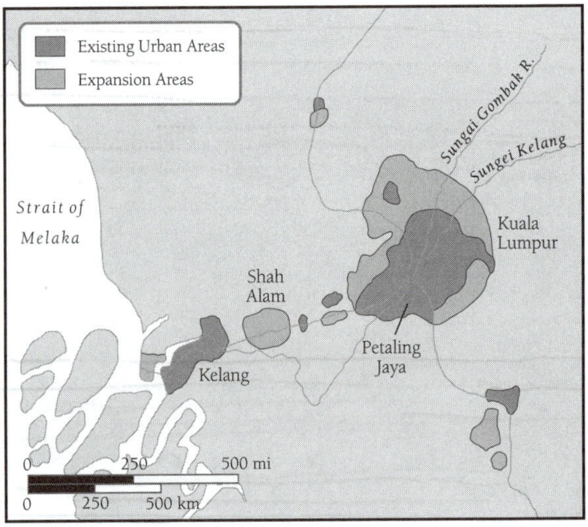

Figure 14-6

The Kuala Lumpur–Kelang development corridor.

Figure 14-5

Kuala Lumpur. The colonial era, Moorish-style Sultan Abdul Samad building, housing city hall and the supreme court, contrasts sharply with the modern office and bank buildings nearby. The tall structure to the right is the Islamic Bank.

Figure 14-7

Selangor's new capital building at Shah Alam is part of the Kuala Lumpur-Kelang development corridor.

tion (to become a hotel), and the 88-story, 1,483 feet (452 m) Petronas "Twin Towers," which were the world's tallest buildings when completed in 1997. A new federal capital is planned at Putrajaya to the south of Kuala Lumpur.

Housing ranges from squatter settlements (fewer than in most of Asia) to condominiums, to suburban homes, to mansions. Because of a housing shortage, the government has built multistory public housing (Figure 14-8). Squatters do not like these structures because it removes them from ground level, where they can garden and watch their children play. Moreover, it is difficult to operate a home-based industry in a high rise. In addition, these projects are frequently distant from employment and services.

Since this area does not have vast expanses of paddy, it does not fit the *desakota* model exactly, although it has some elements of it. ***Kampungs*** (Malay villages) and agricultural land do exist in the interstices of urban places, highways, industrial parks, and the

Figure 14-8

This public housing project in Kuala Lumpur is in Chinatown.

like. Many original inhabitants supplement their income with off-farm jobs.

INDUSTRIALIZATION

Malaysia's national economy is now anchored by high-value manufacturing, including electronics and textiles for domestic consumption and steel and auto assembly for domestic consumption. In the 1980s, the main catalyst in economic growth was an outward-looking industrialization strategy. Rapid industrialization was encouraged by free trade zones and other strategies. Pinang has become Malaysia's "Silicon Island," attracting both foreign and domestic industrial concerns (Figure 14-9). The manufacturing sector is the largest generator of employment.

The Proton Saga

The Proton Saga automobile company is one of several government-owned public enterprises managed by Malays. Its assembly plant is in Shah Alam. A joint venture with Japan's Mitsubishi, the car is replete with the star and crescent symbols of Islam. Initially, the Proton Saga captured over half of the domestic market targeting middle and upper classes. However, this is a low-volume market and the company has looked overseas for potential customers. Britain is the largest thus far. Targeting foreign markets is difficult because of competition and inability to meet safety and emission standards.

Expansion of foreign connections is seen as necessary for the success of Malaysia's auto industry. HICOM, the Proton holding company, is lining up with other foreign companies in Australia and Britain to manufacture certain components such as wheels and catalytic converters. The Malaysian government has turned to Daihatsu to build another car—the Kancil—that is being geared for export markets. Citroen of France is contributing diesel technology to lower vehicles' overall costs.

Foreign investment, to provide capital, technology, and access to foreign distribution networks, was critical to this success. As a result of changing government policies, foreign investment increased tenfold in the 1980s decade. Nearly 50 countries were involved, the major sources being Japan, Taiwan, the United Kingdom, the United States, Hong Kong, Indonesia, and South Korea. The growth of manufacturing has

Figure 14-9

Georgetown, once a British fort, on the island of Pinang, is rapidly industrializing with high-tech industries. Pinang, joined to the mainland by Asia's longest bridge, is becoming Malaysia's "silicon island."

had a tangible effect on urbanization as it generates urban employment.

The Role of Japan and Taiwan

From 1985 to 1990, Japan and Taiwan accounted for 53 percent of total foreign investment. Most of this was in electronics, textile, and apparel industries. Females benefited from these new assembly ventures. They accounted for 60 percent of new employment in the manufacturing sector. Japan has been of particular importance, as Japanese work ethic has been promoted among Malay workers and Japan is a critical source of investment, business skills, and technology transfer.

Some are critical of pervasive Japanese influences. They claim that the Japanese only share outdated technology and that Malaysians are relegated to only middle-level management positions with little hope of advancement. There may be Malay managers, but the important decisions are made by the Japanese. This is different from the U.S. operations, where Malays are often in upper management positions.

With NEP policies, foreign ownership in the corporate arena fell to 21 percent by 1995. Malay involvement increased to 21 percent. Moreover, the Malay employment share in the manufacturing sector rose from 26 percent to 51 percent and in the service sector from 31 to 45 percent. Chinese and Indians experienced parallel share declines.

Beginning in the 1970s, the government targeted a mangrove swamp and an area of aging rubber trees directly east of Johor Baharu for economic development. Much of the early investment was in palm oil refineries and metal-processing plants. By the 1980s the region supported a more diverse industrial base in response to the establishment of a Singaporean-centered growth triangle. In 1988, nearly half of investors were from Singapore, and activities had expanded to include such enterprises as chemicals, textiles, rubber products, and low-value electronic components.

With cheaper land prices, lower labor costs, and access to the Malaysian system of tariffs, Singapore ventures are able to establish the labor-intensive stages of production in Johor prior to adding higher value enhancements in Singapore. The Malaysian government expresses concerns that this evolving core will pull away from the Kuala Lumpur–centered development corridor and that too much Chinese investment will be detrimental to the stated goals of economic equality.

DEFORESTATION AND DEVASTATION

Deforestation is one of Malaysia's most pressing environmental problems. From its origins in early swidden agriculture and colonial plantation, mining, and timber operations, deforestation has reached a zenith with land development schemes and logging over the past two decades. Although devastated, forest cover is greater in East Malaysia than in West Malaysia.

One-quarter of the deforestation is attributable to logging. I recall driving from Kuching, Sarawak, to Brunei along a beautiful, modern road, lined with barren hillsides, stacks of enormous logs, and fires burning off debris. The route is punctuated by Malay *kampungs*, squeezed on the edges of coastal towns that are dominated by Chinese shop houses and fronted with docks piled high with logs and cut lumber. Rivers run brown with eroded material from clear-cut plots deep in the once magnificent rainforest. I thought of the polluted and debris-ridden water, the damage to fish, and the resultant flash floods. All along the route, signs announce the coming of palm oil, pepper, or other estate enterprises, along with new housing and industrial sites.

As I looked upon government-built, tin-roofed long houses lining parts of the highway, I thought of the once-proud Dayak hunters and gatherers whose lives have been disrupted and homes displaced (Figure 14-10). I also thought of the Penan, another group of hunters and gatherers who unsuccessfully have protested the removal of their forest home. Many of these displaced people now work for the logging operations or on plantations.

Figure 14-10

An Iban longhouse along the main highway in Sarawak. As deforestation and modernization ensue, many Iban are forced to seek work in the coastal towns. Although they retain their nuclear family structures and unified housing, the influx of income enables their acquisition of such luxuries as metal roofing and televisions.

A striking note is that women play a leading role in direct action protests. While most men are forced into the cash economy, women are left behind to care for villages and families. Women often work part-time in local industries but still must maintain their subsistence agricultural plots. With their diverse experiences and interests, women are driven to participate in protest events against the logging companies.

A sign of the times is a cultural center near Kuching, which has model villages of all the major tribes of the region. This has been developed for tourists. If you want to see the old ways of doing things, you can go there. Traditional ways of life in Malaysia are sadly disappearing in the wake of development. Both environments and peoples are being devastated.

Brunei: Micro-State

Brunei Darussalam—the country's official name—means "Abode of Peace." Territorially, Brunei is about the size of Luxembourg, a mere 3,577 square miles (5,769 sq km). In 1981, the government purchased a cattle ranch in Australia's Northern Territory. The ranch is larger than Brunei. Brunei is a fragmented state, and people go from one part to the other by boats referred to as "flying coffins." These long-boats are totally enclosed and entertain passengers with screaming programs of Kung Fu genre films.

In the early 1900s, Brunei was known for its exports of *cutch* (or gambier), which is a dye produced from mangrove trees and used for dying leather. Today, the country is known for its oil, with Royal Dutch Shell as the major company. So much wealth derives from oil that Bruneians are the wealthiest people in Asia after the Japanese. Brunei is sometimes called the "Shellfare State," as its citizens do not pay income tax, enjoy free education, medical care, and old-age pensions. Bruneians seldom do manual labor; this is done by Filipino and Thai immigrants. More than half of Brunei citizens hold government jobs. Most Malays are well-housed, although there is a long-lived stilt village in the primate city capital of Bandar Seri Begawan (Figure 14-11).

The Sultan of Brunei is one of the world's richest men, reputedly earning U.S.$100 a

Figure 14-11

Note the contrast between Kampung Ayer and the Omar Ali Saifuddin Mosque in Bandar Seri Begawan, the capital of Brunei. Kampung Ayer is 400 years old and is the world's biggest stilt village, housing around 30,000 people, mostly Malays. In 1987, it was declared a national monument and remains Brunei's most popular tourist attraction.

minute from oil. He lives an extravagant lifestyle and has no problem reconciling his lavish appending and partying with the strictures of Islam. Rock star concerts and alcohol consumption in the palace are generally ignored by the populace. After all, "he is the Sultan." To ensure his safety, he maintains a contingent of (British) Gurkha troops in addition to his own army.

In 1990, Malays were 67 percent of the total population. Most of the rest are Chinese and a handful of indigenous groups. Another 11 percent are Asian and Western expatriates connected to the gas and oil industry. Brunei has one of the fastest-growing populations in the nation. Women continue to have large families, as they are not integrated into industry.

Aside from oil, other primary sectors are not well developed. Brunei is one part of Borneo where most of the forests remain intact. Oil and gas are concentrated at Seria and Kuala Belait. Production began at an onshore field at Seria in 1932, but as of the 1990s, most production is from offshore fields. Reserves of oil and natural gas are expected to last for 40 years. There is little diversification of the oil industry. Downstream processing takes place in facilities at Miri and Bintulu in Sarawak. In terms of patterns of development, Brunei has emerged as a tertiary based economy with a small primary sector and no secondary sector.

Brunei must import about three-quarters of its food needs. Modern grocery stores are dominated by *halal* foods but the main ones have small "non-*halal*" sections for the benefit of the expatriate community. McDonald's and other fast-food chains are prominent in Bandar.

Singapore: The "Intelligent Island"

Geography is central to Singapore's evolution. This 240-square-mile (600 sq km) island, holding 4 million people, is situated at the tip of the Malay peninsula separated from the mainland by a narrow stretch of shallow water. It is a pivotal island in the Straits of Melaka (Figure 14-12). Second only to Japan by development measures, this island city-state's success is contingent on its strategic location, as well as its relationships with its closest neighbors—Malaysia and Indonesia.

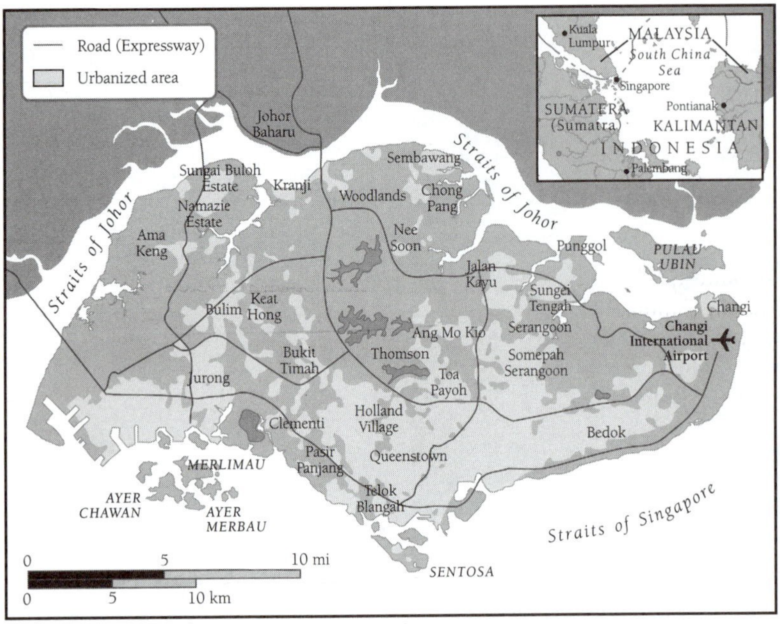

Figure 14-12

Singapore. Note how the urbanized area is oriented toward the sea.

ORIGINS

Countering Dutch power in insular Southeast Asia, the British were seeking strategic ports. Pinang had been acquired in 1786, and the British flag was hoisted over Singapore in 1819 by Sir Stamford Raffles. Now the British East India Company (BEIC) controlled both the northern and southern entries to the Straits of Melaka. At that time, Singapore, with no more than a thousand people, was known as Temasek (Sea Town) and later as Singapura (Lion City).

By the 1830s, Singapore had become the major trading port in Southeast Asia. It was challenged by both Manila and Batavia (Jakarta), but the island had certain advantages. Most ships trading between China, India, and Europe had to pass Singapore. Unlike its competitors, Singapore was a free port. Further, it was enmeshed in the powerful British commercial and industrial empire.

Tin played a major role in Singapore's development. Tin miners in Malaya and Thailand imported their needs from Singapore and used the port to ship their tin to the world. The island attracted traders of all kinds: British, Indians, Malay, Arabs, and Chinese.

The Chinese were the labor force upon which Singapore was built, and the city was the conduit for thousands of Chinese laborers going to Malaya and the Netherlands East Indies. Most Chinese came to Singapore as poor indentured laborers hoping to make a fortune and send money to their ancestral home. The forced opening of the Treaty Ports in China and the annexation of Hong Kong in 1842 accelerated the migration of Chinese—not only to Singapore, but worldwide. Singapore remains a predominantly Chinese country.

By the late nineteenth century, Singapore had become a major transshipment point as well as a commercial and financial center. Here, products from Southeast Asia, such as tin and rubber, were received, packaged, and re-exported. The largest commercial firms were British. Growing numbers of Chinese-owned enterprises were linked to the business web of the Chinese Diaspora. There was little manufacturing in Singapore before 1960; three-quarters of the population were involved in the service sector.

There was also a significant Indian contingent, fluctuating from 6 to 12 percent of the population. The minority was large enough to form its own community known as "Little India." The Indian community was far from united—there were regional, religious, and caste divisions. Thousands were imported from India as convict laborers to build Singapore's infrastructure. Free Indians were primarily merchants or in public employment as teachers, policemen, and the like. There were strong links between the Indian communities of Singapore and Malaya. By the time of World War II, Singapore was a multiracial, multireligious, multilingual society governed by a British elite.

The sea-oriented fortress of Singapore, ill-prepared for a land invasion, fell to the Japanese in February 1942. All communities suffered greatly at the hands of the Japanese, fostering a legacy of distrust of Japan.

After the war, the British reoccupied the island. Although its vision for Malaya was independence, it looked upon Singapore quite differently. There was a fear of the Chinese majority in light of events in China and the insurgence of communism in the region. It was believed that an independent Singapore would come under communist control.

Limited self-government was introduced in 1955. In 1959, the People's Action Party (PAP) began its dominance of Singapore's politics. Led by Lee Kuan Yew, the PAP was the party of a new English-educated elite intent on creating a new, industrialized Singapore.

The creation of Malaysia in 1963 included Singapore. However, unresolved issues surrounding the combined power potential of the Chinese in both Singapore and Malaya led to the creation in 1965 of a separate Republic of Singapore. Under the tutelage of Lee Kwan Yew and the PAP, the island transformed itself from an entrepot economy to an industrial and post-industrial economy in the space of 30 years. Under the strong, stable, and corruption-free PAP, Singapore is a model of planned development.

PLANNING OFFENSIVES

In less than two generations, Singapore has transformed itself from a swampy, overcrowded slum to a modern, gleaming metropolis where rules and regulations control activities and behavior. For example, littering brings a fine of U.S.$500, and chewing gum is forbidden.

The city is a traveler's paradise. As early as 1975, the government introduced a licensing system costing U.S.$3 for cars entering the business district between 7:30 A.M. and 6:30 P.M. on work days. Now there is also an electronic payment system to control traffic on the busiest streets. Moreover, Singapore has an extensive and efficient subway system.

The Land Acquisition Act of 1967 allowed the government to acquire land anywhere, making serious urban planning possible. In 1959, the city was plagued with a soaring population and housing shortages. PAP set up a Housing Development Board (HDB), which cleared slums and built new housing along the Singapore River. Hundreds of Chinese shop-houses gave way to multiunit apartment blocks. Today, in the case of older housing, residents have to pay only 10 to 20 percent of the cost of refurbishing—new plumbing, paint, additions, and so forth.

The HDB then proceeded to build new towns, complete with high-rise apartment blocks, town centers, markets, schools, and light industries to provide jobs. With government loans, apartments (flats) are sold far below free market prices and now house about 90 percent of the city's population.

Population planning is also part of the grand plan. Prior to the 1980s, strict family size controls and incentives were introduced, with a resultant drop in fertility rates, especially among the Chinese majority. Suddenly, it was realized that there were not going to be enough workers for the growing economy. In 1987, new policies were introduced encouraging people to have more children. While Malays and Indians have higher fertility rates, the Chinese remain the majority at 78 percent of the population. Indians comprise 7 percent and the Malays 14 percent.

Many more plans are in the making. A new mass transit line will link more new towns into the system. Deep tunnel sewage systems will be placed 65 to 164 feet (20–50 m) below ground to avoid contamination of usable aquifers. Water desalination plants are intended to reduce the reliance on Malaysia for about half of the city's water. Developers are creating land to join five small islands of the southwest coast into one big island. The new Jurong Island is earmarked for petrochemical industries.

CYBERCITY

Already one of the world's largest hubs for shipping and commerce, this economic tiger is staking its future on becoming a global nerve center for media and communications. The National Computer Board's report titled *IT2000: Vision of an Intelligent Island* proposes an infrastructure that includes both a fiberoptic network and the information services that require this form of digital communication. Singapore is second to Japan in terms of the number of computers it manufactures. In addition, the island makes more than half of the world's hard disc drives, and almost all the sound cards for personal computers.

More than 7,000 multinationals operate out of Singapore, employing 60 percent of its work force and producing 80 percent of its exports. Companies regard the island's infrastructure as superb. It has the world's largest container port, state-of-the-art utilities, and one of the world's most advanced telecommunication networks.

CREATING A HINTERLAND

Building an external economy is considered a national imperative. Foreigners already make up 20 percent of the labor force, and an overseas push is the only place for expansion. Many companies have already relocated in India, Thailand, Vietnam, Russia, Hungary, and Belgium, for example. Twenty-two Singapore companies are constructing a huge industrial park near Suzhou (west of Shanghai) in China. In 1993, nearly 60 percent of Singapore's investment was in Asian countries, with 22 percent in Malaysia. In Malaysia, more than half went into financial-services projects and the rest into manufacturing. Cheaper labor is not the only reason for these investments. Singapore is hoping to tap into large potential markets, as well.

Singapore's goal is for greater interdependence with its neighbors, especially with Malaysia and Indonesia. The adjacent Malay peninsula and Indonesia's Riau islands linking Singapore to Sumatra have potential for a development triangle. Water can be obtained from Sumatra, granite from an island west of Batam, and both Batam and Johor can offer cheap labor. All these areas provide land for expansion. Batam has been developed as an industrial estate, agribusiness site, and tourist haven. It is the third most visited island in Indonesia after Java and Bali. A second causeway has already been built to Johor where spatial patterns and linkages have effected an extended metropolitan region.

The islands of Bintan and Bulan are also being developed. Bintan is becoming a playground for Singaporeans. It is destined to have 20 hotels and 10 golf courses. Bulan is touted as an agribusiness center. Swine production supplies 50 percent of Singapore's demand for pork. Chicken production is expected to supply all Kentucky Fried Chicken outlets. And, the island will have the world's largest orchid farm.

Singapore's growth triangle is neither complete nor proven entirely successful. While these areas are not yet fully integrated, it is still argued that there is more to be gained from cooperation than from competition. This growth triangle is still in the making. The lack of formalization is partly because the growth triangle means different things to each of the participants. According to geographers Scott Macleod and T. G. McGee (1996), "In many ways the strength of the Growth Triangle as a concept lies in its vagueness; it can mean all things to all sides."

FINDING AN IDENTITY

One way in which Singapore is forging a sense of place is through heritage tourism. Certain districts of the city are being refurbished to display their cultural origins (Figure 14-13). "Little India" and "Chinatown" are examples of this phenomenon. However, localism is combined with globalism, as many buildings now host international business chains such as upscale shops and restaurants. In advertising, the city is being portrayed as having "its head in the future and its soul in the past." As geographer T. C. Chang (1999) points out, "On the conceptual front, the power relations between the 'global' and the 'local' in Singapore demonstrate that 'globalization' and 'localization' are mutually constitutive." These relations are revealed in the city's carefully orchestrated heritage landscapes.

Figure 14-13

Singapore's modern skyline rises above Chinese shophouses that have been recently "boutiqued" and now contain trendy restaurants and shops. This is one of the city's deliberately designed heritage landscapes.

Language, as the context for thought, is vital to forging identity. In the 1960s, the PAP stressed learning Malay and English. As Singapore's prosperity grew and the economy became more internationalized, PAP shifted its emphasis to Mandarin and English at the expense of Malay and Chinese dialects. Mandarin and English are now the official state languages.

Singlish

Singlish is a bewildering mixture of English, Mandarin, Chinese dialects, and Malay. There is concern that the rise of this dialect will demote the official languages, which are intended to internationalize the state. The artistic community has embraced the dialect enthusiastically, and Singlish is even popularized on television shows. An example is: "Don't pray pray *lah,*" which means "Don't fool around." Many people feel that Singlish is their own personal language for everyday communication and that "So shiok" ("That's great"). New words and expressions are invented almost on a daily basis fueled by the Internet experience. As the debate over Singlish and English rages, it is clear that Singlish is the language spoken on the street.

Singapore in the twenty-first century will be among the few prosperous countries in the world. Its geographic situation may lessen in importance as it becomes enmeshed in the global economy's computer revolution. In regional terms, Singapore regards itself as the economic engine of the region. The growth triangle is evidence of this view. Nevertheless, Singapore worries about its image and is constantly advocating campaigns regarding what it means to be Singaporean. Its 1990s drive to reveal historic aspects is only part of this continuing search for national identity. However the future is charted, it will be in the context of "Okay can!"

Indonesia: Integration vs. Disintegration

With more than 17,000 islands, Indonesia is the world's largest archipelago (Figure 14-14). With a population of more than 210 million, it is the fourth most populous country in the world. There are more than 350 ethnic groups speaking an estimated 583 languages and dialects. Indonesia also has the world's largest Muslim population.

Indonesia's geography is just as varied. There are vast expanses of tropical swamps in Kalimantan (part of the island of Borneo) and Sumatra and glaciers in

Figure 14-14

Indonesia. Note the concentration of cities on the island of Java as well as the distances between Java and the outer islands.

the central mountains of Irian Jaya (West Papua—the western part of New Guinea). There are more than 300 volcanoes, 200 of which have been active in historical time. Indonesia's motto *Bhinneka Tunggal Ika*—"Unity in Diversity"—is fitting indeed.

Because of Indonesia's cultural diversity, most people first develop a regional identity, only learning the national language *Bahasa Indonesia* when they attend school, and with it a national identity. The Indonesian government is fully aware that national unity and national cultural identity must be created.

Indonesia can be divided into the inner islands and outer islands. The inner islands are Java, Bali, and Madura. The outer islands are the rest, such as Sumatra, Kalimantan, Sulawesi, and the islands of Maluku (Figure 14-15). The political, historical, and economic heart of the country are the islands of Java and Bali. These two islands contain 80 percent of the population but only 7 percent of the land area.

THE EARLY YEARS

As you already know, Southeast Asia lies across the great trading routes from India to China. Chinese, Indian, and Arab traders were common sights in the ports that dotted the region.

In the Indonesian archipelago, there were two types of states. First were the coastal states located at river mouths with secure harbors, dependent on regional and international trade. Second were the inland states, based on agricultural production from the rich volcanic soils and alluvial plains. Central and Eastern Java and Bali were the most prominent of these.

The earliest kingdoms were Hindu/Buddhist states, adopted by local rulers attracted by religious and philosophical ideas as well as court ritual. Yogyakarta is the center of this history and culture. Within 48 miles (30 km) of this city are the great monuments of Borobudur (Buddhist) and Prambanan (Hindu)

Figure 14-15

This woman is a Dayak who lives in the interior of eastern Kalimantan. The Dayaks were once head-hunters. Their traditional way of life in the jungle is rapidly changing with resource exploitation, deforestation, influx of migrants, and government impress from Java. Note the filter cigarette.

(Figures 12-5 and 14-16). The island of Bali remains predominantly Hindu.

Muslim traders arrived as early as the sixth century but the Islamicization of Indonesia began in the thirteenth century with the conversion of the ruler of Aceh in northwestern Sumatra. The nature of Islam in Indonesia varies greatly, ranging from the strict Acehnese to peoples elsewhere who have a more relaxed attitude to the faith.

When the Europeans arrived in the middle of the sixteenth century, they found a series of well-established states across the archipelago. The major ones were at Aceh, Central Java, Bali, the Maluccas (Maluku), and Makassar. There was a constant flow of people and goods across the region using Malay as their means of communication. The Indonesian and Malay languages are very similar.

THE NETHERLANDS EAST INDIES

The Netherlands United East India Company (VOC) began to trade with local kingdoms in the early seventeenth century. In 1619, they attacked Jayakarta (Jakarta), then a major fort in the Javan kingdom of Banten. Jayakarta was renamed **Batavia.** Batavia remained the capital of the Netherlands East Indies until independence in 1945 when it was renamed Jakarta.

Total control of Java was not achieved until 1756 after the bloody Java War whereby the Dutch defeated the Yogyakarta Sultanate. Several years later, in 1796, the VOC was wracked by corruption, and the Netherlands Crown took over what by then was the largest state in the archipelago. Gradually through the nineteenth century, the Dutch extended their control over Sumatra and eastern Indonesia. With the destruction of the Balinese kingdom in 1905 and Aceh in 1911, the colony was complete.

The Dutch maintained a centralized state. Power was vested in Batavia, and the colonial government kept a close eye on religious leaders. Any stirrings of resistance were quickly quashed. The Dutch promoted a Western secular elite built on the children of the pre-colonial elites and made every effort to prevent the formation of an Islamic elite.

Regional trading networks were gradually destroyed. External trade became the preserve of European companies and internal trade became dominated by the Chinese, who were encouraged to migrate from southern China.

Javanese agriculture was transformed by the *Culture System* (see Chapter 12). Compelled to produce designated crops, by the end of the nineteenth century Java was the world's largest sugar producer. Village

Figure 14-16

Evidence of early Hindu kingdoms on Java, these temples at Prambanan have suffered earthquake damage. The (originally 232) temples of Prambanan were built between 900 and 930 A.D. Like Borobudur, they replicate the cosmic Mount Meru. The largest three temples are dedicated to Brahma, Shiva, and Vishnu.

Java's subsistence economies faded to commercialization, and cities and towns emerged to handle the export trade. By the beginning of the twentieth century, most Javanese were tenant farmers, share croppers, or wage laborers.

In Sumatra, vast areas of virgin forest were removed to make room for tobacco and rubber plantations. When oil was discovered in the 1920s, it became the nucleus of what would become the Royal Dutch Shell Oil Company. Much of the labor was Chinese, and although they never made up a large portion of the colony's population, they became dominant in local trade and urban commerce.

The economic transformation of Indonesia led to increased urbanization, and by 1910, Java's cities were unable to cope with the flow of rural to urban migration. Urban places were crowded, without piped water or sanitation systems, and people were suffering from malaria and other water borne diseases. Living conditions for most Indonesians worsened until the 1970s.

The Dutch introduced Western education to provide a needed pool of skilled labor, but entry to these schools was very difficult for ordinary Indonesians. The best schools used Dutch as the medium of instruction. Graduates entered the professions or even attended universities in the Netherlands. Most who benefited were children of the elite or government officials. By the end of the Dutch colonial era, literacy rates were lower than any other European colony in Asia, with the exception of the Portuguese colony of East Timor.

NATIONALISM AND SUKARNO

The term *Indonesia* was first used in the early 1920s, but by 1928, there were enough educated youth with determination to create a modern Indonesia free from Dutch rule. Although several political parties emerged, the most prominent nationalist was a man named **Sukarno,** who could charm the masses with his brilliant oratory. He succeeded in spreading the idea of freedom across all levels of society; he popularized nationalist ideology. An important theme was that people should set aside their ethnic and religious differences to unite in opposition to colonial rule. It was generally agreed that Indonesia should be a secular state. However, none of this pleased the Dutch, and many nationalist leaders were expelled or jailed. Sukarno was imprisoned.

THE JAPANESE AND INDEPENDENCE

The Japanese occupied Indonesia in March 1942. Since they required all males to work for their war effort, and treated people with disdain and brutality, they quickly alienated themselves. However, they removed the Dutch from administrative functions, replacing them with Indonesians. Thousands of Dutch perished in concentration camps from torture, disease, and starvation.

Certain policies would eventually serve the cause of freedom for Indonesia. The Japanese prohibited the use of the Dutch language and encouraged the use of Indonesian. Various schemes of military training were installed for young people. Nationalist leaders, including Sukarno, were freed from jail. As the Japanese realized they were losing the war, they promoted moves toward independence. The Republic of Indonesia was declared on August 17, 1945, in Jakarta. Since the Dutch rejected this notion, a four-year guerilla and diplomatic war ensued. An agreement was reached to end Dutch colonial rule in 1949.

AFTER THE REVOLUTION

The revolution witnesses the rise of a strong army intent on taking the initiative in leading their country to national unity and modernization. The army continues to play a strong role in Indonesian politics to this day. It insists on its *dwifungsi,* "dual function," to defend the nation from external threat and internal subversion and to promulgate economic development.

Despite the perceived power of the army, it did not come into full power until after 1965. Numerous political parties were vying for control, and by 1960, competition was intense. Many were concerned about the influence of the Indonesian Communist Party (PKI). Moreover, in 1965, there were rumors that Sukarno was terminally ill and that the PKI was preparing a coup. On September 30, a group of army officers overthrew the government and took power under the leadership of General Suharto, who proclaimed a government of the "New Order."

Over the next six months, the army vigorously sought out PKI members. At least 400,000 were killed, mostly in Java and Bali. Many of the victims were Chinese. The PKI was destroyed.

Pancasila

With the demise of the PKI, there arose a fear of the power of Islam. The consensus political system has embraced what is called ***pancasila*** democracy. *Pancasila* is the five principles adopted by Sukarno in 1945 as the basis for Indonesian ideology and development: Belief in one God; national unity; humanitarianism; democracy based on consensus and representation; and social justice. Its vagueness allows for many interpretations. All students, members of the civil service, and the armed forces must pass exams concerning *pancasila.*

Indonesia remains a poor country with the largest labor surplus in Southeast Asia—40 percent of its workers are underemployed. Half of its workers are employed in agriculture. The mid-1997 financial crisis dealt the country a devastating blow, with Indonesia's currency falling in value lower than in any other Asian country. This economic crisis generated political turmoil that saw the fall of General Suharto in 1998. This was followed by a period of ***reformasi,*** "reform" that encouraged liberalization. Elections were held in 1999 and a new and more liberal government has been put into place. Also exacerbated by the economic turmoil are Indonesian-Chinese hostilities and criticisms regarding the country's population distribution and transmigration program.

POPULATION DYNAMICS

In the case of Indonesia, economic growth has far outpaced population growth. Women of the 1990s are having only half as many children as women were in the 1960s. In part, this is due to the changing status of women and the decline of the extended family (Figure 14-17). However, family planning programs have

Figure 14-17

These residents of Jakarta have one and two children, respectively. Smaller family size reflects the success of family planning in Indonesia, as well as modernizing trends in terms of the status of women and growing preference for nuclear families.

been in place since the installation of the New Order government.

The program was started in Java and Bali and was progressively extended to the larger Outer Islands, where it gained support from local governments as well as Muslim leaders. The program involved hospitals and family planning services, a battery of field workers, and village leaders. It focused on both reducing family size and improving quality of life.

In the last quarter century, there has been nearly a 50 percent decline in fertility. Mortality levels have also been reduced, a function of the improved nutrition of children. Other populations with poor nutritional status have not fared as well. Life expectancy has risen but still is not as good as in other Asian countries. Maternal mortality remains a problem. Furthermore, there has been an increase in chronic and degenerative diseases such as cancer and heart disease.

Gaining access to medical care is difficult for most Indonesians, especially outside Java. More than a third of Indonesians still do not have access to safe water supplies, and nearly half do not have access to sanitation facilities.

TRANSMIGRATION

Population distribution has long been a concern of Indonesian governments. Even the Dutch had colonization programs to provide pepper estates with labor. ***Transmigrasi*** started up again in 1969 with the first five-year plan of development. In the next two decades, some 1.29 million families (about 5 million people) were moved to correct population imbalances among the islands (Figure 14-18). The proportion of Indonesians living in Java declined from 66 percent at the time of independence to nearly 60 percent in 1995.

The aims of *transmigrasi* are to depopulate crowded areas, to open up new lands, and to Javanize sensitive areas such as Aceh, West Papua, and Kalimantan. People with farming experience were usually selected for the program, but it has been rife with problems.

Lack of funds to prepare areas and assist settlers is one problem. Another is the unsuitability of the land to be developed. For instance, attempts were made to grow wet rice on sandy soil. Some migrants to Sumatra ended up in mangrove swamps. In many cases there were cultural clashes between the immigrants and the original settlers. Madura remains a hot spot in this regard. In another case, pig keepers from Bali insulted the sensitivities of Muslims in Sulawesi, who regard pigs as unclean.

In the 1980s, as government funds became scarce, spontaneous migration was encouraged from Java. Sumatra remains the chief destination of these migrants. However, overall, the number of migrants is decreasing as economic development efforts are in-

Figure 14-18

This is the home of a transmigrasi *in Kalimantan, the destination of thousands of settlers in Indonesia's transmigration programs. The owner derives a living from growing cloves and collecting rubber and other products from the remaining forest. Although typical housing has open windows with shutters these glass windows reflect outside influence and are a sign of relative affluence, unusual for most migrants.*

creasingly focused on Java. In fact, there are many Outer Island peoples who have migrated to Java. One of the most striking patterns of migration is the autonomous movement of women over long distances.

URBANIZATION

The dominant population movement in Indonesia today is rural to urban migration. The proportion of Indonesians living in urban areas has doubled during the last 25 years. In the mid-1990s one in three lived in cities. Many people living in rural areas actually work in urban areas. Improved transportation has fostered commuting on a daily basis. Other people spend part of the year working in the city while maintaining a family in a rural community. These are **circular migrants.** Both commuting and circular migration have increased with expanded transportation systems, advances in education, changes in the role of women, and greater industrial development.

There has been an amazing lateral extension of urban areas. This has occurred along major transportation corridors. New acronyms have become common in planning realms. For example, SIJORI refers to the Singapore, Johor, Riau growth triangle. The most well known of these acronyms is **JABOTABEK,** standing for Jakarta and the contiguous administrative units of Bogor, Tangerang, and Bekasi (Figure 14-19). Another urban agglomeration includes Bandung and its surrounding administrative regions. The Bandung complex is blurring into JABOTABEK.

JABOTABEK

This is the fastest growing area in Indonesia and is becoming a single, integrated urban region with Bandung. JABOTABEK and Bandung consist of nine administrative units. In 1990 the combined population of this region was more than 22 million people. Of the 17.1 million in JABOTABEK, 13.1 million were urban and 4 million were rural. Increased spatial unification is expressed in socioeconomic corridors of activities including agriculture, residences, and retail trade.

An important part of this mega-urban region is housing development. New housing developments have been stimulated by the facts that city center land is very expensive; increasing numbers of middle-income families cannot afford to live in the main cities; there are increasing employment opportunities in urban fringe areas; transportation facilities are better;

Figure 14-19

In Jakarta, modern buildings form the backdrop for part of the Medan Merdeka *or Liberty Square, one of the largest squares in the world.*

and social and economic facilities provided by real estate developments outside the main cities are more prevalent.

Housing development is in the form of large-scale subdivisions. Towns associated with industrial developments are also on the drawing board. Much of this development is associated with foreign investment. Both government and private enterprises are participating in development. However, the government is losing control to private enterprises now dominating construction projects, and many problems are emerging.

Prime agricultural land has been lost to developers. Farmers are reluctant not to sell as demand increases and prices rise. Water supply has become critical—both groundwater and surface water are in short supply. Seawater is encroaching inland as groundwater is depleted. Mangrove habitats are giving way to grand tourism schemes. Displacement of local people is a serious problem as well. While local communities are the last to be told about forthcoming development, they are the first to be removed. So far, the government has not adequately addressed these and other problems of rapid growth in the JABOTABEK region of Java.

ECONOMIC TRANSFORMATION

The Sukarno years were ones of stagnation. Under Suharto, five-year plans were installed and development ensued. Beginning in 1969, economic development actually outpaced population growth. From 1970 to the early 1980s, economic growth was sustained by global increases in oil prices. However, with the 1980s' fall in oil and gas prices, the government was forced to shift its strategy. It proceeded to adopt a market-oriented approach. Liberalization of controls and processes of deregulation allowed the economy to become more diversified, and growth picked up again. Indonesia is now much less reliant on its oil and gas sector.

Although agriculture has decreased as a contributor to Indonesia's GNP, rice is still an important commodity. Paddy land has increased in some regions, but most of the increase in rice production has been achieved through greater inputs other than labor. In the 1990s, the agricultural work force was less than half the overall work force for the first time.

Change in village life is endemic. In the traditional way, all villagers participated in rice cultivation for a share of the harvest. But with commercialization, agricultural work was put on a pay basis, which broke the patron–client relationships between landed and landless peasants.

Growing at more than 10 percent a year, the industrial sector rose to account for 45 percent of the economy in 1995. Manufacturing plays a large role in this rise. Petroleum refineries are a large part of this base. Other main industries include textiles, food products, tobacco, chemicals, and wood products. Indonesia's service sector is also growing, employing 30 percent of the labor force in 1990. Tourism is one of the country's principal sources of foreign exchange (Figure 14-20).

Much of the manufacturing growth is in and around Jakarta. Nevertheless, in terms of numbers of firms, increases were greater in the Outer Islands.

Figure 14-20

Here in a Dayak village in Kalimantan, a chief wears traditional garb specifically for the benefit of tourists. Tourism is an important source of Indonesia's foreign exchange.

Unemployment is low, but is concentrated in the young, educated group. On the other hand, underemployment could be as high as 44 percent of all workers. Another important aspect not to be ignored is the informal sector, which absorbs labor and redistributes income within the larger economic scheme of things.

The Chainsaw Massacre

Indonesia is home to 10 percent of the world's rainforests. Products derived from these forests such as rattan, plywood, sawn timber, and wood furniture are its most profitable export after oil and gas. The biological diversity of these forests is astonishing—more than 10,000 species of trees, 1,500 varieties of birds, and 500 types of mammals. Conservationists claim that the forests are being decimated by logging companies and transmigration settlers.

Of 356 million acres (144 million ha), 74 million acres (30 million ha) are designated as protected; 49 million acres (20 million ha) have been set aside for national parks and wildlife preserves; 77 million acres (31 million ha) are "conversion forests" for settlers; and the remaining 158 million acres (64 million ha) are earmarked for logging. However, millions of acres (ha) have already been deforested or converted for use by smallholders. Others claim that the logging sections are not always suitable for logging.

Indonesian loggers are granted 20-year concessions ranging in size from 49,000 acres (20,000 ha) to 247 million acres (100 million ha). Many of these holdings are connected to Chinese or military interests. Selective cutting is supposed to be practiced in order to ensure a regenerative growth to allow another cutting in 35 years. But the realities may be different.

The World Bank claims that Indonesia's forests are disappearing at the rate of 247 million acres (100 million ha) a year. About a quarter of that amount is attributable to poor logging practices and overlogging. Slash and burn cultivators, forest fires, and a variety of development projects such as palm oil plantations have also contributed to forest losses. There are an estimated 2 million shifting cultivators, although their methods are sustainable when land is freely available. Illegal logging is rampant in all designated forest environments. Many experts have said that Indonesian logging practices are not conducive to sustaining its forests.

Fire has been a major destructive force. As of 1997, according to official estimates, fire had swept through more than 198,000 acres (80,000 ha) in Kalimantan, Sumatra, Sulawesi, and West Papua. In 1982 to 1983 a blaze eradicated more than 9 million acres (3.6 million ha). Vast areas of Kalimantan, with its underground coal deposits, still burn—a blackened wasteland (Figure 14-21). The blazes reach temperatures that roast the subsoil and kill off root systems and microorganisms to a depth of 6.6 feet (2 m). You can walk on this smoldering landscape and have the soles of your shoes melt. This is what happened to me! Forest fires in Indonesia often affect the air as far north as Mainland Southeast Asia and the Philippines where visibility has been cut to 50 feet (15 m) and the sun seems to disappear. Satellite photos reveal that many of these fires are deliberately set to clear the forest by logging and plantation companies.

One well-known endangered animal is the orangutan—"man of the forest." This animal is endemic to the forests of Sumatra and Borneo. They are largely vegetarian, so destruction of the forest is destruction of their habitat and food source. Adult orangutans are frequently shot and their offspring are taken as pets or to be sold to zoos. Conservation efforts exist in both Malaysia and Indonesia, but these are small in relation to the actual devastation of these precious animals.

NATIONAL INTEGRATION

National integration is of primary concern in Indonesia. After Suharto's downfall in 1998, there was a power vacuum and a fear of disintegration. In many outlying areas, local autonomy or independence is seen as the panacea for long-felt ills, and *reformasi* has allowed long-seething discord to surface.

For example, in northern Sumatra, there have been violent clashes over land rights between farmers and security forces protecting the land rights of plantation owners. In southern Sumatra, there has been fighting between the locals and the transmigrants. In

Figure 14-21

This Kalimantan forest fire has been burning since 1983. More than 9 million acres (80,000 ha) of forest and its resident wildlife have been destroyed.

western Kalimantan, the Dayaks are fighting with Madurese migrants. On the island of Ambon in Maluku, Christians have clashed with Muslims.

Autonomy would allow outlying regions to cut their own deals with foreign investors and set their own development priorities. This could spur voluntary migration from overcrowded Java. On the other hand, these areas are typically rich in natural resources and not likely to be granted autonomy.

Take Kalimantan as an example. This province is rich in coal resources, most of which lie under primary forest. Indonesia's oil is expected to run out in the next 20 or 30 years and coal will be an essential fuel. What will happen to the people and the environment of Kalimantan? Kalimantan will remain an essential part of a developing Indonesia.

Three very important regions of dissent are East Timor, Aceh, and West Papua. In 1975, the Portuguese colony of East Timor was occupied by the Indonesian army, which imposed a brutal regime. After decades of struggle and thousands of casualties, East Timor was granted full independence in 1999. Nevertheless, internal strife continues, and thousands of refugees have fled to West Timor. Aceh and West Papua both have separatist movements.

Aceh

There are 25 million Acehnese in the whole of Sumatra. About 5 million live in the northern Sumatran province of Aceh. The region is important for oil, gas, gold, tin, and coal. It also produces coffee, pepper, rice, tobacco, rubber, and timber.

Aceh was where Islam first gained a foothold in Indonesia, and it was the last part of the archipelago to fall to Dutch rule. After a colonial war lasting 69 years, the Dutch withdrew from Aceh in 1942. Aceh was annexed by Indonesia in 1949. Since then, many Javanese have been moved to Aceh in an attempt to integrate the region into the larger Indonesia. Strict Acehnese Muslims resent the more religiously relaxed and alien Javanese.

Complaining that Indonesia was sucking the resources out of Aceh without fair revenue return, the Aceh-Sumatra National Liberation Front was formed in 1976. The movement is also known as *Aceh Merdeka*, or "Free Aceh." In that same year, Aceh declared itself independent. Since then, at least 2,000 Acenese have been killed by the Indonesian army. Since 1992, thousands more have simply disappeared. There are more than 16,000 recorded orphans in Aceh. Furthermore, about 15,000 Javanese migrants have been driven out of areas where they have lived since the 1970s.

Human rights are systematically violated in Aceh. The people are forbidden to use their own language in written form. There is no freedom of expression. Workers are prevented from forming trade unions. They work for foreign companies at below-subsistence wages. Hundreds of schools and other public buildings have been burned and 70,000 refugees driven into impoverished camps along the northern coast. Villagers remain poor, even though Aceh accounts for 15 percent of Indonesia's exports.

Reformasi has made some difference in the situation. The army has backed off, but the *Aceh Merdeka* movement remains strong. Weapons are being smuggled in from mainland Southeast Asia and the movement is well-funded from outside sources. Although independence for Aceh is "nonnegotiable" according to the government, many Acehnese do not recognize the legitimacy of Indonesia. The situation is unresolved.

West Papua

In West Papua, there is a 13,500-foot (4,050 m) mountain thrusting out of the crocodile and malaria infested swamps of the glacier-capped Sundirman Range. Here is a U.S.$50 billion hunk of rock called Grasberg that holds the world's richest copper and gold deposits. The Grasberg mine is owned by Freeport McMoRan Copper and Gold of Louisiana. The mine was one of Suharto's first development projects and is Indonesia's single largest source of tax revenue.

Since Freeport opened, the lives of the indigenous inhabitants have been radically altered. Tribes have received little compensation for their lands. Indonesian law requires indigenous peoples to relinquish customary rights over their land in the national interest. After riots in 1996, Freeport implemented a scheme to parcel out 1 percent of their revenues to the 20,000-strong Melanesian population. But, due to corruption and mismanagement among provincial officials, the scheme collapsed.

The mine has provided 18,000 jobs, but only 1,500 of these are filled by West Papuans (Irian Jayans), and only 400 are filled by local people. In addition, locals are paid only one-seventh as much as other employees. There have been intertribal conflicts. Alcoholism and prostitution abound in the nearby town.

On the mountain's eastern border lies the Lorentz National Park, a 6-million acre (2.5 million ha) expanse whose biotic richness is incomparable. The park's ecological integrity is endangered. The lava flows containing the mineral deposits also exist in the national park. The mountains are rich in timber, and oil and gas deposits lie offshore. How is this area to be protected from further exploitation?

Freeport has one of the world's worse environmental records. It dumps tailings and other wastes into local rivers, and the mining operations leave behind a bare patch large enough to be viewed from space. Tailings devastate lowland vegetation by choking the life out of the plants. Trees are killed by tailings-laden floodwaters. Ninety-seven percent of ore ends up as tailings, which contain arsenic, lead, mercury, and other dangerous metals. These have killed fish, poisoned sago forests (a traditional food source), and made the water unsafe to drink.

Some scientists claim that the vegetation will regenerate on its own. Experiments are taking place to find out what crops might grow in tailings. World pressure has forced Freeport to put more resources into environmental considerations.

Reformasi has allowed organizers to be more vocal about their complaints and desires for self-administration. The Indonesian army has been reigned in, and there is more freedom of expression. There have been meetings between the government and independence activists. But few believe that Indonesia will seriously consider any significant degree of autonomy. Here, too, economic and political issues remain unresolved.

PROGNOSIS

As Indonesia's resources are increasingly eroded and as population growth continues, it will be difficult, if not impossible, to reign in the outer resource regions to the will of the industrial core of Java. In fact, geographer Christine Drake (1992) has suggested that it is really the Outer Islands with their resource wealth that are the foundation for prosperity. It also will be difficult, if not impossible, to suppress the many self-determination movements across this vast region. As Curtis Runyan of *World Watch* (1998) declares: "Indonesia now faces a fundamental choice between developing the country's resources *in spite of* its people, and developing resources *for* its people."

The Philippines: Pearl of the Orient Seas

In bursts of volcanic eruptions and buckling of the earth's crust, the Philippine archipelago began to take shape about 50 million years ago. The archipelago, comprising more than 7,000 islands, stretches 1,100 miles (1,770 km) forming a land chain between the Pacific Ocean and the South China Sea. Only 154 of these islands exceed 5 square miles (14 sq km) in area, and 11 of them make up about 95 percent of the total population. Luzon and Mindanao are the largest islands, and these bracket a regional grouping of islands, known as the Visayas (Figure 14-22).

Rainfall is variable because of the mountainous nature of the islands and shifting wind directions. On Luzon, for instance, rainfall ranges from 35 to 216 inches (84–549 cm) a year, Manila is wettest during the southwest monsoon, which delivers 82 inches (208 cm) from June to November. The islands also straddle the typhoon belt and experience about 15 cyclonic storms yearly. Active volcanoes and earthquakes can also wreak havoc.

In June 1991, Mount Pinatubo erupted and devastated an area of 154 square miles (400 sq km) and blanketed much of Southeast Asia with ash. Fortunately, predictions allowed precautions to be taken, and only 250 people died in the event.

Knowledge of the Philippines' prehistory is sketchy, but Hominid remains estimated to be 30,000 years old have been discovered on Palawan. Malay-related people of Mongoloid descent are predominant today. Other groups are Negritos of obscure origin. Chinese trace their origins back to traders of the tenth

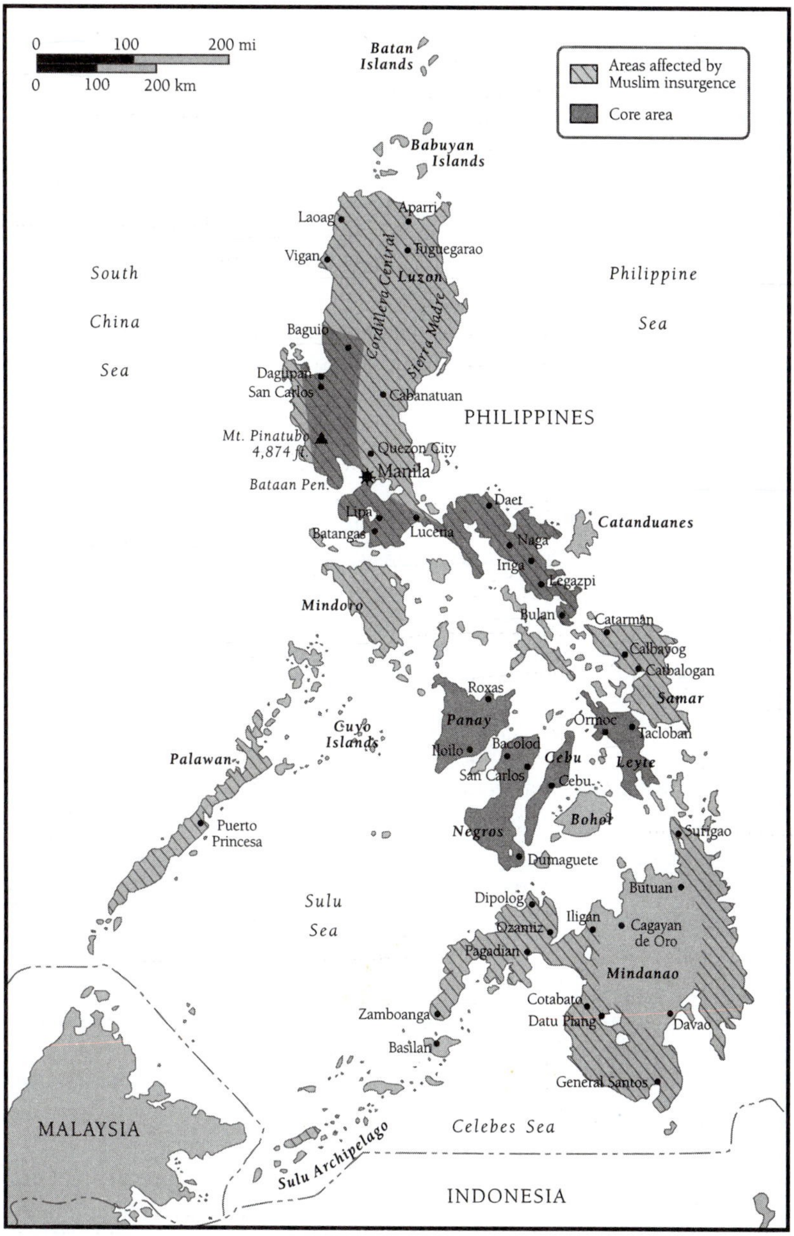

Figure 14-22

The Philippines. Note the focus of activity on Luzon and the Visayan Islands.

century. Five centuries later, Islam from Borneo and the Sulu Islands diffused as far as Luzon, only to be stopped by the arrival of the Spanish Christians in the sixteenth century.

Philippine society is complex, with at least 60 ethnic groups among the Malay majority. Chinese, Americans, other Asians, and indigenous people are also part of the population mix. Several languages are spoken, with **Tagalog**—the language of the people around Manila Bay—being the official language. However, English is the lingua franca throughout the islands.

THE SPANISH PHILIPPINES

Ferdinand Magellan claimed the Philippines for Spain in 1521. Magellan never completed his circumnavigation of the world, as he was murdered in the Philippines. Only one of his ships managed the entire voyage.

Spain controlled the Philippines for the next 377 years, with the familiar twin goals of spreading the Christian faith and setting up trade. The first settlement was on the island of Cebu in 1565, but headquarters were moved to Manila in 1571. The islands were named after the Spanish Crown Prince Felipe.

The economic base of Spanish occupation of the islands was the galleon trade between Macao on the south China coast, Manila, and Acapulco in Mexico. Great galleons left Acapulco for Manila laden with silver. In Manila the silver was exchanged for silk brought from Portuguese Macao. The silk was then taken to Acapulco and on to Europe.

In the seventeenth and eighteenth centuries, Manila became a rival of the major colonial port of Batavia (Jakarta). By this time Manila was in many ways a Chinese city. Chinese traders and ***mestizos*** (Chinese-Filipino or Mexican-Filipino mix) organized the entrepot's trade and became the most influential political and economic force in the Philippines.

Power was vested in Luzon in the north. Efforts to control the Muslim-controlled Sulu archipelago and Mindanao were repeatedly rebuffed by the Islamic sultanate. These regions were not incorporated into the Philippines until the American takeover at the end of the nineteenth century.

Spanish control was based on a close relationship between state and church. Jesuit, Dominican, and Colombian friars were sent into the countryside to proselytize and at the same time establish plantations. Today more than 80 percent of Filipinos are Roman Catholic. Eventually, a state-controlled tobacco monopoly was established in northern Luzon, and local people were forced to provide the labor.

As the galleon trading monopoly came to a close, the islands were open to private investors and traders. In 1785, the Spanish established the Royal Philippine Company to invest in export crops such as sugar, coffee, indigo, and pepper. In all these crops, the Philippines were competing with the Netherlands East Indies. Furthermore, British and American interests had become prominent. By the nineteenth century, British–American export houses dominated the export economy. The volume of trade increased 15 times between 1825 and 1875, with major exports being sugar, coffee, abaca, and tobacco.

The incorporation of the Philippines into the world economy had two chief results. First, it saw the rise of Philippine nationalism, as well as the emergence of a modern nation state. Second, it created regional economic, political, and social forces that would ultimately weaken the country. Powerful regional elites were to become the critical social forces of the twentieth century.

Philippine nationalism was the earliest of its kind in Southeast Asia, but it had little impact on the other colonies. The Chinese mestizos dominated ownership of export plantations. The haciendas (large estates) developed by powerful regional families were worked by tenant farmers. The landowners became richer and the tenants became poorer. From this base arose the powerful landed families that control politics even into the 1990s. This elite educated their children in Spanish schools, seminaries, and universities. These Spanish-educated children, known as ***ilustrados,*** were influenced by liberalized policies in Spain. They were anti-clerical and demanded separation of church and state. As their demands were ignored, they began to call themselves ***Filipinos,*** and *Filipino* became a symbol of nationalism. In 1896, a rebellion erupted.

THE AMERICAN PHILIPPINES

At the same time Spain was dealing with open rebellion in the Philippines, it was at war in Central America and Cuba. U.S. intervention in Cuba resulted in the Spanish-American War. The Americans sailed to Manila, smashed the Spanish fleet, and laid siege to Manila. Philippine opportunists seized the opportunity to declare independence in 1898. The Filipinos were the first people in Asia to fight their colonial oppressors and declare a nation-state. However, the Americans were not done with the Philippines.

In 1899, war broke out between America and the Philippines. Hostilities ended in 1901, when the Americans effectively bought off the *ilustrado* elite, promising to maintain their wealth and power in return for cooperation with American colonial rule.

The United States occupied the Philippines with a "democratic mission." The goal was to lead the Philippines into independence. They stressed the development of health, education, and democratic processes. In 1941, the literacy rate was five times that of Indonesia. In 1935, a Democratic Republic was established with Manuel Quezo as its first president. Leaders were from the landed elite. While espousing complete independence, their political and economic interests were well served by American rule.

Fighting in the Philippines during World War II was more intense than elsewhere in Southeast Asia. However, Japanese slogans such as "Japan, the light of Asia" had little attraction among Filipinos, and resistance to the Japanese was strong. Filipino nationalists welcomed the American liberators, who promised independence by 1946.

INDEPENDENT PHILIPPINES

Independence changed very little in that the landed elite maintained control. Patron–client relations and vested links between rich landowners and poor peasants have continued to underpin the political process. Although rebellions have been crushed, rural discontent and unrest remain problems.

In the 1950s and 1960s, a new industrial class emerged. Some were from the landed elite who were diversifying their capital. Others were professional traders who had merged interests with American or local Chinese businessmen. By the 1960s, the Philippines was the most successful manufacturing country in Southeast Asia. Agribusiness, real estate, and banking were the basis of several multinational conglomerates. Unfortunately subsequent government policies spoiled the record and the economy became stagnant.

Under the rule of Ferdinand Marcos, who proclaimed martial law in 1972, per capital income fell, foreign indebtedness rose to serious levels, and the middle class were protesting corruption and the inability of the government to handle the country's social ills. Moreover, Marcos, friends and relatives, and the military practiced *crony* capitalism—absorbing private businesses and state enterprises—on an unprecedented scale. Family conglomerates controlled 78 percent of all corporate wealth in the country. Land reform and the end of rural poverty remained sheer rhetoric.

By the 1980s, the nation was fragmenting. The Moro Nationalist Liberation Front (MNLF) had 50,000 to 60,000 guerillas fighting for an independent Muslim state in Mindanao. Half the Philippines' army was fighting that battle. Most of the others were fighting communist-controlled guerilla attacks elsewhere in the country. Opposition to Marcos became strident. Marcos called an election in 1986 and lost to Corazon Aquino, wife of assassinated popular politician Benigno Aquino. Marcos ultimately fled the Philippines.

Although the government returned to its 1970s' democratic form, little changed in the social structure. The landed elite retained their riches, land reform was minimal, and political loyalties remained personal rather than institutional. However, there was an impressive economic turnaround.

ECONOMIC PROGRESS

Ferdinand Ramos took power in 1992, and by the mid-decade the Philippines had reached growth levels approaching those of other ASEAN countries. However, the country has a long way to go. Although the agricultural sector has remained relatively constant, with 20 to 25 percent of the total economy, it employs 55 percent of the labor force. In 1993, industry accounted for 33 percent of the economy and services 45 percent. But still, a third of the population are unemployed or underemployed.

Subic Bay

In 1992, the United States gave up the Subic Bay Naval Base and Clark Air Base. Subic Bay alone had employed 32,000 local people and indirectly created 200,000 jobs. However, it also imported some negative aspects of American culture, such as sex work and rampant consumerism. Soon Subic Bay was turned into a free trade zone, and 200 companies—including many from Taiwan—pledged investments in the zone. In 1995, the American company Federal Express established Subic Bay as its air hub for overnight parcel deliveries in the region.

All has not gone as planned for Subic Bay. With the economic crisis in the late 1990s, many Taiwanese, Japanese, and other potential investors backed away from their commitments. Subic's largest investments are tourism related, but poor transport from Manila has limited domestic use. Subic's development is also hampered by a lack of space. The complex is blocked by jungle and the ocean. The airport runways, designed for fighter aircraft, cannot handle large passenger planes, as the runways are lined on three sides by water. Now plans are being made for an expanded harbor facility to provide a more convenient international port than Manila. And, a new superhighway linking Manila with Subic and Clark's faltering freeport zone is on the drawing board.

The traditional industrial base was the processing of agricultural products, but this was altered in the 1990s with a policy of import substitution. For example, the Philippines' largest company, the San Miguel Corporation (of beer fame) diversified into food and soft-drink processing and invested in coconut products, banking, and other commodities.

Recent emphasis has been placed on export-oriented industrialization. Export-processing zones (EPZs) are located near Manila, Cebu, and in Baguio. In addition to Subic's free port, Clark Air Base is now a special economic zone. Numerous private enterprises have been established in other areas, mainly in metro Manila. Today the leading exports of the Philippines are garments and electronics. Gateway Electronics has developed a business park, and numerous chip manufacturers have gravitated to the site. Texas Instruments, Motorola, and Philips have also established industrial sites near Manila.

Manufacturing is highly concentrated in Metro Manila. Its metropolitan area accounts for more than half the total Philippine manufacturing employment. The second largest centers are Cebu and Davao.

Cebu is a world center for the manufacture of rattan products. Nearby locations such as Mactan Island—where Ferdinand Magellan was killed during the first circumnavigation of the globe—include an international airport and an export processing zone. Mactan has also been developed for tourism, with fine hotels and beach resorts. One-third of all foreign tourists, mostly Japanese and Taiwanese, visit Cebu and Mactan. The Japanese are the major foreign investors in the EPZ. One of the zone's first factories was Timex. Now 80 percent of all Timex watches are made in the Mactan plant. Other companies produce semiconductors, cameras, furniture, apparel, and so forth.

Another important manufacturing center is General Santos City on the southern coast of Mindanao. The major industries here are food processing, especially of fish. It is the largest producer of tuna, most of which is exported to Japan as sashimi. The agricultural region surrounding the city accounts for 30 percent of the nation's maize production, 10 percent of its coconuts and copra, 20 percent of its rice, 99 percent of its asparagus, and 40 percent of its pineapples.

One of the largest employers is the Dole Corporation, involved in everything from pineapples and bananas to vegetables.

Another economic scheme is an effort to link Brunei, Indonesia, Malaysia, and the Philippines into an East Asian Growth Area (BIMP-EAGA). These countries are examining ways to enhance trade in the region with the support of the Manila-based Asian Development Bank. A new international airport, expanded harbor facilities, and a new EPZ are part of the future of this new economic growth triangle.

OVERSEAS FILIPINOS

Foreign exchange revenue inflows are an important element of Philippine earnings. The first significant outflow of Philippine workers was to Hawaii to work on plantations. Others have worked in U.S. defense industries in Guam and Vietnam. These ***balikbayans*** (overseas Filipinos) remain important sources of revenue for the Philippines. More than 90 percent (around 1.4 million) live in the United States, Canada, and Australia. Many go to work in the Middle Eastern oil fields of Saudi Arabia, Kuwait, and the United Arab Emirates.

Adjacent Asian countries also offer opportunities for employment, but these are for low-paid domestics and "entertainers." Geographer James A Tyner (1996) has found that marketing and recruiting strategies of the Philippine government, especially in promotional literature, present Filipinos as loyal, disciplined, and obedient and that promotional materials draw attention to gender associated roles and occupations. For example, women are portrayed as nurses and men as construction workers. "Global marketing of workers is organized around specific gendered assumptions of male versus female occupations. . . . The use of gender by government and private institutions as an organizing device provides a gendered context in which migrants and their households must subsequently operate." Gendering of jobs is also most pronounced in Japanese firms. Unfortunately, many Filipino foreign workers find themselves turning to prostitution to pay labor recruitment fees, and otherwise suffer abuse unable to save up enough money to go home.

ONGOING PROBLEMS

Now under new President Estrada, a former movie star, the Philippines has many problems to overcome. In the south, a more militant Moro Islamic Liberation Front is demanding nothing short of complete independence as an Islamic State in western Mindanao and the Sulu archipelago. Environmental damage is severe, with diminishing forests and damaged coral reefs. Little has changed for the poor in either urban or rural areas. Manila is marred with vast squatter settlements and no adequate housing. Wealthy families still control the land, and tenant farmers remain impoverished. Income differentials are huge—some of the largest in Asia—and the middle class remains small. According to the WB, nearly 40 percent of Filipinos still live in poverty.

Conclusion

Clearly, the countries of insular Southeast Asia are plying the path toward modernization. Singapore has achieved this end and is well entrenched in the world of transnationals and cyberspace. However, it is now challenged to preserve its historic landscapes. Malaysia is challenged by the problem of uneven development exacerbated by its fragmented nature. Environmental destruction is taking its toll, especially in Sarawak and Sabah, where traditional lifestyles are meeting their demise. Both Indonesia and the Philippines are fraught with problems deriving from their fragmented nature and their ethnic and religious complexities. Intergroup rivalries are detrimental to a smooth transition to more even economic development in addition to political freedom and self-determination for all concerned. Furthermore, environmental problems overshadow all forms of progress.

Main Point Summary

1. With five nations, this is one of Asia's most complex regions.
2. The course of Malaysia's history has been influenced by its strategic position, which makes it a meeting place between east and west. Sea trade was the basis of Malaysia's history.
3. The earliest inhabitants were the Orang Asli to be followed by the Malays. The Straits of Melaka served as a corridor of interaction between mainland and insular regions.
4. Trade with China had an early history in the region. The Peranakan remind us of the importance of Chinese settlers in the region's commerce.
5. Islam was introduced by Arab traders and became a major influence after 1400.
6. European influence was limited until the coming of the British at the end of the eighteenth century.
7. The British established themselves throughout Malaysia although this political entity did not come into being until the twentieth century.
8. Tin mining and rubber production dominated the economy. Both Indians and Chinese came to work these enterprises. These activities laid the basis for Malaysia's core region on the southwest of the peninsula.
9. East Malaysia's economy was for a long while dominated by indigenous people who dealt in jungle products. By the late nineteenth century, they were transferring these goods to Europeans who were participating in a growing world economy. Development was coastal.
10. Although Malaysia has a plural society, pluralism is a historical construct. Different groups were pigeonholed in specific functions. Modern development in Malaysia stresses the development of all groups, especially the native Malays known as *Bumiputera.*
11. FELDA plans have sought to redistribute population in rural areas. The focus has been on *bumiputeras* and associated with land distribution and plantation crops such as rubber and palm oil. Modern FELDA schemes are more like agrobusiness operations.
12. Malay populations grow at faster rates than other populations such as Indians and Chinese. Malays are now a dominant majority, and more Malays are moving to urban areas.
13. Urban populations are still imbalanced with west coast states accounting for over half. In East Malaysia, populations are mainly coastal yet urbanization is relatively low given the lack of an urban-based manufacturing economy.
14. The paramount urban connurbation includes urban areas from Kuala Lumpur to the port of Klang. This is the core of the country.
15. Kuala Lumpur is a modern landscape with a mix of modern and colonial buildings. Housing is varied and includes many government apartment buildings and few squatter settlements. The city shows many signs of the country's Islamic culture.
16. Malaysia's national economy is now anchored by high-value manufacturing. It has an outward-looking industrialization strategy. Pinang has become the country's "Silicon Island." Foreign investment, especially from Japan and Taiwan, are critical to continued growth.
17. Malaysia is involved in the development of a "growth triangle" involving the state of Johor and Singapore whereby Singapore can garner less expensive land and employ Malaysia's cheap labor.
18. Deforestation is a pressing problem. Much primary forest has been lost to plantations, logging, and land settlement schemes. Logging in East Malaysia has removed one-third of its forests and displaced many indigenous peoples.
19. Brunei is a fragmented micro-state that derives its wealth from oil. A Muslim country, it is ruled by a sultan who is one of the world's richest men. Aside from oil, other industrial sectors are not well developed. Brunei must import most of its food needs.
20. Singapore is a pivotal island in the Straits of Melaka. It is second only to Japan in development in Asia. Its success is contingent on its strategic location relative to other Southeast Asian states.
21. By the 1830s Singapore was a major trading post and was attracting a variety of traders. The Chi-

nese were the labor force upon which the country was built, and they dominate the population today.

22. Singapore had become a major trans-shipment point and financial center by the late nineteenth century. By the time of the Second World War, Singapore was a multiracial, multireligious, and multilingual society governed by a British elite.
23. Singapore was part of Malaysia from 1963 to 1965, when it became an independent republic. It has transformed itself from an entrepot economy to a post-industrial economy in 30 years. It is a model of planned development.
24. Singapore is one of the world's largest hubs for shipping and commerce, and it is staking its future on becoming a global nerve center for media and communications.
25. Building an external economy is a national imperative, and the country has expanded overseas seeking both cheap labor and potential markets. It is also seeking greater interdependence with its neighbors Malaysia and Indonesia. It is central to the growth triangle of Johor and Indonesian islands offshore.
26. Singapore is forging a sense of place through redevelopment and heritage tourism. Both Mandarin and English are official languages, although a linguistic mix called "Singlish" is very popular among young people.
27. Indonesia is the world's largest archipelago and is one of the most complex ethnically and linguistically. It has the world's largest Muslim population. It can be divided into inner and outer islands with Java as the most populated and urbanized island.
28. Because the region is so diverse, people tend to have local identities before national ones. The Indonesian language is supposed to help inculcate a national identity.
29. The earliest kingdoms were Hindu and Buddhist, but Islam started to take over in the thirteenth century.
30. When the Europeans arrived they found a series of well-established states such as Central Java and Aceh. The VOC began to trade with local kingdoms in the early seventeenth century. Not long after, they set up their headquarters at Jayakarta which they renamed Batavia. The Dutch completed their control of the islands in 1911.
31. The Dutch promoted a Western secular elite at the expense of the Malays and encouraged the Chinese to engage in trade.
32. Under the Culture System the islands were compelled to grow designated crops. Cities emerged to handle the export trade but most Javanese were tenant farmers, share croppers or wage laborers. Plantation crops were important to the economy.
33. Oil, discovered in the 1920s, became the nucleus for the development of a modern economy.
34. Java's cities were unable to cope with large flows of in-migrants and living conditions for most were grim. Substantial improvements were not made until the 1970s.
35. Western education benefited the children of the elite but spawned nationalism and a desire to throw off their Dutch rulers. Sukarno became the leader of an independent Indonesia after 1949. He promoted the concept of *Pancasila* with its five goals to make a better and more unified society.
36. The revolution for independence witnessed the rise of the army and its belief in its dual functions to defend the nation and to lead the way in development.
37. In 1965, Sukarno was overthrown and General Suharto took over. He established a government of a "New Order." Many people were killed during the coup, especially Chinese.
38. Indonesia remains a poor country with high levels of unemployment and underemployment.
39. General Suharto was ousted in 1998, and a new government was elected in 1999. The current theme is *reformasi*—reform toward liberalization. Economic and social turmoil is problematic with different ethic and other groups expressing dissatisfaction with the current state of affairs.
40. Indonesia has instituted a successful family planning program, and birth rates have fallen dramatically. This has happened in concert with the facts that the status of women is changing and the extended family is less common.
41. Transmigration programs were instituted in the 1960s to redistribute population from crowded

Java to the Outer Islands. This was also an effort to Javanize these regions. Many of these *transmigrasi* schemes were ill thought out and failed. Others are structurally successful, but the Javanese are resented in the presence of different ethnic groups.

42. The dominant population movement in Indonesia today is rural to urban migration. Improved transportation and other factors have led to people moving or commuting to cities to work. Circular migrants spend part of the year working in cities and the rest at home in rural communities.
43. There has been a dramatic lateral extension of urban areas. JABOTABEK centered on Jakarta is one of these urban connurbations. Housing development is an important element of urban sprawl. Foreign investment plays an important role as well. Prime agricultural land is being lost to developers.
44. Indonesia has adopted a market-oriented approach to development and is attempting to rely less on its oil and gas sector for economic growth. Manufacturing is becoming increasingly important.
45. Fewer people are working in agriculture. Rice remains significant but gains are mostly due to increased inputs. With commercialization, change in village life is endemic.
46. Indonesia is home to 10 percent of the world's rainforests and these forests have the world's greatest biodiversity. Forests are being decimated by logging, settlement schemes, and fire.
47. National integration is of primary concern in Indonesia. It cannot afford to lose the Outer Islands, which hold most of the natural resources such as coal, oil, and gas. But, many of the Outer Islands desire independence from a Java-controlled Indonesia. Aceh in northern Sumatra and West Papua are only two examples of trouble spots.
48. The Philippines archipelago consists of thousands of volcanic islands. Luzon and Mindanao are the largest. The islands straddle a typhoon belt and experience a monsoon climate.
49. The population is predominantly Malay-related people of Mongoloid descent. There are at least 60 ethnic groups.
50. *Tagalog* is the official language, but English is the lingua franca.
51. Spain controlled the islands for more than 300 years after 1521, when Magellan claimed them for Spain. The Spanish were interested in trade and spreading Christianity. The first settlement was on Cebu, but headquarters were established at Manila.
52. The economic base of the Spanish in the Philippines was the silk or silver laden galleon trade between southern China and Mexico.
53. Chinese traders and *mestizos* became the most dominant economic forces in the islands, organizing the entrepot trade. Manila rivaled Jakarta as a port in the seventeenth and eighteenth centuries.
54. Mindanao and the Sulu archipelago, dominated by Muslims, were not brought under control until the nineteenth century.
55. Spanish control was based on a close relationship between state and church. Today more than 80 percent of Filipinos are Roman Catholic.
56. After the galleon trade ended, the Spanish invested in export crops such as tobacco and pepper. By the nineteenth century, British export houses were the most prominent.
57. The incorporation of the Philippines into the world economy saw the rise of nationalism. It also created economic, social, and political forces that saw the rise of regional elites and fostered division.
58. The Spanish-educated *ilustrados* demanded separation of church and state. They began to call themselves *Filipinos,* and the term became a symbol of nationalism. The Filipinos were the first people in Asia to fight their colonial oppressors.
59. In connection with the Spanish-American War, the Americans fought the Filipinos and the country became an American colony in 1901. The United States has a "democratic mission" to lead the Philippines to independence.
60. Independence still left the landed elite in control. Rural rebellions have been crushed but rural discontent remains.
61. A new industrial class emerged in the 1950s and 1960s, and by the 1960s, the Philippines was the

most successful manufacturing country in Southeast Asia. However, subsequent government policies spoiled this record and the economy stagnated. Martial law was proclaimed in 1972 and "crony capitalism" under Ferdinand Marcos became the rule. The country was fragmenting with unrest, and Marcos was defeated in an election in 1986 and fled the country.

62. Even with a new government little reform in terms of land redistribution and reducing the power of the elite was instituted. However, there was an economic turnaround.
63. In the 1990s industry accounted for a third of the economic productivity. Agriculture still employs more than half of workers but contributes about a quarter of the economy. A third of the population are unemployed or underemployed.
64. In 1992, the United States gave up the Subic Bay Naval Base and Clark Air Base. These areas are targeted for development based on foreign investment. Problems with infrastructure have caused many potential investors to shy away from the projects.
65. A policy of import substitution replaced processing of agricultural products in the 1990s. Recent emphasis has been placed on export-oriented industrialization.
66. Manufacturing is highly concentrated in Metro Manila. Cebu, Mactan, and General Santos City are other important nodes of economic activity.
67. The Philippines is planning to participate in a growth triangle with other countries in the region including Brunei, Indonesia, and Malaysia.
68. Many Filipinos live overseas and are important for foreign exchange earnings for the Philippines. Frequently they are "marketed" for specific job types according to gender. Exploitation is common.
69. The country faces many social, economic, and environmental problems. Wealthy families still control the land, and nearly 40 percent of Filipinos continue to live in poverty.

Terms to Know

Orang Asli	Sukarno	*transmigrasi*	*mestizos*
Bumiputera	*dwifungsi*	circular migrants	*ilustrados*
FELDA	*pancasila*	JABOTABEK	Filipinos
Kampung	*reformasi*	Tagalog	*balikbayans*
Batavia			

Bibliography

Airriess, Christopher. 2000. Malaysia and Brunei. In *Southeast Asia: Diversity and Development,* eds. Thomas R. Leinbach and Richard Ulack, pp. 341–78. Upper Saddle River, NJ: Prentice-Hall.

Asian Strategy and Leadership Institute. 1997. *Malaysia Today: Towards the New Millennium.* London: ASEAN.

Balakrishnan, N. 1991. Single-Minded: Census Fails To Allay Concern on Marriage Patterns. *Far Eastern Economic Review* (June 20): 17.

Cartier, Carolyn L. 1993. Creating Historic Open space in Melaka. *The Geographical Review* 83: 359–73.

Chai, Michael. 1992. The "Other Indigenous" Peoples of Sarawak. *Development* 2: 56–61.

Chang, T. C. 1999. Local Uniqueness in the Global Village: Heritage Tourism in Singapore. *The Professional Geographer* 51: 91–103.

Church, Peter, ed. 1995. *Focus on Southeast Asia.* Hong Kong: Allen and Unwin.

Cleary, M. C. 1996. Indigenous Trade and European Economic Intervention in North-West Borneo c. 1860-1930. *Modern Asian Studies* 30: 301–24.

Cleary, Mark, and Brian Shaw. 1994. Ethnicity, Development and the New Economic Policy: The Experience of Malaysia, 1971-1990. *Pacific Viewpoint* 35: 83–107.

Dharmapatni, Ida Ayu Indira, and Tommy Firman. 1995. Problems and Challenges of Mega-Urban Regions in Indonesia: The case of Jabotabek and the Bandung Metropolitan Area. In *The Mega-Urban regions of Southeast Asia,* eds. T. G. McGee and Ira M. Robinson, pp. 296–314. Vancouver: UBC.

Dolven, Ben. 1999 a. Forward March (Singapore). *Far Eastern Economic Review* (March 25): 45.

Dolven, Ben. 1999 b. Singlish: Can or Not? *Far Eastern Economic Review* (September 2): 32–33.

Drake, Christine. 1992. National Integration in China and Indonesia. *The Geographical Review* 82: 295–312.

Hiebert, Murray. 1996. It's A Jungle Out There (Singapore). *Far Eastern Economic Review* (April 25): 58–61.

Hiebert, Murray. 1997. On the Offensive: Singapore's Rules May be Strict—But They Do Curb Urban Blight. *Far Eastern Economic Review* (February 6): 38–39.

Hugo, Graeme. 2000. Indonesia. In *Southeast Asia: Diversity and Development,* eds. Thomas R. Leinbach and Richard Ulack, pp. 304–40. Upper Saddle River: Prentice-Hall.

Leinbach, Thomas R. et al. 1992. Employment Behavior and the Family in Indonesian Transmigration. *Annals of the Association of American Geographers* 82: 23–47.

Johnstone, Bob. 1995. Culture Clash in Cyberspace. *New Scientist* (March 25): 38–41.

Lee, Boon Thong. 1995. Challenges of Superinduced Development: The Mega-Urban Region of the Kuala Lumpur-Klang Valley. In *The Mega-Urban Regions of Southeast Asia,* eds. T. G. McGee and Ira M. Robinson, pp. 315–27. Vancouver: UBC.

Leitner, Helga, and Eric Sheppard. 1987. Indonesia: Internal Conditions, the Global Economy, and Regional Development. *Journal of Geography* 86: 282–91.

Macleod, Scott, and T. G. McGee. 1996. The Singapore-Johore-Riau Growth Triangle: An Emerging Extended Metropolitan Region. In *Emerging World Cities in Pacific Asia,* eds. Fu-chen Lo and Yue-man Yeung, pp. 417–64. New York: United Nations.

Heyzer, Noeleen. 1992. Rainforest Management and Indigenous Livlihoods: A Malaysian Case Study. *Development* 4: 14–17.

Hull, Terence, H. et al. Indonesia's Family Planning Story: Success and Challenge. *Population Bulletin* 32: 1–52.

Jardine, Matthew. 1995. *East Timor: Genocide in Paradise.* Tucson: Odonian.

Madely, John. 1988. People Transplanted, Forests Uprooted. *Geographical Magazine* 60: 22–25.

McBeth, John et al. 1999. Worse to Come (Indonesia). *Far Eastern Economic Review,* July 29: 16–19.

Monbiot, George. 1989. The Transmigration Fiasco. *Geographical Magazine* 61: 26–30.

Murphy, Dan. 1999. The Next Headache (Indonesia). *Far Eastern Economic Review* (April 29): 20–21.

Parsonage, James. 1992. Southeast Asia's "Growth Triangle": a Subregional Response to Global Transformation. *International Journal of Urban and Regional Research* 16: 307–17.

Perry, Martin. 1991. The Singapore Growth Triangle: State, Capital and Labour at a New Frontier in the World Economy. *Singapore Journal of Tropical Geography* 12: 138–51.

Rodan, Gerry. 1989. *The Political Economy of Singapore's Industrialization.* New York: St. Martin's.

Rodrigue, Jean-Paul. 1994. Transportation and Territorial Development in the Singapore Extended Metropolitan Region. *Singapore Journal of Tropical Geography* 15: 56–74.

Runyan, Curtis. 1998. Indonesia's Discontent. *World Watch* 11: 12–23.

Schwarz, Adam. 1992. A Saw Point for Ecology (Indonesia). *Far Eastern Economic Review* (April 19): 60–61.

Sutton, Keith. 1989. Malaysia's FELDA Land Settlement Model in time and Space. *Geoforum* 20: 339–54.

Sutton, Keith, and Amriah Buang. 1995. A New Role for Malaysia's FELDA: From Land Settlement Agency to Plantation Company. *Geography* 80: 125–37.

Taylor, D. M. et al. 1994. The Degradation of Rainforests in Sarawak, East Malaysia, and its Implications for Future Management Policies. *Geoforum* 25: 351–69.

Tiglao, Rigoberto. 1999. About-Face: New Chief Changes Game Plan for Subic Bay. *Far Eastern Economic Review* (February 18): 44–45.

Tyner, James A. 1996. The Gendering of Philippine International Labor Migration. *Professional Geographer* 48: 405–16.

n.a. 1997. *Information Malaysia 1997 Yearbook.* Kuala Lumpur: Berita.

Ulack, Richare. 2000. The Philippines. In *Southeast Asia: Diversity and Development,* eds. T. R. Leinbach and R. Ulack, pp. 408-33. Upper Saddle River: Prentice-Hall.

Vatikiotis, Michael. 1991. Search for a Hinterland. *Far Eastern Economic Review* (January 3): 34–35.

Glossary

38th Parallel Line of latitude dividing the Korean peninsula into Russian and American spheres of influence in 1945.

Activity space The cumulative space employed in the activities of daily life.

Adat Indonesian traditional system of local regulations, rights, and responsibilities in communal context.

Adi Granth The Sikh holy book.

Adivasi Means "original inhabitant." Applied to Scheduled Tribes in India.

Aeolian Referring to wind.

Age of Exploration Refers to European exploration of the world beginning in the fifteenth century.

Age of Imperialism (1870–1914) The acquisition, control, and exploitation of colonies by competing European powers.

Ainu Aboriginal people of Japan. Now confined mainly on Hokkaido.

Allah "God" in Islam.

Alluvial soil Fertile, water-deposited soil.

Altitudinal zonation The variation in plant communities caused by climatic change associated with increasing altitude.

Amenity-migration The movement of people to places offering amenities such as rest and recreation or a more comfortable climate.

Americanization The promotion and acceptance of American culture.

Ammonite A flat, spiral fossil shell from the ancient Tethys Sea.

Analects of Confucius The ideals of Confucius recorded by his students.

Annam Means "the pacified south." Chinese name for Vietnam during the Tang Dynasty. Referred to central Vietnam under the French.

APEC Asia Pacific Economic Cooperation. A regional trading block.

Aquaculture The cultivation of sea products such as shrimp, oysters, seaweed, and pearls.

Aquifer A permeable subsurface rock layer that can store, transmit, and supply water.

Archipelago A collection of islands.

Arithmetic density The number of people per given area of land.

Artesian basin An aquifer in which water is held under pressure. Drilling a well would allow water to flow freely to the surface without artificial pumping.

Aryans Relatively light-skinned nomadic people who invaded India from Central Asia.

Asia-Pacific Similar to the economic concept of the Pacific Rim.

Asoka The most important Mauryan ruler in India.

Aung San Suu Kyi Leader of the National League for Democracy in Burma. Has been under house arrest since 1989.

Azad Kashmir A revolutionary government in northern Kashmir.

Bagasse Sugar cane fiber.

Bangla The Bangladeshi name for the language Bengali.

Balikbayans Overseas Filipinos.

Bao Dao The name for Taiwan as portrayed on ancient Chinese maps.

Barani Dry, non-irrigated regions of Pakistan.

Bari The basic spatial unit in a Bangladeshi village.

Base population The actual population from which growth or decline occurs.

Batavia Dutch name for Jakarta. Batavia was renamed Jakarta in 1945.

Benazir Bhutto The youngest and first female leader of a Muslim state. Served as Prime Minister of Pakistan.

Bilateral aid Government to government transfers of cash, low-interest loans, food, and the like.

Biodiversity The range of living organisms defined in terms of genes, species, and ecosystems.

Biological Oxygen Demand (BOD) How much oxygen is required to support microorganisms that break down organic wastes.

Bodhisattva A divine being in Buddhism such as Avalokitesvara meaning "the lord who looks in every direction."

Brahman The one ultimate reality in Hinduism.

Brahmin Hindu priest.

Buddha The enlightened one in Buddhism.

Buffer state A country situated between two other opposing countries.

Bumiputera Means "sons of the soil." The Malay peoples of Malaysia.

Bungalow A low-lying, thick-walled, one-story house.
Burakumin Social outcasts in Japan.
Burgher Descendant of Portuguese or Dutch in Sri Lanka.
Buttress Giant, supportive base of a shallow-rooted tree.
Canopy The upper layer of the rainforest. The forest's biomass.
Cao Dai Means "high tower." A Vietnamese religion that is a mixture of Buddhism, Christianity, Confucianism, and Daoism.
Cardamom A highly valued spice commonly used in South Asian cuisine.
Caste system The division of Hindu society according to social ranking and occupation.
Castle town Fortified town along a transportation corridor in traditional Japan.
Ch'I The vital force or cosmic breath. The most important force of *feng shui.*
Chaebol A giant, corporate conglomerate in South Korea.
Char As rivers sway between valley walls, they assume a braided pattern separated by small, sandy islands called *chars* in Bangladesh.
Chawl Slum in Mumbai.
Chiang Kai-shek Leader of the Nationalist forces in the war against Mao Zedong and the Communists. Fled to Taiwan in 1949. Ruled Taiwan until his death in 1975.
China frontier The rugged and arid environment of western China. Sparsely populated.
China proper The relatively low relief, humid environment of eastern China. Densely populated.
Chipko Movement A grass-roots, women's movement to save trees in the Indian Himalaya.
Cilao Portuguese name for Ceylon (Sri Lanka).
Cinchona Source of quinine extracted from tree bark; it was the only treatment for malaria in historic times.
Circular migrant A person who lives and works in the city for part of the year and spends the remainder of the year in his or her village.
Civil line British civilian district socially and spatially separated from Indian society.
Cloud Forest Forest existing in fog and mist characteristic of elevations between 8,000 and 13,000 feet (2,400–39,000 m) in the Himalaya.
Cochin China A southern Vietnam colony established in 1867 in French Indochina.
Collectivization Put an end to land ownership in China. Land was worked collectively.
Colonizer's Model The idea that civilization and progress diffuse from Europe to the rest of a culturally barren world.
Commoditization The division of a particular market into two sectors: big-name brand marketing and sales, and hundreds of no-name component suppliers. The garnering of the no-name component sector is known as the Original Equipment Manufacture market.
Confucius A Chinese teacher of ethics, ritual, and philosophy.
Congress of Vienna 1815 A meeting at which European boundaries were defined and Europe was demarcated from Asia by the Ural Mountains.
Congress Party The Indian National Congress was formed in 1865 to provide a forum for nationalist sentiments. The Congress Party was run by Brahmin Hindus.
Continental drift Millions of years ago, a continent named Pangaea broke into pieces forming separate continents.
Convectional rainfall Precipitation brought about by immediate rising, cooling, and condensing of moisture-carrying air.
Coral reef A biological entity formed in sea water by small marine animals living in symbiotic relationships with algae.
Core-periphery The concept that there are core regions of development surrounded by regions of lesser development. Applicable at any scale.
Cultural convergence Takes place when life ways and belief systems blend in a landscape.
Cyberspace The sphere of information technology.
Cyclone A low pressure center. Can be associated with cyclonic storms having high velocity winds and copious precipitation.
Dalit Refers to India's untouchables and means "oppressed."
Dao de Jing Means "classic of the way." A book of Daoist philosophy.
Dasht Expanses of black pebbles in Baluchistan, Pakistan.
Deciduous Refers to trees that lose all their leaves in a cold or dry season.
Decolonization The process of gaining of independence from colonial powers.
Demographic Transition Model Sequential changes in population growth or decline beginning with high births and high deaths and ending with low births and low deaths.
Deng Xiaoping A colleague of Mao Zedong and survivor of the Long March. Deng ruled China from 1976 until 1989.
Dependency Theory The idea that core regions can grow only through exploitation of their dependent peripheries.
Deposition The depositing of material through the process of erosion.

become a center of tissue culture; laboratories commercially mass-produce plant clones for customers around the world. Further, Thai concerns have been exporting agribusiness technology to developing countries (e.g., China, Indonesia, and Vietnam).

New Theory Farming

New Theory Farming is a form of land and agricultural management considered by various government agencies as a means of introducing a community-based economy. In the initial stage, farmers divide their 6-acre (2.4-ha) plot of land into three 30 percent and one 10 percent portion allotted to rice, field crop or fruit, water supply, and residence. This initial stage should enable farmers to become self-sufficient and to pay off their debts.

In subsequent stages, farmers join to form cooperatives and then link up with financial and energy sources such as commercial banks and oil companies. In this third stage, farmers will manage their own mills and coops and take on joint ventures.

In Thailand New Theory Farming has been launched as "The Community Self-Sufficient Economy Project." With farmers focusing on their families and communities, networks such as savings groups will emerge and production groups will network with groups in other provinces. A local marketing and civil society will thereby develop.

In the first year of the project, 1998, thousands of villages participated in various aspects of the program. Community markets, networks, water source development, agricultural service centers, and civil society activities were put into place. Nearly 5,000 families have participated in the New Theory self-sufficiency project with positive results. It is the hope of the government that these programs will help farmers to pay off their debt and ease themselves out of poverty.

Tourism

Since 1987, tourism has become a leading earner of foreign exchange, outdistancing even textiles. Ecotourism has increased substantially among the northern hill tribes. Sex tourism continues in Bangkok and some of the southern resorts such as Pattaya. Unfortunately, many of the hill tribe traditions have been compromised to suit tourist desires, leaving substantial numbers culturally marginalized. Now, much of hill tribe life is "staged" for the tourist, its timing and presentation directed by the majority of Thais to fit in with tourist schedules. Traditional lives are disrupted, and the tourist "objects" of attention are merely players on a constructed stage.

Thailand is noted for its sex tourism. Refer to Chapter 3 for a discussion of the sex industry and AIDS in Thailand.

Industry

Manufactured goods have become increasingly important as a source of foreign revenue. Textiles, cement, and electronics lead the way. Car and truck manufacturing are on the rise since several international manufacturers such as Toyota, Mitsubishi, Mazda, Ford, and GM have established themselves along the eastern seaboard. Thailand now has the second largest pickup truck market in the world after the United States.

The 1980s and 1990s have witnessed a surge of foreign investment from such countries as Japan, South Korea, and Taiwan, where labor costs had risen. Thailand had numerous attractions for relocating industrial operations. Unskilled, trainable, and cheap labor was available. It had a decent infrastructure of roads, airlines, and port facilities, as well as sufficient power and water supplies for industry. Tax incentives, infrastructure provision, and other favorable policies encouraged foreign investment. Bureaucratic procedures were streamlined. Moreover, there was political stability and lack of racial conflicts. Finally, there was no adversity toward Chinese business, an important consideration for firms coming from Hong Kong and Taiwan.

Much foreign capital has flowed toward primary industries such as rubber, tiger prawns, livestock, non-

Figure 13-8

Wat Pho, originating in the sixteenth century, is the oldest and largest in Bangkok. It is also the earliest center of public education.

THE ECONOMY

Agriculture

Sixty percent of Thailand's exports are agricultural. Thailand ranks first in the world in rice production, second in tapioca (manioc) and sugar, and fifth in coconuts. It is the world's largest producer of natural rubber. Other exports include corn, jute, pineapple, palm oil, and canned or frozen shrimp, tuna, and chicken. Chickens are grown American style in large, industrial feed houses, and they are exported frozen. Pigs are also raised in industrial fashion, with the side effect of epidemics of Japanese encephalitis. Freshwater aquaculture is also important. Thailand exports mainly to the United States, Japan, and Singapore.

About 50 percent of Thailand's work force is engaged in agriculture, but the sector is fraught with problems. One of these involves the availability of water, which has dwindled in the dry season. Some worry that decreased flow in the Chao Praya will draw salt water from the Gulf of Thailand to foul waterworks north of Bangkok. Climatic change has resulted in less rainfall. Water shortage is also attributed to deforestation. Thailand has lost 50 percent of its forests over the last 30 years, and there is now a moratorium on logging. Thailand is procuring its wood from Burma and Laos.

Farmers are used to expanding space rather than intensifying inputs to increase production. This is no longer possible. The industrial boom has sent harvest wages soaring, and many people have left the farm for work in industry.

Rising competition is another problem. Sugar, the only crop subsidized by the government, faces stiff competition in the newly established ASEAN Free Trade Area. Vietnamese rice exports, especially lower grades, are hurting thousands of subsistence farmers. Chicken and shrimp exports face competition from China. The rubber industry is threatened by cheaper labor in Indonesia. Thailand is especially vulnerable to world market prices as 70 percent of its land is in only six commodities: rice, sugar, rubber, maize, cassava, and sorghum.

Some changes are taking place. Improvements are being made in mechanization and intensification, along with land consolidation and better merchandising. A dairy industry is taking off with cows from New Zealand and the Netherlands being cross bred with other breeds to withstand Thailand's climate. Agribusiness firms are entering into new areas such as fresh and processed fruit and cut flowers. The country has the Four Noble Truths, or focusing on how to be an upstanding human being. Even in the whirling commercial world of Bangkok, the role of monks and *wats* and the belief system of Theravada Buddhism is central.

In Thai thought, Buddhism encompasses virtue and wisdom, which can liberate one from the common order of life. It transcends and relates to the ultimate order of morality and goodness symbolized by the home, the mother, and the female symbols of Mother Earth and Mother Rice. The Buddha and these female entities constitute the domain of moral goodness.

Figure 13-7

This is a phrapbuum, *a Thai spirit house. It is believed that if it is well-treated and respected, it will protect homes and businesses.*

Next comes the less tenuous order of the realm of supernatural powers outside the home. This area of *saksit* is the focus of religious preoccupation, with rules for dealing with it.

Beyond the area of *saksit* lies the area of chaos represented by wickedness and immorality. Forces of goodness can intermingle with forces of evil. Pure order can transcend into chaos; the Buddha can emerge into evil; pure virtue can become deeply immoral; and safety can transform into danger.

Each Thai house and business possesses a spirit shrine of the ***phrapbuum***—"lord of the place" (Figure 13-7). If respected, well-treated, and given offerings, it will be protective and care for the safety of the place. If things do not go well, it is not the fault of the *phrapbuum;* the cause is always elsewhere.

Just as individual households have lords of place, so do temples, villages, and provinces, where the lord resides at the city pillar. These rulers have no power outside their local territories and must be vested with appropriate rights and offerings.

What's a Wat?

Wat is a Thai word from a Pali-Sanskrit word meaning "dwelling for pupils and ascetics." A *wat* is a Buddhist compound where men or women can be ordained as monks or nuns (Figure 13-8). Virtually every village has a *wat*, while towns and cities may have several. Each *wat* has a number of buildings such as a *bot* where ordinations are held; a *sala* for community meetings and lectures; a *tripitaka* or library for the Buddhist scriptures; and various *chedis* and *stupas* housing the ashes of worshippers. *Wats* are multifunctional and are fundamental in villages—they operate schools, clinics, community centers, old age homes, and so forth. The typical *wat* is a locus of social activity. Especially important are the temple fairs that take place on auspicious dates such as Buddha's birthday.

written language. This is the predecessor of today's Thai writing system. Ultimately, Sukothai faded and in 1351 a new kingdom emerged to the south. This was Ayudhya, which eventually became known as Siam.

Ayudhya

Ayudhya lasted until 1757, prospering from its strategic position only 43 miles (70 km) up the Chao Praya from the sea. Commanding the vast Chao Praya plain, it provided rice for export. Ayudhya was seen by the Europeans as a great trading power. However, Ayudhya had to face a challenge from the Burmans who crushed the empire by 1767. Ayudhya's ruling class was decimated. Thousands of people and their wealth were taken away. The city was burned and vast tracts subjected to a scorched earth policy.

The Rise of Bangkok

By 1785, new Thai rulers had emerged to smash the Burmese and establish new kingdoms. These eventually became consolidated into an area larger than Ayudhya had ever known. The new empire included all of mainland Southeast Asia excluding Burma and Vietnam. It also included some of the northern Malay states. A new capital was established at Bangkok in 1782.

Not experiencing any intense Western pressures, Bangkok remained traditional until the mid-nineteenth century. At this time, many Thai nobles were studying Western languages, science, and mathematics, and Bangkok was prepared to orient toward the West. In 1855, the Bowring trade treaty was signed with Britain giving Britain extraterritorial rights. Treaties with other countries followed.

Eventually the Thai empire was stripped of some of its holdings. It had to abandon its Cambodian and Laotian territory to the French and the Malay states to the British. In 1896, a treaty between England and France guaranteed the independence of most of the territory that today forms Thailand. Siam became Thailand in 1939, although it was still referred to as Siam during World War II.

Modernization proceeded with the building of railways and Western-style education for royal and upper-class children. Rice exports boomed, as did teak and rubber. But there was no significant industrialization. Western and Chinese interests controlled the economy.

The government stimulated Chinese assimilation into Thai society through business partnerships and intermarriage. The Chinese also adopted the Thai language, education, and culture.

European predominance in Southeast Asia was challenged by the Japanese, who invaded Thailand in 1941. When Western forces collapsed in the area, Thailand nominally allied itself with Japan, although effectively, it was an occupied country.

The defeat of Japan was followed by an era of close relations with the United States in an alignment against the spread of Communism in the region. Thailand has served as a refugee center for hundreds of thousands fleeing the conflicts of French Indochina.

CHANGE, FAITH, AND ORDER

America's involvement in Vietnam was a powerful factor in opening up Thai society. In the 1970s, international tourism struck the country in a big way, and labor migration began in earnest. This stimulated consumer culture and the emergence of a plethora of "lifestyles." What does this mean for Thai culture?

The Thais are confident that future developments will be adapted to the Thai way of life. One important aspect of Thai culture is order. They like a well-regulated life with a bent for aestheticism and appreciation for harmonious social order. Implicit in this is respect for interpersonal relationships and hierarchy in concert with ensuing rights and obligations. Even with modernization and social transformation, the expectation that society should be orderly remains.

The role of the *Sangha* in the knowledge of *Theravada*—original teachings—is predominant. The monks are fed twice a day by the common people before noon. This gives people a chance to acquire merit. Almost every Thai male makes merit for his family by going into the monkhood as a novice for a rainy season. Traditionally, few Thai women would consider marrying a man who had not spent time thinking about ethics, learning to read, meditating on

Non-Governmental Organization (NGO) Non-governmental aid organization.

North-South Dichotomy The difference between the rich, industrialized countries of the north and the poorer, developing countries of the south.

Nucleated village Clustered settlement.

Nutritional density The number of people per given area of cultivated land.

Occident The West.

Oracle Bones Bones, especially scapulae, heated and their cracks interpreted by ancient Chinese rulers and sages. Also inscribed with questions and interpretations.

Orang Asli Means "man of the land." Earliest of the current inhabitants of the Malay Peninsula.

Orient The East.

Orographic precipitation Precipitation brought about by the presence of a topographic barrier which forces a moisture carrying air mass to rise, cool, and condense.

Oustee Uprooted person who flees from war and violence, oppression, disasters, economic stagnation, and human-induced environmental displacement.

Out of Africa Theory Postulates that humankind originated in Africa.

Overseas Chinese Chinese living outside of China including both individuals and corporations.

Pacific rim An economic concept referring to all countries facing the rim of the Pacific Ocean. These countries' primary trading partners are each other.

Pacific Ring of Fire The region of seismic activity circling the Pacific Rim.

Pancasila The five principles adopted by Sukarno in 1945 as the basis for Indonesian ideology and development.

Pan-gu A primeval being in China whose body parts transmuted into various parts of the cosmos.

Parsis Zoroastrians who fled Persia to India. Believe that fire is the essence of purity.

Pat Clay desert in Pakistan.

Pathet Lao Communist guerilla fighters in Laos.

Pax Mongolica A period of peace and prosperity under the Mongols from 1280 to 1300.

People's communes A collection of cooperatives housing thousands of people organized under China's Great Leap Forward of 1958.

Percentage urbanized The percentage of a country's population living in cities.

Phrapbuum Means "lord of the place" and refers to the spirit inhabiting houses, office buildings, and other private and commercial establishments in Thailand.

Pinyin A simplified character and transliteration system for Chinese language.

Plate tectonics A theory that rigid plates in the upper portion of the lithosphere float on the more plastic upper layer of the earth's mantle. Movement along the plates' fracture zones cause seismic activity.

Playas Salt-encrusted dry lakes.

Plural society A society in which two or more culture groups live adjacent to each other in the same country without mixing.

Pol Pot Leader of the Khmer Rouge in Cambodia. Responsible for the death of millions.

Polytheistic Having many gods.

Popular culture Usually refers to American popular culture that is diffusing via transnationals and media into all corners of the world.

Population composition The age and sex structure of a particular population.

Population pyramid Pyramidal graphs showing percentage of males and females in various age cohorts.

Power relationships The fact that men and women are accorded unequal roles in virtually every society.

Primate city The largest city in context of urban primacy.

Privatization Changing from public control to private control of development.

Protestantism A major branch of Christianity.

Purdah The Muslim custom of secluding women.

Quality of life Human well-being.

Quoc Ngu Means "national language." Transliteration of Vietnamese spoken language into the Roman alphabet.

Rabi In South Asia, crops sown after onset of the wet monsoon and harvested in autumn.

Rainshadow The leeward side of a topographic barrier which is the dry side in the process of orographic precipitation.

Raj Means "the rule" and refers to the period of British rule in India.

Rajputs Highly militarized tribe of Rajasthan, India.

Ramadan A month of fasting. One of the Five Pillars of Islam.

Reformasi A period of economic reform and liberalization in Indonesia following the end of General Suharto's rule in 1998.

Regional state A natural economic zone shaped by the global economy in which it participates. Guangdong province in China is an example.

Regionalism Focusing on the local rather than the global.

Region-based urbanization The continual growth of cities over large areas.

Reincarnation Rebirth in the cycle of birth and death.

Religion A set of sacred beliefs and practices.

Restructuring The rapid change in local, national, and international social, political, and economic systems wrought from the dynamics of global interaction.

Rich–poor gap The difference between rich and poor between parts of the world as well as between countries and within individual countries.

Roman Catholicism A major branch of Christianity.

Sangha Monastic orders of Theravada Buddhism.

Sati Being burned alive on a husband's funeral pyre.

Sawah Rice paddies in Indonesia.

Second World Refers to communist-bloc nations.

Semi-deciduous Refers to trees that lose some of their leaves in the dry months of the monsoon.

Sepoy A British-trained Indian soldier.

Serendip Arab name for Ceylon (Sri Lanka).

Sex city A city that is known for its sex trade.

Sex ratio The numerical ratio between men and women in a population.

Sex-selective abortion The deliberate abortion of males or females (typically females) after sex determination of the unborn fetus.

Shadow economy Illegal economic activities such as black market enterprises.

Shatter-zone A region of multiple, independent cultural groups often generated by fragmented terrain.

Sherpas Herders and traders of Nepal's Himalaya who are now dependent on tourism. Many serve as guides for climbing expeditions.

Shifting cultivation A migratory, field rotation system of agriculture.

Shiites A breakaway Islamic group with roots in Persia (Iran).

Shinto Traditional religion of Japan.

Sikh A follower of Sikhism.

Sikhism Belief system founded in fifteenth century by Guru Nanak in India. Believes that all people are the children of God. Opposes the caste system.

Simhala Dvipa Island of the Lion Tribe (Sri Lanka).

Sinhala Language of the Aryan, Buddhist, Sinhalese in Sri Lanka.

Sinicize To make Chinese.

Sinocentric A world view that places China at the central and superior position relative to the rest of the world.

Slash and burn method Cutting and burning to clear an area for planting.

Social construction The construction of landscapes by internal and external political, social, and economic forces.

Sodic land Land where crusts of salt or other alkalis prevent nutrients and water from going below the surface.

South Asian Association for Regional Cooperation (SAARC) Formed to disassociate political conflicts from economic cooperation.

Southward policy Taiwan's policy of increasing investment in Southeast Asian countries to offset investments in China.

Spatial duality The differences between the landscapes of core and peripheral regions.

Special Economic Zone (SEZ) A zone set aside by China's government to attract foreign technology and investment.

Species diversity Variation in plant or animal species.

Squatter settlement Makeshift housing in and around cities associated with rural to urban migration.

State capitalism Where the government plays a strong role in economic development such as in South Korea.

Steppe Short grassland.

Storm surge A cyclone-driven tide that is shaped into astronomical proportions by an increasingly narrow channel across low, flat terrain.

Subak Agricultural cooperative in Bali.

Sudra Serf in India's caste system.

Sukarno Indonesian nationalist who ruled from 1945 until 1965.

Sun Yat-sen Asia's first Asian nationalist. Proclaimed a Chinese Republic in 1912.

Sunderbans The ever-shifting mangrove, mudflats, islands, and distributary channels of the Indus and Brahmaputra delta in the Bay of Bengal.

Sunni The Islamic faith of the majority of Muslims.

Sustainable development Development that sustains environmental integrity.

Swidden Shifting cultivation in Southeast Asia.

Tagalog The official language of the Philippines.

Taiping Rebellion Rebellion against the corrupt and ineffective Manchus in China. Began in 1850 and ended in 1864 with the loss of 20 million lives.

Tais Principle ancestors of the Thai, Lao, and Shan peoples.

Taiwanese Everyone who lives in Taiwan.

Tamil Language of the Dravidian Hindus in Sri Lanka.

Tamil Eelam The proposed name for an independent state being fought for by the Tamil Tigers in northern Sri Lanka.

Tamil Tigers A militant Tamil group of northern Sri Lankans who are agitating for an independent state called Tamil Eelam.

Tantric Buddhism Vajrayana Buddhism developed in north India in the seventh century and spread to the Himalaya to be influenced by Tibetan shamanism. Official religion of Bhutan and Sikkim.

Techno-Archipelago A series of technopolises linked by information and communication channels.

Technobelt Zone selected for research and development and linked by communication and telecommunication.

Technopole A place where information and technology disseminates from.

Technopolis A network of science cities centering on a "mother city" and connected to a major metropolis.

Technoscapes Vast, interlocking channels of communication create their own landscapes in the form of global transactions and transnational or supranational corporations, organizations, and agencies.

Temporary resident In China, a person who obtains permission to stay in a city for a fixed length of time to work in a factory or construction site.

Terai A marshy plain in southern Nepal at 900 feet (270 m).

Theravada Buddhism The school of thought practiced in Sri Lanka and most of Southeast Asia.

Third World Refers to nations other than those of the First and Second Worlds.

Three Gorges Dam World's largest water conservation project on the Yangzi river in China. Will have the world's most powerful hydroelectric dam and improve shipping. Will also displace thousands of acres of farmland and people in addition to numerous important cultural sites.

Tied Aid The linking of foreign aid to structural adjustment or specific activities or purchases.

Time-space convergence With improvements in transportation and communication, more space can be covered in less time.

Timor Gap The disputed ocean and seabed between Australia and Indonesia.

Tokaido Megalopolis Urbanized, industrialized region from Tokyo to Osaka. Named for the old Tokaido Road.

Tonkin Region of the Red River delta in north Vietnam.

Topographic barrier An area of raised relief preventing the passage of wind or other phenomena.

Topography Landforms.

Total Fertility Rate (TFR) The number of children a woman would have if age-specific birth rates remained constant throughout her childbearing years.

Trans-Border Development Region Includes an economic development zone in adjacent political units.

Transgenic plant Human-engineered plant.

Transmigrasi People who move from crowded areas of Indonesia to open up and populate frontier regions.

Transmigration of souls The changing of people's life status through rebirth according to the merits of their behavior.

Transnational enterprise An economic, social, or political entity with ties to several nations.

Tree line The line above which growing season and soil depth do not support trees.

Tropical cyclone A violent storm created by a deep, circular, low-pressure cell. Called a typhoon in the Pacific.

Tsukuba Science City Conceived in the 1960s, it is an extensive research and development center northeast of Tokyo.

Tsunami Seismic sea wave.

Tundra A sub-arctic environment above the tree line that is a zone of low-lying flora, mosses, and lichens.

Typhoon A violent storm created by a deep, circular, low-pressure cell. Widely known as a tropical cyclone.

Underemployment A western notion regarding the underutilization of labor.

Untouchables Non-caste individuals associated with "polluting" occupations in India.

Upanishads Hindu literature meaning "sitting down near one's spiritual advisor."

Upazila A sub-district in Bangladesh created in the 1980s.

Urban agglomeration A cluster of cities.

Urban primacy When the population of the largest city is greater than the combined population of the next three largest cities.

Vaishya Peasant farmer in India's caste system.

Varna Means "color" and refers to classes of people such as priests or warriors in Hinduism.

Vedas The earliest known Hindu literature meaning "knowledge."

Vedda An aboriginal person of Sri Lanka.

Vegeculture The cultivation of tubers and tree crops.

Viet Cong National Liberation Front in southern Vietnam in opposition to the American supported South Vietnamese government.

Viet Minh Front organization of the Indochina Communist Party in North Vietnam.

Wallace Line The line of demarcation between the distinct zoogeographic regimes of Malaysia and Indonesia.

Well-field system Land divided into nine squares. Eight farmed by families, and the ninth farmed in common. Landscape representation of proper hierarchical relationships in China.

We-they syndrome Ethnocentric peoples (we) apply a sense of superiority to other cultures (they) that are considered inferior.

White man's burden Europeans' zealous mission to convert all people to Christianity.

White revolution The transformation of India's dairy industry to produce more milk.

Wind-chill factor The role that wind plays in reducing temperature.

World city A city at the top of a global urban hierarchy such as Tokyo or New York.

Xenophobia Fear of foreigners.

Xerophytic Drought tolerant.

Yang With *yin*, one of two opposing yet complementary forces that govern the universe.

Yi jing A Chinese literary work in which hexagrams depict the harmonies of complementary opposites.

Yin With *yang*, one of two opposing yet complementary forces that govern the universe.

Zaibatsu Japanese military-industrial conglomerates. Disbanded after World War II.

Zeylon Dutch name for Ceylon (Sri Lanka).

Index

Page references in *italic* type indicate illustrations. Page numbers followed by italic *t* indicate material in tables.

T